The PRINCETON FIELD GUIDE *to* DINOSAURS

2ND EDITION

PRINCETON FIELD GUIDES

Rooted in field experience and scientific study, Princeton's guides to animals and plants are the authority for professional scientists and amateur naturalists alike. **Princeton Field Guides** present this information in a compact format carefully designed for easy use in the field. The guides illustrate every species in color and provide detailed information on identification, distribution, and biology.

Albatrosses, Petrels, and Shearwaters of the World, by Derek Onley and Paul Scofield
Birds of Aruba, Curaçao, and Bonaire by Bart de Boer, Eric Newton, and Robin Restall
Birds of Australia, Eighth Edition, by Ken Simpson and Nicolas Day
Birds of Borneo: Brunei, Sabah, Sarawak, and Kalimantan, by Susan Myers
Birds of Botswana, by Peter Hancock and Ingrid Weiersbye
Birds of Central Asia, by Raffael Ayé, Manuel Schweizer, and Tobias Roth
Birds of Chile, by Alvaro Jaramillo
Birds of the Dominican Republic and Haiti, by Steven Latta, Christopher Rimmer, Allan Keith, James Wiley, Herbert Raffaele, Kent McFarland, and Eladio Fernandez
Birds of East Africa: Kenya, Tanzania, Uganda, Rwanda, and Burundi, by Terry Stevenson and John Fanshawe
Birds of East Asia: China, Taiwan, Korea, Japan, and Russia, by Mark Brazil
Birds of Europe, Second Edition, by Lars Svensson, Dan Zetterstrom, and Killian Mullarney
Birds of the Horn of Africa, by Nigel Redman, Terry Stevenson, and John Fanshawe
Birds of India, Pakistan, Nepal, Bangladesh, Bhutan, Sri Lanka, and the Maldives, by Richard Grimmett, Carol Inskipp, and Tim Inskipp
Birds of Kenya and Northern Tanzania: Field Guide Edition, by Dale A. Zimmerman, Donald A. Turner, and David J. Pearson
Birds of Melanesia: Bismarcks, Solomons, Vanuatu, and New Caledonia, by Guy Dutson
Birds of the Middle East, by R. F. Porter, S. Christensen, and P. Schiermacker-Hansen
Birds of Mongolia, by Sundev Gombobaatar, Axel Bräunlich, and Sh. Boldbaatar
Birds of Nepal, by Richard Grimmett, Carol Inskipp, and Tim Inskipp
Birds of New Guinea, by Thane K. Pratt and Bruce M. Beehler
Birds of Northern India, by Richard Grimmett and Tim Inskipp
Birds of Peru, by Thomas S. Schulenberg, Douglas F. Stotz, Daniel F. Lane, John P. O'Neill, and Theodore A. Parker III
Birds of the Seychelles, by Adrian Skerrett and Ian Bullock
Birds of Southeast Asia, by Craig Robson
Birds of Southern Africa, Fourth Edition, by Ian Sinclair, Phil Hockey, Warwick Tarboton, and Peter Ryan
Birds of Thailand, by Craig Robson
Birds of the West Indies, by Herbert Raffaele, James Wiley, Orlando Garrido, Allan Keith, and Janis Raffaele
Birds of Western Africa, by Nik Borrow and Ron Demey
Carnivores of the World, by Luke Hunter
Caterpillars of Eastern North America: A Guide to Identification and Natural History, by David L. Wagner
Common Mosses of the Northeast and Appalachians, by Karl B. McKnight, Joseph Rohrer, Kirsten McKnight Ward, and Warren Perdrizet
Coral Reef Fishes, by Ewald Lieske and Robert Meyers
Dragonflies and Damselflies of the East, by Dennis Paulson
Dragonflies and Damselflies of the West, by Dennis Paulson
Mammals of Europe, by David W. Macdonald and Priscilla Barrett
Mammals of North America, Second Edition, by Roland W. Kays and Don E. Wilson
Marine Mammals of the North Atlantic, by Carl Christian Kinze
Minerals of the World, by Ole Johnsen
Nests, Eggs, and Nestlings of North American Birds, Second Edition, by Paul J. Baicich and Colin J. O. Harrison
Palms of Southern Asia, by Andrew Henderson
Parrots of the World, by Joseph M. Forshaw
The Princeton Field Guide to Dinosaurs, Second Edition, by Gregory S. Paul
Raptors of the World, by James Ferguson-Lees and David A. Christie
Seeds of Amazonian Plants, by Fernando Cornejo and John Janovec
Sharks of the World, by Leonard Compagno, Marc Dando, and Sarah Fowler
Stars and Planets: The Most Complete Guide to the Stars, Planets, Galaxies, and the Solar System, Fully Revised and Expanded Edition, by Ian Ridpath and Wil Tirion
Trees of Eastern North America, by Gil Nelson, Christopher J. Earle, and Richard Spellenberg
Trees of Panama and Costa Rica, by Richard Condit, Rolando Perez, and Nefertaris Daguerre
Trees of Western North America, by Richard Spellenberg, Christopher J. Earle, and Gil Nelson
Whales, Dolphins, and Other Marine Mammals of the World, by Hadoram Shirihai and Brett Jarrett

The PRINCETON FIELD GUIDE *to* DINOSAURS

2ND EDITION

GREGORY S. PAUL

Princeton University Press

Princeton and Oxford

Published by Princeton University Press, 41 William Street, Princeton, New Jersey 08540
In the United Kingdom: Princeton University Press, 6 Oxford Street, Woodstock, Oxfordshire OX20 1TR
nathist.press.princeton.edu

ISBN 978-0-691-16766-4
Library of Congress Control Number: 2016933929
British Library Cataloging-in-Publication Data is available

This book has been composed in Galliard, Goudy and Optima

Printed on acid-free paper. ∞

Designed by D & N Publishing, Baydon, Wiltshire, UK

Printed in China

10 9 8 7 6 5 4 3 2 1

CONTENTS

PREFACE

If I were, at about age twenty as a budding paleoresearcher and artist, handed a copy of this book by a mysterious time traveler, I would have been shocked as well as delighted. The pages would reveal a world of new dinosaurs and ideas that I barely had a hint of or had no idea existed at all. My head would spin at the revelation of the therizinosaurs such as the wacky feathered *Beipiaosaurus* and at the biplane flying dromaeosaurids—or at the oversized shoulder spines of *Gigantspinosaurus*, the neck spines of *Amargasaurus*, the brow horns and atrophied arms of bulldog-faced *Carnotaurus*, the furry adornments of *Tianyulong* and *Kulindadromeus*, the bristly tail of *Psittacosaurus*, the bat-like membranous wings of scansoriopterygids, and the often psychedelic frilled horns of the new stable of centrosaurine and chasmosaurine ceratopsids. Even *Triceratops* has proven to have strange skin. It is a particular pleasure to at long last be able to restore the skeleton of the once mysterious *Deinocheirus*, long known from only its colossal arms—the skull and the rest of its peculiar skeleton do not disappoint. And who would have imagined it would become possible to figure out the colors of some feathered dinosaurs? I would note the new names for some old dinosaurs, including my favorite, *Giraffatitan*. And that old *Brontosaurus* is back! There would be the dinosaur-bearing beds with the familiar yet often exotic names Tendaguru, Morrison, Nemegt, Great Oolite, Hell Creek, and Lance. Plus there are the novel formations, at least to my eyes and ears, Yixian, Tiouraren, Dinosaur Park, Anacleto, Fangyan, Portezuelo, and Maevarano. The sheer number of new dinosaurs would demonstrate that an explosion in dinosaur discoveries and research, far beyond anything that had previously occurred, and often based on new high technologies, marked the end of the twentieth century going into the twenty-first.

Confirmed would be the paradigm shift already under way in the late 1960s and especially the 1970s that observed that dinosaurs were not so much reptiles as they were near birds that often paralleled mammals in form and function. Dinosaurs were still widely seen as living in tropical swamps, but we now know that some lived through polar winters so dark and bitterly cold that low-energy reptiles could not survive. Imagine a small dinosaur shaking the snow off its hairy body insulation while the flakes melt on the scaly skin of a nearby titanic sauropod whose body, oxygenated by a birdlike respiratory complex and powered by a high-pressure four-chambered heart, produces the heat needed to prevent frostbite.

In just the six years since the appearance of the first edition of this book, the number of dinosaur species named has expanded about 15 percent relative to some 190 years of research. Producing this second edition has been satisfying in that it has given me yet more reason to more fully achieve a long-term goal, to illustrate the skeletons of almost all dinosaur species for which sufficiently complete material is available. These have been used to construct the most extensive library of side-view life studies of dinosaurs in print to date. The result is a work that covers what is fast approaching two centuries of scientific investigation into the group of animals that ruled the continents for over 150 million years. Enjoy the travel back in time.

The author's paleozoological website can be found at www.gspauldino.com, and includes a complete list of his technical papers and other publications.

Acknowledgments

A complaint back in the last decade on the online Dinosaur List by Ian Paulsen about the absence of a high-quality dinosaur field guide led to the production of the first edition of this dinosaur guide, and its exceptional success, combined with the continuous flux of new discoveries and research, led to production of the second. Many thanks to those who have provided the assistance over the years that has made this book possible, including Peter Galton, Kenneth Carpenter, James Kirkland, Michael Brett-Surman, Philip Currie, Alex Downs, John Horner, Xu Xing, Robert Bakker, Saswati Bandyopadhyay, Rinchen Barsbold, Frank Boothman, David Burnham, Thomas Carr, Daniel Chure, Kristina Curry Rogers, Steven and Sylvia Czerkas, Peter Dodson, David Evans, James Farlow, Tracy Ford, Catherine Forster, John Foster, Mike Fredericks, Peter Galton, Roland Gangloff, Donald Glut (whose encyclopedia supplements made this work much easier), Mark Hallett, Jerry Harris, Scott Hartman, Thomas Holtz, Nicholas Hotton, Hermann Jaeger, Peter Larson, Guy Leahy, Nicholas Longrich, James Madsen, Jordan Mallon, Charles Martin, Teresa Maryanska, Octavio Mateus, John McIntosh, Carl Mehling, Ralph Molnar, Markus Moser, Darren Naish, Mark Norell, Fernando Novas, Halszka Osmólska, Kevin Padian, Armand Ricqles, Timothy Rowe, Dale Russell, Scott Sampson, John Scannella, Mary Schweitzer, Masahiro Tanimoto, Michael Taylor, Robert Telleria, Michael Triebold, David Varricchio, Matthew Wedel, David Weishampel, Jeffrey Wilson, Lawrence Witmer, and many others. I would also like to thank all those who worked on this book for Princeton University Press: Robert Kirk, Samantha Nader, Kathleen Cioffi, Laurel Anderton, and Namrita and David Price-Goodfellow.

INTRODUCTION

The spectacular plated dinosaur *Stegosaurus*

HISTORY OF DISCOVERY AND RESEARCH

Dinosaur remains have been found by humans for millennia and probably helped form the basis for belief in mythical beasts including dragons. A few dinosaur bones were illustrated in old European publications without their true nature being realized. In the West the claim in the Genesis creation story that the planet and all life were formed just two thousand years before the pyramids were built hindered the scientific study of fossils. At the beginning of the 1800s the numerous three-toed trackways found in New England were attributed to big birds. By the early 1800s the growing geological evidence that Earth's history was much more complex and extended back into deep time began to free researchers to consider the possibility that long-extinct and exotic animals once walked the globe.

Modern dinosaur paleontology began in the 1820s in England. Teeth were found, and a few bones of the predatory *Megalosaurus* and herbivorous *Iguanodon* were published and named. For a few decades it was thought that the bones coming out of ancient sediments were the remains of oversized versions of modern reptiles. In 1842 Richard Owen recognized that many of the fossils were not standard reptiles, and he coined the term "Dinosauria" to accommodate them. Owen had pre-evolutionary concepts of the development of life, and he envisioned dinosaurs as elephantine versions of reptiles, so they were restored as heavy-limbed quadrupeds. This led to the first full-size dinosaur sculptures for the grounds of the Crystal Palace in the 1850s, which helped initiate the first wave of dinomania as they excited the public. A banquet was actually held within one of the uncompleted figures. These marvelous examples of early dinosaur art still exist.

The first complete dinosaur skeletons, uncovered in Europe shortly before the American Civil War, were those of small examples, the armored *Scelidosaurus* and the birdlike *Compsognathus*. The modest size of these fossils limited the excitement they generated among the public. Found shortly afterward in the same Late Jurassic Solnhofen sediments as the latter was the "first bird," *Archaeopteryx*, complete with teeth and feathers. The remarkable mixture of avian and reptilian features preserved in this little dinobird did generate widespread interest, all the more so because the publication of Charles Darwin's theory of evolution at about the same time allowed researchers to put these dinosaurs in a more proper scientific context. The enthusiastic advocate of biological evolution Thomas Huxley argued that the close similarities between *Compsognathus* and *Archaeopteryx* indicated a close link between the two groups. In the late 1870s Belgian coal miners came across the complete skeletons of iguanodonts that confirmed that they were three-toed semibipeds, not full quadrupeds.

At this time, the action was shifting to the United States. Before the Civil War, incomplete remains had been found on the Eastern Seaboard. But matters really got moving when it was realized that the forest-free tracts of the West offered hunting grounds that were the best yet for the fossils of extinct titans. This quickly led to the "bone wars" of the 1870s and 1880s in which Edward Cope and Charles Marsh, having taken a dislike for one another that was as petty as it was intense, engaged in a bitter and productive competition for dinosaur fossils that would produce an array of complete skeletons. For the first time it became possible to appreciate the form of classic Late Jurassic Morrison dinosaurs such as agile predatory *Allosaurus* and *Ceratosaurus*, along with *Apatosaurus*, *Brontosaurus*, *Diplodocus*, and *Camarasaurus*—which were really elephantine quadrupeds—the protoiguanodont *Camptosaurus*, and the bizarre plated *Stegosaurus*. Popular interest in the marvelous beasts was further boosted.

By the turn of the century, discoveries shifted to younger deposits such as the Lance and Hell Creek, which produced classic dinosaurs from the end of the dinosaur era including duck-billed *Edmontosaurus*, armored *Ankylosaurus*, horned *Triceratops*, and the great *Tyrannosaurus*. As paleontologists moved north into Canada in the early decades of the twentieth century, they uncovered a rich collection of slightly older Late Cretaceous dinosaurs including *Albertosaurus*, horned *Centrosaurus*, spiked *Styracosaurus*, and the crested duckbills *Corythosaurus* and *Lambeosaurus*.

Inspired in part by the American discoveries, paleontologists in other parts of the world looked for new dinosaurs. Back in Europe abundant skeletons of German *Plateosaurus* opened a window into the evolution of early dinosaurs in the Late Triassic. In southeastern Africa the colonial Germans uncovered at exotic Tendaguru the supersauropod *Giraffatitan* (was *Brachiosaurus*) and spiny *Kentrosaurus*. In the 1920s Henry Osborn at the American Museum in New York dispatched Roy Andrews to Mongolia in a misguided search for early humans that fortuitously led to the recovery of small Late Cretaceous dinosaurs, parrot-beaked *Protoceratops*, the "egg-stealing" *Oviraptor*, and the advanced, near-bird theropod *Velociraptor*. Dinosaur eggs and entire nests were found, only to be errantly assigned to *Protoceratops* rather than the oviraptorid that had actually laid and incubated them. As it happened, the Mongolian expeditions were somewhat misdirected. Had paleontologists also headed northeast of Beijing, they might have made even more fantastic discoveries that would have dramatically altered our view and understanding of dinosaurs, birds, and their evolution, but that event would have to wait another three-quarters of a century.

The mistake of the American Museum expeditions in heading northeast contributed to a set of problems that seriously damaged dinosaur paleontology as a science between the twentieth-century world wars. Dinosaurology became rather ossified, with the extinct beasts widely portrayed as sluggish, dim-witted evolutionary dead ends doomed to extinction, an example of the "racial senescence" theory that was widely held among researchers who preferred a progressive concept of evolution at odds with more random Darwinian natural selection. It did not help matters when artist/paleontologist Gerhard Heilmann published a seminal work that concluded that birds were not close relatives of dinosaurs, in part because he thought dinosaurs lacked a wishbone furcula that had just been found, but misidentified, in *Oviraptor*. The advent of the Depression, followed by the trauma of World War II—which led to the loss of some important specimens on the continent as a result of Allied and Axis bombing—brought major dinosaur research to a near halt.

Even so, public interest in dinosaurs remained high. The paleoart of Charles Knight made him famous. The *Star Wars–Jurassic Park* of its time, RKO's *King Kong* of 1933 amazed audiences with its dinosaurs seemingly brought to life. Two major film comedies, 1938's *Bringing Up Baby*, starring Cary Grant and Katherine Hepburn, and 1949's *On the Town*, featuring Gene Kelly and Frank Sinatra, involved climactic scenes in which sauropod skeletons at a semifictional New York museum collapsed because of the hijinks of the lead characters. Unfortunately, the very popularity of dinosaurs gave them a circus air that convinced many scientists that they were beneath their scientific dignity and attention.

Despite the problems, discoveries continued. In an achievement remarkable for a nation ravaged by the Great Patriotic War and suffering under the oppression of Stalinism, the Soviets mounted postwar expeditions to Mongolia that uncovered the Asian version of *Tyrannosaurus* and the enigmatic arms of enormous clawed *Therizinosaurus*. Equally outstanding was how the Poles took the place of the Soviets in the 1960s, discovering in the process the famed complete skeleton of *Velociraptor* engaged in combat with *Protoceratops*. They too found another set of mysterious long arms with oversized claws, *Deinocheirus*.

In the United States, Roland Bird studied the trackways of herds of Texas-sized Cretaceous sauropods before World War II. Shortly after the global conflict, the Triassic Ghost Ranch quarry in the Southwest, packed with complete skeletons of little *Coelophysis*, provided the first solid knowledge of the beginnings of predatory dinosaurs. Also found shortly afterward in the Southwest was the closely related but much larger crested theropod *Dilophosaurus* of the Early Jurassic.

What really spurred the science of dinosaur research were the Yale expeditions to Montana in the early 1960s that dug into the little-investigated Early Cretaceous Cloverly Formation. The discovery of the *Velociraptor* relative *Deinonychus* finally made it clear that some dinosaurs were sophisticated, energetic, agile dinobirds, a point reinforced by the realization that it and the other sickle claws, the troodontids, as well as the ostrichlike ornithomimids, had fairly large, complex brains. These developments led John Ostrom to note and detail the similarities between his *Deinonychus* and *Archaeopteryx* and to conclude that birds are the descendants of energetic small theropod dinosaurs.

Realizing that the consensus dating back to their original discovery that dinosaurs were an expression of the reptilian pattern was flawed, Robert Bakker in the 1960s and 1970s issued a series of papers contending that dinosaurs and their feathered descendants constituted a distinct group of archosaurs whose biology and energetics were more avian than reptilian. Eventually, in the article "Dinosaur Renaissance" in a 1975 *Scientific American*, Bakker proposed that some small dinosaurs themselves were feathered. In the late 1970s, Montana native John Horner found baby hadrosaurs and their nests, providing the first look at how some dinosaurs reproduced. At the same time, researchers from outside paleontology stepped into the field and built up the evidence that the impact of an asteroid over six miles in diameter was the long-sought great dinosaur killer. This extremely controversial and contentious idea turned into the modern paradigm on the finding of a state-sized meteorite crater in southeastern Mexico dating to the end of the dinosaur era.

These radical and controversial concepts greatly boosted popular attention on dinosaurs, culminating in the *Jurassic Park* novels and films that sent dinomania to unprecedented heights. The elevated public awareness was combined with digital technology in the form of touring exhibits of robotic dinosaurs. This time the interest of paleontologists was elevated as well, inspiring the second and ongoing golden age of dinosaur discovery and research, which is surpassing that which has gone before. Assisting the work are improved scientific techniques in the area of evolution and phylogenetics, including cladistic genealogical analysis, which has improved the investigation of dinosaur relationships. A new generation of artists has portrayed dinosaurs with a "new look" that lifts tails in the air and gets feet off the ground to represent the more dynamic gaits that are in line with the more active lifestyles the researchers now favor. I noticed that the sickle-clawed dromaeosaurs and troodonts, as well as the oviraptorosaurs, possessed anatomical features otherwise found in flightless birds and suggested that these dinosaurs were also secondarily flightless.

Dinosaurs are being found and named at an unprecedented rate as dinosaur science goes global, with efforts under way on all continents. In the 1970s the annual Society of Vertebrate Paleontology meeting might have seen a half-dozen presentations on dinosaurs; now it is in the area of a couple of hundred. Especially important has been the development of local expertise made possible by the rising economies of many second-world nations, reducing the need to import Western expertise.

In South America, Argentine and American paleontologists collaborated in the 1960s and 1970s to reveal the first Middle and Late Triassic protodinosaurs, finally showing that the very beginnings of dinosaurs started among surprisingly small archosaurs. Since then, Argentina has been the source of endless remains from the Triassic to the end of the Cretaceous that include the early theropods *Eoraptor* and *Herrerasaurus*, supertitanosaur sauropods such as *Argentinosaurus*, *Futalongnkosaurus*, and *Dreadnoughtus*, and the oversized theropods such as

The dinobird *Deinonychus*

Giganotosaurus that preyed on them. Among the most extraordinary finds have been sauropod nesting grounds that allow us to see how the greatest land animals of Earth's history reproduced themselves.

In southern Africa excellent remains of an Early Jurassic species of *Coelophysis* verified how uniform the dinosaur fauna was when all continents were gathered into Pangaea. Northern Africa has been the major center of activity as a host of sauropods and theropods have filled in major gaps in dinosaur history. Australia is geologically the most stable of continents, with relatively little in the way of tectonically driven erosion to either bury fossils or later expose them, so dinosaur finds have been comparatively scarce despite the aridity of the continent. The most important discoveries have been of Cretaceous dinosaurs that lived close to the South Pole, showing the climatic extremes dinosaurs were able to adapt to. Glacier-covered Antarctica is even less suitable prospecting territory, but even it has produced the Early Jurassic crested avepod *Cryolophosaurus* as well as other dinosaur bones.

At the opposite end of the planet, the uncovering of a rich Late Cretaceous fauna on the Alaskan North Slope confirms the ability of dinosaurs to dwell in latitudes cold and dark enough in the winter that lizards and crocodilians are not found in the same deposits. Farther south, a cadre of researchers have continued to plumb the great dinosaur deposits of western North America as they build the most detailed sample of dinosaur evolution from the Triassic until their final loss. We now know that armored ankylosaurs were roaming along with plated stegosaurs in the Morrison Formation, a collection of sauropods has been exposed from the Early Cretaceous, and one new ceratopsian and hadrosaur after another is coming to light in the classic Late Cretaceous beds.

Now Mongolia and especially China have become the great frontier in dinosaur paleontology. Even during the chaos of the Cultural Revolution, Chinese paleontologists made major discoveries, including the first spectacularly long-necked mamenchisaur sauropods. As China modernized and Mongolia gained independence, Canadian and American researchers have worked with their increasingly skilled resident scientists, who have become a leading force in dinosaur research. It was finally realized that the oviraptors found associated with nests at the Flaming Cliffs were not eating the eggs but brooding them in a pre-avian manner. Almost all of China is productive when it comes to dinosaurs, and after many decades paleontologists started paying attention to the extraordinary fossils being dug up by local farmers from Early Cretaceous lake beds in the northeast of the nation.

In the mid-1990s, complete specimens of small compsognathid theropods labeled *Sinosauropteryx* began to show up with their bodies covered with dense coats of bristly protofeathers. More recently it has been argued that it is often possible to determine the color of the feathers! This was just the start: the Yixian beds are so extensive and productive that they have become an inexhaustible source of beautifully preserved material as well as of strife, as the locals contend with the authorities for the privilege of excavating the fossils for profit—sometimes altering the remains to "improve" them—rather than for rigorous science. The feathered dinosaurs soon included the potential oviraptorosaur *Caudipteryx*. Even more astonishing have been the Yixian dromaeosaurs. These small sickle claws bear fully developed wings not only on their arms but on their similarly long legs as well. This indicates not only that dromaeosaurs first evolved as fliers but that they were adapted to fly in a manner quite different from the avian norm. The therizinosaur *Beipiaosaurus* looks like a refugee from a Warner Brothers cartoon. But the Yixian is not just about confirming that birds are dinosaurs and that some dinosaurs were feathered. One of the most common dinosaurs of the Early Cretaceous is the parrot-beaked *Psittacosaurus*. Although it was known from numerous skeletons across Asia found over the last eighty years, no one had a clue that its tail sported large, arcing, bristly spines until a complete individual with preserved skin was found in the Yixian. To top things off, the Yixian has produced the small ornithischian *Tianyulong*, which suggests that insulating fibers were widespread among small dinosaurs. There are new museums in China packed with enormous numbers of undescribed dinosaur skeletons on display and in storage.

On a global scale, the number of dinosaur trackways that have been discovered is in the many millions. This is logical in that a given dinosaur could potentially contribute only one skeleton to the fossil record but could make innumerable footprints. In a number of locations, trackways are so abundant that they form what have been called "dinosaur freeways." Many of the trackways were formed in a manner that suggests their makers were moving in herds, flocks, packs, and pods. A few may record the attacks of predatory theropods on herbivorous dinosaurs.

The history of dinosaur research is not just one of new ideas and new locations; it is also one of new techniques and technologies. The turn of the twenty-first century has seen paleontology go high tech with the use of computers for processing data and high-resolution CT scanners to peer inside fossils without damaging them. Dinosaurology has also gone microscopic and molecular in order to assess the lives of dinosaurs at a more intimate level, telling us how fast they grew, how long they lived, and at what age they started to reproduce. Bone isotopes are being used to help determine dinosaur diets and to state that some dinosaurs were semiaquatic. And it turns out that feather pigments can be preserved well enough to restore original colors. Meanwhile the *Jurassic World* franchise helps sustain popular interest in the group even as it presents an obsolete, prefeather image of the birds' closest relations.

The evolution of human understanding of dinosaurs has undergone a series of dramatic transformations since they were scientifically discovered almost two hundred years ago. This is true because dinosaurs are a group of "exotic" animals whose biology was not obvious from the start, unlike fossil mammals or lizards. It has taken time to build up the knowledge base needed to resolve their true form and nature. The latest revolution is still young. When I was a youth, I learned that dinosaurs were, in general, sluggish, cold-blooded, tail-dragging, slow-growing, dim-witted reptiles that did not care for their young. The idea that some were feathered and that birds are living descendants was beyond imagining. Dinosaur paleontology has matured in that it is unlikely that a reorganization of similar scale will occur in the future, but we now know enough

The flying dinosaur *Sinornithosaurus* attacking *Psittacosaurus*

about the inhabitants of the Mesozoic to have the basics well established. Sauropods will not return to a hippo-like lifestyle, and dinosaurs' tails will not be chronically plowing through Mesozoic muds. Dinosaurs are no longer so mysterious. Even so, the research is nowhere near its end. To date, over seven hundred valid dinosaur species in about five hundred genera have been discovered and named. This probably represents at most a quarter, and perhaps a much smaller fraction, of the species that have been preserved in sediments that can be accessed. And, as astonishingly strange as many of the dinosaurs uncovered so far have been, there are equally odd species waiting to be unearthed. Reams of work based on as-yet-undeveloped technologies and techniques will be required to provide further details about both dinosaur biology and the world in which they lived. And although a radical new view is improbable, there will be many surprises.

WHAT IS A DINOSAUR?

To understand what a dinosaur is, we must first start higher in the scheme of animal classification. The Tetrapoda are the vertebrates adapted for life on land—amphibians, reptiles, mammals, birds, and the like. Amniota comprises those tetrapod groups that reproduce by laying hard-shelled eggs, with the proviso that some have switched to live birth. Among amniotes are two great groups. One is the Synapsida, which includes the archaic pelycosaurs, the more advanced therapsids, and mammals, which are the only surviving synapsids. The other is the Diapsida. Surviving diapsids include the lizard-like tuataras, true lizards and snakes, crocodilians, and birds. The Archosauria is the largest and most successful group of diapsids and includes crocodilians and dinosaurs. Birds are literally flying dinosaurs.

Archosaurs also include the basal forms informally known as thecodonts because of their socketed teeth, themselves a diverse group of terrestrial and aquatic forms that include the ancestors of crocodilians and the flying pterosaurs, which are not intimate relatives of dinosaurs and birds.

The great majority of researchers now agree that the dinosaurs were monophyletic in that they shared a common ancestor that made them distinct from all other archosaurs, much as all mammals share a single common ancestor that renders them distinct from all other synapsids. This consensus is fairly recent—before the 1970s it was widely thought that dinosaurs came in two distinct types that had evolved separately from thecodont stock, the Saurischia and Ornithischia. It was also thought that birds had evolved as yet another group independently from thecodonts. The Saurischia and Ornithischia still exist, but they are now the two major parts of the Dinosauria, much as living Mammalia is divided mainly into marsupials and placentals. Dinosauria is formally defined as the phylogenetic clade that includes the common ancestor of *Triceratops* and birds and all their descendants. Because different attempts to determine the exact relationships of the earliest dinosaurs produce somewhat different results, there is some disagreement about whether the most primitive, four-toed theropods were dinosaurs or lay just outside the group. This book includes them, as do most researchers.

In anatomical terms, one of the features that most distinguish dinosaurs centers on the hip socket. The head of the femur is a cylinder turned in at a right angle to the shaft of the femur that fits into a cylindrical, internally open hip socket. This allows the legs to operate in the nearly vertical plane characteristic of the group, with the feet directly beneath the body. You can see this system the next time you have chicken thighs. The ankle is a simple fore-and-aft hinge joint that also favors a vertical leg posture. Dinosaurs were "hind-limb dominant" in that they were either bipedal or, even when they were quadrupedal, most of the animal's weight was borne on the legs, which were always built more strongly than the arms. The hands and feet were generally digitigrade, with the wrist and ankle held clear of the ground. All dinosaurs shared a trait also widespread among archosaurs in general, the presence of large and often remarkably complex sinuses and nasal passages.

Aside from the above basic features, dinosaurs, even when we exclude birds, were an extremely diverse group of animals, rivaling mammals in this regard. Dinosaurs ranged in form from nearly bird-like types such as the sickle-clawed dromaeosaurs to

A basal archosaur,
Euparkeria

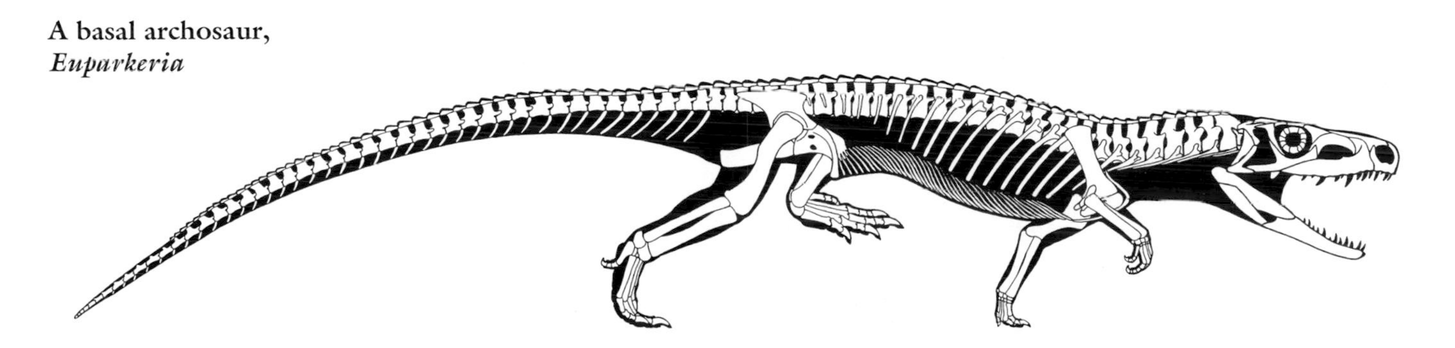

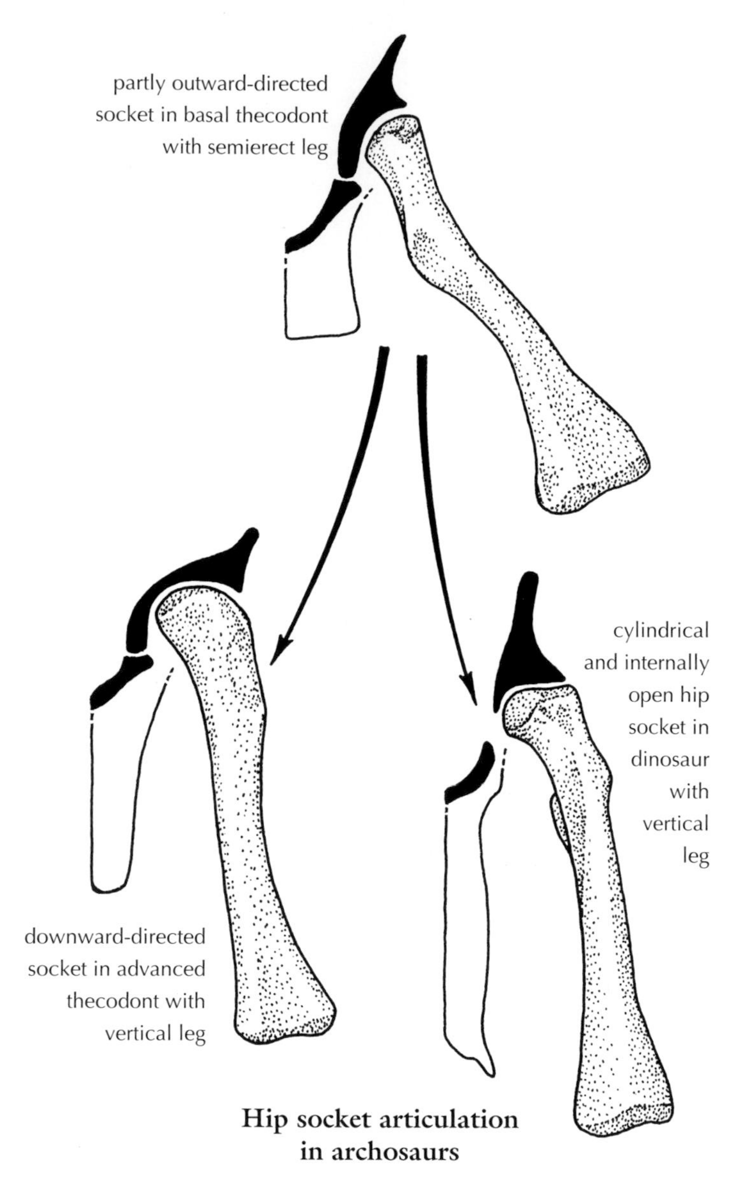

Hip socket articulation in archosaurs

rhino-like horned ceratopsians to armor-plated stegosaurs to elephant- and giraffe-like sauropods and dome-headed pachycephalosaurs. They even took to the skies in the form of birds. However, dinosaurs were limited in that they were persistently terrestrial. Although some dinosaurs may have spent some time feeding in the water like moose or fishing cats, at most a few became strongly amphibious in the manner of hippos, much less marine like seals and whales. The only strongly aquatic dinosaurs are some birds. The occasional statement that there were marine dinosaurs is therefore incorrect—these creatures of Mesozoic seas were various forms of reptiles that had evolved over the eons.

Because birds are dinosaurs in the same way that bats are mammals, the dinosaurs aside from birds are sometimes referred to as "nonavian dinosaurs." This usage can become awkward, and in general in this book dinosaurs that are not birds are, with some exceptions, referred to simply as dinosaurs.

Dinosaurs seem strange, but that is just because we are mammals biased toward assuming the modern fauna is familiar and normal, and past forms are exotic and alien. Consider that elephants are bizarre creatures with their combination of big brains, massive limbs, oversized ears, a pair of teeth turned into tusks, and noses elongated into hose-like trunks. Nor were dinosaurs part of an evolutionary progression that was necessary to set the stage for mammals culminating in humans. What dinosaurs do show is a parallel world, one in which mammals were permanently subsidiary and the dinosaurs show what largely diurnal land animals that evolved straight from similarly day-loving ancestors should actually look like. Modern mammals are much more peculiar, having evolved from nocturnal beasts that came into their own only after the entire elimination of nonavian dinosaurs. While dinosaurs dominated the land, small nocturnal mammals were just as abundant and diverse as they are in our modern world. If not for the accident of the later event, dinosaurs would probably still be the global norm.

DATING DINOSAURS

How can we know that dinosaurs lived in the Mesozoic, first appearing in the Late Triassic over 230 million years ago and then disappearing at the end of the Cretaceous 66 million years ago?

As gravels, sands, and silts are deposited by water and sometimes wind, they build up in sequence atop the previous layer, so the higher in a column of deposits a dinosaur is, the younger it is relative to dinosaurs lower in the sediments. Over time sediments form distinct stratigraphic beds that are called formations. For example, *Apatosaurus*, *Brontosaurus*, *Diplodocus*, *Barosaurus*, *Stegosaurus*, *Camptosaurus*, *Allosaurus*, and *Ornitholestes* are found in the Morrison Formation of western North America, which was laid down in the Late Jurassic, from 156 to 147 million years ago. Deposited largely by rivers over an area covering many states in the continental interior, the Morrison Formation is easily distinguished from the marine Sundance Formation lying immediately below as well as from the similarly terrestrial Cedar Mountain Formation above, which contains a very different set of dinosaurs. Because the Morrison was formed over millions of years, it can be subdivided into lower (older), middle, and upper (younger) levels. So a fossil found in the Sundance is older than one found in the Morrison, a dinosaur found in the lower Morrison is older than one found in the middle, and a dinosaur from the Cedar Mountain is younger still.

Geological time is divided into a hierarchical set of names. The Mesozoic is an era—preceded by the Paleozoic and followed by the Cenozoic—that contained the three progressively younger periods called the Triassic, Jurassic, and Cretaceous. These are then divided into Early, Middle, and Late, except that the Cretaceous is split only into Early and Late despite being considerably longer than the other two periods (this was not known when the division was made in the 1800s). The periods are further subdivided into stages. The Morrison Formation, for example, began to be deposited during the last part of the Oxfordian, continued through the entire Kimmeridgian, and the top part was formed at the beginning of the Tithonian.

The absolute age of recent fossils can be determined directly by radiocarbon dating. Dependent on the ratios of carbon isotopes, this method works only on bones and other specimens going back up to fifty thousand years, far short of the dinosaur era. Because it is not possible to directly date Mesozoic dinosaur remains, we must instead date the formations that the specific species are found in. This is viable because a given dinosaur species lasted only a few hundred thousand to a few million years.

The primary means of absolutely determining the age of dinosaur-bearing formations is radiometric dating. Developed by nuclear scientists, this method exploits the fact that radioactive elements decay in a very precise manner over time. The main nuclear transformations used are uranium to lead, potassium to argon, and one argon isotope to another argon isotope. This system requires the presence of volcanic deposits that initially set the nuclear clock. These deposits are usually in the form of ashfalls similar to the one deposited by Mount Saint Helens over neighboring states that leave a distinct layer in the sediments. Assume that one ashfall was deposited 144 million years ago, and another one higher in the sediments 141 million years ago. If a dinosaur is found in the deposits in between, then we know that the dinosaur lived between 144 and 141 million years ago. As the technology advances and the geological record is increasingly better known, radiometric dating is becoming increasingly precise. The further back in time one goes, the greater the margin of error, and the less exactly the sediments can be dated.

Volcanic deposits are often not available, and other methods of dating must be used. Doing so requires biostratigraphic correlation, which can in turn depend in part on the presence of "index fossils." Index fossils are organisms, usually marine invertebrates, that are known to have existed for only geologically brief periods of time, just a few million years at most. Assume a dinosaur species is from a formation that lacks datable volcanic deposits. Also assume that the formation grades into marine deposits laid down at the same time near its edge. The marine sediments contain small organisms that lasted for only a few million years in time. Somewhere else in the world, the same species of marine life was deposited in a marine formation that includes volcanic ashfalls that have been radiometrically dated to between 84 and 81 million years. We can then conclude that the dinosaur in the first formation is also 84 to 81 million years old.

A number of dinosaur-bearing formations lack both volcanic deposits and marine index fossils. It is not possible to accurately date the dinosaurs in these deposits. It is only possible to broadly correlate the level of development of the dinosaurs and other organisms in the formation with faunas and floras in better-dated formations, and this produces only approximate results. This situation is especially common in central Asia. The reliability of dating therefore varies. It can be very close to the actual value in formations that have been well studied and contain volcanic deposits; these can be placed in specific parts of a stage. At the other extreme are those formations that, because they lack the needed age determinants, and/or because they have not been sufficiently well examined, can only be said to date from the early, middle, or late portion of one of the periods, an error that can span well over 10 million years. North America currently has the most robust linkage of the geological time scale with its fossil dinosaurs of anywhere on Earth.

THE EVOLUTION OF DINOSAURS AND THEIR WORLD

Dinosaurs appeared in a world that was both ancient and surprisingly recent—it is a matter of perspective. The human view that the age of dinosaurs was remote in time is an illusion that results from our short life span. A galactic year, the time it takes our solar system to orbit the center of the galaxy, is 200 million years. Only one galactic year ago the dinosaurs had just appeared on planet Earth. When dinosaurs first appeared, our solar system was already well over 4 billion years old, and 95 percent of the history of our planet had already passed. A time traveler arriving on Earth when dinosaurs first appeared would have found it both comfortingly familiar and marvelously different from our time.

As the moon slowly spirals out from the Earth because of tidal drag, the length of each day grows. When dinosaurs first evolved, a day was about 22 hours and 45 minutes long, and the year had 385 days; when they went largely extinct, a day was up to 23 hours and over 30 minutes, and the year was down to 371 days. The moon would have looked a little larger and would have more strongly masked the sun during eclipses—there would have been none of the rare annular eclipses in which the moon is far enough away in its elliptical orbit that the sun rings the moon at maximum. The "man on the moon" leered down upon the dinosaur planet, but the prominent Tycho crater was not blasted into existence until toward the end of the Early Cretaceous. As the sun converts an increasing portion of its core from hydrogen into helium, it becomes hotter by nearly 10 percent every billion years, so the sun was about 2 percent cooler when dinosaurs first showed up and around a half percent cooler than it is now when most went extinct.

At the beginning of the great Paleozoic Era over half a billion years ago, the Cambrian Revolution saw the advent of complex, often hard-shelled organisms. Also appearing were the first, simple vertebrates. As the Paleozoic progressed, first plants and then animals, including tetrapod vertebrates, began to invade the land, which saw a brief Age of Amphibians in the late Mississippian followed by the classic Age of Reptiles in the Pennsylvanian and much of the Permian. By the last period in the Paleozoic, the Permian, the continents had joined together into the supercontinent Pangaea, which straddled the equator and stretched nearly to the poles north and south. With the majority of land far from the oceans, most terrestrial habitats were harshly semiarid, ranging from extra hot in the tropics to sometimes glacial at high latitudes. The major vertebrate groups had evolved by that time. Among synapsids, the mammal-like therapsids, some up to the size of rhinos, were the dominant large land animals in the Age of Therapsids of the Late Permian. These were apparently more energetic than reptiles, and those living in cold climates may have used fur to conserve heat. Toward the end of the period, the first archosaurs appeared. These low-slung, vaguely lizard-crocodilian creatures were a minor part of the global fauna. The conclusion of the Permian saw a massive extinction that has yet to be entirely explained and that, in many regards, exceeded the extinction that killed off the terrestrial dinosaurs 185 million years later.

At the beginning of the first period of the Mesozoic, the Triassic, the global fauna was severely denuded. As it recovered, the few remaining therapsids enjoyed a second evolutionary radiation and again became an important part of the wildlife. Again, they never became truly enormous or tall. This time they had competition, as the archosaurs also underwent an evolutionary explosion, first expressed as a wide variety of thecodonts, some of which reached a tonne in mass. One group evolved into aquatic, armored crocodile mimics. Others became armored land herbivores. Many were terrestrial predators that moved on erect legs achieved in a manner different from dinosaurs. The head of the femur did not turn inward; instead, the hip socket expanded over the femoral head until the shaft could be directed downward. Some of these erect-legged archosaurs were nearly bipedal. Others became toothless plant eaters. It is being realized that in many respects the Triassic thecodonts filled the lifestyle roles that would later be occupied by dinosaurs. Even so, these basal archosaurs never became gigantic or very tall. Also coming onto the scene were the crocodilians, the only group surviving today that reminds us what the archosaurs of the Triassic were like. Triassic crocodilians started out as small, long-legged, digitigrade land runners. Their sophisticated liver-pump lung systems may have evolved to help power a highly aerobic exercise ability. Crocodilians, like many of the thecodonts, had a very undinosaurian feature. Their ankles were complex, door-hinge-like joints in which a tuber projecting from one of the ankle bones helped increase the leverage of the muscles on the foot, rather as in mammals. At some time in the period, the membrane-winged, long-tailed pterosaurs evolved. Because pterosaurs had the same kind of simple-hinge ankle seen in dinosaurs, it has been suggested that the two groups are related. The energetic pterosaurs were insulated; we do not yet know whether other nondinosaurian archosaurs were also covered with thermal fibers, but the possibility is substantial.

In the Anisian and Ladinian, the two stages of the Middle Triassic, quite small predatory archosaurs appeared that exhibited many of the features of dinosaurs. Although the hip socket was still not internally open, the femoral head was turned inward, allowing the legs to operate in a vertical plane. The ankle was a simple hinge. The skull was lightly constructed. At first known only from South America, these protodinosaurs have since been found on other continents. These early dinosaurian forms would survive only until the Norian. Protodinosaurs show that dinosaurs started out as little creatures; they did not descend from the big basal archosaurs.

From small things big things can evolve, and very quickly. In the Carnian stage of the Late Triassic the fairly large-bodied, small-hipped, four-toed herrerasaur theropods were on the global stage. These bipeds dwelled in a world still dominated by complex-ankled archosaurs and would not last beyond the Norian or maybe the Rhaetian stage, perhaps because these early dinosaurs did not have an aerobic capacity high enough to vie with their new competitors. The Norian saw the appearance of the great group that is still with us, the bird-footed avepod theropods, whose large hips and beginnings of the avian-type respiratory system imply a further improvement in aerobic performance and thermoregulation. At about the same time, the first members of one of the grand groups of herbivorous dinosaurs are first recorded in the fossil record, the small-hipped, semibipedal prosauropods, followed almost immediately by the quadrupedal and bigger-hipped sauropods. These new dinosaurs gave thecodonts increasing competition as they rapidly expanded in diversity as well as size. Just 15 or 20 million years after the evolution of the first little protodinosaurs, prosauropods and sauropods weighing 2 tonnes had developed. In only another 10 million years, sauropods as big as elephants, the first truly gigantic land animals, were extant. These long-necked dinosaurs were also the first herbivores able to browse at high levels, many meters above the ground. Dinosaurs were showing the ability to evolve enormous dimensions and bulk on land, an attribute otherwise seen only among mammals. In the Carnian the first of the beaked herbivorous ornithischians arrived. These little semibipeds were not common, and they, as well as small prosauropods, may have dug burrows as refuges from a predator-filled world. By the last stage of the Triassic the saurischian dinosaurs were becoming the ascendant land animals, although they still lived among thecodonts and some therapsids. From the latter, at this time, evolved the first mammals. Mammals and dinosaurs have, therefore, shared the planet for over 200 million years—and for 140 million of those years, mammals would remain small.

Because animals could wander over the entire supercontinent with little hindrance from big bodies of water, faunas

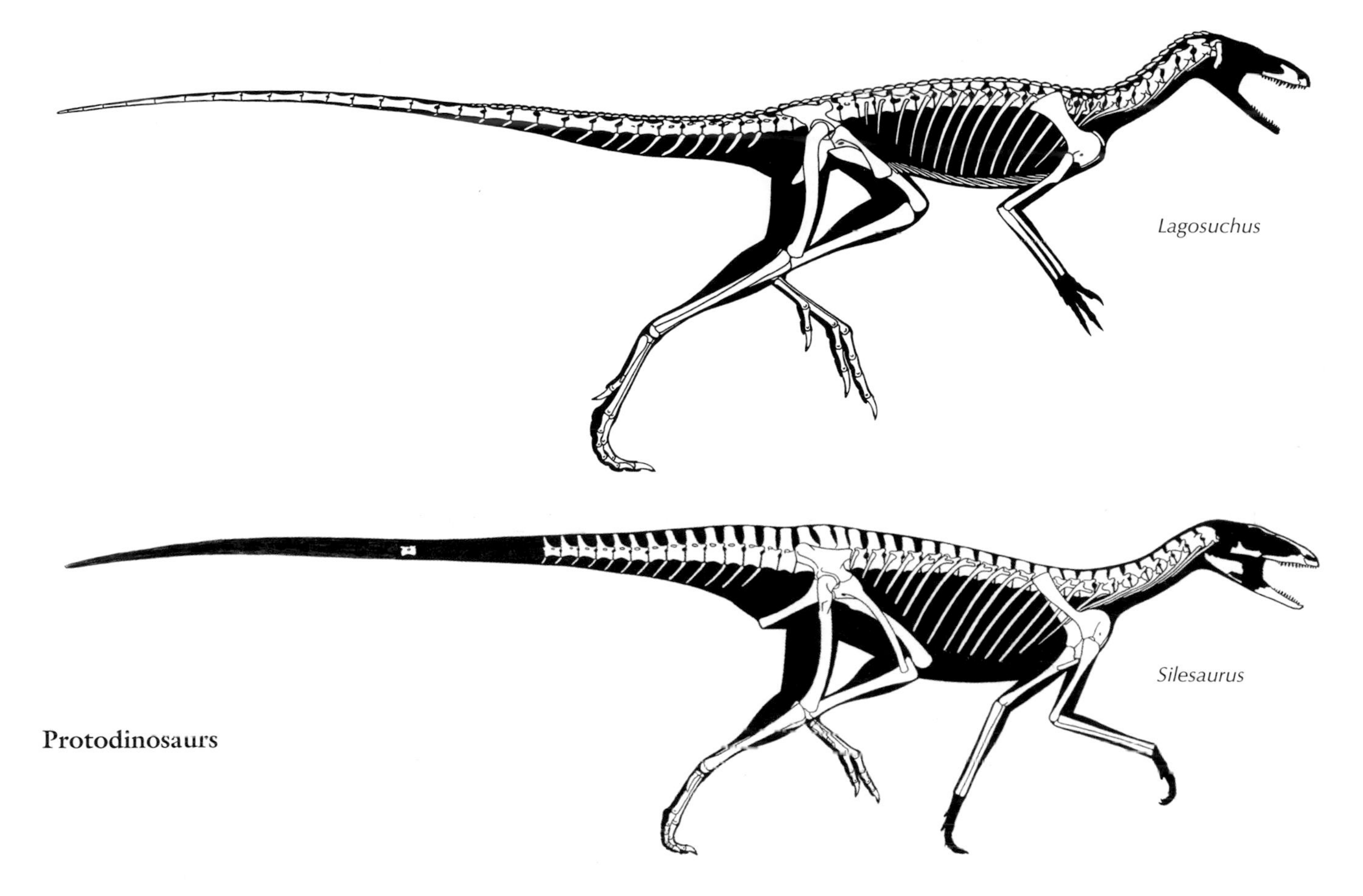

Protodinosaurs

tended to exhibit little difference from one region to another. And with the continents still collected together, the climatic conditions over most of the supercontinent remained harsh. It was the greenhouse world that would prevail through the Mesozoic. The carbon dioxide level was two to ten times higher than it is currently, boosting temperatures to such highs—despite the slightly cooler sun of those times—that even the polar regions were fairly warm in winter. The low level of tectonic activity meant there were few tall mountain ranges to capture rain or interior seaways to provide moisture. Hence, there were great deserts, and most of the vegetated lands were seasonally semiarid, but forests were located in the few regions of heavy rainfall and groundwater created by climatic zones and rising uplands. It appears that the tropical latitudes were so hot and dry that the larger dinosaurs, with their high energy budgets, could not dwell near the equator and were restricted to the cooler, wetter, higher latitudes. The flora was in many respects fairly modern and included many plants we would be familiar with. Wet areas along watercourses were the domain of rushes and horsetails. Some ferns also favored wet areas and shaded forest floors. Other ferns grew in open areas that were dry most of the year, flourishing during the brief rainy season. Large parts of the world may have been covered by fern prairies, comparable to the grasslands and shrublands of today. Tree ferns were common in wetter areas. Even more abundant were the fernlike or palmlike cycadeoids, similar to the cycads that still inhabit the tropics. Taller trees included water-loving ginkgoids, of which the maidenhair tree is the sole—and, until widely planted in urban areas, the nearly extinct—survivor. Dominant among plants were conifers, most of which at that time had broad leaves rather than needles. Some of the conifers were giants rivaling the colossal trees of today, such as those that formed the famed Petrified Forest of Arizona. Flowering plants were completely absent.

The end of the Triassic about 200 million years ago saw another extinction event whose cause is obscure. A giant impact occurred in southeastern Canada, but it was millions of years before the extinction. The thecodonts and therapsids suffered the most: the former were wiped out, and only scarce remnants of the latter survived along with mammal relatives. In contrast, crocodilians, pterosaurs, and especially dinosaurs sailed through the crisis into the Early Jurassic with little disruption. Avepod theropods such as *Coelophysis* remained common and little changed, as did prosauropods. Sauropods just got bigger. For the rest of the Mesozoic, dinosaurs would enjoy almost total dominance on land except for some semiterrestrial crocodilians; there simply were no competitors above a few kilograms in weight. Such extreme superiority was unique in Earth's history. The Jurassic and Cretaceous combined were the Age of Dinosaurs.

The Late Triassic *Coelophysis*

As the Jurassic progressed, the prosauropods appear to have been unable to compete with their more sophisticated sauropod relatives and were gone by the end of the Early Jurassic. The larger hip muscles and the beginnings of a birdlike respiratory system suggest that sauropods had the higher aerobic capacity and higher-pressure circulatory system needed to achieve truly great height and bulk. Although some theropods were getting moderately large, the much more gigantic sauropods enjoyed a period of relative immunity from attack. Ornithischians remained uncommon, and one group was the first set of dinosaurs to develop armor protection. Another group of ornithischians was the small, chisel-toothed, semibipedal heterodontosaurs, which established that fiber coverings had evolved in some small dinosaurs by this time if not earlier. On the continents, crocodilians remained small and fully or semiterrestrial, while other groups became marine giants.

Partly splitting Pangaea into northern Laurasia and southern Gondwanaland like a marine wedge was the great Tethys tropical ocean, the only surviving remnant of which is the Mediterranean. Farther west, the supercontinent was beginning to break up, creating African-style rift valleys along today's Eastern Seaboard of North America that presaged the opening of the Atlantic. More importantly for dinosaur faunas, the increased tectonic activity in the continent-bearing conveyor belt formed by the mantle caused the ocean floors to lift up, spilling the oceans onto the continents in the form of shallow seaways that began to isolate different regions from one another, encouraging the evolution of a more diverse global wildlife. The expansion of so much water onto the continents also raised rainfall levels, although most habitats remained seasonally semiarid. The moving land masses also produced more mountains able to squeeze rain out of the atmosphere.

Beginning 175 million years ago, the Middle Jurassic began the Age of Sauropods, whose increasingly sophisticated respiratory and circulatory systems allowed them to match medium-sized whales in bulk and trees in height. Sauropods thrived even in dry habitats by feeding on the forests that lined watercourses as well as the fern prairies in the wet season. In China, partly isolated by seaways, some sauropods evolved slender necks so long that they could feed 10 meters (over 30 feet) high. A few sauropods had tail spikes or clubs. Also appearing were the first small, armored stegosaur ornithischians that also introduced tail spikes. Even smaller were the little ornithopods, the beginnings of a group of ornithischians whose respiratory systems—which may have paralleled those of mammals—and dental batteries gave them great evolutionary potential. Although the increasingly sophisticated tetanuran, avetheropod, and coelurosaur theropods evolved and featured highly developed avian-type respiratory systems, for reasons that are obscure, they continued to fail to produce true giants. There is new evidence that flowering plants were present by the middle of the Jurassic, but even if so they were not yet common.

The Late Jurassic, which began 160 million years ago, was the apogee of two herbivorous dinosaur groups, the sauropods and the stegosaurs. Sauropods, which included haplocanthosaurs, mamenchisaurs, dicraeosaurs, diplodocines, apatosaurines, camarasaurs, and the first titanosaurs, would never again be so diverse. Some neosauropods rapidly enlarged to 50 to 75 tonnes, and a few may have greatly exceeded 100 tonnes, rivaling the biggest baleen whales. The tallest sauropods could feed over 20 meters (70 feet) high. But it was a time of growing danger for the sauropods: theropods had finally evolved hippo-sized yangchuanosaurs and allosaurs that could tackle the colossal herbivores. Meanwhile, some sauropods isolated on islands underwent dwarfing to rhino size to better accommodate to the limited resources (the same would happen to elephants and hippos). The rhino- and sometimes elephant-sized stegosaurs were at their most diverse. But the future of the other group of big armored dinosaurs, the short-legged ankylosaurs, was beginning to develop. Also entering the fauna were the first fairly large ornithopods, sporting thumb spikes. Asia saw the development of small semibipedal ceratopsians.

The still-small ancestors of tyrannosaurs seem to have been developing at this time, and assorted gracile maniraptor coelurosaurs were numerous. Also present by the Late Jurassic were the curious alvarezsaurs, whose stout and short arms and hands were adapted for breaking into insect nests. But it is the advent of the highly birdlike and probably partly arboreal aveairfoilans at the end of the mid-Jurassic going into the late part of the period that was a major event. Dinosaur flight appears to have come in two versions. One experiment was the bat-winged scansoriopterygids of Asia; these apparently soon disappeared, perhaps because of competition from the bird-winged aveairfoilans. The Chinese deinonychosaur *Anchiornis* is the earliest dinosaur known to have had large feathers on its arms, and on its legs too. Because the moderately long, symmetrical feathers were not proper airfoils despite the great length of the arms, this apparent climber may be the first example of a reduction of flight abilities from an ancestor with superior aerial prowess. A few million years later, when Europe was still a nearshore

The Late Jurassic *Giraffatitan* and *Dicraeosaurus*

The Late Cretaceous *Tyrannosaurus* and *Therizinosaurus*

extension of northeastern North America, the first "bird," the deinonychosaur *Archaeopteryx*, was extant. Preserved in lagoonal deposits on the northwestern edge of the then great Tethys Ocean, it had a combination of very large arms and long, asymmetrical wing feathers indicating that it was part of the process of developing the early stages of powered flight. The advent of the little aveairfoilans also heralded the first major increase in dinosaurian mental powers, as brain size and complexity rose to the lower avian level. Pterosaurs, which retained smaller brains, remained small bodied, and most still had long tails. Although some crocodilians were still small runners, the kind of highly amphibious crocodilians of the sort we are familiar with were appearing. Their liver-pump lung systems readapted into buoyancy control devices. Although small, mammals were undergoing extensive evolution in the Jurassic. Many were insectivorous or herbivorous climbers, but some were burrowers, and others had become freshwater-loving swimmers weighing a few kilograms.

During the Middle and Late Jurassic, carbon dioxide levels were incredibly high, with the gas making up between 5 percent and 10 percent of the atmosphere. As the Jurassic and the Age of Sauropods ended, the incipient North Atlantic was about as large as today's Mediterranean. Vegetation had not yet changed dramatically from the Triassic. Wetter areas were dominated by conifers similar to cypress. A widespread and diverse conifer group of the time was the araucarians. Some appear to have evolved a classic umbrella shape in which most of the adult trunk was as bare of foliage as a telephone pole, with all of the branches concentrated at the top. Still seen in a few South American examples, this odd shape may have evolved as a means of escaping browsing by the sauropods, which should have had a profound impact on floral landscapes as they heavily browsed and wrecked trees to an extent that probably exceeded that of elephants. What happened to the fauna at the end of the Jurassic is not well understood because of a lack of deposits. Some researchers think that there was a major extinction, but others disagree.

The Cretaceous began 145 million years ago. This period would see an explosion of dinosaur evolution that surpassed all that had gone before as the continents continued to split, the south Atlantic began to open, and seaways crisscrossed the continents. Greenhouse conditions became less extreme as carbon dioxide levels gradually edged downward, although never down to the modern preindustrial level. Early in the Cretaceous, the warm Arctic oceans kept conditions up there balmy even in the winter. At the other pole, continental conditions rendered winters frigid enough to form permafrost. General global conditions were a little wetter than they were earlier in the Mesozoic, but seasonal aridity remained the rule in most places, and true rain forests continued to be scarce at best.

Sauropods remained abundant and often enormous, but they were less diverse than before, as a few small-bodied, short-necked diplodocoids—some with broad, square-ended mouths specialized for grazing—tall brachiosaurs, and especially the broad-bellied titanosaurs predominated.

To a fair extent the Cretaceous was the Age of Ornithischians. Ornithopods small and especially large flourished. Thumb-spiked iguanodonts soon became common herbivores in the Northern Hemisphere. Their well-developed dental batteries may have been a key to their success. A few evolved tall sails formed by their vertebral spines. Until recently it was thought that the heterodontosaur clade had failed well back in the Jurassic, but we now know that they made it into at least the early Cretaceous in Asia with little change in form. Among ceratopsians, the small Asian chisel-toothed psittacosaurs first proliferated, and their relatives, the big-headed protoceratopsids, appeared in the same region. So did the first of the dome-headed pachycephalosaurs. Stegosaurs, however, soon departed the scene, the final major dinosaur group to become totally extinct since the prosauropods. This reveals that over time the dinosaurs tended to add new groups without losing the old ones, building up their diversity over the Mesozoic. In the place of stegosaurs, the low-slung and extremely fat-bellied armored ankylosaurs became a major portion of the global fauna, their plates and spikes providing protection from the big Laurasian allosauroids and the snub-nosed, short-armed abelisaurs in Gondwana. Another group of giant theropods, the croc-snouted spinosaurs, apparently adapted to catch fish as part of their diet. Bone isotopes indicate that spinosaurs were semiaquatic like hippos. Some of them also evolved great sail backs.

It was among the smaller theropods that dinosaur evolution really went wild in the Early Cretaceous. The first of the ostrich-mimicking ornithomimids were present, as were the initial, not yet titantic, tyrannosaurs with similarly long running legs and reduced arms. But the focus of events was among the nearly avian aveairfoilans. As revealed by the spectacular lake deposits of northeastern China, deinonychosaurs developed into an array of flying and flightless forms, with the latter possibly secondarily flightless descendants of the fliers. The famous sickle-clawed dromaeosaurs appear to have begun as small aerialists with two sets of wings, the normal ones on the arms and an equally large set on the hind legs. From these appear to have evolved bigger terrestrial dromaeosaurs that hunted large game. The other major sickle-clawed deinonychosaur group, the more lightly built and swifter-running troodonts, also thrived.

At the same time, birds themselves not only descended from deinonychosaur dinosaurs, the Chinese deposits show they had already undergone a spectacular evolutionary radiation by 125 million years ago. Some retained teeth; others were toothless. Some had long tails; most did not. None were especially large. Among these early birds were the toothed, long-skulled, and long-tailed herbivorous jeholornithiformes. It is possible that they were the ancestors of the enigmatic, potbellied, land-bound therizinosaur dinosaurs. The short-tailed, deep-beaked omnivoropterygid birds bear a striking resemblance to the caudipterygid and protarchaeopterygid oviraptorosaurs from the same formations. It is possible that the short-tailed oviraptorosaurs were another group of secondarily flightless dinosaur-birds, ones more advanced than the archaeopterygian-dromaeosaur-troodont deinonychosaurs, and the therizinosaurs. The conventional view held by most researchers is that flightless therizinosaurs and oviraptorosaurs happened to be convergent with the flying jeholornithiformes and omnivoropterygids, respectively.

Pterosaurs, most of them now short tailed and consequently more dynamic fliers, were becoming large as they met increasing competition from birds. Also fast increasing in size were the freshwater crocodilians, making them an increasing threat for dinosaurs coming to water to drink or for other purposes. Some large crocodilians were semiterrestrial and able to attack big dinosaurs on land as well as in the water. Still scampering about were a few small running crocodilians. Some carnivorous mammals were big enough, about a dozen kilograms, to catch and consume the smallest dinosaurs and their babies. Even gliding mammals had evolved by this time.

During the late Early Cretaceous a major evolutionary event occurred, one that probably encouraged the rapid evolution of dinosaurs. Flowering plants began to become an important portion of the global flora. The first examples were small shrubs growing along shifting watercourses where their ability to rapidly colonize new territory was an advantage. Others were more fully aquatic, including water lilies. Their flowers were small and simple. The fast growth and strong recovery potential of flowering plants may have encouraged the development of low-browsing ankylosaurs and ornithopods. Conversely, the browsing pressure of dinosaurs may have been a driving force behind the evolution of the fast-spreading and fast-growing new plants. Also appearing about this time were South American conifers with monkey-puzzle foliage, their umbrella shape encouraged by the ever-hungry sauropods.

In the Late Cretaceous, which began 100 million years ago, the continental breakup was well under way, with interior seaways often covering vast tracts of land. As carbon dioxide levels continued to drop, the dark Arctic winters became cold enough to match the conditions seen in today's high northern forests, and glaciers crept down high-latitude mountains. Mammals were increasingly modern, and small. Pterosaurs, marine and terrestrial, became gigantic to a degree that stretches credulity. Oceanic pteranodonts had wings stretching 8 meters (over 25 feet). Toward the end of the Cretaceous, the freshwater-loving

***Tyrannosaurus* biting off the horn of a *Triceratops*, based on a fossil of the latter**

azhdarchids sported wings of 11 meters (over 35 feet) and outweighed ostriches. Small running crocodilians remained extant, and a few even became herbivorous. As for the conventional freshwater crocodilians, in some locales they become colossi up to 12 meters (close to 40 feet) long and approaching 10 tonnes, as large as the biggest flesh-eating theropods. Although these monsters fed mainly on fish and smaller tetrapods, they posed a real threat to all but the largest dinosaurs. The hazard should not be exaggerated, however, because these supercrocs do not appear to have been very numerous in many locations and were absent at higher latitudes. Even so, their existence may have discouraged the evolution of highly aquatic dinosaurs.

Although sauropods soon became limited to the titanosaurs, they diversified and proliferated across most of the globe, being especially diverse in the Southern Hemisphere, wrapping up the 150 million years that made them the most successful herbivore group in Earth's history. Sauropods disappeared from North America for part of the Late Cretaceous, only to reappear in the drier regions toward the end. Some sauropods were armored; this may have been a means to protect the juveniles against the increasing threat posed by a growing assortment of predators. A few small titanosaurs had the short necks and square, broad mouths suited for grazing. Others were titanic, exceeding 50 and perhaps 100 tonnes up to the end of the dinosaur era. These were subject to attack from abelisaur and allosauroid theropods, some matching bull elephants in bulk. Perhaps even larger were the African sail-backed spinosaurs of the early Late Cretaceous; unlike the abelisaurs and allosauroids, this group did not make it to the end of the Mesozoic.

The ultrawide-bodied ankylosaurs continued their success, especially in the Northern Hemisphere. One group of the armored herbivores developed tail clubs with which to deter and if necessary damage their enemies, as well as settle breeding and perhaps feeding disputes within the species. The iguanodonts faded from the scene to be replaced by their descendants, the duck-billed hadrosaurs, which evolved the most complex grinding dental batteries among dinosaurs and often used elaborate head crests to identify the variety of species. The most common herbivores in much of the Northern Hemisphere, hadrosaurs may have been adapted in part to browse on the herbaceous shrubs and ground cover that were beginning to replace the fern prairies as well as to invade forest floors. Small ornithopods, not all that different from the bipedal ornithischians that had appeared back near the origins of the dinosaurs, continued to dwell over much of the globe. In the Northern Hemisphere the protoceratopsids, small in body and big in head, were common in many locales. It was from this stock that some of the most spectacular dinosaurs evolved—the rhino- and elephant-sized ceratopsids whose oversized heads sported horns, neck frills, great parrot-like beaks, and slicing dental batteries. These remarkable dinosaurs flourished for just the last 15 million years of the dinosaur era, limited largely to the modest-sized stretch of North America that lay west of the interior seaway; for some reason their presence in Asia was very limited.

Birds, some still toothed, continued to thrive. One group of oceanic birds lost flight to the point that they evolved into fully marine divers. By the late Cretaceous the classic short-armed coelurosaurs were no longer extant. The small predatory theropods consisted of the intelligent and sickle-clawed swift troodonts and leaping dromaeosaurs, some of which were still able to fly. Also successful were the short-tailed nonpredatory aveairfoilans, those being the deep-headed omnivorous oviraptorosaurs, many exhibiting dramatic head crests, as well as the small-headed, big-clawed herbivorous therizinosaurs. In both groups some species became quite large, as did some ornithomimosaurs. But among the latter group the long- and slender-legged ornithomimids became perhaps the fastest of all dinosaurs, although they were closely matched by the colonial insect-eating alvarezsaurs.

Culminating the over 150 million years of theropod history were the great tyrannosaurids, the most sophisticated and powerful of the gigantic predators. The classic great tyrannosaurids came into existence only some 15 million years before the end of the Mesozoic and were limited to Asia and North America. Apparently they wandered, along with other theropods, hadrosaurs, and ankylosaurs, across the subpolar Bering land bridge, where some became specialized for the winter climate. In North America a size race occurred as tyrannosaurids, ceratopsids, ankylosaurids, and pachycephalosaurids reached unprecedented sizes for their groups in the final few million years of the Cretaceous, resulting in the classic *Tyrannosaurus*, *Triceratops*, *Ankylosaurus*, and *Pachycephalosaurus* fauna; the ornithomimids got bigger too. This may have been the result of a predator-prey arms race, or expansion of the resource base as the retreating interior seaway linked the eastern and western halves of the continent into a larger land area, or a combination of both. It is interesting that the hadrosaurs did not get bigger—some earlier edmontosaurs were if anything larger than those that followed, some of the latter being well adapted for grazing. This pattern indicates that the enormous size and firepower of the American *Tyrannosaurus* was a specialization for hunting the equally oversized contemporary horned dinosaurs rather than just dispatching the easier-to-kill edmontosaurs. Nor did the armored nodosaurids enlarge at this time.

By the end of the Cretaceous the continents had moved far enough that the world was beginning to assume its modern configuration. At the terminus of the period a burst of uplift and mountain building had helped drain much of the seaways. Flowering plants were fast becoming an ever more important part of the flora, and the first hardwood trees—among them the plane tree commonly planted in cities—evolved near the end of the period and were evolving into the first large hardwood trees. Conifers remained dominant, however, among them the deciduous, moisture-dependent dawn redwoods that barely survived to modern times. Also common were the classic redwoods, which reached towering heights as they do today. In South America the browsing pressure of the towering titanosaurs may have continued to encourage the evolution of the umbrella-topped monkey-puzzle araucarians. Classic rain forests, however, still did not exist. Grasses had evolved: they tended to be water-loving forms and did not yet form dry grassland prairies.

Then things went catastrophically wrong.

EXTINCTION

The mass extinction at the end of the Mesozoic is generally seen as the second most extensive in Earth's history, after the one that ended the Paleozoic. However, the earlier extinction did not entirely exterminate the major groups of large land animals. At the end of the Cretaceous all nonavian dinosaurs, the only major land animals, were lost, leaving only flying birds as survivors of the group. Among the birds, all the toothed forms, plus a major Mesozoic bird branch, the enantiornithines, as well as the flightless birds of the time, were also destroyed. So were the last of the superpterosaurs and the most gigantic of the crocodilians.

It is difficult to exaggerate how remarkable the loss of the dinosaurs was. If dinosaurs had repeatedly suffered the elimination of major groups and experienced occasional diversity squeezes in which the Dinosauria was reduced to a much smaller collection that then underwent another evolutionary radiation until the next squeeze, then their final loss would not be so surprising. But the opposite is the case. A group that had thrived for over 150 million years over the entire globe, rarely suffering the destruction of a major group and usually building up diversity in form and species over time as they evolved into an increasingly sophisticated group, was in short order completely expunged. The small dinosaurs went with the large ones, predators along with herbivores and omnivores, and intelligent ones along with those with reptilian brains. It is especially notable that even the gigantic dinosaurs did not suffer repeated extinction events. Sauropods were always a diverse and vital group for almost the entire reign of dinosaurs. The same was true for giant theropods once they appeared, as well as ankylosaurs and the iguanodonts/hadrosaurs. Only the stegosaurs had faded away well into the dinosaur era. In contrast, many of the groups of titanic mammals appeared, flourished relatively briefly, and then went extinct. Dinosaurs appear to have been highly resistant to large-scale extinction. Rendering their elimination still more remarkable is that one group of dinosaurs, the birds, did survive, as well as aquatic crocodilians, lizards, snakes—the latter had evolved by the Late Cretaceous—amphibians, and mammals that proved able to weather the same crisis.

It has been argued that dinosaurs were showing signs of being in trouble in the last few million years before the final extinction. Whether they were in decline has been difficult to verify or refute even in those few locations where the last stage of the dinosaur era was recorded in the geological record, such as western North America. Even if true, the decline was at most only modest. At the Cretaceous/Paleocene (K/Pg), formerly the Cretaceous/Tertiary (K/T), boundary, the total population of juveniles and adult dinosaurs should have roughly matched those of similar-sized land mammals before the advent of humans, numbering in the billions and spread among many dozens or a few hundred species on all continents and many islands.

A changing climate has often been offered as the cause of the dinosaurs' demise. But the climatic shifts at the end of the Cretaceous were neither strong nor greater than those already seen in the Mesozoic. And dinosaurs inhabited climates ranging from tropical deserts to icy winters, so yet another change in the weather should not have posed such a lethal problem. If anything, reptiles should have been more affected. The rise of the flowering plants has been suggested to have adversely impacted dinosaurs, but the increase in food sources that the fast-growing seed- and fruit-producing plants provided appears to have been so much to the dinosaurs' benefit that it spurred the evolution of late Mesozoic dinosaurs. Mammals consuming dinosaur eggs are another proposed agent. But dinosaurs had been losing eggs to mammals for nearly 150 million years, and so had reptiles and birds without long-term ill effects. The spread of diseases as retreating seaways allowed once-isolated dinosaur faunas to intermix is not sufficient because of their prior failure to crash the dinosaur population, which was too diverse to be destroyed by one or a few diseases and which would have developed resistance and recovered its numbers. Also unexplained is why other animals survived.

The solar system is a shooting gallery full of large rogue asteroids and comets that can create immense destruction. There is widespread agreement that the K/P extinction was caused largely or entirely by the impact of at least one meteorite, a mountain-sized object that formed a crater 180 km (over 100 miles) across, located on the Yucatán Peninsula of Mexico. The evidence strongly supports the object being an asteroid rather than a comet, so speculations that a perturbation of the Oort cloud as the solar system traveled through the galaxy and its dark matter are at best problematic. The explosion of 100 teratons surpassed the power of the largest H-bomb detonation by a factor of 20 million and dwarfed the total firepower of the combined nuclear arsenals at the height of the Cold War. The blast and heat generated by the explosion wiped out the fauna in the surrounding vicinity, and enormous tsunamis cleared off many coastlines. On a wider scale, the cloud of high-velocity debris ejected into space glowed hot as it reentered the atmosphere in the hours after the impact, creating a global pyrosphere that may have been searing enough to bake animals to death as it ignited planetary wildfires. The initial disaster would have been followed by a solid dust pall that plunged the entire world into a dark, cold winter lasting for years, combined with severe air pollution and acid rain. As the aerial particulates settled, the climate then flipped as enormous amounts of carbon dioxide—released when the impact hit a tropical marine carbonate platform—created an extreme greenhouse effect that baked the planet for many thousands of years. Such a combination of agents appears to solve the mystery of the annihilation of the dinosaurs. Even so, some problems remain.

It is not certain whether the pyrosphere was as universally lethal as some estimate. Even if it was, heavy storms covering a small percentage of the land surface should have shielded a few million square kilometers, equal in total to the size of India, creating scattered refugia. In other locations dinosaurs that happened to be in burrows, caves, and deep gorges, as well as in water, should have survived the pyrosphere. So should many of the eggs buried in covered nests. Birds and amphibians, which are highly sensitive to environmental toxins, survived the acid rain and pollution. Because dinosaurs were rapidly reproducing animals whose self-feeding young could survive without the care of the parents, at least some dinosaurs should have made it through the crisis, as did some other animals, recolonizing the planet as it recovered.

Massive volcanism occurred at the end of the Cretaceous as enormous lava flows covered 1.5 million square kilometers, a third of the Indian subcontinent. It has been proposed that the massive air pollution produced from the repeated supereruptions damaged the global ecosystem so severely in so many ways that dinosaur populations collapsed in a series of stages, perhaps spanning tens or hundreds of thousands of years. This hypothesis is intriguing because extreme volcanic activity also occurred during the great Permo-Triassic extinction; those eruptions were in Siberia. Although the K/Pg Deccan Traps were being extruded before the Yucatán impact, evidence indicates that the latter—which generated earthquakes of magnitude 9 over most of the globe (11 at the impact site)—greatly accelerated the frequency and scale of the eruptions. If this is correct, then the impact was responsible for the extinction not just via its immediate, short-term effects, but by sparking a level of extended supervolcanism that prevented the recovery of dinosaurs. It is also possible that the Yucatán impactor was part of an asteroid set that hit the planet repeatedly, further damaging the biosphere. Even so, the combined impact/volcanic hypothesis does not fully explain why dinosaurs failed to survive problems that other continental animals did.

Although extraterrestrial impact(s), perhaps indirectly linked with volcanism, is the leading explanation, the environmental mechanisms that destroyed all of the nonflying dinosaurs while leaving many birds and other animals behind remain incompletely understood.

AFTER THE AGE OF DINOSAURS

Perhaps because trees were freed from chronic assault by sauropods, dense forests, including rain forests, finally appeared. After the extinction of the nonavian dinosaurs, there were no large land animals, and only large freshwater crocodilians could make a living feeding on fish. The loss of dinosaurs led to a second, brief Age of Reptiles as superboa snakes as long as the biggest theropods and weighing over a tonne quickly evolved in the tropics. Their main prey was probably a diverse array of crocodilians, some semiterrestrial, as well as mammals, which were also swiftly expanding in size. By 40 million years ago, about 25 million years after the termination of large dinosaurs, some land and marine mammals were evolving into giants rivaling the latter. Among the survivors of the Dinosauria, a number of birds lost flight and soon became large land runners and marine swimmers. But the main story of Cenozoic dinosaurs has been their governance of the daylight skies, while the night has been dominated by the mammalian fliers, the bats. The greatest success story of modern flying dinosaurs? The marvelous diversity and numbers of the little but sophisticated passerine songbirds that fill field guides.

BIOLOGY

General Anatomy

Dinosaur heads ranged from remarkably delicately constructed to massively built. In all examples the nasal passages or the sinuses or both were very well developed, a feature common to archosaurs in general. Many dinosaurs retained a large opening immediately in front of the orbits; in others this opening was almost entirely closed off. Unlike mammals with their extensive facial musculature, dinosaurs, like reptiles and birds, lacked facial muscles, so the skin was directly appressed to the skull. This feature makes dinosaur heads easier to restore than those of mammals. The external nares are always located far forward in the nasal depression no matter how far back on the skull the nasal openings extend. In some sauropods the nasal openings are set far back on the skull, above the eye sockets. It was once thought that this allowed these dinosaurs to snorkel when submerged. More recently it has been suggested that the retracted nostrils evolved to avoid irritation from needles as sauropods fed on conifers. Most conifers at that time, however, had soft leaves. In any case, the fleshy nostrils extended far forward so that the external nares were in the normal position near the tip of the snout. There is no anatomical evidence that any dinosaur had a proboscis. The skin covering the large openings in front of the orbits of many dinosaurs probably bulged gently outward. Jaw muscles likewise bulged gently out of the skull openings aft of the eye sockets.

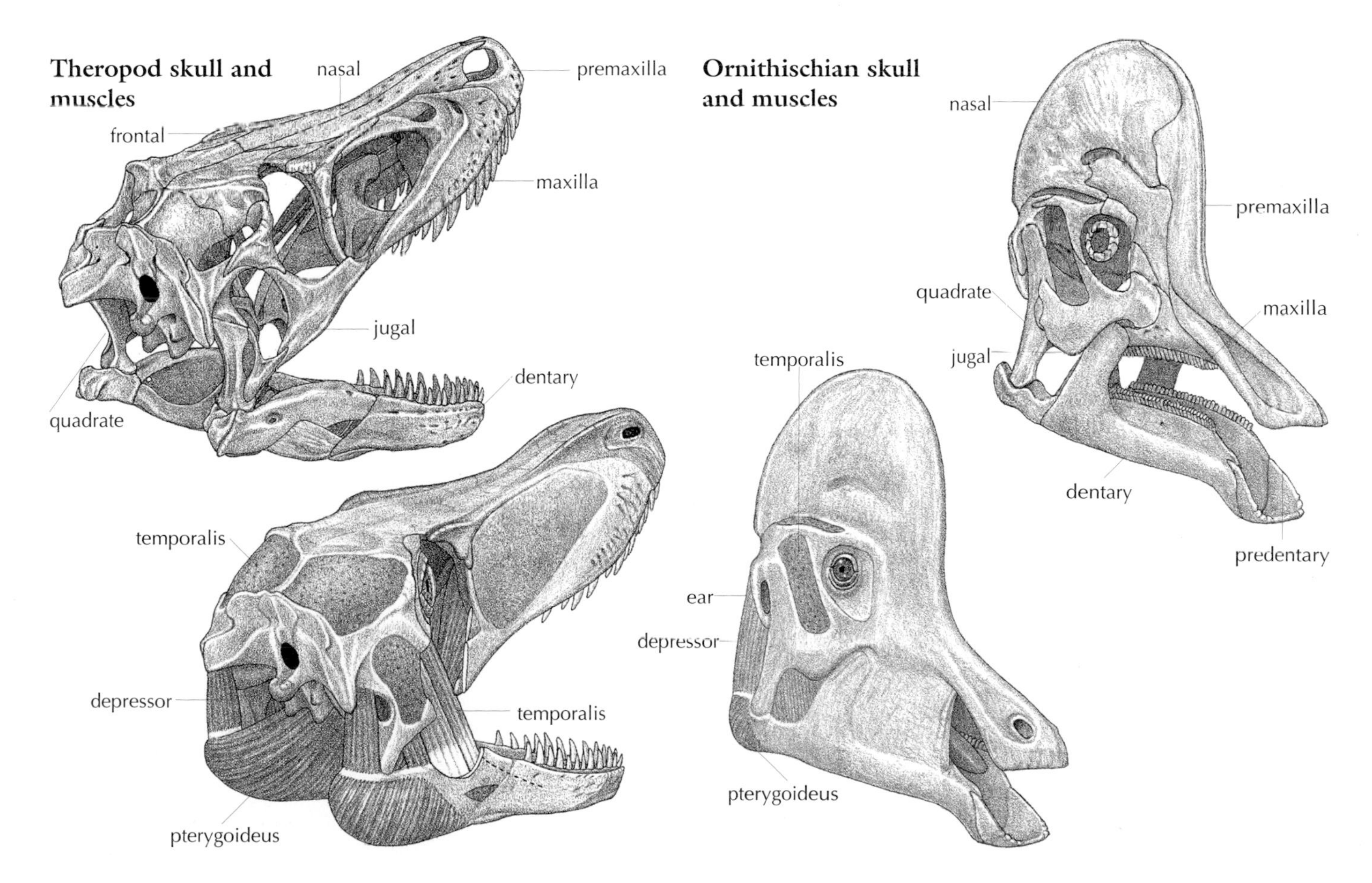

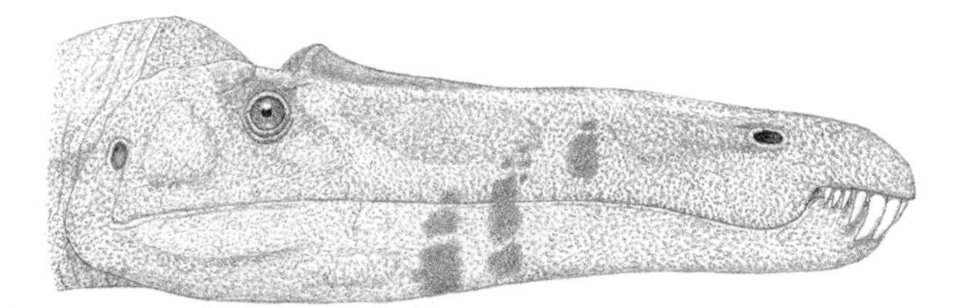

Baryonyx with exposed front teeth

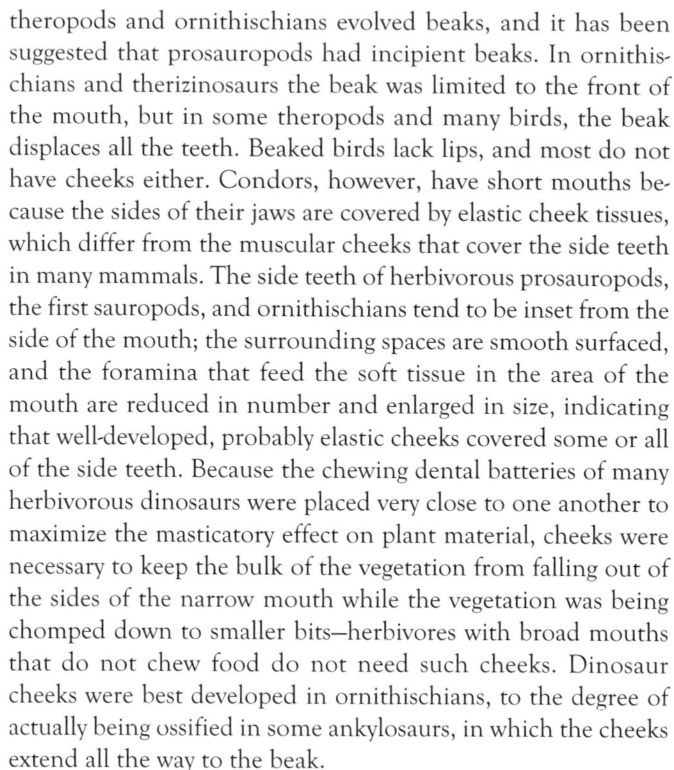

crocodilian without lips

theropod restored with lips

Archosaur lip anatomy

Among amphibians, tuataras, lizards, and snakes, the teeth tend to be set close to one another along fairly sharp-rimmed jaws, and the mouth is sealed and the teeth covered by nonmuscular lips when closed. This arrangement appears to be true of most theropods, and sauropods as well. An exception among theropods would be the spinosaurs, which have a more crocodilian arrangement in which at least the front teeth are widely spaced in separate sockets, so they may have been lipless and their snaggly teeth exposed when the jaws were closed. Some theropods and ornithischians evolved beaks, and it has been suggested that prosauropods had incipient beaks. In ornithischians and therizinosaurs the beak was limited to the front of the mouth, but in some theropods and many birds, the beak displaces all the teeth. Beaked birds lack lips, and most do not have cheeks either. Condors, however, have short mouths because the sides of their jaws are covered by elastic cheek tissues, which differ from the muscular cheeks that cover the side teeth in many mammals. The side teeth of herbivorous prosauropods, the first sauropods, and ornithischians tend to be inset from the side of the mouth; the surrounding spaces are smooth surfaced, and the foramina that feed the soft tissue in the area of the mouth are reduced in number and enlarged in size, indicating that well-developed, probably elastic cheeks covered some or all of the side teeth. Because the chewing dental batteries of many herbivorous dinosaurs were placed very close to one another to maximize the masticatory effect on plant material, cheeks were necessary to keep the bulk of the vegetation from falling out of the sides of the narrow mouth while the vegetation was being chomped down to smaller bits—herbivores with broad mouths that do not chew food do not need such cheeks. Dinosaur cheeks were best developed in ornithischians, to the degree of actually being ossified in some ankylosaurs, in which the cheeks extend all the way to the beak.

Set in sockets, all dinosaur teeth were constantly replaced through life in the manner of reptile teeth. Teeth ranged from blunt, leaf-shaped dentition suitable for crushing plants to serrated blades adapted to piercing flesh. Like the teeth of today's carnivores, those of predatory theropods were never razor sharp, as is often claimed: one can run a finger firmly along the serrations without harm. The teeth of iguanodonts and especially hadrosaurs and ceratopsids were concentrated into compact dental batteries made up of hundreds of teeth, although only a minority formed the plant-processing pavement at a given time. A few sauropods also evolved fast-replacement dental arrays, in their case at the front edge of the jaws where the teeth cropped

Ornithischian cheeks

ankylosaur with ossified cheeks covering dental battery

side view

cross section

Californian condor with elastic cheek tissues

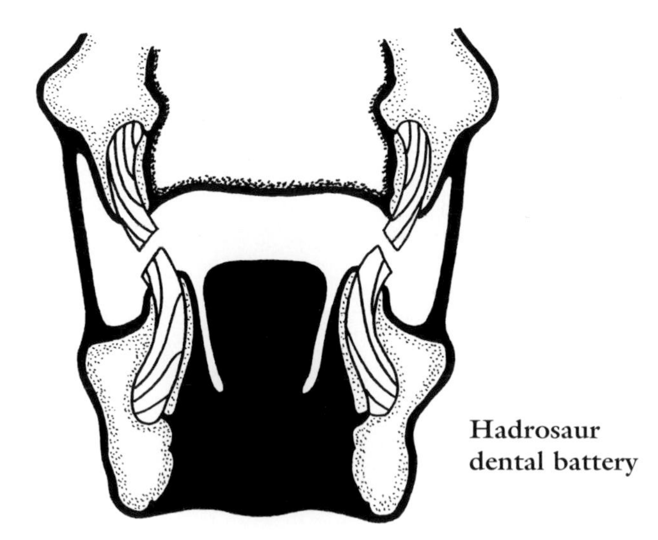

Hadrosaur dental battery

plant material. Because dinosaurs were not lizards or snakes, they lacked flickering tongues. Dinosaurs had well-developed hyoids, suggesting that the tongues they supported were similarly developed. In predatory theropods the tongue was probably simple and inflexible. The tongues of herbivorous dinosaurs were likely to have been more supple and complex in order to help manipulate and, in the case of ornithischians, chew fodder.

In some large dinosaurs the eyes were in the upper part of the orbit. Bony eye (sclerotic) rings often show the actual size of the eye both in total and indirectly in that the diameter of the inner ring tends to closely match the area of the visible eye when the eyelids are open. Most dinosaurs had large eyes, yet relative eye size decreases as animals get bigger. Although the eyes of giant theropods were very large, they looked small compared to the size of their heads. Even the eyes of ostriches, the biggest among living terrestrial animals, do not appear that large on the living animal. In the predatory daylight raptors, a bony bar running above the eyeball provides the fierce "eagle look." Interestingly, the flesh-eating theropods lacked this bar, but it was present in some of the smaller ornithischians, giving these plant eaters a more intimidating appearance than that of equally peaceful, doe-eyed herbivorous mammals. The purpose of the eye bar is not well understood. It may shade the eyes from glare, it may strengthen the skull during feeding and chewing, and it may have protected the eyes of burrowing ornithopods from dirt and dust. Whether the pupils of dinosaur eyes were circular or slit shaped is not known. The latter are most common in nocturnal animals, and either may have been present in different species. The eyes of birds and reptiles are protected by both lids and a nictitating membrane, so the same was presumably true in dinosaurs.

The outer ear is a deep, small depression between the quadrate and jaw-closing muscles at the back of the head. The eardrum was set in the depression and was connected to the inner ear by a simple stapes rod. The orientation of the semicircular canals of the inner ears is being used to determine the posture of dinosaur heads. For example, short-necked diplodocoid heads pointed straight down according to this method, implying that they grazed ground cover. The situation may, however, be more complicated, reducing the reliability of the method. In living animals the relationship between the orientation of the canals and the normal carriage of the head is not all that uniform. That animals position their heads in different manners depending on what they are doing does not help. Giraffes feed with the head pointing straight down when browsing on low shrubs, or horizontally, or straight up when reaching as high as possible, so the orientation of the semicircular canals is not particularly informative. It is widely thought that the broad-beaked, duck-billed hadrosaurs were grazers, so their heads should often have been held directed straight down. Yet their semicircular canals favor a horizontal head posture. The semicircular canals of at least some prosauropods seem to show that they typically held the nose tilted somewhat upward, an odd pose not normal to large herbivores. It seems that the posture of the semicircular canals is determined as much by the orientation of the braincase with the rest of the skull and does not reflect the orientation of the head as well as has been thought.

The necks of many dinosaurs tend to articulate in a birdlike S curve, as they do in most theropods and ornithopods. The beveling of the vertebrae is especially strong in some theropods. If anything, animals tend to hold their necks more erect than the articulations indicate. In other groups, such as ankylosaurs and ceratopsids, the necks were straighter. There has been a tendency to make dinosaur necks too short by placing the shoulder girdle too far forward. Even ankylosaur necks were long enough to accommodate two or three well-spaced armor rings. The flexibility of dinosaur necks ranged from low—the first few vertebrae of the short-necked ceratopsids were even fused together—to fairly high in longer-necked examples, but no dinosaur had the special adaptations that make bird necks exceptionally mobile.

The posture and function of the long necks of sauropods have become controversial. Some researchers propose a simplistic model in which the necks of all sauropods were held nearly straight and horizontally and, in a number of cases, could not be raised much above shoulder level. This was most true of one group, the short-necked diplodocoids. Otherwise the situation is complex and in many regards is not well understood. Many of the sauropod necks that have been restored in a straight line show obvious misarticulations or are based on vertebrae that are too distorted and incomplete to be reliably articulated. The vertebrae of the necks of different giraffe individuals do not articulate in a consistent manner: they can range from arcing strongly downward to strongly erect. This reflects the differing thickness of the cartilage pads between the vertebrae and demonstrates that the cartilage as well as the bones must be present to articulate necks properly. This is an obvious problem in that cartilage is rarely preserved in fossils. In many dinosaur skeletons the vertebrae are found jammed tightly together, probably because the intervening cartilage disks dried out after death and pulled the bones together.

In some articulated dinosaur skeletons, the vertebrae are still separated by the substantial gap that had been filled by the cartilage. The only example of the cartilage between the vertebrae being preserved in a sauropod neck is in two neck-base vertebrae of an old camarasaur that fused together before death. Contrary to the prediction based on horizontal-necked sauropods, the vertebrae are flexed upward as though the neck was held above shoulder level. Because sauropod necks had so many vertebrae, just 10 degrees of upward flexion between each pair allowed most of the sauropods to raise most of their necks nearly vertically, with the head far above shoulder level. Ostriches and giraffes hold their necks at different angles, and it is possible that sauropods did not really have specific neutral neck postures. There is no reason to assume that sauropods did not hold their necks higher than the bones may seem to indicate, and a growing number of researchers favor the probability that many sauropods held their heads high.

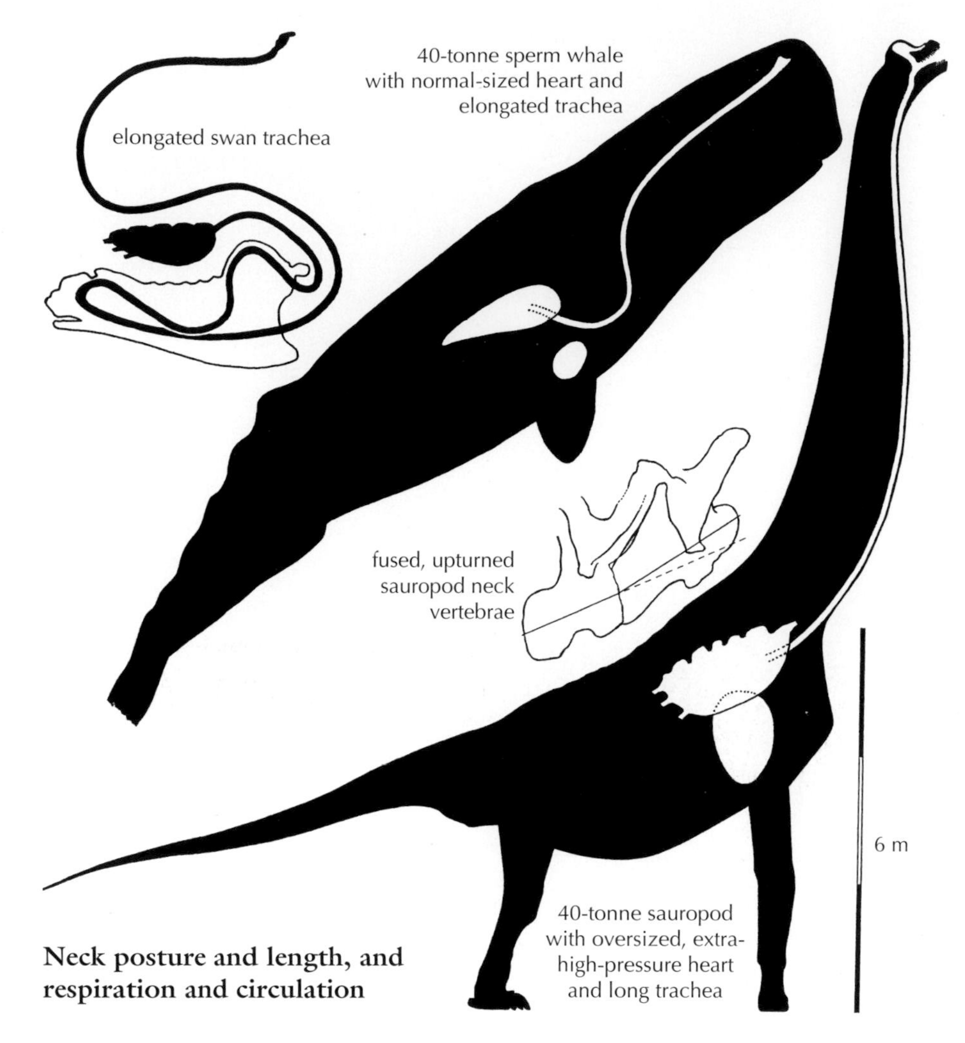

Neck posture and length, and respiration and circulation

Giraffe necks are not heavily muscled despite their having solid vertebrae that had to support a large head. Sauropod necks held up much smaller heads and were highly pneumatic, so they should not have been heavily muscled either. In some sauropods tall shoulder spines indicate that a fairly deep set of nuchal tendons helped to support the neck. In a number of other sauropods the neural spines were doubled in order to improve neck support. The upper neck muscles of big-headed pachycephalosaurs and ceratopsians should have been powerfully built, and some ceratopsids had the tall shoulder withers that indicate the presence of deep nuchal tendons. Mummies show that the hadrosaurs, whose neck vertebrae appear slender relative to their rather large heads, had deep nuchal tendons to help hold them up. The predatory theropods probably had the strongest neck muscles, which helped drive the teeth deeper into the flesh of their prey.

The trunk vertebrae of dinosaurs articulated either in a straight line or, more often, in a dorsally convex arch that varied from subtle to very strong. The nature of the vertebral articulations, and in many cases ossified interspinal tendons, indicates that dinosaurs had stiffer backs than lizards, crocs, and most mammals, although the trunk vertebrae of dinosaurs were not normally fused the way they often are in birds. As in lizards, crocodilians, and birds, the front ribs are strongly swept back in articulated dinosaur skeletons of all types; they are not vertical as they are in many mammals. Dinosaur belly ribs tend to be more vertical, but this condition is variable. The bellies and hips of the flesh-eating theropods were narrow, reflecting the small size of their digestive tracts as well as their athletic form. Big-game-hunting predators gorged after a kill and then fasted until the next one, so their bellies were hollow when they were on the hunt. The same should have been true of flesh-eating theropods, although abdominal air sacs, if present, may have filled out some the space of the gut even when the animals were hungry. The abdomens and hips of herbivorous dinosaurs were broader in order to accommodate more capacious digestive tracts.

Some plant-eating dinosaurs, therizinosaur theropods, titanosauriform sauropods, pachycephalosaurs, most stegosaurs, and especially ankylosaurs took the broadening of the belly and hips to an extreme, to a degree that seems absurd in the fattest of the armored dinosaurs. The shoulder blades of ankylosaurs were even twisted along their long axis to fit onto the rapid shift from the narrow shoulders to the fat abdomen. Because dinosaur trunk vertebrae and ribs formed a short, fairly rigid body with the shoulder and hip girdles close together, the trunk musculature was rather light, like that of birds. Theropods and prosauropods retained gastralia, a series of flexible bony rods in the skin of the belly. Each segment of the gastralia was made of multiple pieces. This may have been necessary in prosauropods because they flexed their trunks while running on all fours. Theropods needed flexible gastralia because their bellies changed dramatically in size as they gorged and fasted between hunts. In therizinosaurs the gastralia became more rigid, probably because these rigid-trunked herbivores always kept their abdomens full of fermenting fodder. These structures were absent in sauropods and ornithischians.

The tails of dinosaurs were highly flexible in most stegosaurs, theropods, and sauropodomorphs, especially the titanosaur sauropods, whose ball-and-socket joints may have allowed the tail to be arced directly over the back. In the sickle-clawed dromaeosaurid theropods, club-tailed ankylosaurids, and ornithopods, part or all of the tail was stiffened by ossified tendons, with the tails of iguanodonts and hadrosaurs being especially inflexible.

In most dinosaurs the hip vertebrae and tail were in much the same line as the trunk vertebrae. Because tail drag marks are very rare among the immense number of trackways known for all the major dinosaur groups, the old-style convention of persistently tail-dragging dinosaurs cannot be correct. This is true even in those dinosaurs whose tail base was swept downward. In therizinosaurs and some sauropods the hips and tail were flexed upward relative to the trunk vertebrae. This allowed the trunk to be held strongly pitched up while the hips and tail remained horizontal, increasing the vertical reach of the head while the dinosaur retained the ability to move on the hind legs. Because all dinosaurs bore most of their weight on their hind legs and usually had long tails that acted as counterweights to the body, all of them could rear up, even the few that had arms that were longer than their legs.

Unlike many mammals, no dinosaur had hands that looked like its feet. The hands always lacked a heavy central pad, even in the giant quadrupeds. Sauropods, stegosaurs, iguanodonts, and hadrosaurs united their short fingers into a hoof-like hand by encasing them in single, tight pad. A very distinctive character of theropods, prosauropods, some sauropods, and some ornithischians was the big-clawed, inwardly directed thumb weapon, which could be held clear of the ground when walking with the arms. The palms of dinosaurs always faced partly or strongly inward, especially in bipedal examples. In some of the larger dinosaurs—iguanodonts and hadrosaurs, armored dinosaurs, ceratopsids, and sauropods—the hind feet were underlain by a large central pad similar to those of rhinos and elephants.

The front of the rib cage of dinosaurs was narrow from side to side in order to accommodate the shoulder girdle, both sides of which nearly met one another on the chest, and the shoulder joint was immediately in front of the rib cage. This differs from mammals, in which the shoulder joint is on the side of the chest. In theropods, including birds, the shoulder girdle is fixed in place, partly by a fused furcula that braces both scapula blades. Many reptiles and mammals have mobile shoulder girdles that help increase the stride length of the arms. This appears to have been true of quadrupedal dinosaurs because their clavicles are not fused together, or they do not contact one another, or they are lost. In side view, the scapula blade of most dinosaurs was subvertical as in most tetrapods, not horizontal. The exceptions are the most birdlike theropods and birds themselves, whose scapula blades are horizontal.

In flying birds the shoulder joint faces sideways and upward so the arms can be held out to the side and raised vertically for flapping. In many predatory theropods the arms could also be swung laterally to grapple with prey. But even in winged protobirds like *Archaeopteryx* the arms could not be directed straight up. When dinosaurs were walking or running, trackways show that neither their arms nor legs were sprawled sideways like those of lizards. It is difficult to restore the precise posture of dinosaur limbs because in life the joints were formed by thick cartilage pads similar to those found on store-bought chickens, which are immature. Even so, some basics can be determined. The shoulder joints of quadrupedal dinosaurs faced down and backward so that the arm could swing below the shoulder joints, and the cylindrical hip joints forced the legs to work below the hips. But this does not mean that the erect limbs worked in simple, entirely vertical fore-and-aft planes. The elbows and knees, for instance, were bowed somewhat outward to clear the body, a feature common to many mammals as well. Trackways show that unlike the hands of mammals, which are often near the body midline when walking, the hands of dinosaurs were almost always separated by at least two hand widths, the hands were rarely placed closer to the midline than the feet, and the hands were often farther from the midline than the feet. This was because the arms were oriented so that the hands were either directly beneath the shoulder joints or a little farther apart. The hind feet of dinosaurs often did fall on the midline, even among some of the largest quadrupeds, and were never separated by much more than the width of a single hind print, even among the broadest-hipped sauropods and armored dinosaurs.

Dinosaur hands and feet were digitigrade, with the wrists and ankles held clear of the ground. Most dinosaurs retained the strongly flexed shoulder, elbow, hip, knee, and ankle joints that provided the springlike limb action needed to achieve a full run in which all feet were off the ground at some point in each complete step cycle. In addition, the ankle remained highly flexible, allowing the long foot to push the dinosaur into the ballistic stride. This was true of even the most gigantic theropods, ornithopods, ankylosaurs, and ceratopsids, which reached 5 to 15 tonnes. The knee joints of flexed-limbed dinosaurs were not fully articulated if they were straightened. Humans have vertical legs with straight knees because our vertical bodies place the center of gravity in line with the hip socket. In bipedal dinosaurs, because the head and body were held horizontally and were well forward of the hips, the center of gravity was ahead of the hip socket even with the long tail acting as a counterbalance, so the femur had to slope strongly forward to place the feet beneath the center of gravity. This arrangement is taken to an extreme in short-tailed birds, whose femur is nearly horizontal when they are walking in order to place the knees and feet far enough forward; in running, the femur of birds swings more strongly backward.

That dinosaurs normally retained thick cartilage pads in their limb joints throughout their entire lives, no matter how fast or big they became, is a poorly understood difference between them and birds and mammals that have well-ossified limb joints. The manner in which dinosaurs grew up and matured may explain the divergence. In terms of locomotory performance it does not seem to have done dinosaurs any harm and may have had advantages in distributing weight and stress loads.

Two groups of dinosaurs, the stegosaurs and sauropods, evolved elephantine, more columnar, straighter-jointed limbs. The configuration of the knee was altered so that it remained fully articulated when straight. In addition, the ankle was less mobile, and the hind feet were very short. This suite of

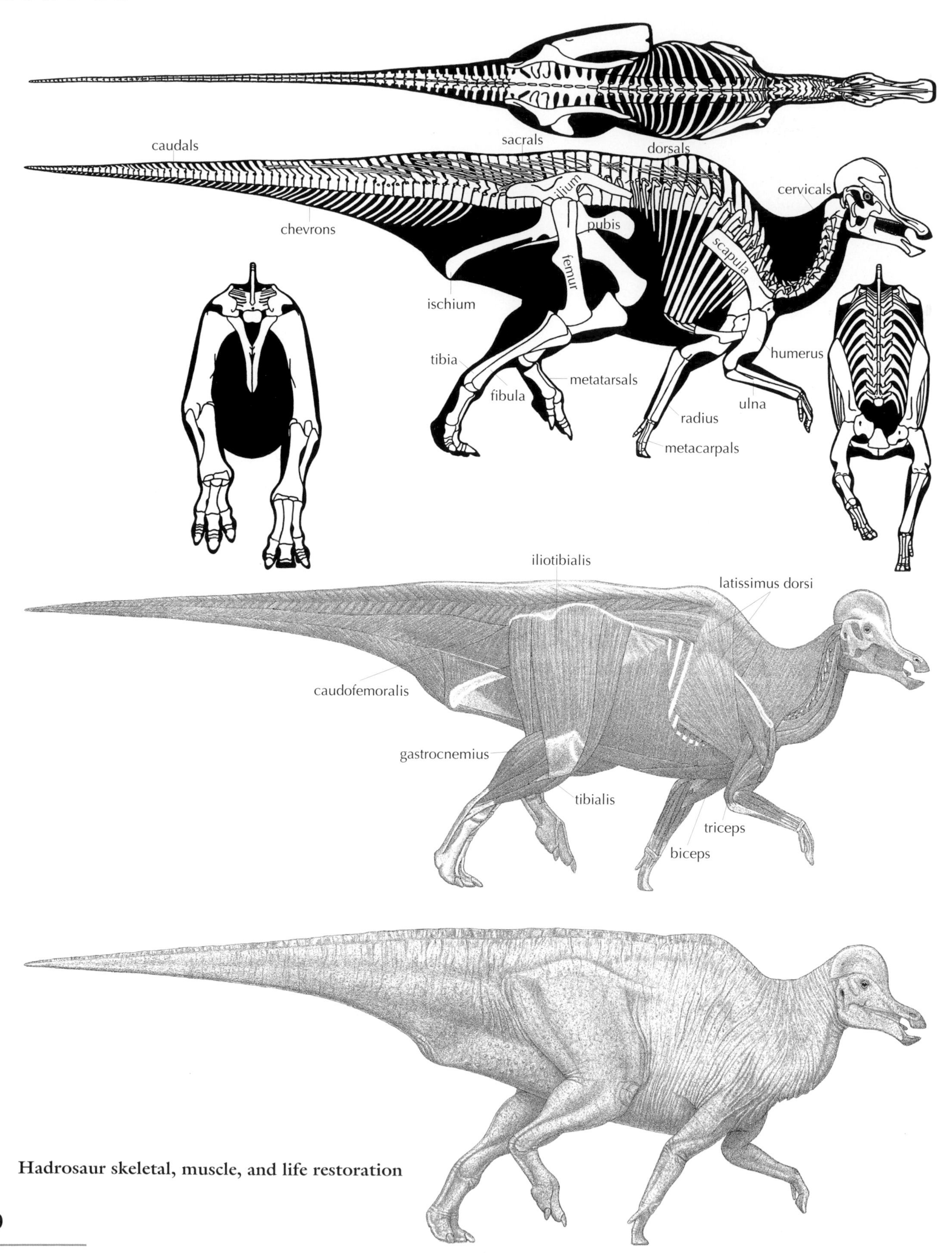

Hadrosaur skeletal, muscle, and life restoration

Dinosaur trackways, drawn to same stride length

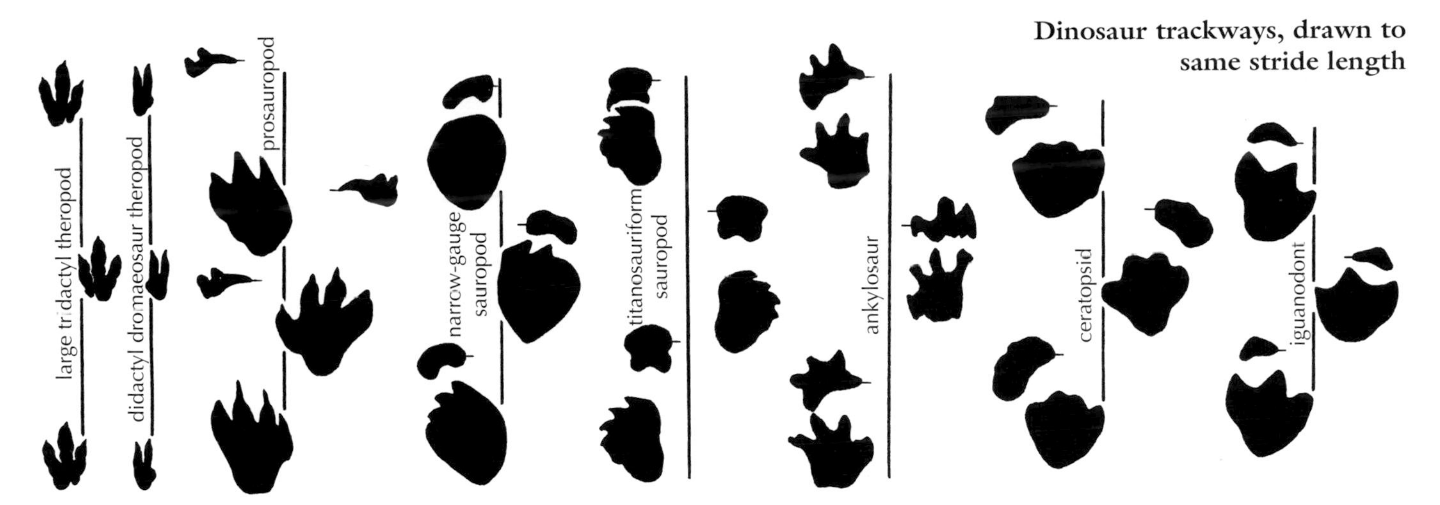

adaptations prevents the body from being propelled into a true run regardless of size: juvenile elephants cannot move any faster than their parents. Instead, at least one foot remains in contact with the ground at the highest speed.

The straight-limbed dinosaurs should not have been able to move faster than elephants, which cannot exceed 25 km/h (15 mph). Nor is it a problem to conclude that small and medium-sized dinosaurs with long, slender, flexed legs were able to run at speeds comparable to those of similar-sized ground birds and galloping mammals, which can reach 40–60 km/h (25–40 mph). Difficulties arise when trying to estimate the top speeds of flexed-limbed dinosaurs weighing many tonnes. Computer analysis has calculated that *Tyrannosaurus* could reach a top speed ranging from no better than that of a similar-sized elephant up to 40 km/h, the speed of a sprinting human. Because big-hipped, birdlike *Tyrannosaurus* was much better adapted for running than are elephants, it is unlikely that it was similarly slow, and other estimates suggest that giant theropods could run almost twice as fast as elephants, matching rhinos and nonthoroughbred horses.

The computer analyses to date are not able to fully simulate important aspects of animal locomotion, including the energy

Dinosaur limb articulations and posture

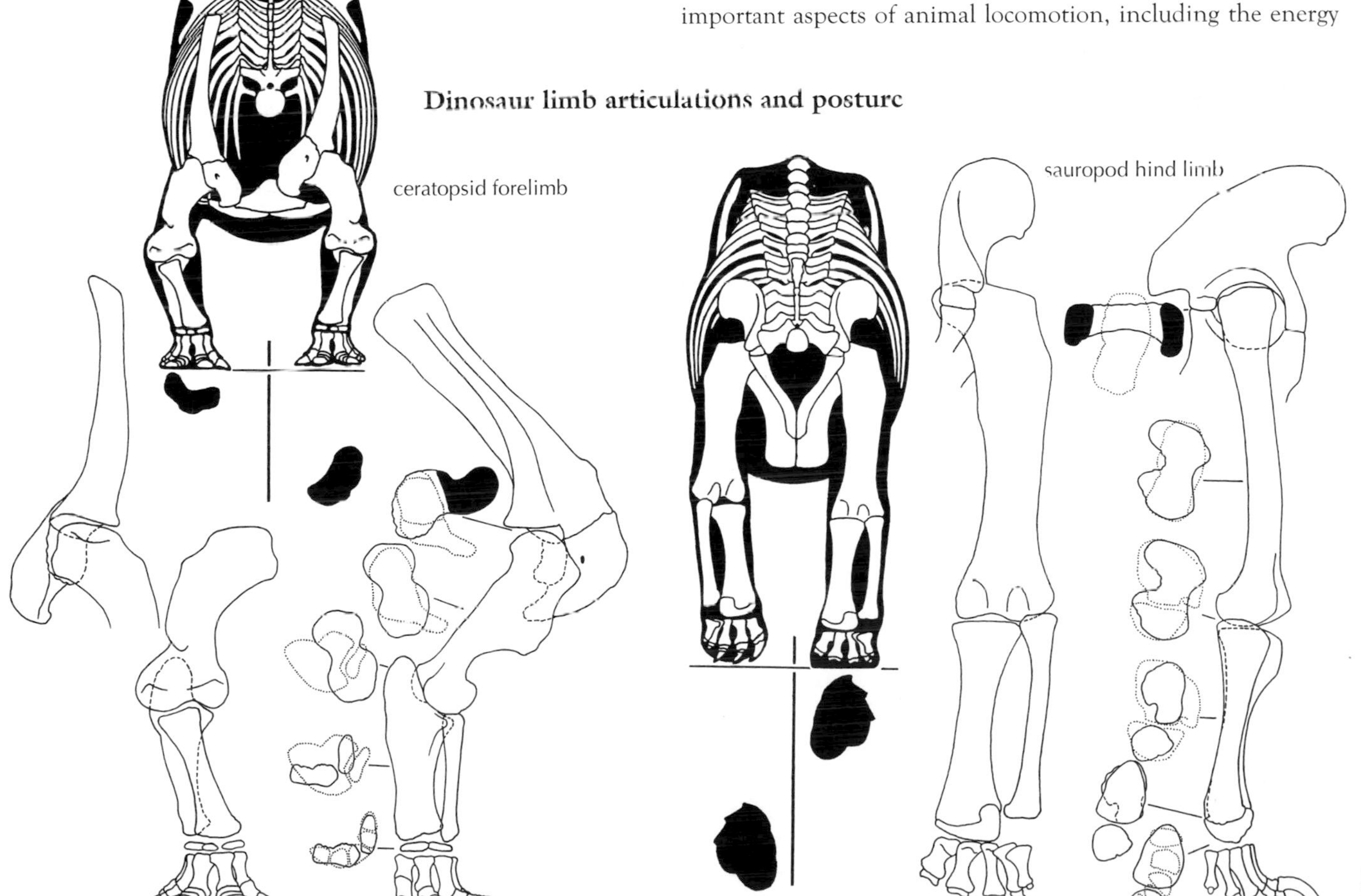

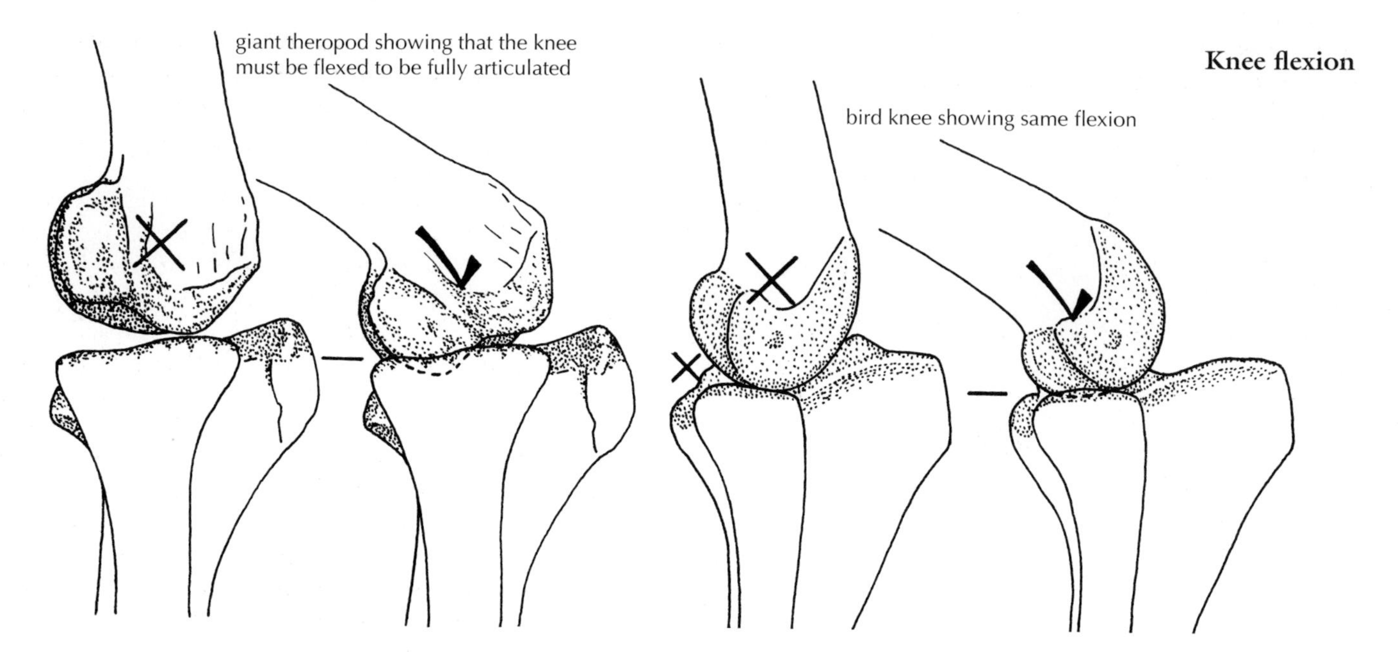

Knee flexion

storage of prestretched elastic leg tendons and the resonant springlike effect of the torso and tail. Nor has the ability of these programs to successfully calculate the performance of extreme animals been better established by showing how the most extreme of all dinosaurs, the supersauropods, managed to even stand upright much less move their whale-sized bodies. This important yet unanswered question is all the more pressing because trackways prove that the greatest sauropods walked without the support of water, yet they appear no better proportioned to support their mass than are the slow-moving elephants, which are ten or more times smaller. Did the supersauropods not need "super" adaptations beyond those seen in elephants to move about the Mesozoic landscapes, or did special adaptations such as stronger muscle fibers and pretensed tendons evolve to solve the problem? If the latter proves true, then other giant running dinosaurs may have used special adaptations to move faster than our computer models are indicating.

An important aspect of assessing dinosaur speed and power is the mass of the limb muscles, which tend to make up a larger percentage of the total mass in fast runners than in slower animals. Because the muscles are not preserved in dinosaur fossils, it is not possible to accurately restore the speed of a given dinosaur—at best it can only be approximated. The complex limb muscles of living mammals are the heritage of the unusual history of the early members of the group. Dinosaurs retained the simpler muscle patterns of reptiles, which are still seen in birds. A major muscle present in many reptiles and most dinosaurs, but not in birds and mammals, was the tail-based caudofemoralis, which helped pull the hind limb back during the propulsive stroke.

Although the absolute size of dinosaur muscles cannot be exactly determined, their relative size between the different groups can be approximated. In reptile hips the ilium is so short that the thigh muscles have to be narrow, limiting their size. The much longer ilia of birds and mammals anchor a broad and powerful set of thigh muscles. The ilium of the early herrerasaurs and prosauropods was short, so they must have had narrow thigh muscles. In other dinosaurs the ilium was longer and deeper, anchoring a larger set of thigh muscles able to produce more sustainable power. This trend was taken to an extreme in some dinosaurs. In the ostrichlike ornithomimids and tyrannosaurids, the oversized pelvis indicates the presence of exceptionally large leg muscles able to power high speeds. The ceratopsid dinosaurs had even longer hips, which probably supported the big leg muscles required to propel the fast charges needed to fend off the similarly strong-muscled tyrannosaurs. It is interesting that the enormous sauropods did not have especially large ilia. That is because they did not need large muscles to move at a fast pace. The same is true of elephants, which also lack large muscles below the knees because the feet that the shank muscles help operate are very short and nearly immobile. A similar situation was true of sauropods and stegosaurs. Faster animals have a large bundle of shank muscles that operate the long, mobile foot via long tendons. In bipedal dinosaurs, including birds, the large, drumstick-shaped collection of muscles below the knee is anchored on the cnemial crest projecting forward of the knee joint.

Fossil burrows indicate that some small dinosaurs dug burrows. This explains why small, bipedal ornithischians often had oversized shoulder girdles—they anchored the powerful upper arm muscles needed for digging with their broad hands.

Restorations of dinosaurs commonly simplify their surface contours, making their necks, tails, and legs into rather simple tubes and smoothing over the topography of the body. In sauropods the bulge of each neck vertebra was probably visible on the side of the neck, as it is in giraffes. Because the trachea and esophagus of sauropods were probably tucked up between their cervical ribs, the bottom of their necks should have been fairly flat, unlike giraffe necks, which lack well developed ribs. In dinosaurs with large arms, the upper end of the humerus bulged out a little, and in many but not all dinosaurs, a very large crest of the humerus formed a prominent contour along the upper front edge of the arm. The elbow joint formed a large bulge in front view, especially in the dinosaurs with massive arms—the ceratopsids, armored dinosaurs, and diplodocoids. The upper edge of the ilium was visible in living dinosaurs, especially the herbivores, in the same way that the pelvic bones of a cow can be seen under the skin.

Skin, Feathers, and Color

Most dinosaurs are known from their bones alone, but we know a surprising amount about dinosaur body coverings from a rapidly growing collection of fossils that record their integument. It has long been known that large, and some small, dinosaurs were covered with mosaic-patterned scales. These are usually preserved as impressions in the sediments before the skin rotted away, but in some cases traces of keratin are still preserved. Footprints sometimes preserved the shape of the bottom scales as well as the foot pads. The large dinosaurs whose skin is best known are the duck-billed hadrosaurs, for which some almost complete "mummies" are known. Lizard-like overlapping scales were not common among dinosaurs, although birdlike examples like those on the tops of some bird feet may have been present in birdlike dinosaurs, and overlapping plates have been found along the tail of a small ornithischian. Dinosaur mosaic scales were commonly semihexagonal in shape, with larger scales surrounded by a ring of smaller scales, forming rosettes that were themselves set in a sea of small scales. These scales were often flat, but some were more topographic, ranging from small beads on up. Because dinosaur scales were usually not large, they tend to disappear from visual resolution when viewed from a dozen feet or more away. However, in some cases the center scale in a rosette was a large, projecting, subconical scale; these were often arranged in irregular rows. On a given dinosaur the size and pattern of the scales varied depending on their location. The most spectacular scales yet known are those that adorned *Triceratops*. As big as the palm of a large person's hand, they were strongly subconical and may have borne a large central bristle.

The backs of some dinosaurs were adorned with nonarmor display tissue. This took the form of large, prominent scales, spines, and segmented and smooth-edged frills. At least some psittacosaur tails were adorned with a comblike set of very long bristles. In heterodontosaurs the dorsal bristles were denser and finer and ran along the back as well as the tail. Prominent skin folds like those seen on lizards are sometimes preserved and may have been fairly common in various dinosaurs. Soft crests, combs, dewlaps, wattles, and other soft display organs may have been more widespread than we realize. A pelican-like throat pouch has been found under the jaws of an ornithomimosaur theropod, and the throat pouches of stegosaurs and ankylosaurs, which started at the front of the lower jaws, were armored with a dense pavement of small ossicles. Armor plates were covered with hard keratin; when the plates were erect, the horn coverings probably enlarged them. Also lengthened by keratin sheaths were beaks, horns, and claws; in a few cases these have been preserved. Keratin typically lengthens a bony horn core by a third up to twofold; I usually add half.

Until recently neither scales nor any other kind of body covering had been discovered on small ornithischians. This data gap has finally been dramatically reduced by the discovery of a fiber coat on two examples. Feathers have long been known on the fossils of birds preserved in fine-grained lake or lagoon bottom sediments, including *Archaeopteryx*. In the last two decades a growing array of small theropod dinosaurs have been found covered with bristle protofeathers or fully developed pennaceous feathers in the Yixian beds. Some researchers have claimed that the simpler bristles are really degraded internal collagen fibers. This idea is untenable for a number of reasons, including the discovery of pigmentation—either visible to the naked eye or in microscopic capsules—in the fibers that allows their actual color to be approximated. Some small nonflying theropods also had scales at least on the tail and perhaps legs, and some small ornithischians such as psittacosaurs were largely scale covered. This suggests that the body covering of small dinosaurs was variable—ostriches lack feathers on the legs, and many mammals from a small bat through a number of suids and humans to rhinos and elephants are essentially naked. Ironically, some paleoartists are going too far with feathering dinosaurs, giving many the fully developed aeroshells in which contour feathers streamline the head, neck, and body of most flying birds. But most Mesozoic dinosaurs did not fly, and like those birds whose ancestors lost flight long ago, flightless dinosaurs would have had shaggier, irregular coats for purposes of insulation and display. Also, modern birds have hyperflexible necks that allow many but not all fliers to strongly U-curve the neck to the point that the head and heavily feathered neck aerodynamically merge with the body. Dinosaurs and even early-flying dinobirds like *Archaeopteryx* and microraptors could not do this, so their less flexible necks stuck out ahead of the shoulders like those of a number of modern long-necked flying birds.

Because fibers covered basal ornithischians, it is a good scientific bet that dinosaur insulation evolved once, in which case they were all protofeathers. The absence to date of protofeathers in Triassic and Early Jurassic theropods is the kind of negative evidence that is no more meaningful than their lack of fossil scales, the kind that long led to the denial of insulation in any dinosaurs and is likely to be corrected by the eventual discovery of insulation in basal examples. However, it cannot be ruled out that insulation evolved more than once in dinosaurs. A question is why dinofur and feathers appeared in the first place. The first few bristles must have been too sparse to provide insulation, so

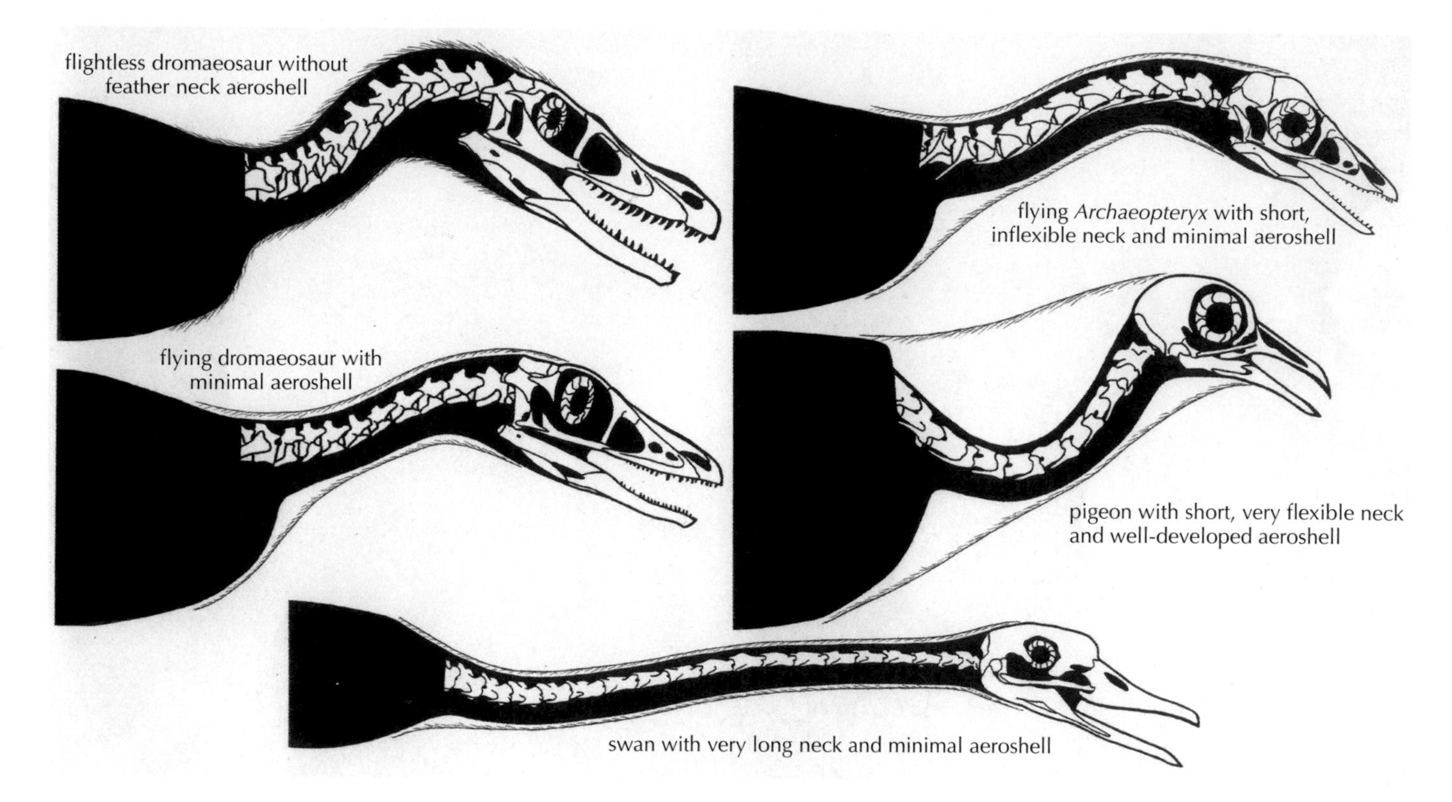

their initial appearance should have been for nonthermoregulatory reasons. One highly plausible selective factor was display, as in the visually striking tail bristles of psittacosaurs. As the bristles increased in number and density to improve their display effect, they became thick enough to help retain the heat generated by the increasingly energetic archosaurs. The display-to-insulation hypothesis is supported by how the fibers and feathers found on some flightless dinosaurs such as heterodontosaurs functioned as both prominent display organs on some parts of the body and as insulation cover on others.

A number of researchers argue that the pigment organelles of feathers preserve well, and their shape varies according to color, so they are being used to restore the actual colors of feathered dinosaurs. Although some researchers have challenged the reliability of this method, it appears to be sound, so this book uses the colors determined by this technique—doing so maximizes the probability of achieving correct coloration, whereas not doing so essentially ensures incorrect results. It appears that the feathers of some dinosaurs were, as might be expected, iridescent, using refraction rather than pigmentation to achieve certain color effects. There is no known method to restore the colors of scales. The hypothesis offered by some researchers that the differing scale patterns on a particular species of dinosaur correspond to differences in coloration is plausible, but some reptiles are uniformly colored regardless of variations in the scales. Dinosaur scales were better suited to carry bold and colorful patterns like those of reptiles, birds, tigers, and giraffes than is the dull gray, nonscaly skin of big mammals, and the color vision of dinosaurs may have encouraged the evolution of colors for display and camouflage. Dinosaurs adapted to living in forested areas may have been prone to using greens as stealth coloring. On the other hand, big reptiles and birds tend to be earth tinged despite their color vision. Small dinosaurs are the best candidates for bright color patterns like those of many small lizards and birds. Archosaurs of all sizes may have used specific color displays for intraspecific communication or for startling predators. Crests, frills, skin folds, and taller neural spines would be natural bases for vivid, even iridescent, display colors, especially in the breeding season. Because dinosaur eyes were bird- or reptile-like, not mammal-like, they lacked white surrounding the iris. Dinosaur eyes may have been solid black or brightly colored, like those of many reptiles and birds.

Respiration and Circulation

The hearts of turtles, lizards, and snakes are three-chambered organs incapable of generating high blood pressures. The lungs, although large, are internally dead-end structures with limited ability to absorb oxygen and exhaust carbon dioxide and are operated by rib action. Crocodilian hearts are incipiently four chambered but are still low pressure. Their lungs are internally dead end, but they may have unidirectional airflow, and the method by which they are ventilated is sophisticated. Muscles attached to the pelvis pull on the liver, which spans the full height and breadth of the rib cage, to expand the lungs. This action is facilitated by an unusually smooth ceiling of the rib cage that allows the liver to easily glide back and forth, the presence of a rib-free lumbar region immediately ahead of the pelvis, and, at least in advanced crocodilians, a mobile pubis in the pelvis that enhances the action of the muscles attached to it.

Birds and mammals have fully developed four-chambered, double-pump hearts able to propel blood in large volumes at high pressures. Mammals retain fairly large dead-end lungs, which are internally very intricate, greatly expanding the gas-exchange surface area. The lungs are operated by a combination of rib action and the vertical, muscular diaphragm. The presence of the diaphragm is indicated by the existence of a well-developed, rib-free lumbar region, preceded by a steeply plunging border to the rib cage on which the vertical diaphragm is stretched.

It is widely agreed that all dinosaurs probably had fully four-chambered, high-capacity, high-pressure hearts. Their respiratory complexes appear to have been much more diverse.

It is difficult to reconstruct the respiratory systems of ornithischians because they left no living descendants, and because their rib cages differ not only from those of all living tetrapods but among differing ornithischian groups. It is not possible to determine the complexity of their lungs; it can only be said that if ornithischians had high aerobic capacity, then their lungs should have been internally intricate. Because no ornithischian shows evidence of pneumatic bones, it can be assumed that they retained high-volume, dead-end lungs, although airflow may have been partly unidirectional. Nor were their ribs highly mobile—in ankylosaurs most of the ribs were actually fused to the vertebrae. The belly ribs of ceratopsids were packed tightly together and attached to the pelvis, so they could not move either. It can be speculated that in most ornithischians abdominal muscles anchored on the ventral pelvis were used to push the viscera forward, expelling stale air from the lungs; when the muscles were relaxed the lungs expanded. One group of ornithischians had a different arrangement. In ornithopods there was a large rib-free lumbar region with a steeply plunging rib cage immediately ahead. This is so similar to the mammalian lumbar region that it is probable that a diaphragm, perhaps muscular, had evolved in the group.

Restoring the respiratory complexes of saurischians, especially theropods, is a much more straightforward process because birds are living members of the group and retain the basic theropod system. Birds have the most complex and efficient respiratory system of any vertebrate. Because the lungs are rather small, the chest ribs that encase them are fairly short, but the lungs are internally intricate so they have a very large gas-exchange area. The lungs are also rather stiff and set deeply into the strongly corrugated ceiling of the rib cage. The lungs do not dead end; instead, they are connected to a large complex of air sacs whose flexibility and especially volume greatly exceed those of the lungs. Some of the air sacs invade the pneumatic vertebrae and other bones, but the largest sacs line the sides of the trunk; in most birds the latter air sacs extend all the way back to the pelvis, but in some, especially flightless examples, they are limited to the rib cage. The chest and abdominal sacs are operated in part by the ribs; the belly ribs tend to be extra long in birds with well-developed abdominal air sacs. All the ribs are highly mobile because they attach to the trunk vertebrae via well-developed hinge articulations. The hinging is oriented so that the ribs swing outward as they swing backward, inflating the air sacs within the rib cage, and then deflating the sacs as they swing forward and inward. In most birds the movement of the ribs is enhanced by ossified uncinate processes that form a series along the side of the rib cage. Each uncinate process acts as a lever for the muscles that operate the rib the process is attached to. In most birds the big sternal plate also helps ventilate the air sacs. The sternum is attached to the ribs via ossified sternal ribs that allow the plate to act as a bellows on the ventral air sacs. In those birds with short sternums, the flightless ratites, and in active juveniles, the sternum is a less important part of the ventilation system.

The system is set up in such a manner that most of the fresh inhaled air does not pass through the gas-exchange portion of the lungs but instead goes first to the air sacs, from where it is injected through the lungs in one direction on its way out. Because this unidirectional airflow eliminates the stale air that remains in dead-end lungs at the end of each breath and allows the blood and airflow to work in opposite, countercurrent directions that maximize gas exchange, the system is very efficient. Some birds can sustain cruising flight at levels higher than Mount Everest and equaling those of jet airliners.

Neither the first theropods nor prosauropods show evidence that they possessed air sacs, and aside from their lungs being dead-end organs, little is known about their respiration. In the first avepod theropods some of the vertebrae are pneumatic, indicating the presence of some air sacs. Also, the hinge jointing of the ribs increased, indicating that they were probably helping to

Respiratory complexes of archosaurs

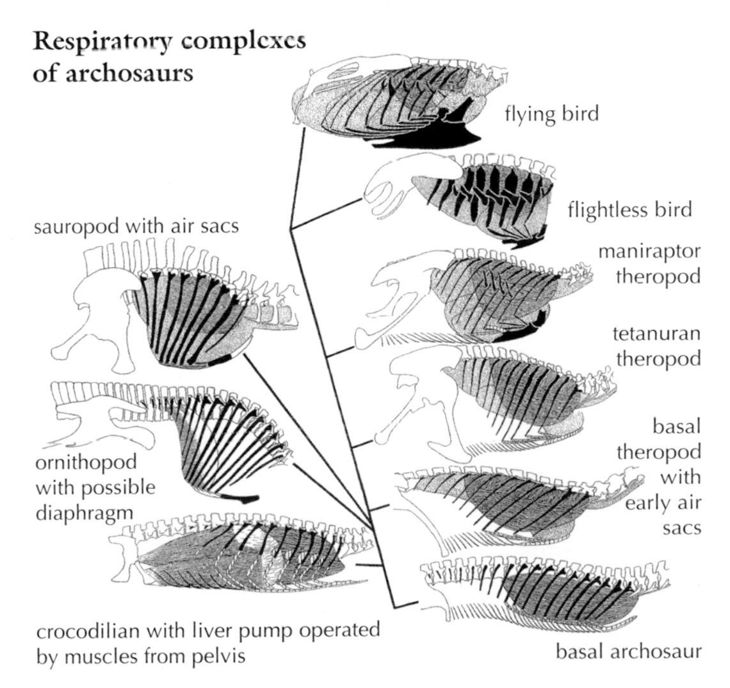

ventilate the lungs by inflating and deflating air sacs. As theropods evolved, the hinge jointing of the ribs further increased, as did the invasion of the vertebrae by air sacs until it reached the hips. Also, the chest ribs began to shorten, probably because the lungs were becoming smaller and stiffer as the air sacs did more of the work. By this stage the air-sac complex was probably approaching the avian condition, and airflow in the lungs should have been largely unidirectional. The sternum was still small, but the gastralia may have been used to help ventilate the ventral, belly air sacs. Alternatively the air sacs were limited to the rib cage as they are in some flightless birds—the extra-long belly ribs of birds with big abdominal air sacs are absent in theropods. In many aveairfoilan theropods the ossified sternum was as large as it is in ratites and juvenile birds and was attached to the ribs via ossified sternal ribs, so the sternal plate was combining with the gastralia to inflate and deflate the air sacs. Also, ossified uncinate processes are often present, indicating that the bellows-like action of the rib cage was also improved. At this stage the respiratory complex was probably about as well developed as it is in some modern birds.

The few researchers who think birds are not dinosaurs deny that theropods breathed like birds. Some propose that theropod dinosaurs had a crocodilian liver-pump system. Aside from theropods not being close relatives of crocodilians, they lacked the anatomical specializations that make the liver-pump system possible—a smooth rib cage ceiling, a lumbar region, and a mobile pubis. Instead, some of the theropods' adaptations for the avian air-sac system—the corrugated rib cage ceiling created by the hinged rib articulations, the elongated belly ribs—would have prevented the presence of a mobile liver. Advocates of the liver pump point to the alleged presence of a deep liver within the skeletons of some small theropods. The fossil evidence for these large livers is questionable, and in any case, predators tend to have big livers, as do some birds. The existence of a crocodilian liver-pump lung ventilation system in dinosaurs can be ruled out.

Sauropods show strong evidence that they independently evolved an air-sac system. The vertebrae were usually highly pneumatic. Also, all the ribs were hinge jointed, even the belly ribs, which one would expect to instead be solidly anchored in order to better support the big belly. Most researchers agree that the air-sac-filled vertebrae and mobile belly ribs of sauropods are strong signs that they had an air-sac-driven respiratory complex that probably involved unidirectional airflow. Because sauropods lacked gastralia, the air sacs should have been limited to the rib cage. Sauropods pose an interesting respiratory problem because most of them had to breathe through very long tracheas, which created a large respiratory dead space that had to be overcome with each breath. Presumably the great air capacity of the air sacs helped them to completely flush the lungs with fresh air during each breath.

Mammal red blood cells lack a nucleus, which increases their gas-carrying capability. The red blood cells of reptiles, crocodilians, and birds retain a nucleus, so those of dinosaurs should have as well.

Digestive Tracts

In a number of dinosaur specimens from a number of groups, gastroliths, or gizzard stones, are preserved within the rib cage, often as bundles of stones. In some dinosaur formations large numbers of polished stones are present even though geological forces that could explain their presence appear to be absent. This evidence indicates that many if not all dinosaurs had gizzards.

The digestive tracts of predatory theropods were relatively short, simple systems that quickly processed the easily digested chunks of flesh bolted down by the simple scissors action of the serrated-toothed jaws. Coprolites attributable to large theropods often contain large amounts of undigested bone, confirming the rapid passage of food through the tract. Some vegetarian theropods used numerous gastroliths to break up the plant material. Like herbivorous birds, most sauropods lacked the ability to chew the plant materials they ingested. The fodder was physically broken down in the gizzard, which may have used stones to help stir it up. Sauropods had large rib cages that contained the long, complex digestive tracts needed to ferment and chemically break down leaves and twigs. The system was taken to an extreme in the broad-bellied titanosaurs.

The cheeks that appear to have been present on at least some prosauropods, early sauropods, and therizinosaurs should have allowed them to pulp food before swallowing. But it was the ornithischians that fully exploited this system. After cropping food with their beaks, they could break up plant parts with their dental batteries. As some of the food fell outside the tooth rows, it was held in the elastic cheek pouches until the tongue swept it up for further processing or swallowing. Hadrosaurs took the evolution of the dental complexes the furthest and had modest-sized abdomens to further process the well-chewed fodder. Some ornithischians had relatively weakly developed tooth complexes and used massive digestive tracts contained in enormous bellies to ferment and break down food. In pachycephalosaurs the expansion of the digestive tract was further accomplished by broadening the base of the tail in order to accommodate an enlargement of the intestines behind the pelvis. A few ornithischians supplemented plant processing with dense gastrolith bundles.

There is no evidence that any dinosaur evolved a highly efficient ruminant-like system in which herbivores chew their own cuds. Such a system works only in animals of medium size in any case, and it was not suitable for the most titanic dinosaurs.

Senses

The large eyes and well-developed optical lobes characteristic of most dinosaurs indicate that vision was usually their primary sensory system, as it is in all birds. Reptiles and birds have full color vision extending into the ultraviolet range, so dinosaurs

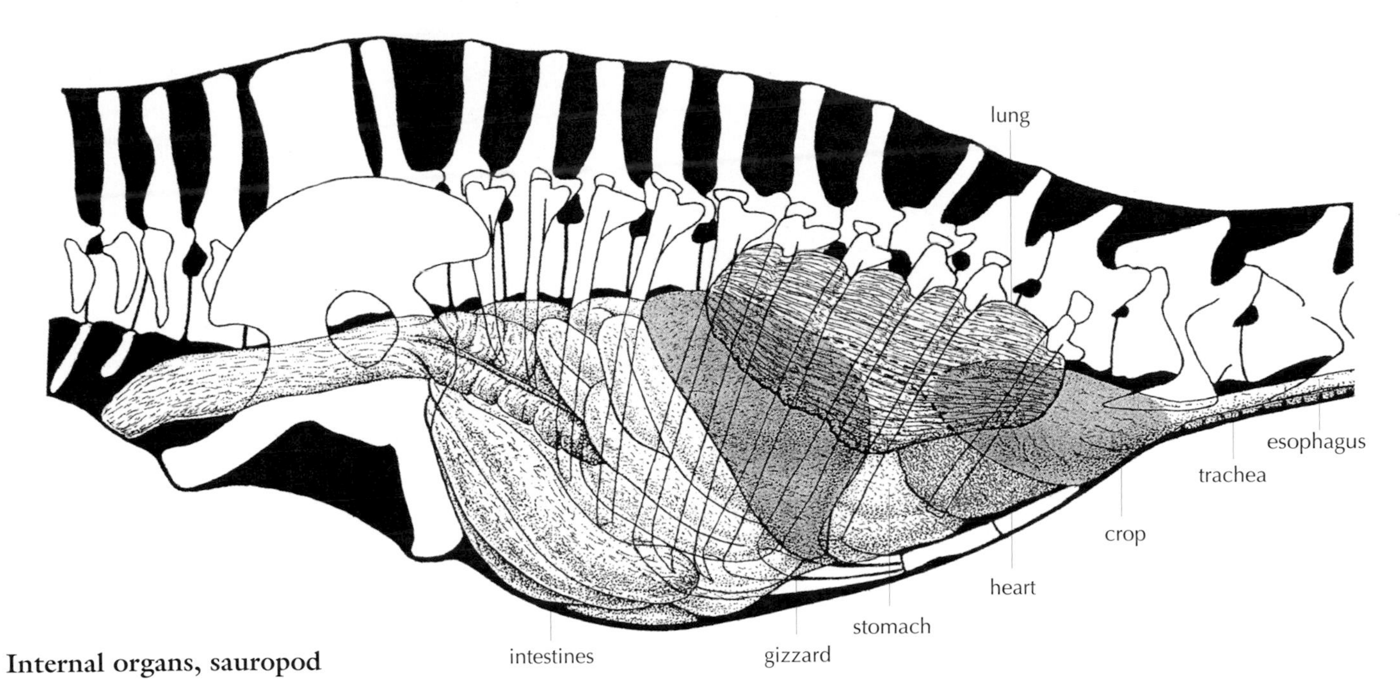

Internal organs, sauropod

probably did too. The comparatively poorly developed color vision of most mammals is a heritage of the nocturnal habits of early mammals, which reduced vision in the group to the degree that eyesight is often not the most important of the senses. Reptile vision is about as good as that of mammals, and birds tend to have very high-resolution vision both because their eyes tend to be larger than those of reptiles and mammals of similar body size and because they have higher densities of light-detecting cones and rods than mammals. The cones and rods are also spread at a high density over a larger area of the retina than in mammals, in which high-density light cells are more concentrated at the fovea (so our sharp field of vision covers just a few degrees). Some birds have a secondary fovea. Day-loving raptors can see about three times better than people, and the sharp field of vision is much more extensive, so birds do not have to point their eye at an object as precisely as mammals to focus on it. Birds can also focus over larger ranges, 20 diopters compared to 13 diopters in young adult humans. The vision of the bigger-eyed dinosaurs may have rivaled this level of performance. The dinosaurs' big eyes have been cited as evidence for both daylight and nighttime habits. Large eyes are compatible with either lifestyle—it is the (in this case unknowable) structure of the retina and pupil that determines the type of light sensitivity.

Birds' eyes are so large relative to the head that they are nearly fixed in the skull, so looking at specific items requires turning the entire head. The same was likely to have been true of smaller-headed dinosaurs. Dinosaurs with larger heads should have had more mobile eyeballs that could scan for objects without rotating the entire head. The eyes of most dinosaurs faced to the sides, maximizing the area of visual coverage at the expense of the view directly ahead. Some birds and mammals—primates most of all—have forward-facing eyes with overlapping fields of vision, and in at least some cases vision includes a binocular, stereo effect that provides depth perception. Tyrannosaurid, ornithomimid, and many aveairfoilan theropods had partly forward-facing eyes with overlapping vision fields. Whether vision was truly stereo in any or all of these dinosaurs is not certain; it is possible that the forward-facing eyes were a side effect of the expansion of the back of the skull to accommodate larger jaw muscles in tyrannosaurids.

Most birds have a poorly developed sense of smell, the result of the lack of utility of this sense for flying animals, as well the lack of space in heads whose snouts have been reduced to save weight. Exceptions are some vultures, which use smell to detect rotting carcasses hidden by deep vegetation, and grub-hunting kiwis. As nonfliers with large snouts, many reptiles and mammals have very well-developed olfaction, sometimes to the degree that it is a primary sensory system, canids being a well-known example. Dinosaurs often had extremely well-developed, voluminous nasal passages, with abundant room at the back of the passages for large areas of olfactory tissues. In many dinosaurs the olfactory lobes are large, verifying an effective sense of smell. Herbivorous dinosaurs probably had to be approached from downwind to avoid their fleeing from an attack, and it is possible that olfaction was as important as vision in the smaller-eyed ankylosaurs. Among theropods the tyrannosaurs and dromaeosaurs had excellent olfaction, useful for finding both live prey and dead carcasses.

Mammals have exceptional hearing, in part because of the presence of large, often movable outer ear pinnae that catch and direct sounds into the ear opening, and especially because of the intricate middle ear made up of three elements that evolved from jaw bones. In some mammals hearing is the most important sense, bats and cetaceans being the premier examples. Reptiles and birds lack fleshy outer ears, and there is only one inner ear bone. The combination of outer and complex inner ears means that mammals can pick up sounds at low volume. Birds partly compensate by having more auditory sensory cells per unit length of the cochlea, so sharpness of hearing and discrimination of frequencies are broadly similar in birds and mammals. Where mammalian hearing is markedly superior is in high-frequency sound detection. In many reptiles and birds the auditory range is just 1–5 kHz; owls are exceptional in being able to pick up from 250 Hz to 12 kHz, and geckos go as high as 10 kHz. In comparison, humans can hear 20 kHz, dogs up to 60 kHz, and bats 100 kHz. At the other end of the sound spectrum, some birds can detect very low frequencies: 25 Hz in cassowaries, which use this ability to communicate over long distances, and just 2 Hz in pigeons, which may detect approaching storms. It has been suggested that cassowaries use their big, pneumatic head crests to detect low-frequency sounds, but pigeons register even bassier sounds without a large organ.

In the absence of fleshy outer and complex inner ears, dinosaur hearing was in the reptilian-avian class, and they could not detect very high frequencies. Nor were the auditory lobes of dinosaur brains especially enlarged, although they were not poorly developed either. Nocturnal, flying, rodent-hunting owls are the only birds that can hear fairly high-frequency sounds, so certainly most and possibly all dinosaurs could not hear them either. Oviraptorosaurs had hollow head crests similar to those of cassowaries, hinting at similar low-frequency sound detection abilities. The big ears of large dinosaurs had the potential to capture very low frequencies, allowing them to communicate over long distances. It is unlikely that hearing was the most important sense in any dinosaur, but it was probably important for detection of prey and of predators, and for communication, in all species.

Vocalization

No reptile has truly sophisticated vocal abilities, which are best developed in crocodilians. Some mammals do, humans most of all. A number of birds have limited vocal performance, but many have evolved a varied and often very sophisticated vocal repertoire not seen among other vertebrates outside of people. Songbirds sing, and a number of birds are excellent mimics, to the point that some can imitate artificial sounds such as bells and sirens, and parrots can produce understandable humanlike speech. Some birds, such as swans, possess elongated tracheal loops in the chest that they use to produce high-volume vocalizations. Cassowaries call one another over long ranges with very low-frequency sounds, and so do elephants. Some or many dinosaurs may have had limited vocal abilities, although it is very doubtful that any had vocal abilities to match the more sophisticated examples seen in birds and mammals. Still, the sound-generating performance of the group probably exceeded that of reptiles. The long trachea of long-necked dinosaurs may have been able to generate powerful low-frequency sounds that could be broadcast over long ranges. Vocalization is done through the open mouth rather than through the nasal passages, so complex nasal passages acted as supplementary resonating chambers. This system was taken to an extreme in the lambeosaurine hadrosaurs. Although we will never know what dinosaurs sounded like, there is little doubt that the Mesozoic forests, prairies, and deserts were filled with their voices.

Disease and Pathologies

Dinosaurs lived in a world filled with diseases and other dangers to their health. The disease problem was accentuated by the global greenhouse effect, which maximized the tropical conditions that favored disease organisms, especially bacteria and parasites. Biting insects able to spread assorted diseases were abundant during the Mesozoic; specimens have been found in amber and fine-grained sediments. Reptile and bird immune systems operate somewhat differently from those of mammals; in birds the lymphatic system is particularly important. Presumably the same was true of their dinosaur ancestors.

Dinosaur skeletons often preserve numerous pathologies. Some appear to record internal diseases and disorders. Fused vertebrae are fairly common. Also found are growths that represent benign conditions or cancers. Most pathologies are injuries caused by stress or wounds; the latter often became infected, creating long-term, pus-producing lesions that affected the structure of the bone. Injuries tell us a lot about the activities of dinosaurs.

The predaceous theropods are, not surprisingly, especially prone to show signs of combat-related injury. One *Allosaurus* individual shows evidence of damage to its ribs, tail, shoulder, feet, and toes as well as chronic infections of its foot, finger, and a rib. The tail injury, probably caused by a kick or fall, had occurred early in life. Some of the injuries, including those to the feet and ribs, look severe enough that they may have limited its activities and contributed to its death. A wound in another *Allosaurus* tail appears to have been inflicted by the spike of a stegosaur. The famous *Tyrannosaurus* "Sue" had problems with its face, a neck rib, tail, finger, and a fibula. The head and neck wounds appear to have been caused by other *Tyrannosaurus* and in one case had undergone considerable healing. The sickle-claw-bearing toes of dromaeosaurs and troodonts frequently show signs of stress damage.

Among herbivorous dinosaurs, stegosaur tail spikes are often damaged or even broken and then healed, verification that they were used for combat. The horn of a *Triceratops* was bitten off by a *Tyrannosaurus*, according to the tooth marks, and then healed during the following years, indicating that the prey survived face-to-face combat with the great predator. Healed bite marks in the tails of sauropods and duck-billed hadrosaurs indicate that they too survived attacks by pursuing allosaurs and tyrannosaurs, respectively. Sauropods, despite or perhaps because of their size and slow speeds, show relatively little evidence of injury.

BEHAVIOR

Brains, Nerves, and Intelligence

The brains of the great majority of dinosaurs were reptilian both in size relative to the body and in structure. There was some variation in the size compared to body mass: the giant tyrannosaurids had unusually large brains for dinosaurs of their size, and so did the duck-billed hadrosaurs they hunted. However, even the diminutive brains of sauropods and stegosaurs were within the reptilian norm for animals of their great mass.

The small, fairly simple brains common to most dinosaurs indicate that their behavioral repertoire was limited compared to those of birds and mammals, being more genetically programmed and stereotypical. Even so, small-brained animals can achieve remarkable levels of mental ability. Fish and lizards can retain new information and learn new tasks. Many fish live in organized groups. Crocodilians care for their nests and young. Social insects with tiny neural systems live in organized collections that rear the young, enslave other insects, and even build large, complex architectural structures.

The major exception to dinosaurian reptile brains appeared in the birdlike aveairfoilan theropods. Their brains were proportionally larger, falling into the lower avian zone, as did their complexity. It is possible that the expanded and upgraded brains of aveairfoilans evolved in the context of the initial stages of dinosaurian flight. Presumably the bigger-brained dinosaurs were capable of more sophisticated levels of behavior than other dinosaurs.

The enlarged spinal cavity in the pelvic region of many small brained dinosaurs was an adaptation to better coordinate the function of the hind limbs and is paralleled in big ground birds. The great length of some dinosaurs posed a potential problem in terms of the time it took for electrochemical impulses to travel along the nerves. In the biggest sauropods, a command to the end of the tail and the response back could have to travel as much as 75 meters (250 feet) or more. Synaptic gaps where chemical reactions transmit information slow down the impulses, so this problem could have been minimized by growing individual nerve cords as long as possible.

Social Activities

Land reptiles do not form organized groups. Birds and mammals often do, but many do not. Most big cats, for instance, are solitary, but lions are highly social. Some, but not all, deer form herds.

That dinosaurs often formed social groups is supported by bone beds, some containing hundreds, thousands, or tens of thousands of individuals, and smaller collections that include a single species. Some accumulations of dinosaur skeletons can be attributed to death traps that accrued specimens over time or to droughts that compelled numerous individuals to gather at a water source where they starved to death as the vegetation ran out. Other accumulations, however, appear to have been the result of sudden events caused by volcanic ashfalls, by flash floods, by drownings when large numbers of dinosaurs crossed fast-flowing streams, or by dune slides. Such bone beds, which in some cases suggest the existence of very large herds, usually consist of large, herbivorous hadrosaurs or ceratopsids.

The presence of a number of individuals of a single species of theropod in association with the skeleton of a potential prey animal has been cited as evidence that predatory dinosaurs sometimes killed and fed in packs. It is, however, often difficult to explain why so many theropods happened to die at the same time while feeding on a harmless carcass. It is more probable that the theropod skeletons represent individuals killed by other theropods in disputes over feeding privileges, an event that often occurs when large carnivorous mammals compete over a kill.

Trackways are the closest thing we have to motion pictures of the behavior of fossil animals. A significant portion of the trackways of a diverse assortment of dinosaurs are solitary, indicating that the maker was not part of a larger group. It is also very common for multiple trackways of a variety of dinosaur species to have been laid close together on parallel paths. In some cases this may be because the track makers were forced to follow the same path along a shoreline even if they were moving independently of one another. But many times the parallel trackways are crisscrossed by the trackways of other dinosaurs that appear to have been free to travel in other directions. The large number of parallel trackways is therefore evidence that many species of predatory and herbivorous dinosaurs of all sizes often formed collectives that moved as pods, flocks, packs, and herds.

The degree of organizational sophistication of dinosaur groups was probably similar to that in fish schools and less developed than that in organized mammal herds and packs. Suggestions that the trackways of sauropods show that the juveniles were ringed by protective adults have not been borne out. Nor is it likely that theropod packs employed tactics as advanced as those attributed to canid packs or lion prides.

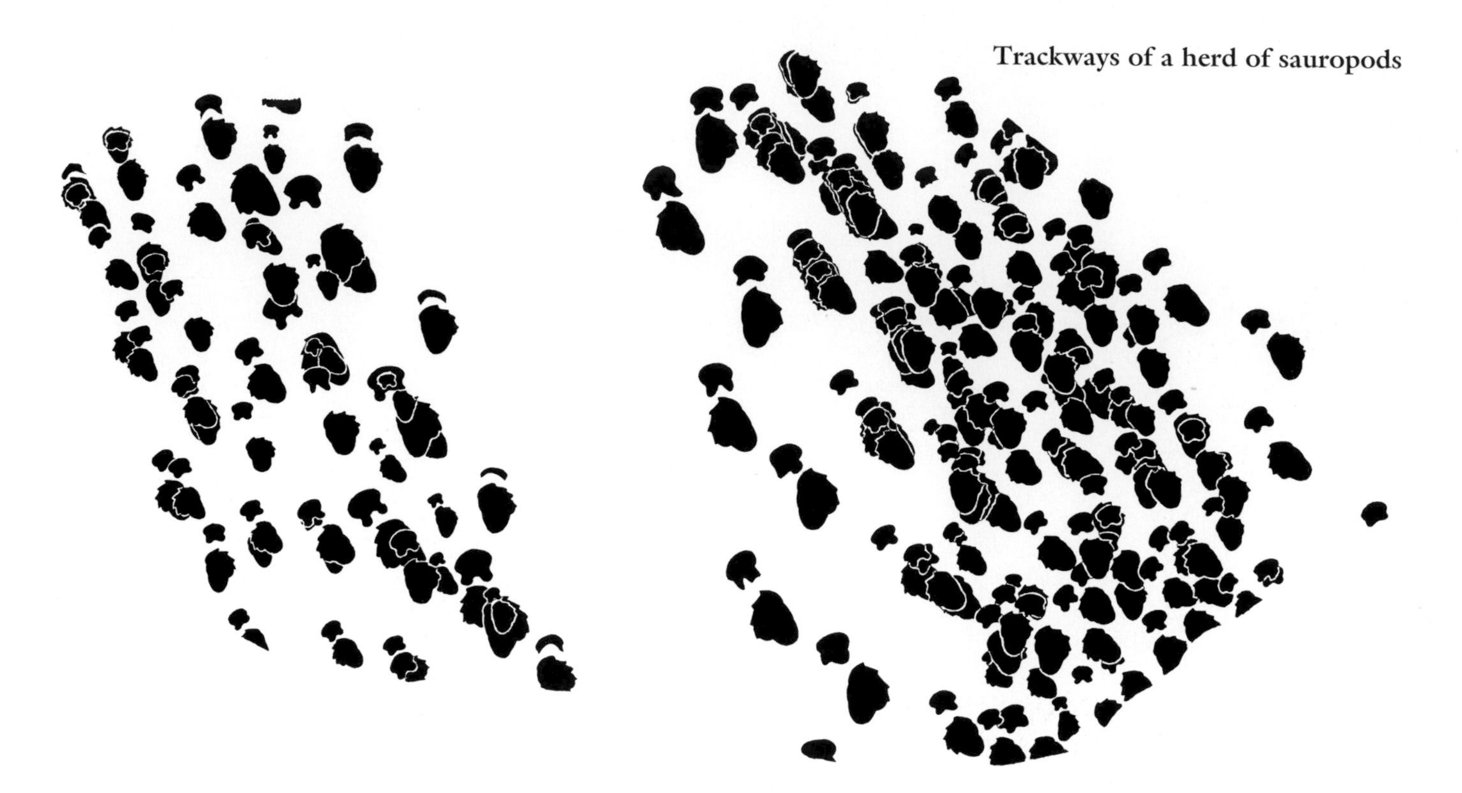

Trackways of a herd of sauropods

Reproduction

It has been suggested that some dinosaur species exhibit robust and gracile morphs that represent the two sexes. It is difficult to either confirm or deny many of these claims because it is possible that the two forms represent different species. Males are often more robust than females, but there are exceptions. Female raptors are usually larger than the males, for instance, and the same is true of some whales. Attempts to use the depth of the chevron bones beneath the base of the tail to dinstiguish the males from the females have failed because the two factors are not consistent in modern reptiles. Heterodontosaurs appear to come both with small tusks and without, and the former may be the males. Head-crested oviraptorosaurs and dome-headed pachycephalosaurs may be males if they are not mature individuals of both sexes. The robust form of *Tyrannosaurus* has been tentatively identified as the female on the basis of the inner bone tissues associated with egg production in birds, but the distribution of the stout and gracile morphs of this genus over stratigraphic time is more compatible with different species than with sexes.

Reptiles and some birds and mammals including humans achieve sexual maturity before reaching adult size, but most mammals and extant birds do not. Females that are producing eggs deposit special calcium-rich tissues on the inner surface of their hollow bones. The presence of this tissue has been used to show that a number of dinosaurs began to reproduce while still immature in terms of growth. The presence of still-growing dinosaurs brooding nests confirms this pattern. Most dinosaurs probably became reproductive before maturing. Exceptions may have been ceratopsids and hadrosaurs, whose display organs did not become completely developed until they approached adult proportions.

The marvelous array of head and body crests, frills, horns, hornlets, spikes, spines, tail clubs, bristles, and feathers evolved by assorted dinosaurs shows that many were under strong selective pressure to develop distinctive display organs and weapons to identify their species to other members of the species, and to achieve success in sexual competition. The organs we find preserved record only a portion of these visual devices—those consisting of soft tissues and color patterns are largely lost. How these organs were used varied widely. Females used display organs to signal males of the species that they were suitable and fertile mates. Males used them both to intimidate male rivals and to attract and inseminate females.

Healthy animals in their reproductive prime are generally able to dedicate more resources to grow superior-quality displays. Use of display organs in sexual attraction and competition was a relatively peaceful affair, and this system was taken to its dinosaurian height among the hadrosaurs with their spectacular head crests. Many dinosaurs probably engaged in intricate ritual display movements and vocalizations during competition and in courtship that have been lost to time. The head and body display surfaces of many dinosaurs were oriented to the sides so they had to turn themselves to best flaunt their display. The ceratopsians, whose head frills were most prominent in front view with the frill tilted up, were a major exception. The domes adorning pachycephalosaur heads were at least as prominent in front as in side view and may also have been tilted forward to intimidate opponents. Among the predatory theropods, the transverse head crest of *Cryolophosaurus* and the horns and domes of some abelisaurs provided unusual frontal displays. The same was true of the crest of the rather small-headed brachylophosaur hadrosaurs.

Intraspecific competition is often forceful and even violent in animals that bear weapons. Sauropods could have reared up and assaulted one another with their thumb claws. The iguanodonts' thumb spikes were potentially even more dangerous intraspecific weapons. Domeheads may have battered each others' flanks with their heads. Whether they used their rounded domes to head butt like bighorn sheep is controversial. Male ankylosaurids are quite likely to have pummeled one another with their tail clubs, and other ankylosaurs probably locked their shoulder spines and engaged in strength-testing shoving matches. Multihorned ceratopsids may have interlocked their horns and done the same. Healed wounds indicate that ceratopsids also used their horns to injure one another. Tusked male heterodontosaurs may have done the same thing. The conceit that males have evolved means to avoid lethally injuring one another in reproductive contests is true in many cases but not in others. Male hippos and lions suffer high mortality from members of their own species, and the same may have been true of theropods, ceratopsians, and big-thumb-spiked iguanodonts.

In reptiles and birds the penis or paired penises (if either are present) and the testes are internal, and this was the condition in dinosaurs. Most birds lack a penis, but whether any dinosaur shared this characteristic is unknown. Presumably copulation was a quick process that occurred with the female lowering her shoulders and swinging her tail aside to provide clearance for the male, which reared behind her on two legs or even one leg while placing his hands on her back to steady them. The need of sauropods to copulate supports the ability of these giants to stand on the hind legs alone. The vertical armor plates of stegosaurs probably required a modification, with the male resting his hands on one side of the female's pelvis.

As far as is known, dinosaurs produced hard-shelled eggs like those of birds rather than the softer-shelled eggs of reptiles and crocodilians. The evolution of calcified shells may have precluded live birth, which is fairly common among reptiles and is absent in birds, but fossil remains of dinosaur eggs remain surprisingly scarce through much of the Mesozoic. For example, not a single eggshell fragment attributable to the many sauropod species that inhabited the enormous Morrison Formation has yet been found; so far only some small eggs laid by ornithopods have been discovered in its sediments. A fast-growing and diverse collection of eggs and nests is now known for a wide variety of Cretaceous dinosaurs, especially from the latter half of the period. Firmly identifying the producer of a given type of egg requires the presence of intact eggs within the articulated trunk skeleton, or identifiable embryo skeletons within the eggs, as well as adults found atop their nests in brooding posture. Because each dinosaur group produced distinctive types of eggshells and shapes, the differences can be used to further identify their origin, although the producers of many types remain obscure. Dinosaur eggs ranged from near-perfect spheres to highly elongated and in some cases strongly tapered. The surface texture of the egg was crenulated in some, and bumpy in others. The arrangement of eggs within dinosaur bodies and in their nests shows that they were formed and deposited in pairs as in reptiles, rather than singly as in birds. Even small reptiles lay small eggs relative to the size of the parent's body, whereas birds lay proportionally larger eggs. The eggs of small dinosaurs are intermediate in size between those of reptiles and birds. It is interesting that no known dinosaur egg matches the size of the gigantic, 12-kilogram (25-pound) eggs laid by the flightless elephant bird *Aepyornis*, which, as big as it was at nearly 400 kilograms (800 pounds), was dwarfed by many dinosaurs. The eggs of the huge sauropods, for instance, weighed less than a kilogram (2 pounds). The largest dinosaur eggs discovered so far weighed 5 kilograms and probably belonged to 1-tonne-plus oviraptors.

There are two basic reproductive stratagems, r-strategy and K-strategy. K-strategists are slow breeders that produce few young; r-strategists produce large numbers of offspring that offset high losses of juveniles. Rapid reproduction has an advantage. Producing large numbers of young allows a species to quickly expand its populations when conditions are suitable, so r-strategists are "weed species" able to rapidly colonize new territories or to promptly recover their population after it has crashed for one reason or another. As far as we know, dinosaurs were r-strategists that typically laid large numbers of eggs in the breeding season, although herbivorous dinosaurs isolated on predator-free islands might have been slow breeders. This may explain why dinosaurs laid smaller eggs than birds, most of which produce a modest number of eggs and provide the chicks with considerable parental attention. One r-strategist bird group is the big modern ratites, which produce numerous eggs. Sauropods appear to have placed the largest number of eggs into a single nest, up to a few dozen. Giant dinosaurs were very different in this respect from giant mammals, which are K-strategists that produce few calves that then receive extensive care over a span of years. Nor did any dinosaur nurse its young via milk-producing mammary glands. It is possible that some dinosaurs produced a "milk"-like substance in the digestive tract that was regurgitated to their young, as pigeons do, but there is no direct evidence of this.

It was long tacitly assumed that, like most reptiles, dinosaurs paid little or no attention to their eggs after burying them. A few lizards do stay with the nest, and pythons actually incubate their eggs with muscle heat. Crocodilians often guard their nests and the hatchlings. All birds lavish attention on their eggs. Nearly all incubate the eggs with body heat; the exception is megapode fowl that warm eggs in mounds that generate heat via fermenting vegetation. The fowl carefully regulate the temperature of the nest by adding and removing vegetation to and from the mound. But when megapode chicks hatch they are so well developed that the precocial juveniles quickly take off and survive on their own. The newly hatched chicks of ratites are also precocial, but they remain under the guardianship of adults that guide them to food sources and protect them from attack. Most bird chicks are altricial: they are so poorly developed when they break out of the egg that they have to be kept warm and fed by adults.

A spate of recent discoveries has revealed that the manner in which dinosaurs deposited eggs and then dealt with them and the offspring varied widely, and in various regards was both similar to and distinctive from this behavior in living tetrapods.

Some dinosaur eggs whose makers have yet to be identified were buried in a manner that implies they were immediately abandoned. This was probably true of the eggs of sauropods. The large, vegetation-covered nests that can be attributed to the giants were structured in a rather irregular manner that differs from the more organized nature of nests that are tended by adults. There is evidence that at least some sauropods deposited their eggs near geothermal heat sources. Because large numbers of nests were created at the same time and place, the adult sauropods would have risked denuding the local vegetation as well as trampling their own eggs if they remained to guard their nests. Also in danger of being trampled were the hatchlings, which were thousands of times less massive than their parents. Laying so many eggs in so many nests made it possible for the adults to overwhelm the ability of the local predators to find and eat all the hatchlings, although a fossil shows a large snake feeding on a just-emerged hatchling. Trackways indicate that small juvenile sauropods formed their own pods, independent of multitonne adults. Other trackways further indicate that sauropod calves joined up with full-sized adults only after a few years, when they had reached about a tonne, large enough to keep up and to not be stepped on. The mature sauropods probably paid the young ones no particular notice and were unlikely to have even been closely related to them. In this scenario, the juveniles were seeking the statistical safety of being in the vicinity of aggressive grown-ups able to battle the biggest predators. A mystery is why the enormous Morrison Formation, home to an array of sauropod species, has yet to produce any trace of their eggs, even though the shells of smaller dinosaur eggs have been found.

Also apparently forming juvenile pods were at least some ankylosaurs. The intact skeletons of over a dozen large juvenile *Pinacosaurus* skeletons have been found grouped together, apparently killed at the same moment by a dune slide. The absence of an adult suggests that the growing armored dinosaurs were moving together as an independent gang.

The compact nests of duck-billed hadrosaurs have a structural organization that suggests they were monitored by the adults. Hadrosaurs may have regulated the temperatures of their mound nests like megapode fowl. The nests seem to form colonies in at least some cases, and breeding hadrosaurs were not so large that they would have stripped the local flora if they remained to care for their young. In many hadrosaur nests the eggs are so thoroughly broken up that they seem to have been trampled on over time, and the skeletons of juveniles considerably larger than the hatchlings have been found in the nests, so the young hadrosaurs did not immediately abandon their nests. The heads of baby hadrosaurs had the short snouts and large eyes that encourage parental behavior. These factors suggest that the parents opened the mounds as the eggs hatched and then brought food to the altricial juveniles while they remained in the nest. This arrangement would have avoided the problem of stepping on the tiny hatchlings, would have provided them protection from predators, and would have improved growth rates by supplying the nestlings with plenty of food while the youngsters saved energy by remaining immobile. What happened when hadrosaur juveniles left the nests after a few weeks or months is not certain—the still-extreme size disparity between the parents and their offspring favors the latter forming independent pods until they were large enough to join the adult herds.

Little is known about the nesting of large predatory theropods. Tyrannosaurid chicks were vulnerable to being killed by the adults either accidentally or cannibalistically. Juvenile tyrannosaurids were unusual in having elongated snouts, which are the opposite of the short faces of juveniles cared for by their parents. This suggests that growing tyrannosaurids hunted independently of the adults. Suggestions that the gracile juvenile tyrannosaurids hunted prey for their parents are implausible; when food is exchanged between juveniles and adults, it is the latter who feed the former.

Because smaller dinosaurs did not face the problem of accidentally crushing their offspring, they had the potential to be more intensely parental. The best evidence for dinosaur brooding and incubating is provided by the birdlike aveairfoilan theropods, especially oviraptors. The large number of eggs, up to a few dozen in some cases, could not have been produced by a single female, so the nests were probably communalistic. The big ratites also nested communally. Oviraptors laid their elongated eggs in two-layered rings with an open center. Laid flat, the eggs were partly buried and partly exposed. Because eggs left open to the elements would die from exposure or predation, eggs were not left exposed unless they were intended to be protected and incubated by adults. A number of oviraptor nests have been found with an adult in classic avian brooding posture atop the eggs, the legs tucked up alongside the hips, the arms spread over the eggs. The egg-free area in the center of the ring allowed the deep pelvis to rest between the eggs without crushing them; flatter-bellied birds do not need this space between their eggs. Presumably the arm and other feathers of oviraptorosaurs completely covered the eggs in order to protect them from inclement conditions and to retain the incubator's body heat. It is thought that brooding oviraptors were killed in place by sandstorms or more likely duneslides. The giant eggs appear to be of the type laid by oviraptors, and they too are laid in rings, in their case of enormous dimensions (up to 3 meters or 10 feet across). These are the largest incubated nests known and were apparently brooded by oviraptors weighing a tonne or two. In troodont nests the less-elongated eggs were laid subvertically in a partial spiral ring, again with the center open to accommodate the brooder's pelvis. The size of the adult troodonts found in brooding posture atop their nests is as small as 1 pound. The half-buried, half-incubated nesting habits of aveairfoilans ideally represent the near-avian arrangement expected in the dinosaurs closest to birds.

A problem that all embryos that develop in hard-shelled eggs face is getting out of that shell when the time is right. The effort to do so is all the harder when the egg is large and the shell

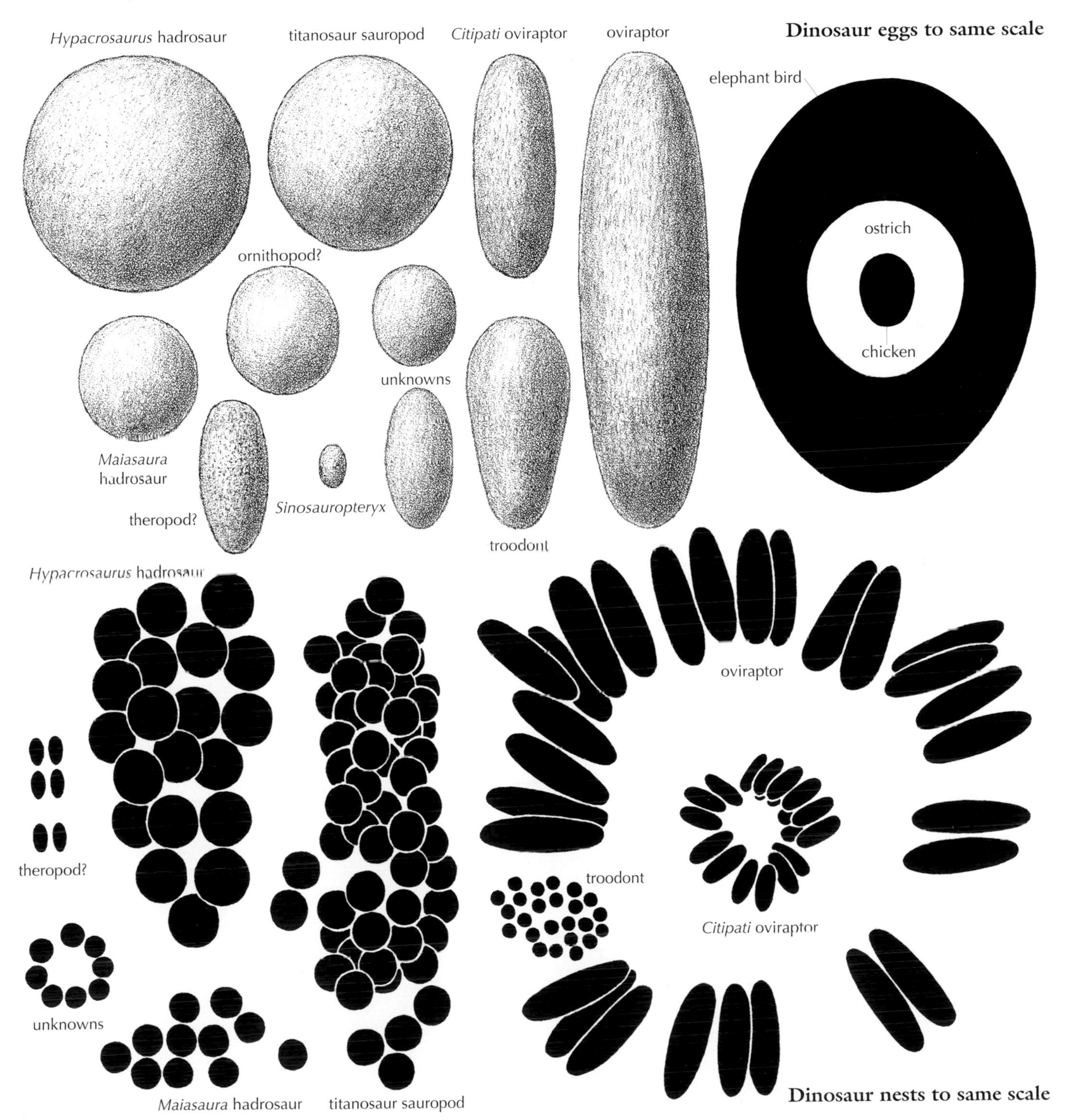

Dinosaur eggs to same scale

Dinosaur nests to same scale

Oviraptor *Citipati* incubating a nest, based on a fossil, with feathers drawn short enough to show eggs

correspondingly thick. Fortunately, some of the shell is absorbed and used to help build the skeleton of the growing creature. Baby birds use an "egg tooth" to break out of the shell. The same has been found adorning the nose of titanosaur sauropod embryos, and this may have been true of other Mesozoic dinosaurs.

Whether small birdlike theropods and many other dinosaurs continued to care for their young after they hatched is not known but is plausible. The best evidence for dinosaur parenting of juveniles found so far is among small ornithischians. A compact clutch of nearly three dozen articulated juvenile (about a tenth

Titanosaur hatchling with egg tooth

of a kilogram, or a fifth of a pound) *Psittacosaurus* skeletons was found in intimate association with the remains of an adult over a dozen times heavier. There is dispute over whether this is a true association between a parent and its offspring; if it is, the situation seems to parallel that of some ratite birds, which gather the offspring of a number of females into a large crèche that is tended by a set of adults. It is possible that the tightly packed collection of psittacosaurs was entombed in a fossil burrow. Parental care probably ranged from minimal to extensive in dinosaurs and in a number of cases probably exceeded that seen in reptiles or even crocodilians, and rivaled that of birds. However, no dinosaur lavished its offspring with the parenting typical of mammals, and because dinosaurs did not nurse, it is likely that most of them could grow up on their own.

GROWTH

All land reptiles grow slowly. This is true even of giant tortoises and big, energetic (by reptilian standards) monitors. Land reptiles can grow most quickly only in perpetually hot equatorial climates, and even then they are hard pressed to reach a tonne. Aquatic reptiles can grow more rapidly, probably because the low energy cost of swimming allows them the freedom to acquire the large amounts of food needed to put on bulk. But even crocodilians, including the extinct giants, which reached nearly 10 tonnes, do not grow as fast as many land mammals. Reptiles tend to continue to grow slowly throughout their lives.

Some marsupials and large primates including humans grow no faster or only a little faster than the fastest-growing land reptiles. Other mammals, including other marsupials and a number of placentals, grow at a modest pace. Still others grow very rapidly; horses are fully grown in less than two years, and aquatic whales can reach 50 to 100 tonnes in just a few decades. Bull elephants take about thirty years to mature. All living birds grow rapidly; this is especially true of altricial species and of the big ratites. No extant bird takes more than a year to grow up, but some of the recently extinct giant island ratites may have taken a few years to complete growth. The secret to fast growth appears to be having an aerobic exercise capacity high enough to allow the growing juvenile, or its adult food provider, to gather the large amounts of food needed to sustain rapid growth.

High mortality rates from predation, disease, and accidents make it statistically improbable that unarmored, nonaquatic animals will live very long lives, so they are under pressure to grow rapidly. On the other hand, starting to reproduce while still growing tends to slow down the growth process as energy and nutrition resources are diverted to producing offspring. Few mammals and no living birds begin to breed before they reach adult size. So even elephants do not live much more than half a century, and most medium-sized and large mammals and birds live for only a few years or decades. No bird continues to grow once it is mature. Nor do most mammals, but some marsupials and elephants never quite cease growing.

At the microscopic scale the bone matrix is influenced by the speed of growth, and the bones of dinosaurs tend to be more similar to those of birds and mammals, which grow at a faster pace than those of reptiles. Bone ring counts are being used to estimate the growth rate and life span of a growing number of extinct dinosaurs, but this technique can be problematic because some living birds lay down more than one ring in a year, so ring counts can overestimate age and understate growth rate. There is also the problem of animals that do not lay down growth rings; it is probable that they grow rapidly, but exactly how fast is difficult to pin down. There are additional statistical issues, because as animals grow the innermost growth rings tend to be destroyed, leading to controversy over the correct estimates of ring numbers. Almost all dinosaurs sampled so far appear to have grown at least somewhat faster than land reptiles. The possible exception is a very small birdlike troodont theropod whose bone rings seem to have been laid down multiple times in a year, perhaps because it was reproducing while growing. Small dinosaurs fall along the lower end of the mammalian zone of growth, perhaps because they were reproducing while immature. Most gigantic dinosaurs appear to have been growing as fast as similar-sized land mammals, with the hadrosaurs and ceratopsids—which did not lay down growth rings when juveniles—apparently being particularly quick to mature for the group. None grew with the spectacular swiftness seen in the big rorqual whales. The growth achievement of the greatest sauropods is astonishing. Giant mammals get a head start, being born as large calves only a few dozen times smaller than the adults, and then being nourished with enormous amounts of nutrient-dense milk. Hatchling sauropods had to expand their mass tens of thousands-fold in just a few decades and with little or no nourishment provided by the adults. Armored dinosaurs appear to have grown less quickly than the others.

There is no evidence that dinosaurs lived longer than mammals or birds of similar size. In fact, the giant theropods appear to have normally died after just three decades. This was probably the result of lives of extreme danger that involved attacking large and dangerous adult prey; the small-brained dinosaurs were throwaway organisms, unlike large, big-brained mammals, which are major investments requiring extensive parental care and resources. The short life spans of these great dinosaurs were acceptable because they were expendable creatures, being early and fast-breeding r-strategists that could readily replace their losses. The cessation of significant growth of the outer surface of many adult dinosaur bones indicates that most species did not grow throughout life the way many reptiles do.

Comparison of growth between same-scale 6-tonne African elephant and 50-tonne sauropod

ENERGETICS

Vertebrates can utilize two forms of power production. One is aerobiosis, the direct use of oxygen taken in from the lungs to power muscles and other functions. This system has the advantage of producing power indefinitely but is limited in its maximum power output. An animal that is walking at a modest speed for a long distance, for instance, is exercising aerobically. The other is anaerobiosis, in which chemical reactions that do not immediately require oxygen are used to power muscles. This system has the advantage of being able to generate about ten times more power per unit of tissue and time. But it cannot be sustained for an extended period and produces toxins that can lead to serious illness if sustained at too high a rate for too long. Anaerobiosis also builds up an oxygen debt that has to be paid back during a period of recovery. An animal that is running near its top speed is exercising anaerobically.

Most fish and all amphibians and reptiles have low resting metabolic rates and low aerobic exercise capacity. They are therefore bradyenergetic, and even the most energetic reptiles, including the most aerobically capable monitor lizards, are unable to sustain truly high levels of activity for extended periods of time. Many bradyenergetic animals are, however, able to achieve very high levels of anaerobic burst activity, such as when a monitor lizard or crocodilian suddenly dashes toward and captures prey. Because bradyenergetic animals do not have high metabolic rates, they are largely dependent on external heat sources, primarily the ambient temperature and the sun, for their body heat, so they are ectothermic. As a consequence, bradyenergetic animals tend to experience large fluctuations in body temperature, rendering them heterothermic. The temperature at which reptiles normally operate varies widely depending on their normal habitat. Some are adapted to function optimally at modest temperatures of 12°C (52°F). Those living in hot climates are optimized to function at temperatures of 38°C (100°F) or higher, so it is incorrect to generalize reptiles as "cold blooded." In general, the higher the body temperature is, the more active an animal can be, but even warm reptiles have limited activity potential.

Most mammals and birds have high resting metabolic rates and high aerobic exercise capacity. They are therefore tachyenergetic and are able to sustain high levels of activity for extended periods of time. The ability to better exploit oxygen for power over time is probably the chief advantage of being tachyenergetic. Tachyenergetic animals also use anaerobic power to briefly achieve the highest levels of athletic performance, but they do not need to rely on this as much as reptiles and can recover more quickly. Because tachyenergetic animals have high metabolic rates, they produce most of their body heat internally, so they are endothermic. As a consequence, tachyenergetic animals can achieve more stable body temperatures. Some, like humans, are fully homeothermic, maintaining a nearly constant body temperature at all times when healthy. Many birds and mammals, however, allow their body temperatures to fluctuate to varying degrees on a daily and/or seasonal basis, so they are heterothermic. The ability to keep the body at or near its optimal temperature is another advantage of having a high metabolic rate. Normal body temperatures range from 30°C to 44°C (86°F–105°F), with birds always at least at 38°C. High levels of energy production are also necessary to do the cardiac work that creates the high blood pressures needed to be a tall animal.

Typically, mammals and birds have resting metabolic rates and aerobic exercise capacities about ten times higher than those of reptiles, and differences in energy budgets are even higher. However, there is substantial variation from these norms in tachyenergetic animals. Some mammals, among them monotremes, some marsupials, hedgehogs, armadillos, sloths, and manatees, have modest levels of energy consumption and aerobic performance, in some cases not much higher than those seen in the most energetic reptiles. In general, marsupials are somewhat less energetic than their placental counterparts, so kangaroos are about a third more energy efficient than deer. Among birds, the big ratites are about as energy efficient as similar-sized marsupials. At the other extreme, some small birds share with similarly tiny mammals extremely high levels of oxygen consumption.

Widely different energy systems have evolved because they permit a given species to succeed in its particular habitat and lifestyle. Reptiles enjoy the advantage of being energy efficient, allowing them to survive and thrive on limited resources. Tachyenergetic animals are able to sustain much higher levels of activity that can be used to acquire even more energy that can then be dedicated to the key factor in evolutionary success, reproduction. Tachyenergy has allowed mammals and birds to become the dominant large land animals from the tropics to the poles. But reptiles remain very numerous and successful in the tropics and, to a lesser extent, in the temperate zones.

As diverse as the energy systems of vertebrates are, there appear to be things that they cannot do. All insects have low, reptile-like resting metabolic rates. When flying, larger insects use oxygen at very high rates similar to those of birds and bats. Insects can therefore achieve extremely high maximal/minimal metabolic ratios, allowing them to be both energy efficient and aerobically capable. Insects can do this because they have a dispersed system of tracheae that oxygenate their muscles. No vertebrate has both a very high aerobic exercise capacity and a very low resting metabolism, probably because the centralized respiratory-circulatory system requires that the internal organs work hard even when resting in tachyenergetic vertebrates. An insect-like metabolic arrangement should not, therefore, be applied to dinosaurs. However, it is unlikely that all the energy systems that have evolved in land vertebrates have survived until today, so the possibility that some or all dinosaurs were energetically exotic in their nature needs to be considered.

The general assumption until the 1960s was that dinosaur energetics was largely reptilian in nature, but most researchers now agree that their power production and thermoregulation were closer to those of birds and mammals. It is also widely agreed that because dinosaurs were such a large group of diverse forms, there was considerable variation in their energetics, as there is in birds and especially mammals.

Reptiles have nonerect, sprawling legs that are suitable for the slow walking speeds of 1–2 km/h (0.5–1 mph) that their low aerobic capacity can power over extended periods of time. Sprawling limbs also allow reptiles to easily drop onto their belly and rest if they become exhausted. No living bradyenergetic animal has erect legs. Walking is always energy expensive—it is up to a dozen times more costly than swimming the same distance—so only aerobically capable animals can easily walk faster than 3 km/h. The long, erect legs of dinosaurs matched those of birds and mammals and favored the high walking speeds of 3–10 km/h (2–6 mph) that only tachyenergetic animals can sustain for hours at a time. The speed at which an animal of a given size is moving can be approximately estimated from the length of its stride—an animal that is walking slowly steps with shorter strides than it does when it picks up the pace. The trackways of a wide variety of dinosaurs show that they normally walked at speeds of over 3 km/h, much faster than the slow speeds recorded in the trackways of prehistoric reptiles. Dinosaur legs and the trackways they made both indicate that their sustained aerobic exercise capacity exceeded the reptilian maximum.

Even the fastest reptiles have slender leg muscles because their low-capacity respirocirculatory systems cannot supply enough oxygen to a larger set of locomotory muscles. Mammals and birds tend to have large leg muscles that propel them at a fast pace over long distances. As a result, mammals and birds have a large pelvis that supports a broad set of thigh muscles. It is interesting that protodinosaurs, the first theropods, and the prosauropods had a short pelvis that could have anchored only a narrow thigh. Yet their legs are long and erect. Such a combination does not exist in any modern animal. This suggests that the small-hipped dinosaurs had an extinct metabolic system, probably intermediate between those of reptiles and mammals. All other dinosaurs had the large hips able to support the large thigh muscles typical of more aerobically capable animals. Among the big-hipped dinosaurs, the relatively sluggish therizinosaurs, stegosaurs, and armored ankylosaurs were likely to have had lower energy budgets than their faster-moving relatives.

That many dinosaurs could hold their brains far above the level of their hearts indicates that they had the high levels of power production seen in similarly tall birds and mammals.

An intermediate metabolism is compatible with the unsophisticated lungs that protodinosaurs, early theropods, and prosauropods appear to have had. Too little is known about the respiration of ornithischians to relate them to metabolic level, except that the possible presence of a mammal-like diaphragm in ornithopods hints that they had a mammalian level of oxygen intake. The highly efficient, birdlike, air-sac-ventilated respiratory complex of avepod theropods and sauropods is widely seen as evidence that elevated levels of oxygen consumption evolved in these dinosaurs. Sauropods probably needed a birdlike breathing complex in order to oxygenate a high metabolic rate through their long trachea. Some reptiles with low energy levels had long necks, some marine plesiosaurs among them, but because they had low metabolic rates, they did not need air sacs to help pull large volumes of air into their lungs.

Many birds and mammals have large nasal passages that contain respiratory turbinals. These are used to process exhaled air in a manner that helps retain heat and water that would otherwise be lost during the high levels of respiration associated with high metabolic rates. Because they breathe more slowly, reptiles do not need or have respiratory turbinals. Some researchers point to the lack of preserved turbinals in dinosaur nasal passages, and the small dimensions of some of the passages, as evidence that dinosaurs had the low respiration rates of bradyenergetic reptiles. However, some birds and mammals lack well-developed respiratory turbinals, and in a number of birds they are completely cartilaginous and leave no bony traces. Some birds do not even breathe primarily through their nasal passages: California condors, for example, have tiny nostrils. The space available for turbinals has been underestimated in some dinosaurs, and other dinosaurs had very large passages, able to accommodate very large, unossified examples of these structures. Overall, the turbinal evidence does not seem to be definitive.

The presence of a blanket of hollow fibers in a growing array of small dinosaurs is strong evidence of elevated metabolic rates. Such insulation hinders the intake of environmental heat too much to allow ectotherms to quickly warm themselves and is never found adorning bradyenergetic animals. The evolution of insulation early in the group indicates that high metabolic rates also evolved near the beginning of the group or in their ancestors. The uninsulated skin of most dinosaurs is compatible with high metabolic rates, as in mammalian giants, many suids, human children, and even a small naked bat. The tropical climate most dinosaurs lived in reduced the need for insulation, and the bulk of large dinosaurs eliminated any need for it.

The low exercise capacity of land reptiles appears to prevent them from being active enough to gather enough food to grow rapidly. In an expression of the principle that it takes money to make money, tachyenergetic animals are able to eat the large amounts of food needed to produce the power needed to gather the additional large amounts of food needed to grow rapidly. Tachyenergetic juveniles either gather the food themselves or are fed by their parents. That dinosaurs, large and small, grew at rates faster than those seen in land reptiles of similar size indicates that the former had markedly higher aerobic capacity and energy budgets.

Bone isotopes have been used to help assess the metabolism of dinosaurs. These can be used to examine the temperature

The feathered theropod *Sinosauropteryx*

fluctuations that a bone experienced during life. If the bones show evidence of strong temperature differences, then the animal was heterothermic on either a daily or seasonal basis. In this case the animal could have been either a bradyenergetic ectotherm or a tachyenergetic endotherm that hibernated in the winter. The results indicate that most dinosaurs, large and small, were more homeothermic, and therefore more tachyenergetic and endothermic, than crocodilians from the same formations. An ankylosaur shows evidence of being heterothermic. Because the armored dinosaur lived at a high latitude, it is possible that it hibernated in the dark winter, perhaps bedding down in dense brush where it was protected by its armor against the chill as well as predators.

The presence of a diverse array of dinosaurs, from small species to titanic sauropods, in polar regions that are known to have experienced freezing conditions during the winter provides additional evidence that dinosaurs were better able to generate internal heat than reptiles, which were scarce or even absent in the same habitats. It was not practical for land-walking dinosaurs to migrate far enough toward the equator to escape the cold; it cost too much in time and energy, and in some locations oceans barred movement toward warmer climes—that polar dinosaurs appear to have been distinct from those of lower latitudes also contradicts epic migrations. The presence of sauropods in wintry habitats directly refutes the hypothesis that big dinosaurs used their bulk to keep warm by retaining the small amount of internal heat produced by a reptilian metabolism; only a higher level of energy generation could have kept the body core balmy and the skin from freezing. The discovery of probable dinosaur burrows in then-polar Australia suggests that some small ornithopods did hibernate through the winter in a manner similar to bears.

Because the most primitive and largest of living birds, the ratites, have energy budgets similar to those of marsupials, it is probable that most dinosaurs did not exceed this limit. This fits with some bone isotope data that seem to indicate that dinosaurs had moderately high levels of food consumption, somewhat lower than seen in most placentals of the same size. Possible exceptions are the tall sauropods with their high circulatory pressures, and polar dinosaurs that remained active in the winter and needed to produce lots of warmth. At the opposite extreme, early dinosaurs, slower-growing armored forms, and the awkward therizinosaurs probably had modest energy budgets like those of the less-energetic mammals. It is likely that dinosaurs, like birds, were less prone to controlling their body temperatures as precisely as do many mammals. This is in accord with their tendency to lay down bone rings. Because they lived on a largely hot planet, it is probable that most dinosaurs had high body temperatures of 38°C or more to be best able to resist overheating. The possible exception was again high-latitude dinosaurs, which may have adopted slightly lower temperatures and saved some energy if they were active during the winter.

GIGANTISM

Although dinosaurs evolved from small protodinosaurs, and many were small—birds included—dinosaurs are famous for their tendency to develop gigantic forms. The average mammal is the size of a dog, whereas the average dinosaur was bear sized. But those are just averages. Predatory theropods reached as much as 10 tonnes, as big as elephants and dwarfing the largest carnivorous mammals by a factor or ten or more. Sauropods exceeded the size of the largest land mammals, mammoths, and the long-legged indricothere rhinos of 15 to 20 tonnes, by a factor of at least four to five.

Among land animals whose energetics are known, only those that are tachyenergetic have been able to become gigantic on land. The biggest fully terrestrial reptiles, some oversized tortoises and monitors, have never much exceeded a tonne. Land reptiles are probably not able to grow rapidly enough to reach great size in reasonable time. Other factors may also limit their size. It is possible that living at 1 g, the normal force of gravity, without the support of water, is possible only among animals that can produce high levels of sustained aerobic power. The inability of the low-power, low-pressure reptilian circulatory system to pump blood far above the level of the heart probably helps limit the size of bradyenergetic land animals. Conversely, the extreme height of sauropods indicates that their hearts could push blood many meters up against the gravity well at pressures up to two or three times higher than the 200 mm Hg giraffes need to oxygenate their brains. And it is unlikely that such tall and massive animals in danger of fatal injury from falling could risk a moment of hypoxic wooziness from an oxygen-deprived brain. If so, then sauropods had oversized hearts whose high energy demands would have required a very high level of oxygen consumption. It may not be possible for a land animal to get much over 20 meters (65 feet) tall, both because of the great pressures needed to pump blood up to the brain and because of the very high pressures produced in the feet by the liquid column of such height. Supertall animals would have needed, like giraffes, special vascular adaptations to cope with the problems associated with fluctuating pressures as the animal stood or lay down, and raised and lowered its head from drinking level to the maximum vertical reach.

The hypothesis that only tachyenergetic animals can grow to enormous dimensions on land is called terramegathermy. An alternative concept, gigantothermy, proposes that the metabolic systems of giant reptiles converge with those of giant mammals, resulting in energy efficiency in all giant animals. In this view, giants rely on their great mass, not high levels of heat production, to achieve thermal stability. This idea reflects a misunderstanding of how animal power systems work. A consistently high body temperature does not provide the power needed to sustain high levels of activity; it merely allows a tachyenergetic animal, and only an animal with a high aerobic exercise capacity, to sustain high levels of activity around the clock. A gigantic reptile with a high body temperature would still not be able to remain highly athletic for extended periods of time. Measurements show that the metabolic rates and aerobic capacity of elephants and whales are as high as expected in mammals of their size and are far higher than those of the biggest crocodilians and turtles, which have the low levels of energy production typical of reptiles.

It has long been questioned how sauropods fed themselves with their small heads, all the more so if they had the high rates of food consumption expected in tachyaerobic animals of their size. However, the small head of a sauropod was like the small head of an emu or ostrich—it was basically all mouth. Most of the head of herbivorous mammals consists of the dental batteries used to chew food after it has been cropped with the mouth, which is restricted to the front end of the jaws. Also, sauropod heads are not as small as they look—the mouths of the biggest sauropods could engulf the entire head of a giraffe. The breadth of the mouth of sauropods is the same as that of herbivorous mammals of the same body mass. If a tachyenergetic sauropod of 50 tonnes ate as much as expected in a mammal of its size, then it needed to consume over half a tonne of fresh fodder a day. But that is only 1 percent of its own body mass, and if the sauropod fed for fourteen hours each day and took one bite per minute, then it needed to bite off only about half a kilogram of plant material each time. That would have been easy for the sauropod's head, which weighed as much as a human body and had a mouth about half a meter (1.5 feet) wide.

Some researchers are concerned that giant dinosaurs would have overheated in the Mesozoic greenhouse if they had avian- or mammalian-like levels of energy production. However, the largest animals dwelling in the modern tropics, including deserts, are big birds and mammals. Some of the largest elephants live in the Namib Desert of the Skeleton Coast of southwestern Africa, where they often have to tolerate extreme heat and sun without the benefit of shade. It is widely thought that elephants use their ears to keep themselves cool when it is really hot, something dinosaurs could not do. However, elephants flap their ears only when the ambient temperature is below that of their bodies. When the air is as warm as the body, heat can no longer flow out, and flapping the ears actually picks up heat when the air is warmer than the body. Nor was the big-eared African elephant the main savanna elephant until fairly recently; the dominant savanna elephant used to be one of the biggest land mammals ever, *Palaeoloxodon recki*. A relative of the Asian elephant, it probably had small ears of little use for shedding body heat at any temperature. It is actually small animals that are most in danger from suffering heat exhaustion and stroke, because their small bodies pick up heat from the environment very quickly. The danger is especially acute in a drought, when water is too scarce to be used for evaporative cooling. Because they have a low surface area/mass ratio, large animals are protected by their bulk against the high heat loads that occur on very hot days, and they can store the heat they generate internally. Large birds and mammals retain the heat they produce during the day by allowing their body temperatures to climb a few degrees above normal and then dump it into the cool night sky, preparing for the cycle of the next day.

Another, and subtle, reason that dinosaurs could become so enormous has to do with their mode of reproduction. Because big mammals are slow-breeding K-strategists that lavish attention and care on the small number of calves that they produce, there always has to be a large population of adults present to raise the next generation. A healthy herd of elephants has about as many breeding adults as it does juveniles, which cannot survive without parental care. Because there always has to be a lot of grown-ups, the size of the adults has to be limited in order to avoid

Dinosaur giants compared to mammals

overexploiting their habitat's food resources; doing the latter will cause the population to collapse. This constraint appears to limit slow-reproducing mammalian herbivores from exceeding 10–20 tonnes. Flesh eaters live off an even smaller resource base because they are preying on the surplus herbivores, and it seems that carnivorous mammals cannot maintain a viable population if they are larger than between half a tonne and 1 tonne.

Because giant dinosaurs were fast-breeding r-strategists that produced large numbers of offspring that could care for themselves, their situation was very different from that of big mammals. A small population of adults was able to produce large numbers of young each year. Even if all adults were killed off on occasion, their eggs and offspring could survive and thrive, keeping the species going over time. Because dinosaurs could get along with smaller populations of adults, the grown-ups were able to grow to enormous dimensions without overexploiting their resource base. This evolutionary scheme allowed plant-eating dinosaurs to grow to 20 to perhaps more than 100 tonnes. It is notable that supersauropods were relatively rare, indicating that they had small populations. Because the bulk of the biomass of adult herbivorous dinosaurs was tied up in oversized giants, the theropods needed to evolve great size themselves in order to be able to fully access the nutrition tied up in the huge adults—the idea that theropods grew to 6 to 10 tonnes only to "play it safe" by consistently hunting smaller juveniles is not logical—and the fast-breeding and fast-growing predators could reach tremendous size. The existence of oversized predators in turn may have resulted in a size race in which sauropods evolved great size in part as protection against their enemies, which later encouraged the appearance of supersized theropods that could bring them down.

Very tall necks like those of sauropods and giraffes evolve in an evolutionary feedback loop that involves two factors. Increasing height serves as a dominance display that enhances reproductive success by intimidating rivals and impressing mates. This is similar to other reproductive displays such as the tails of peacocks and the giant antlers of big cervids. As the head gets higher the herbivore has a competitive feeding advantage over shorter herbivores that provides the power source needed to pump blood to the brain held far above the heart. Lacking dental batteries and big brains, sauropod heads were relatively small, so sauropods were able to evolve extremely tall necks that required enormous bodies to anchor them upon and to contain the big hearts they needed.

In the 1800s Edward Cope proposed what has become known as Cope's Rule, the tendency of animal groups to evolve gigantism. The propensity of dinosaurs to take this evolutionary pattern to an extreme means that the Mesozoic saw events on land that are today limited to the oceans. In modern times combat between giants occurs between orcas and whales. In the dinosaur era it occurred between orca-sized theropods and whale-sized sauropods, hadrosaurs, and ceratopsids.

MESOZOIC OXYGEN

Oxygen was absent from the atmosphere for much of the history of the planet, until the photosynthesis of single-celled plants built up enough oxygen to overwhelm the processes that tend to bind it to various elements such as iron. Until recently it was assumed that oxygen levels then became stable, making up about a fifth of the air for the last few hundred million years. It has now been proposed that oxygen levels have instead fluctuated strongly over time. The methods used to estimate past oxygen levels suggest that they reached a uniquely high level of about a third of the atmosphere during the late Paleozoic, when the great coal forests were forming and, because of the high oxygen levels, often burning. It is notable that this is when many insects achieved enormous dimensions by the standards of the group, including dragonfly relatives with wings over half a meter (2 feet) across. Because insects bring oxygen into their bodies by a dispersed set of tracheae, the size of their bodies may be tied to the level of oxygen.

Soon afterward, oxygen levels may have plunged precipitously, sinking to a little over half the current level by the Triassic and Jurassic. In this case oxygen availability at sea level would have been as poor as it is at high altitudes today. Making matters worse were the high levels of carbon dioxide. Although not high enough to be directly lethal, the combination of low oxygen and high carbon dioxide would have posed a serious respiratory challenge. Reptiles subjected to low-oxygen conditions become more sluggish and grow more slowly, whereas some birds can fly higher than Mount Everest. If oxygen was scarce in the Mesozoic, then the ability of dinosaurs to achieve high levels of sustained activity and grow rapidly was all the more remarkable and is evidence that they evolved systems able to efficiently take in and utilize oxygen at high levels while coping with excess carbon dioxide. In this context, the evolution and success of saurischian dinosaurs in the Late Triassic and Jurassic may have resulted from the development of the efficient air-sac-driven respiratory systems in avepod theropods and sauropods, which would have allowed them to breathe as easily at low altitudes as birds do today at high levels. This allowed them to normally walk at 3–10 km/h without running out of breath. It also allowed a group of small theropods to evolve into powered fliers despite the absence of abundant oxygen. There is evidence that pterosaurs likewise evolved an air-sac system of their own, allowing them to power fly beginning in the Triassic. Because of their less efficient dead-end lungs, the evolution of reptiles, mammals, and ornithischians may have been hindered in the Jurassic. The oxygen problem may have restricted the habitation of highlands, and again the saurischians would have been best suited for the conditions.

During the Cretaceous, oxygen levels are estimated to have crept upward toward modern levels, although they never reached the current concentration in the Mesozoic. This rise in oxygen may have allowed the ornithischians to finally evolve large size and great diversity, helping them to partly displace the sauropods. It is interesting that the most athletic of the big ornithischians, the ceratopsids and hadrosaurids, as well as the exceptionally fast-moving tyrannosaurids, appeared in the closing stages of the Cretaceous, when oxygen levels were at their Mesozoic maximum. It is similarly notable that the biggest pterosaurs also evolved at this time.

But there is a problem. A different method of estimating oxygen levels agrees that there was a big dip in levels at the beginning of the Mesozoic but soon has the level soaring to the present level early in the Triassic and then edging up higher, perhaps much higher, in the later portion of the Mesozoic. If so, then most of the above discussion is moot, and dinosaurs would have been able to easily exploit oxygen to power their active lives. Figuring out the actual oxygen content of the atmosphere in the dinosaur days remains an important challenge.

THE EVOLUTION—AND LOSS—OF AVIAN FLIGHT

Powered flight has evolved repeatedly among animals, including numerous times in insects in the late Paleozoic and three times in tetrapods—pterosaurs in the Triassic, birds in the Jurassic, and bats in the early Cenozoic. In all cases among vertebrates, flight evolved rapidly by geological terms, so much so that the earliest stages have not yet been found in the fossil record for pterosaurs and bats. The means by which flight evolved in pterosaurs remain essentially unknown. The fact that bats evolved from tiny insectivorous mammals, and the recent discovery of an early fossil bat with smaller wings than those of more modern forms, show that mammalian flight evolved in arboreal forms.

The origin of birds and their flight is much better understood than it is for pterosaurs and bats. This knowledge extends back to the discovery of Late Jurassic *Archaeopteryx* in the mid-1800s and is rapidly accelerating with the abundance of new fossils that have come to light in recent years, especially from the Early Cretaceous, and also from the middle of the Jurassic before *Archaeopteryx*. However, a major gap still exists because little is known about what was happening in the Early and Middle Jurassic, well before *Archaeopteryx*.

When it was assumed that birds did not evolve from dinosaurs, it was correspondingly presumed that their flight evolved among climbers that first glided and then developed powered flight. This has the advantage that we know that arboreal animals can evolve powered flight with the aid of gravity, as in bats. When it was realized that birds descended from deinonychosaurs, many researchers switched to the hypothesis that running dinosaurs learned to fly from the ground up. This has the disadvantage that it is not certain whether it is practical for tetrapod flight to evolve among ground runners working against gravity.

The characteristics of birds indicate that they evolved from dinosaurs that had first evolved as bipedal runners, and then evolved into long-armed climbers. If the ancestors of birds had been entirely arboreal, then they should have been semiquadrupedal forms whose sprawling legs were integrated into the main airfoil, like bats. That birds are bipeds whose erect legs are separate from the wings indicates that their ancestors evolved to run. Conversely, how and why ground animals would directly develop the long, strongly muscled arms and wings necessary for powered flight has not been adequately explained. The hypothesis that running theropods developed the ability to fly as a way to enhance their ability to escape up tree trunks itself involves a degree of arboreality. Small theropods, with their grasping hands and feet, were inherently suited for climbing. Some avepod theropods show specializations for climbing, especially *Scansoriopteryx*, microraptorine dromaeosaurs, *Anchiornis*, and *Archaeopteryx*.

Avian flight may have evolved among predatory dinosaurs that spent time both on the ground and in the trees, evolving long arms that facilitated the latter. Leaps between branches could have been lengthened by developing aerodynamically asymmetric pennaceous feathers that turned the leaps into short glides. As the feathers lengthened, they increased the length of the glides. When the protowings became large enough, flapping would have added power, turning the glides into a form of flight.

An early bird, *Confuciusornis*

The same flapping motion would have aided the rapid climbing of trees. Selective pressures then promoted further increases in arm muscle power and wing size until the level seen in *Archaeopteryx* was present. The flying deinonychosaur had an oversized furcula and large pectoral crest on the humerus, which supported an expanded set of muscles for flapping flight. The absence of a large sternum shows that its flight was weak by modern standards. As bird flight further developed, the sternum became a large plate like those seen in dromaeosaurs. Fixed on the rib cage with ossified sternal ribs, the plate anchored large wing-depressing muscles and later sported a keel that further expanded the flight muscles. Adaptations at and near the shoulder joint improved the ability of the wing to elevate, increasing the rate of climb in flight. At the same time, the hand was stiffened and flattened to better support the outer wing primaries, and the claws were reduced and lost. The tail rapidly shortened in most early birds until it was a stub. This means that birds quickly evolved a dynamic form of flight, much more rapidly than pterosaurs, which retained a long tail stabilizer through most of the Jurassic. The above adaptations were appearing in early Cretaceous birds, and the essentially modern flight system had evolved by the early Cretaceous.

For all its advantages, flight has its downsides, including all the energy that is absorbed by the oversized wing tissues, especially the enormous flight muscles. Nor can flying birds be especially large. A number of birds have lost flight, and dinobirds with only modest flight abilities and clawed hands that could be used for multiple purposes would have been more prone to losing the ability to take to the air. Evidence for the loss of flight includes the presence of flight features such as large sternal plates supported by bony sternal ribs, bony uncinate processes on the ribs, folding arms, and stiffened, pterosaur-like tails in animals whose arms were too small for flight. These features typify nonflying dromaeosaurs, whose early examples appear to have been better adapted for flight than *Archaeopteryx*. The large dromaeosaurs were almost certainly neoflightless like big ground birds. *Anchiornis* suggests that deinonychosaurs began to lose flight in the Late Jurassic. Therizinosaurs and oviraptorosaurs show signs that some level of flight was present early in their evolution. In the Cretaceous, birds themselves lost flight on occasion, most famously the widely distributed marine hesperornithiform divers, as well as some chicken- to ratite-sized European birds of uncertain relationships known from near the end of the period.

DINOSAUR SAFARI

Assume that a practical means of time travel has been invented, and, *Field Guide to Dinosaurs* in hand, you are ready to take a trip to the Mesozoic to see the dinosaurs' world. What would such an expedition be like? Here we ignore some practical issues that might preclude such an adventure, such as the problem of cross-contaminating different time periods with exotic diseases. Then there is the classic time paradox issue that plagues the very concept of time travel. What would happen if a time traveler to the dinosaur era did something that changed the course of events to such a degree that humans never evolved?

One difficulty that might arise could be the lack of modern levels of oxygen and extreme greenhouse levels of carbon dioxide (which can be toxic for unprepared animals), especially if the expedition traveled to the Triassic or Jurassic. Acclimation could be necessary, and even then, supplemental oxygen might be needed at least on an occasional basis. Movement and activities would be constrained if oxygen levels were well below modern standards. Work at high altitudes would be even more difficult. Another problem would be the chronically high levels of heat in most dinosaur habitats. Relief would be found at high latitudes, at least during the perpetually dark winters, as well as on mountains.

Assuming that the safari were to one of the classic Mesozoic habitats that included gigantic dinosaurs, the biggest problem would be the sheer safety of the expedition members. The bureaucratic protocols developed for a Mesozoic expedition would emphasize safety, with the intent of keeping the chances of losing any participants to a bare minimum. Modern safaris in Africa require the presence of a guard armed with a rifle when visitors are not in vehicles in case of an attack by big cats, cape buffalo, rhinos, or elephants. Similar weaponry is needed in tiger country, in areas with large populations of grizzlies, and in Arctic areas inhabited by polar bears. The potential danger level would be even higher in the presence of flesh-eating dinosaurs as big as rhinos and elephants and easily able to run down a potentially out-of-breath human. It is possible that theropods would not recognize humans as prey, but it is at least as likely that they would, and the latter would have to be assumed. Aside from the desire to not kill members of the indigenous fauna, rifles, even automatic rapid-fire weapons, might not be able to reliably bring down a 5-tonne allosauroid or tyrannosaur, and heavier weapons would be impractical to carry about. Nor would the danger come from just the predators. A herd of whale-sized sauropods would pose a serious danger of trampling or impact from tails, especially if they were spooked by humans and either attacked them as a possible threat or stampeded in their direction. Sauropods would certainly be more dangerous than elephants, whose high level of intelligence allows them to better handle situations involving humans. The horned ceratopsids, even less intelligent than rhinos, and probably with the attitudes of oversized pigs, would pose another major danger.

Travel by foot would, therefore, probably be largely precluded in habitats that included big theropods, sauropods, and ceratopsids. Expedition members would have to move about on the ground in vehicles sufficiently large and strong to be immune from attacks by colossal dinosaurs. Movement away from the vehicles would be possible only when aerial vehicles, drones perhaps, could show that the area was safe. Nor would it be feasible to simply set up tents in a clearing. The camp would have to be a protected space, ringed by a fence, wall, or ditch able to fend off the giant predators as well as a panicked herd of supersauropods. In places lacking giant dinosaurs, such stringent levels of protection would not be necessary. Even so, medium-sized dinosaurs would still pose significant risk. An attack by sickle-clawed dromaeosaurs, for instance, could result in serious casualties. So could assault by a pack of parrot-beaked peccary-like protoceratopsids. Defensive weapons would be necessary. If the expedition protocol required minimal risk to the fauna, then transport in vehicles under most circumstances would be standard. Yet another danger in some Cretaceous habitats would be elephant-sized crocodilians that would undoubtedly be willing to snap up and gulp down whole a still-living human unwary enough to go near or in the water. One way or another, dinosaur watching would pose a series of difficult problems not seen in dealing with modern animals.

IF DINOSAURS HAD SURVIVED

Assume that the K/P impact is what killed off the dinosaurs, but also assume that the impact did not occur and that nonavian dinosaurs continued into the Cenozoic. What would the evolution of land animals have been like in that case?

Although much will always be speculative, it is likely that the Age of Dinosaurs would have persisted—indeed the Mesozoic Era would have endured—aborting the Cenozoic Age of Mammals. Thirty million years ago western North America probably would have been populated by great dinosaurs rather than the rhino-like titanotheres. The continuation of sauropods should have inhibited the growth of dense forests. But the flowering angiosperms would have continued to evolve and to produce a new array of food sources including well-developed fruits that herbivorous dinosaurs would have needed to adapt to in order to exploit.

What is not certain is whether mammals would have remained diminutive or would have begun to compete with dinosaurs for the large-body ecological niches. By the end of the Cretaceous sophisticated marsupial and placental mammals were appearing, and they may have been able to begin to mount a serious contest with dinosaurs as time progressed. Eventually, southward-migrating Antarctica would have arrived at the South Pole and formed the enormous ice sheets that act as a giant air conditioning unit for the planet. At the same time, the collision of India and Asia that closed off the once-great Tethys Ocean built up the miles-high Tibetan Plateau, which has also contributed to the great planetary cool-off of the last 20 million years that eventually led to the current ice age despite the rising heat production of the sun. This should have forced the evolution of grazing dinosaurs able to crop the spreading savanna, steppe, and prairie grasslands that thrive in cooler climates. In terms of thermoregulation, dinosaurs should have been able to adapt, but the also energetic mammals may have been able to exploit the decreasing temperatures. Perhaps big mammals of strange varieties would have formed a mixed dinosaur-mammal fauna, with the former perhaps including some big birds. Mammals may also have proven better able to inhabit the oceans than nonavian dinosaurs.

The birdlike dinosaurs evolved brains larger and more complex than those of reptiles toward the end of the Jurassic and beginning of the Cretaceous, but they never exceeded the lower avian range, and they did not exhibit a strong trend toward larger size and intricacy in the Cretaceous similar to the startling increase in neural capacity in Cenozoic mammals. We can only wonder whether dinosaurs would have eventually undergone their own expansion in brain power had they not gone extinct. Perhaps the evolution of large-bodied, big-brained mammals would have compelled dinosaurs to upgrade thinking performance as well. Or perhaps smarter mammals would have outcompeted dinosaurs still stuck with inferior mental capacity.

The specific species *Homo sapiens* would not have evolved if not for the extinction of dinosaurs, but whether some form of highly intelligent, language- and tool-using animal would have developed is another matter. Modest-sized, bipedal, birdlike predatory theropods with their grasping hands might have been able to do so. Or perhaps arboreal theropods with stereo color vision would have become fruit eaters whose evolution paralleled that of the increasingly brainy primates that spawned humans. It is possible that actual primates would have appeared and evolved above the heads of the great dinosaurs, producing at some point bipedal ground mammals able to create and use tools. On the other hand, the evolution of superintelligent humans may have been a fluke and would not have been repeated in another world.

DINOSAUR CONSERVATION

Taking the above scenario to its extreme, assume that some group of smart dinosaurs or mammals managed to survive and thrive in a world of great predatory theropods and became intelligent enough to develop agriculture and civilization as well as an arsenal of lethal weapons. What would have happened to the global fauna? The fate of large dinosaurs would probably have been grim. We actual humans may have been the leading factor in the extinction of a large portion of the megafauna that roamed much of the Earth toward the end of the last glacial period, and matters continue to be bad for most wildlife on land and even in the oceans. The desires and practical needs of our imaginary sapients would have compelled them to wipe out the giant theropods, whose low adult populations would have rendered them much more susceptible to total loss than the big mammal carnivores. If whale-sized herbivorous dinosaurs were still extant, their low populations would also have made them more vulnerable than elephants and rhinos. By the time the sapients developed industry, the gigantic flesh and plant eaters would probably already have been part of historical lore. If superdinosaurs had instead managed to survive in an industrial world, they would have posed insurmountable problems for zoos. Feeding lions, tigers, and bears is not beyond the means of zoos, but a single tyrannosaur-sized theropod (assuming it were tachyenergetic) would break the budget by consuming a couple of thousand cattle-sized animals over a few decades. How could a zoo staff handle a 50-foot-tall sauropod weighing 30 or 50 tonnes and eating ten times as much as an elephant?

WHERE DINOSAURS ARE FOUND

Because the big dinosaurs are long gone and time travel probably violates the nature of the universe, we have to be satisfied with finding the remains they left behind. With the possible exception of very high altitudes, dinosaurs lived in all places on all continents, so where they are found is determined by the existence of conditions suitable for preserving their bones and other traces, eggs and footprints especially, as well as by conditions suitable for finding and excavating the fossils. For example, if a dinosaur habitat lacked the conditions that preserved fossils, then that fauna has been totally lost. Or, if the fossils of a given fauna of dinosaurs are currently buried so deep that they are beyond reach, then they are not available for scrutiny.

All but a very small percentage of carcasses are destroyed soon after death. Many are consumed by predators and scavengers, and others rot or are weathered away. Even so, the number of animals that have lived over time is immense. Because at any given time a few billion dinosaurs were probably alive, mostly juveniles and small adults, and the groups existed for most of the Mesozoic, the number of dinosaur fossils that still exist on the planet is enormous, probably in the hundreds of millions or low billions of individuals.

Of these only a fraction of 1 percent have been found at or near the small portion of the dinosaur-bearing formations that are exposed on the surface where the fossils can be accessed, or in the mines that allow some additional remains to be reached. Even so, the number of dinosaur fossils that have been scientifically documented to at least some degree is considerable. Some dinosaur bone beds contain the remains of thousands of individuals, and the total number of dinosaur individuals known in that sense is probably in the tens or hundreds of thousands. The question is where to find them.

Much of the surface of the planet at any given time is undergoing erosion. This is especially true of highlands. In erosional areas, sediments that could preserve the bones and other traces are not laid down, so highland faunas are rarely found in the geological record. Fossilization has the potential to occur in areas in which sediments are being deposited quickly enough, and in large enough quantities, to bury animal remains before they are destroyed. Animals can be preserved in deep fissures or caves in highland areas; this is fairly rare but not unknown when dealing with the Mesozoic. Areas undergoing deposition tend to be lowlands downstream of uplifting highlands that provide abundant sediment loads carried in streams, rivers, lakes, or lagoons that settle out to form beds of silt, sand, or gravel. Therefore, large-scale formation of fossils occurs only in regions experiencing major tectonic activity. Depositional lowlands can be broad valleys or large basins of varying size in the midst of highlands, or coastal regions. As a result, most known dinosaur habitats were flatlands, with little in the way of local topography. In some cases the eroding neighboring highlands were visible in the distance from the locations where fossilization was occurring; this was especially true in ancient rift valleys and along the margins of large basins. In deserts, windblown dunes can preserve bones and trackways, and also when they slump when wetted by rains. So can ashfalls, but lava flows tend to incinerate and destroy animal remains. Also suitable for preserving the occasional dinosaur carcass as drift are sea and ocean bottoms.

Most sediment deposition occurs during floods, which may also drown animals that are then buried and preserved. The great majority of remains, however, died before a given flood. Once burial occurs, the processes that preserve remains are complex and in many regards poorly understood. It is being realized, for instance, that bacterial activity is often important in preserving organic bodies. Depending on the circumstances, fossilization can be rapid or very slow to the point that it never really occurs even after millions of years. The degree of fossilization therefore varies and tends to be more extensive the further back in time the animal was buried. The most extreme fossilization occurs when the original bone is completely replaced by groundwater-borne minerals. Some Australian dinosaur bones have, for instance, been opalized. Most dinosaur bones, however, retain the original calcium structure. The pores have been filled with minerals, converting the bones into rocks much heavier than the living bones. In some locations, such as the Morrison Formation, bacterial activity encouraged the inclusion of uranium in many bones, leading to a significant radiation risk from stored bones. In other cases the environment surrounding dinosaur bones has been so stable that little alteration has occurred, leading to the partial retention of some soft tissues near the core of the bones.

Late Triassic (Rhaetian–Norian–Carnian)

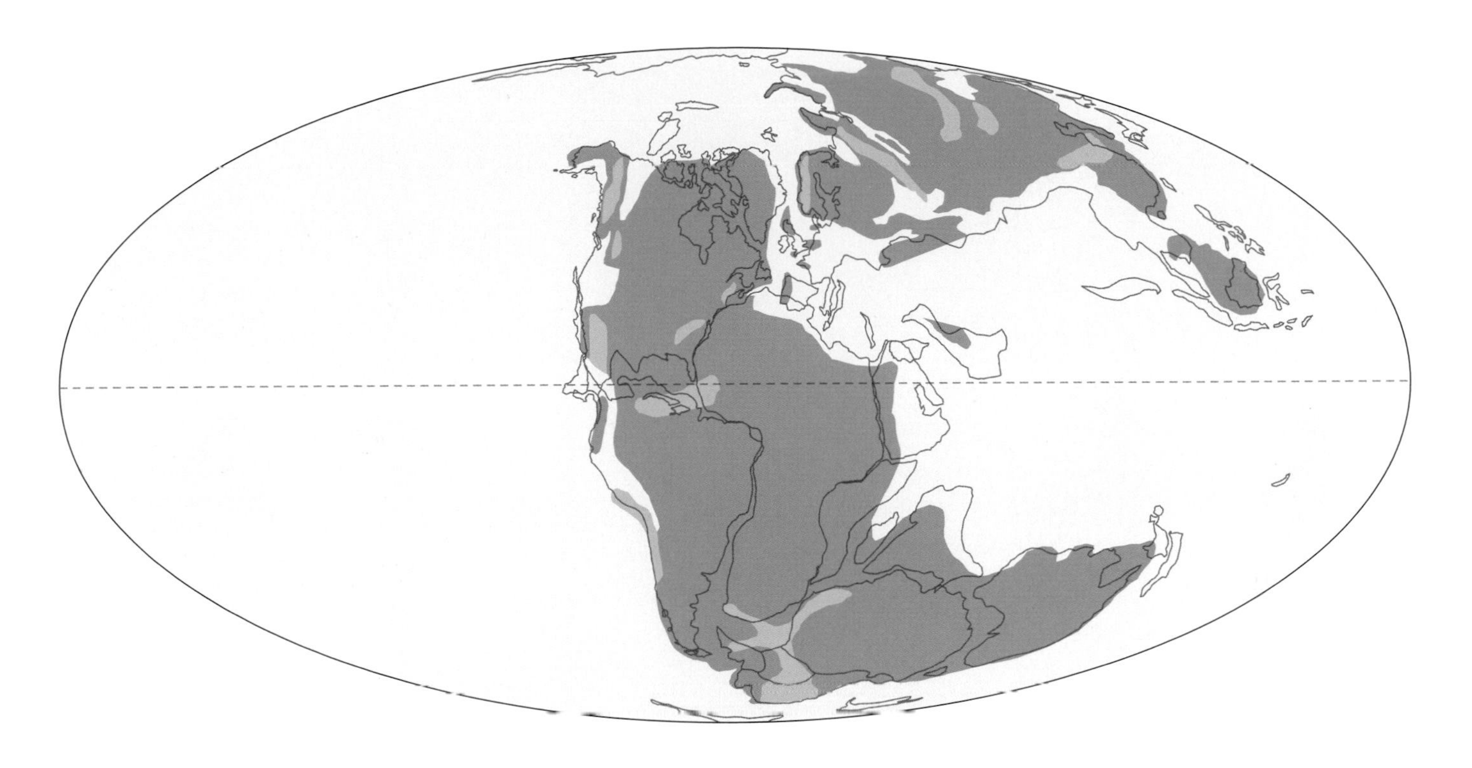

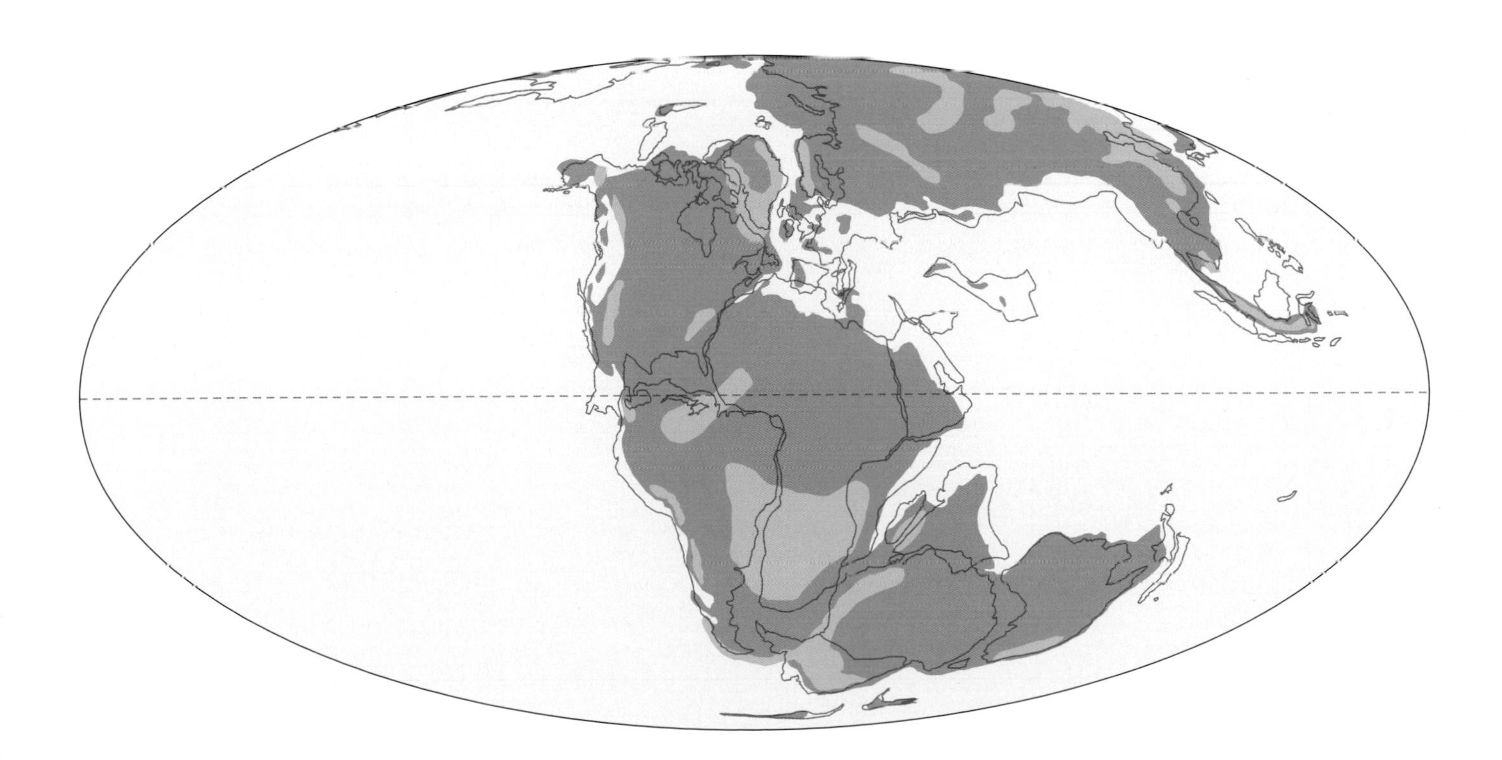

Early Jurassic (Sinemurian)

Middle Jurassic (Callovian)

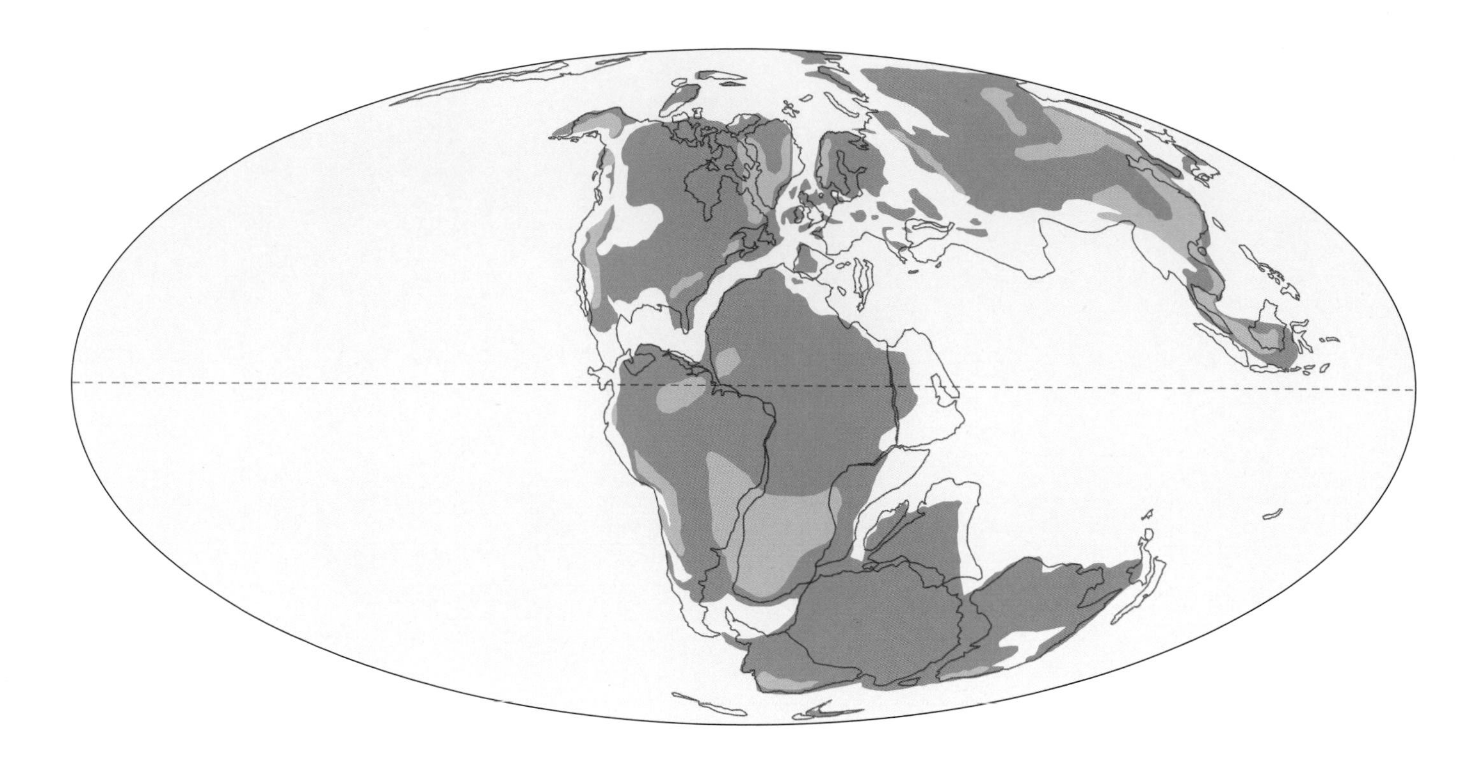

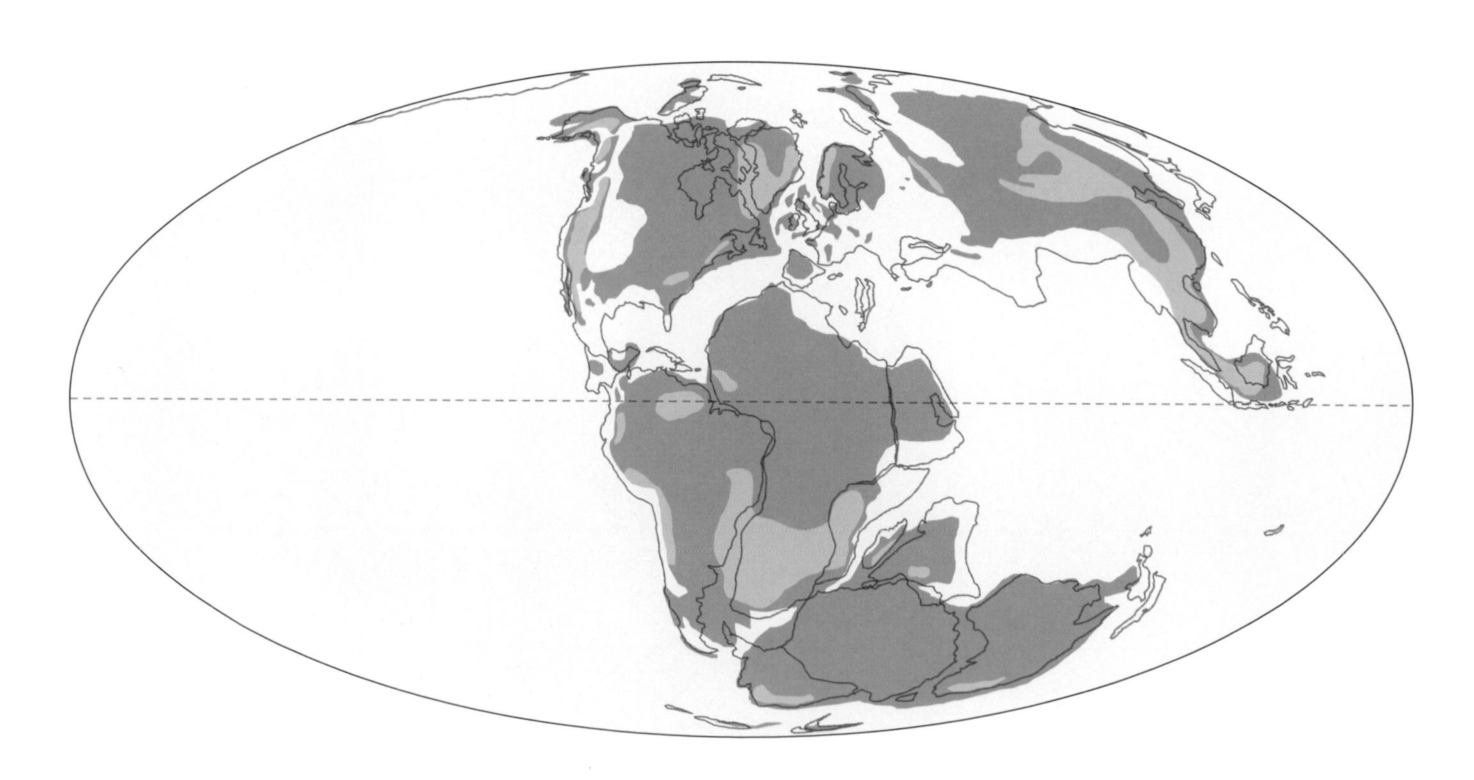

Late Jurassic (Kimmeridgian)

Early Cretaceous (Valanginian–Berriasian)

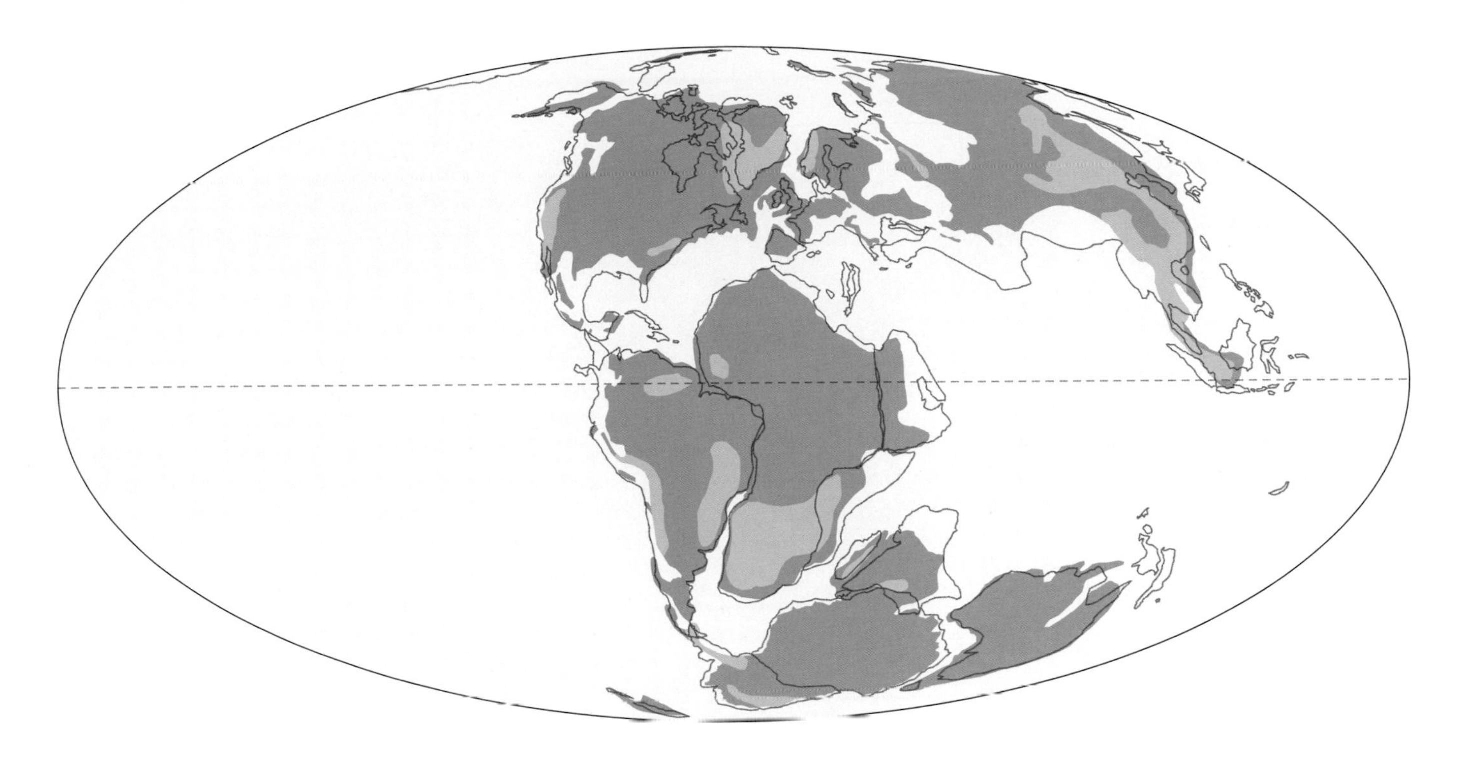

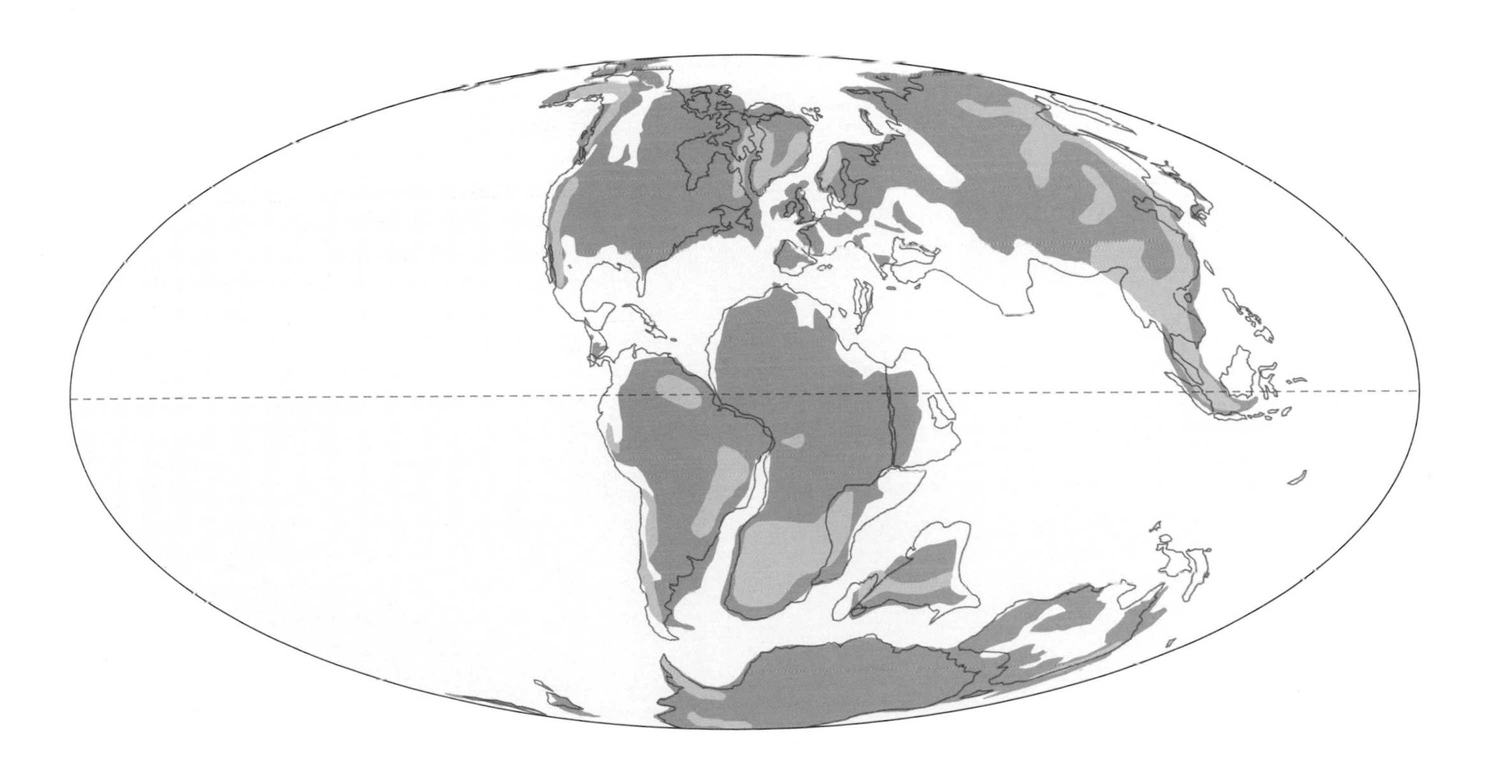

Early Cretaceous (Aptian)

Late Cretaceous (Coniacian)

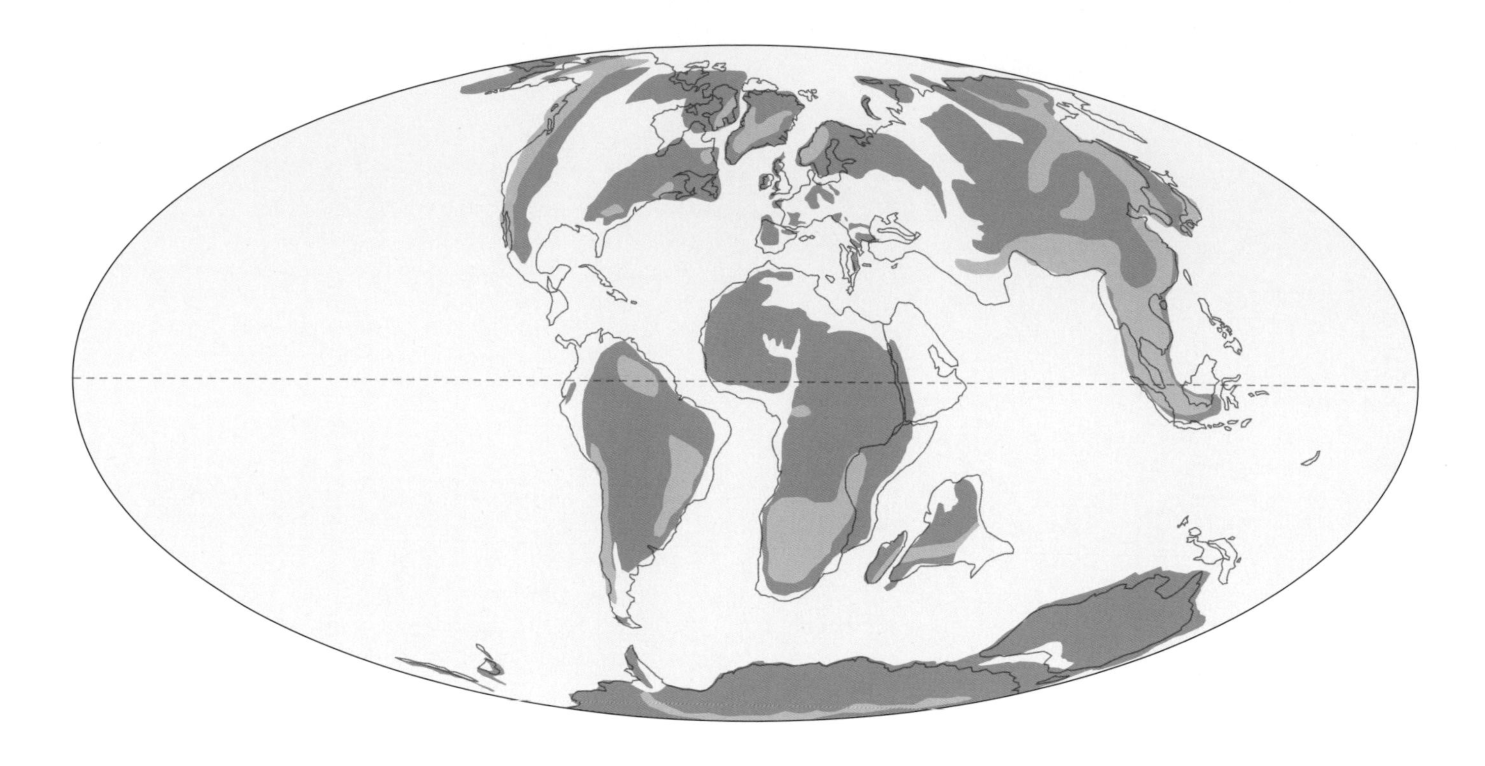

Late Cretaceous (Campanian)

Although the number of dinosaur bones and trackways that lie in the ground is tremendous, all but a tiny fraction are for practical reasons out of reach. Nearly all are simply buried too deeply. The great majority of fossils that are found are on or within a few feet of the surface. Occasional exceptions include deep excavations such as construction sites and quarries, or mining operations. Even if deposits loaded with dinosaur fossils are near the surface, their discovery is difficult if a heavy cover of well-watered vegetation and soil hides the sediments. For example, large tracts of dinosaur-bearing Mesozoic sediments lie on the Eastern Seaboard, running under some major cities such as Washington, DC, and Baltimore. But the limited access to the sediments hinders discoveries, which are largely limited to construction sites made available by willing landowners. Coastal cliffs made up of Mesozoic deposits are another location for dinosaur hunting in forested areas.

Prime dinosaur real estate consists of suitable Mesozoic sediments that have been exposed and eroded over large areas that are too arid to support heavy vegetation. This includes short-grass prairies, badlands, and deserts. There are occasional locations in which dinosaur bone material is so abundant that their remains are easily found with little effort, especially before they have been picked over. Dinosaur Provincial Park in southern Alberta is a well-known example. In some locations countless trackways have been exposed. In most cases dinosaur bones are much less common. Finding dinosaurs has changed little since the 1800s. It normally consists of walking slowly, stooped over, usually under a baking sun, often afflicted by flying insects, looking for telltale traces. If really small remains are being looked for, such as fragmented egg shells, crawling on (padded) knees is necessary. Novices often miss the traces against the background of sediments, but even amateurs soon learn to mentally key in on the characteristics of fossil remains. Typically, broken pieces of bones on the surface indicate that a bone or skeleton is eroding out. One hopes that tracing the broken pieces upslope will soon lead to bones that are still in place. In recent years GPS has greatly aided in determining and mapping the position of fossils. Ground-penetrating radar has sometimes been used to better map out the extent of a newly found set of remains, but usually researchers just dig and see what turns up.

Now it becomes a matter of properly excavating and removing the fossil without damaging it while scientifically investigating and recording the nature of the surrounding sediments in order to recover the information they may contain. These basic methods have also not changed much over the years. On occasion thick overburden may be removed by heavy equipment or even explosives. But usually it is a job for jackhammers, sledgehammers, picks, or shovels, depending on the depth and hardness of the sediments and the equipment that can be brought in. When the remains to be recovered are reached, more-careful excavation tools, including trowels, hammers and chisels, picks, and even dental tools and brushes, are used. It is rare to be able to simply brush sand off a well-preserved specimen as in the movies, although this happy circumstance does occur in some ancient dune deposits in Mongolia. Usually sediments are cemented to some degree and require forceful action. At the same time, the bones and other remains are fragile, and care must be taken to avoid damaging them. And their position has to be documented by quarry maps, photography, or laser scanning before removal. Individual bones can be removed, or blocks of sediment including multiple bones or articulated skeletons may have to be taken out intact. Again, these operations are usually conducted under conditions that include flying insects, dust, heat, and sun, although tarps can provide shade. In Arctic locations heat is not a problem, but insect swarms are intense during the summer field season.

After exposure, especially fragile bones may be soaked with glue to harden them. On the other hand, the increasingly sophisticated techniques being applied to bones in the laboratory discourage alteration and contamination of bones. Before removal, most remains are quickly covered with tissue paper that is wetted in place, followed by heavier paper, and over that a thick layer of plaster to form a protective jacket. Wood is usually used to brace the jacket. When the top is so protected, the remains are undermined and then flipped—a process often requiring considerable exertion and entailing some risk to both excavators and the fossils. Then the other side is papered and plastered, forming a protective cocoon. If the jacketed block is very heavy and not accessible to heavy equipment, a helicopter may be brought in to lift it out. On occasion this requires a heavy lift copter; the US Army is sometimes willing to conduct such operations gratis as part of dissimilar cargo training that provides its crews with the opportunity to learn how to cope with challenging objects rather than standard pallets.

Because dinosaur paleontology is not a high-priority science backed by large financial budgets, and because the number of persons searching for and excavating dinosaurs in the world in a given year is only a few thousand—far more than in the past—the number of dinosaur skeletons that now reside in museums is still just a few thousand. A growing exception is China, where government funding is filling warehouses and new museums with material.

In the lab, preparators remove part or all of the jacket and use fine tools to eliminate some or all of the sediment from the bones and any other remains. Most bones are left intact, and only their surface form is documented. In some cases chemical treatment is required to stabilize bones; this is especially true if the bones are impregnated with pyrite, which gradually swells with moisture. Increasingly, certain bones are opened to reveal their internal structure for various purposes: sectioning to examine bone histology and microstructure, to count growth rings, to search for traces of soft tissues, and to sample bone isotopes and proteins. It is becoming the norm to conduct CT scans on skulls and complex bones as a means to determine the three-dimensional structure without invasive preparation, as well as to reduce costs. These can be published as conventional hard copies and in digital form. There is increasing reluctance to put original bones in mounted skeletons in display halls because delicate fossils are better conserved when properly restored. Instead, the bones are molded, and lightweight casts are used for the display skeleton.

There has never been as much dinosaur-related activity as there is today. At the same time, there is the usual shortage of funding and personnel. The happy result is that there are plenty of opportunities for amateurs to participate in finding and preparing dinosaurs. If nonprofessionals search for fossils on their own, they need to pay attention to laws and to paleontological ethics. In some countries all dinosaur fossils are regulated by the state—this is true in Canada, for instance. In the United States, fossils found on privately held land are entirely the property of the landowner, who can dispose of any prehistoric remains as he or she sees fit. Any search for and retention of fossils on private property is therefore by permission of the owner, many but not all of whom are interested in the fossils on their land. Because dinosaur remains in the eastern states usually consist of teeth and other small items, nonhazardous construction sites are often available for exploration on nonwork days. In the West, ranches with cooperative owners are primary sources of dinosaur remains. Unfortunately, the rising sums of money to be made by selling fossils are making it more difficult for scientific teams to access such lands. The religious opinions of some landowners are also an occasional barrier. Fortunately, dinosaur fossils are a part of western lore and heritage, so many locals favor paleontological activities, which contribute to the tourist trade. All fossils on federal government land are public property and are heavily regulated. Removal can occur only with official permission, which is limited to accredited researchers. Environmental concerns may be involved because dinosaur excavations are in effect small-scale mining operations. Fossils within Indian reservations may likewise be regulated, and collaboration with resident natives is indispensable. Dinosaur fossils found by nonprofessionals searching on their own should not be disturbed. Instead they should be reported to qualified experts, who can then properly document and handle the remains. In such cases the professionals are glad to do so with the assistance of the discoverer.

A growing number of museums and other institutions offer courses to the public on finding, excavating, and preparing dinosaurs and other fossils. Most expeditions include unpaid volunteers who are trained, often on-site, to provide hands-on assistance to the researchers. Participants are usually expected to pay for their own transportation and general expenses, although food and in some cases camping gear as well as equipment may be provided. In order to tap into the growing number of dinosaur enthusiasts, commercial operations led by qualified experts provide a dinosaur-hunting experience for a fee, usually in the western states and Canada. Those searching for and digging up dinosaurs need to take due precautions to protect themselves from sunlight and heat, in terms of UV exposure, dehydration, and hyperthermia, as well as biting and stinging insects and scorpions. Rattlesnakes are often common in the vicinity of dinosaur fossils. Steep slopes, cliffs, and hidden cavities are potential dangers. In many dinosaur formations gravel-like caliche deposits formed in the ancient, semiarid soils form roller-bearing-like surfaces that undermine footing. Flash floods can hit quarries or ill-placed campsites. The use of mechanical and handheld tools when excavating fossils poses risks, as does falling debris from quarry walls. When impact tools are used on hard rock, eye protection may be necessary. Chemicals used while working with fossils require proper handling.

Back in the museums and other facilities, volunteers can be found helping to prepare specimens for research and display, and cataloging and handling collections. This is important work because, in addition to the constant influx of new specimens with each year's harvest, many dinosaur fossils found as long as a century ago have been sitting on shelves, sometimes still in their original jackets, without being researched.

Landowners who allow researchers onto their land sometimes get a new species found on their property named after them, informally or formally. So do volunteers who find new dinosaurs. Who knows, you may be the next lucky amateur.

USING THE GROUP AND SPECIES DESCRIPTIONS

About seventeen hundred dinosaur species have been named, but a large portion of these names are invalid. Many are based on inadequate remains, such as teeth or one or a few bones, that are taxonomically indeterminate. Others are junior synonyms for species that have already been named. *Dynamosaurus*, for instance, proved to be the same as previously named *Tyrannosaurus*, which had been named shortly before, so the former is no longer used. This guide includes those species that are generally considered valid and are based on sufficient remains. A few exceptions are allowed when a species based on a single bone or little more is important in indicating the existence of a distinctive type or group of dinosaurs in a certain time and place. Most of the group and species entries have been changed very little if at all from the first edition; corrections and new information have caused a substantial minority to be more heavily revised and in a few cases dropped; at the same time, a large array of new species have been added.

The species descriptions are listed hierarchically, starting with major groups and working down the level of rankings to the genera and species. Because many researchers have abandoned the traditional Linnaean system of classes, orders, suborders, and families, there is no longer a standard arrangement for the dinosaurs—many dinosaur genera are no longer placed in official families—so none is used here. In general the taxa are arranged phylogenetically. This presents a number of problems. It is more difficult for the general reader to follow the various groupings. Although there is considerable consensus concerning the broader relationships of the major groups, at lower levels the incompleteness of the fossil record hinders a better understanding. The great majority of dinosaur species are not known, many of those that are known are documented by incomplete remains, and it is not possible to examine dinosaur relationships with genetic analysis. Because different cladistic analyses often differ substantially from one another, I have used a degree of personal choice and judgment to arrange the groups and species within the groups. Some of these placements reflect my considered opinion, while others are arbitrary choices between competing research results. Most of the phylogeny and taxonomy offered here is not a formal proposal, but a few new group labels were found necessary and are coined and defined here for future use by others if it proves efficacious. Disputes and alternatives concerning the placement of dinosaur groups and species are often but not always mentioned.

Under the listing for each dinosaur group, the overall geographic distribution and geological time span of its members are noted. This is followed by the anatomical characteristics that apply to the group in general, which are not repeated for each species in the group. The anatomical features usually center on what is recorded in the bones, but other body parts are covered when they have been preserved. The anatomical details are for purposes of general characterization and identification and reflect as much as possible what a dinosaur watcher might use; they are not technical phylogenetic diagnoses. The type of habitat that the group favored is briefly listed, and this varies from specific in some types to very generalized in others. Also outlined are the restored habits that probably characterize the group as a whole. The reliability of these conclusions varies greatly. There is, for example, no doubt that theropods with bladed, serrated teeth consumed flesh rather than plants. There is also little doubt that the sickle-clawed *Velociraptor* regularly attacked the similar-sized herbivore *Protoceratops*—there is even an example of two skeletons still locked in combat. Less certain is exactly how *Velociraptor* used its sickle claw to dispatch prey on a regular basis. It is not known whether or not *Velociraptor* packs attacked the much larger armored *Pinacosaurus* that lived in the same desert habitat.

The naming of dinosaur genera and species is often problematic, in part because of a lack of consistency. In some cases genera are badly oversplit. For example, every species of tyrannosaurid receives a generic title despite the high uniformity of the group. A similar situation is true of chasmosaurine and centrosaurine ceratopsids, and of kritosaur and hypacrosaur hadrosaurs, in which the core skulls and skeletons are often very similar; the characters that vary are the crests, frills, and horns used for species-level identification. A number of species are probably one of the growth stages of another species. Meanwhile, a large number of species have been appropriately placed in the genus *Psittacosaurus*, and iguanodont species have tended to be placed in too few genera and species. The inconsistency in naming genera is seen when comparing the great variation between *Psittacosaurus gobiensis* and *P. sibiricus* versus the very close similarity of *Corythosaurus* and *Lambeosaurus*. Among sauropods it has been recognized that the apatosaurs were overlumped, resulting in the revival of *Brontosaurus*. In this work I have attempted to apply more uniform standards to generic and species designations, with the divergences from the conclusions of others noted.

The entry for each species first cites the dimensions and estimated mass of the taxon. The total length is for the combined skull and skeleton; any tail feathers that further lengthened the dinosaur are not included. The values represented are a general figure for the size of the largest known adults of the species and do not necessarily apply to the value estimated for specific specimens, most of which can be found via http://press.princeton.edu/titles/10851.html, which includes the mass estimates for each included specimen. Because the number of specimens for a particular species is a small fraction of those that lived, the largest individuals are not measured; "world record" specimens can be a third or more heavier than is typical. The sizes of species known only from immature specimens are not estimated. All values are, of course, approximate, and their quality varies depending on the completeness of the remains for a given species. If the species is known from sufficiently complete remains, the dimensions and mass are based on the skeletal restoration. The last are used to estimate the volume of the dinosaur, which can then be used to calculate the mass with the portion of the volume that was occupied by lungs and any air sacs taken into account. For dinosaurs without air sacs, the density, or specific gravity, is set at 0.95 relative to the density of water. For those with air sacs, the specific gravity is up to 0.85, except that sauropods' necks are 0.60. When remains are too incomplete to make a direct estimate of dimensions and mass, they are extrapolated from those of relatives and are considerably more approximate. Both metric and English measurements are included except for metric tonnes, which equal 1.1 English tons; all original calculations are metric, but because they are often imprecise, the conversions from metric to English are often rounded off as well. Most length and mass estimates have not changed from those of the first edition, but a significant number have.

The next line outlines the fossil remains, whether they are skull or skeletal material or both, that can be confidently assigned to the species to date; the number of specimens varies from one to thousands. The accuracy of the list ranges from exact to a generalization. The latter sometimes results from recent reassignment of specimens from one species to another, leaving the precise inventory uncertain. Skeletal and/or skull restorations have been rendered for those species that are known from sufficiently complete remains that were available as the book was being produced to execute a reconstruction—the pace of discovery is so fast that some new finds could not be included—or for species that are of such interest that a seriously incomplete restoration is justified (the rather poorly known but oversized new oviraptorosaur *Gigantoraptor* being an example, as is the long-known but still very incomplete *Ankylosaurus*). A few complete skeletons are so damaged or distorted that a restoration is not feasible; this is true of the flattened Yixian *Psittacosaurus* with skin impression and bristles, and *Chasmosaurus irvinensis*. When existing examples of some major dinosaur types are not sufficient, the remains of multiple species have been used to construct a composite representation, such as for the derived alvarezsaurid and therizinosaurid.

A number of species known from good remains have yet to be made available for research, in some cases decades after their discovery. In some cases only oblique-view photographs unsuitable for a restoration of reasonably complete skulls and skeletons are obtainable. Despite the absences, this is by far the most extensive skull-skeletal library yet published in print. The core skeletal specimens that have been restored can be found via a link at http://press.princeton.edu/titles/10851.html. The restorations show the bones as solid white set within the solid black restored muscle and keratin profiles; cartilage is not included. When the skull is very small relative to the rest of the dinosaur, it is also shown at a larger scale. In many cases only the skull is available for illustration. In most cases the skull and skeletal restorations are of adults, but some juveniles have been included, sometimes as a growth series. The skeletons are posed in a common basic posture, with the right hind leg pushing off at the end of its propulsive stroke, in order to facilitate cross comparisons. The accuracy of the restorations ranges from very good for those that are known from extensive remains and for which a detailed description and/or good photographs of the skeleton are on hand, down to approximate if much of the species remains are missing or have not been well illustrated. The restorations have been prepared over three and a half decades to fulfill differing requirements. A number of skeletons and skulls show only those bones that are known, ranging from a large fraction to nearly all, whereas others have been filled out to represent a complete skeleton, and in others only major sections, such as a skull, or hand, etc. are not drawn in. Reliable information about exactly which bones have and have not been preserved is often not available, so the widely used term "rigorous" restoration for incomplete skeletals is best avoided in favor of "known bone." A representative sample of top views of skeletons has been included, or skulls are shown in top view when available. Some representative examples of shaded skull restorations have been included with some of the major groups, but preparing these for every case was not practical timewise. The same has been done with a sample of muscle studies, whose detailed nature is no less or more realistic than are the particulars found in full-life restorations, which, if anything, involve additional layers of speculation.

The color plates are based on species for which the fully or nearly adult skeletal restorations or skulls are deemed of sufficient quality for a full-life restoration. If it is unlikely that new information will significantly alter the skeletal plan in the near future, a life restoration has been prepared; if a color restoration is absent for a species illustrated by a skeleton, it is because the latter is not sufficiently reliable. Most of the skull as well as the skeleton needs to be present to justify the life restoration of the entire animal—although *Majungasaurus* bones are exquisitely preserved, there are too many uncertainties about its unusual proportions to warrant a life drawing. In a few cases color restorations were executed despite significant questions about the skeletal study because the species is particularly important or interesting for one reason or another, the spectacularly feathered but incomplete *Beipiaosaurus* being an example. In a few cases only the skull is good enough to warrant a life restoration to the exclusion of the overall body. The colors and patterns are entirely speculative except in the few cases where feather coloration has been restored—there is not a consistent effort to coordinate speculative color patterns for a species between the life profiles and full-color scenes. Extremely vibrant color patterns have generally not been used to avoid giving the impression that they are identifying features. Those who wish to use the skeletal, muscle, and life restorations herein as the basis for commercial and other public projects are reminded to first contact the copyright illustrator.

The particular anatomical characteristics that distinguish the species are listed, but these too are for purposes of general identification by putative dinosaur watchers, not for technical species diagnoses. These differ in extent depending on the degree of uniformity versus diversity in a given group as well as the completeness of the available fossil remains. In some cases the features of the species are not different enough from those of the group to warrant additional description. In other cases not enough is known to make a separate description possible.

Listed next is the formal geological time period and, when available, the stage that the species is known from. As discussed earlier, the age of a given species is known with a precision of within a million years in some cases, or as poorly as an entire period in others. The reader can refer to the time scale on the timeline chart to determine the age, or age range, of the species in years (see pp. 66–67). Most species exist for a few hundred thousand years to a couple of million years before either being replaced by a descendant species or going entirely extinct. In some cases it is not entirely clear whether a species was present in just one time stage or crossed the boundary into the next one. In those cases the listing includes "and/or," such as Late Santonian and/or Early Campanian.

Next the geographic location and geological formation that the species is known from so far are listed. The paleomaps of coastlines (see pp. 57–60) can be used to geographically place a species in a world of drifting continents and fast-shifting seaways, with the proviso that no set of maps is extensive enough to show the exact configuration of the ancient lands when each species was extant. I have tended to be conservative in listing the presence of a specific species only in those places and levels where sufficiently complete remains are present. Some dinosaur species are known from only a single location, whereas others have been found in an area spread over one or more formations. In some cases formations have yet to be named, even in areas that are well studied. Many formations were formed over a time span that was longer than that of some or all of the species that lived within them, so when possible the common procedure of simply listing just the overall formation a given species is from is avoided. For example, a host of large herbivorous dinosaurs is often thought to have lived at the same time in the famous Dinosaur Park Formation in Alberta, including *Centrosaurus apertus*, *Styracosaurus albertensis*, *Chasmosaurus belli*, *Hypacrosaurus casuarius*, *Lambeosaurus lambei*, and *Parasaurolophus walkeri*. The actual situation over the million and a half years the formation was laid down is more complicated. In the lower, earlier portion of the formation dwelled *Centrosaurus apertus*, *Chasmosaurus russelli*, *Hypacrosaurus intermedius*, and the uncommon *Parasaurolophus walkeri*. In the middle of the formation dwelled *Centrosaurus nasicornis*, *Chasmosaurus belli*, *Hypacrosaurus intermedius*, and *Hypacrosaurus clavinitialis*. The upper or later sections of Dinosaur Park held the species *Styracosaurus albertensis*, *Chasmosaurus irvinensis*, *Hypacrosaurus intermedius*, *Hypacrosaurus lambei*, and later *Hypacrosaurus magnicristatus*. It has been assumed that one species of *Triceratops* and *Tyrannosaurus* lived during the 1.5 million years it took to deposit the Hell Creek, Lance, and other formations, but the evidence favors species evolution over this period. Because the Morrison Formation was deposited over a span of 8 million years in the Late Jurassic, there was extensive change over time in the allosaurs, apatosaurs, diplodocines, camarasaurs, and stegosaurs that dwelled in the area. Many of the familiar particular species such as *Allosaurus fragilis*, *Stegosaurus stenops*, *Camptosaurus dispar*, and *Ceratosaurus nasicornis* are known from the earlier period of the formation, and later species of these genera are known from higher parts of the Morrison. I have therefore listed the level of the formation that each species comes from when this information is available. The reader can get an impression of what dinosaur species constituted a given fauna in a particular bed of sediments by using the formation index. In a few cases the geology of the sediments a dinosaur is from is not yet well enough known to name a formation, and the geological group may be named instead.

Noted next are the basic characteristics of the dinosaur's habitat in terms of rainfall and vegetation, as well as temperature when it is not generally tropical or subtropical year-round. Environmental information ranges from well studied in heavily researched formations to nonexistent in others. If the habits of the species are thought to include attributes not seen in the group as a whole, then they are outlined. Listed last are special notes about the species when they are called for. In many cases other dinosaurs that the species shared its habitat with are listed. Possible ancestor-descendant relationships with close older or younger relatives are sometimes noted, but these are always tentative. This section is also used to note alternative hypotheses and controversies that apply to the species.

Edmontonid

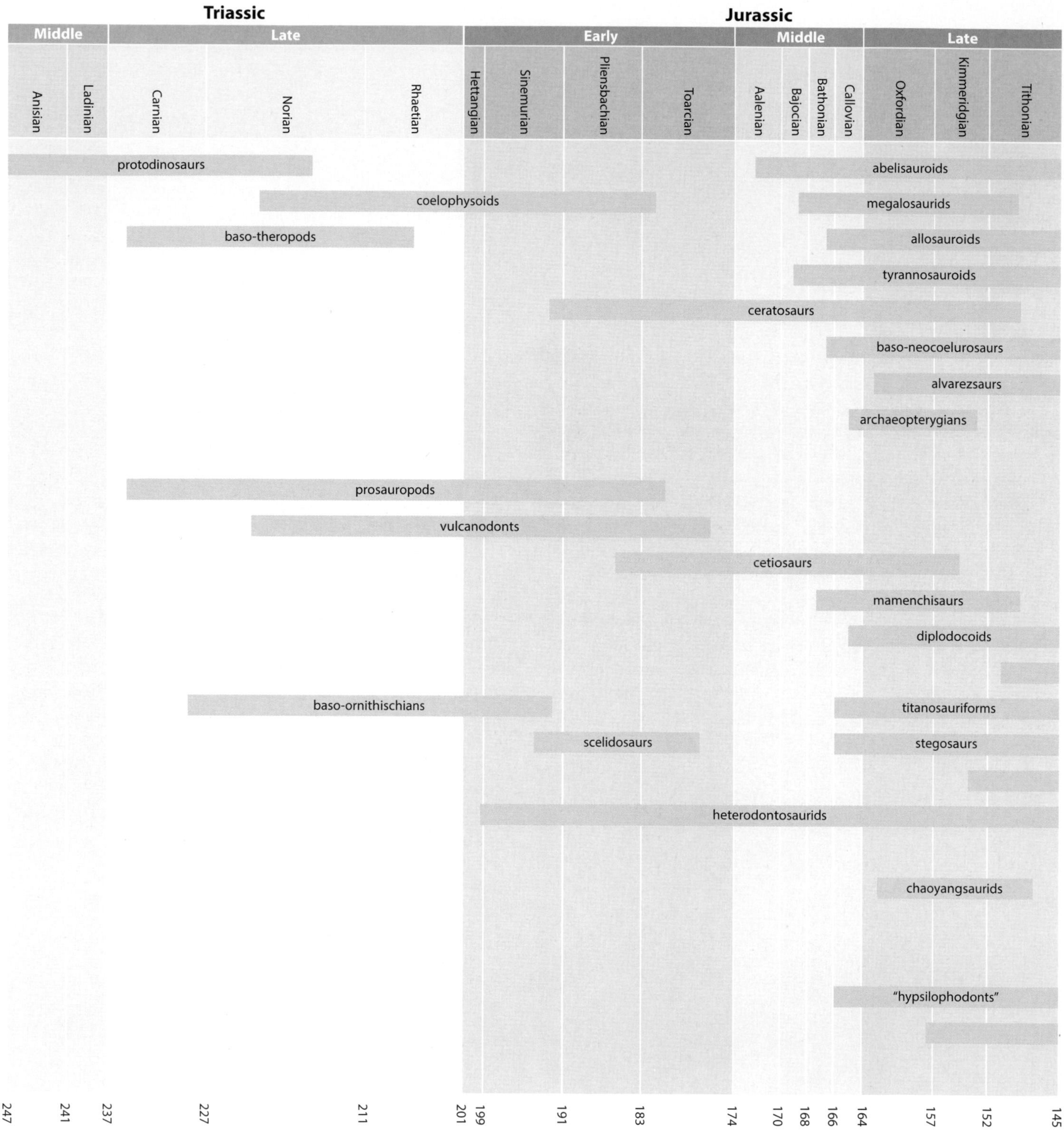

Triassic

Jurassic

Middle

Late

Early

Middle

Late

Anisian

Ladinian

Carnian

Norian

Rhaetian

Hettangian

Sinemurian

Pliensbachian

Toarcian

Aalenian

Bajocian

Bathonian

Callovian

Oxfordian

Kimmeridgian

Tithonian

protodinosaurs

abelisauroids

coelophysoids

megalosaurids

baso-theropods

allosauroids

tyrannosauroids

ceratosaurs

baso-neocoelurosaurs

alvarezsaurs

archaeopterygians

prosauropods

vulcanodonts

cetiosaurs

mamenchisaurs

diplodocoids

baso-ornithischians

titanosauriforms

scelidosaurs

stegosaurs

heterodontosaurids

chaoyangsaurids

“hypsilophodonts”

247 241 237 227 211 201 199 191 183 174 170 168 166 164 157 152 145

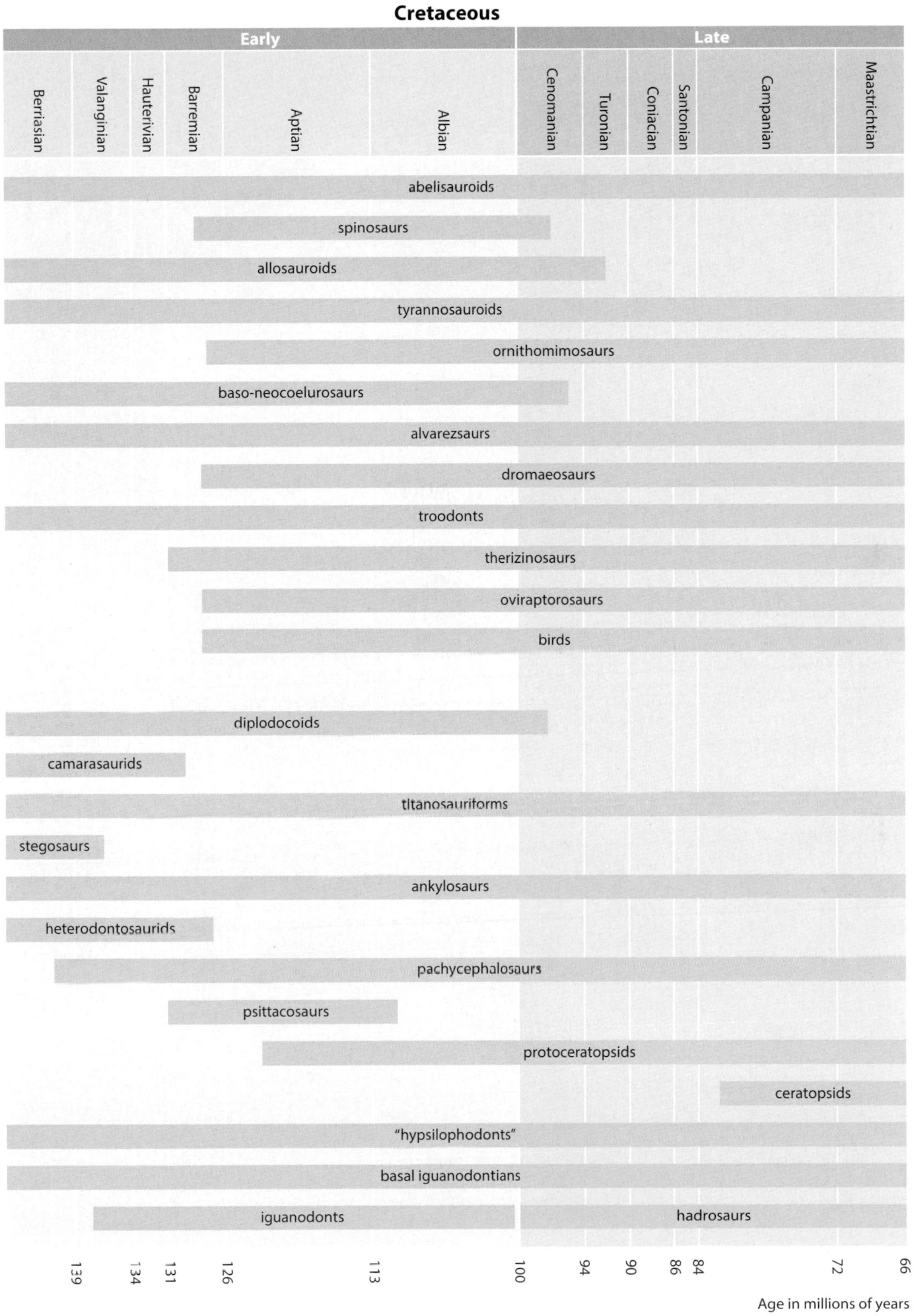

Cretaceous

Early | Late

Berriasian | Valanginian | Hauterivian | Barremian | Aptian | Albian | Cenomanian | Turonian | Coniacian | Santonian | Campanian | Maastrichtian

abelisauroids

spinosaurs

allosauroids

tyrannosauroids

ornithomimosaurs

baso-neocoelurosaurs

alvarezsaurs

dromaeosaurs

troodonts

therizinosaurs

oviraptorosaurs

birds

diplodocoids

camarasaurids

titanosauriforms

stegosaurs

ankylosaurs

heterodontosaurids

pachycephalosaurs

psittacosaurs

protoceratopsids

ceratopsids

"hypsilophodonts"

basal iguanodontians

iguanodonts

hadrosaurs

139 134 131 126 113 100 94 90 86 84 72 66

Age in millions of years

Deinocheirus mirificus

DINOSAURS

SMALL TO GIGANTIC ARCHOSAURS FROM THE LATE TRIASSIC TO THE END OF THE MESOZOIC, ALL CONTINENTS.

ANATOMICAL CHARACTERISTICS Erect leg posture achieved by cylindrical femoral head fitting into a perforated hip socket and a simple hinge-jointed ankle. All are hind-limb dominant in that legs are sole locomotory organs in walking and running and/or are more strongly built than the arms. Hands and feet digitigrade, with wrist and ankle held clear of ground. Trackways show that when quadrupedal, hands always at least as far or farther apart from midline as feet; never hopped, and tail normally held clear of ground. Body scales, when known and present, usually form a nonoverlapping mosaic pattern.
REPRODUCTION AND ONTOGENY Most rapid breeders, probably all laid hard-shelled eggs in pairs, nests on the ground; growth rates moderate to rapid; usually reached sexual maturity while still growing.
HABITS AND HABITATS Strongly terrestrial; although all were able to swim, none were marine; otherwise highly variable.

THEROPODS

SMALL TO GIGANTIC SAURISCHIAN DINOSAURS, MOST PREDATORS, FROM THE LATE TRIASSIC TO THE END OF THE DINOSAUR ERA, ALL CONTINENTS.

ANATOMICAL CHARACTERISTICS All obligatory bipeds, otherwise very variable. Head size and shape variable, skull bones usually somewhat loosely attached to one another, extra joint usually at middle of lower jaw, eyes normally large, usually if not always supported by internal bone ring, teeth from large, bladed, and serrated to absent. Neck long to fairly short, usually S-curved to greater or lesser extent, moderately flexible. Series of trunk vertebrae short and stiff. Tail from long and very flexible to very short and stiff. Arm very long to severely reduced, fingers four to one, fingers long and slender to short, sharp claws from large to reduced. Pelvis moderate in size to very large, leg flexed at all sizes, long, main toes four to three; footprints confirm that trackway gauge was very narrow. Brains vary from reptilian in size and form to similar to birds.
HABITATS Very diverse, from sea level to highlands, from tropics to polar winters, from arid to wet.
HABITS Diets ranged from classic hunting with opportunistic scavenging to full herbivory. Small and juvenile theropods with long arm and hook-clawed fingers were probably able to climb. Enormous numbers of trackways laid down along watercourses show that many theropods of all sizes spent considerable time patrolling shorelines and using them to travel.
NOTES The only dinosaur group that includes archpredators. Already somewhat birdlike at the beginning, generally became increasingly so with time, especially among some advanced groups that include the direct ancestors of birds.

BASO-THEROPODS

SMALL TO MODERATELY LARGE PREDATORY THEROPODS, LIMITED TO THE LATE TRIASSIC.

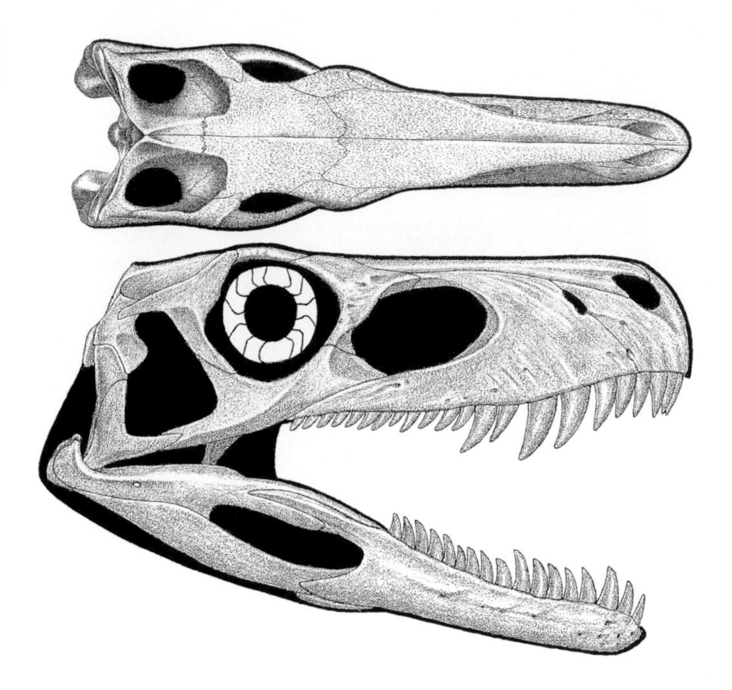

Herrerasaurus shaded skull

ANATOMICAL CHARACTERISTICS Fairly uniform. Generally lightly built. Head moderately large, usually long and shallow, subrectangular, fairly robustly constructed, fairly narrow, teeth usually serrated blades. Neck moderately long, only gently S-curved. Tail long. Arm and four-fingered hand moderately long, claws well developed. Pelvis short but deep. Four load-bearing toes. Beginnings of birdlike respiratory system possibly present. Brains reptilian.
HABITS Pursuit predators, some also omnivorous. Head and arms primary weapons. Jaws and teeth probably delivered slashing wounds to disable muscles and cause bleeding, shock, and infection. Arms used to hold onto and control prey, possibly delivered slashing wounds. Prey items include prosauropods, possibly sauropods, especially juveniles, small ornithischians, herbivorous thecodonts, small game.
ENERGETICS Thermophysiology probably intermediate, energy levels and food consumption probably low compared to more-derived dinosaurs.
NOTES The most primitive dinosaurs, the briefly existing baso-theropods were apparently not able to compete with the more sophisticated avepods. It is possible that some of these very early dinosaurs were not theropods. Absence from at least some other continents probably reflects lack of sufficient sampling.

Eoraptor lunensis

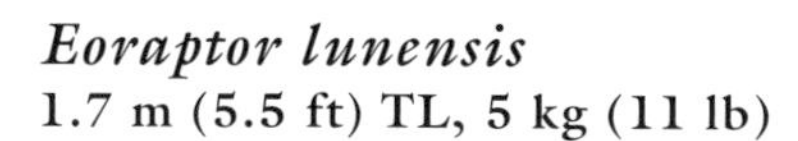

1.7 m (5.5 ft) TL, 5 kg (11 lb)

FOSSIL REMAINS Two nearly complete skulls and skeletons, almost completely known.
ANATOMICAL CHARACTERISTICS Back teeth bladed and serrated, front teeth more leaf shaped.
AGE Late Triassic, Carnian.

Eoraptor lunensis

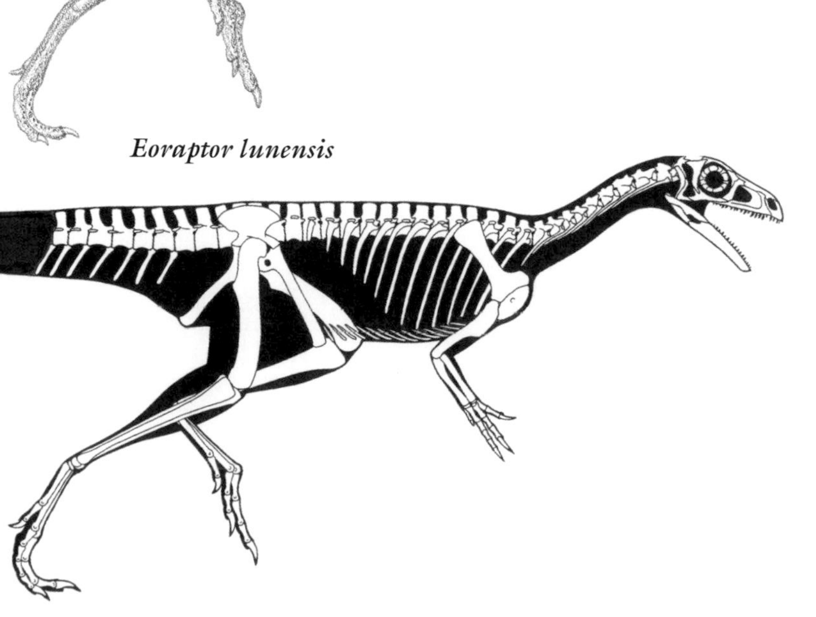

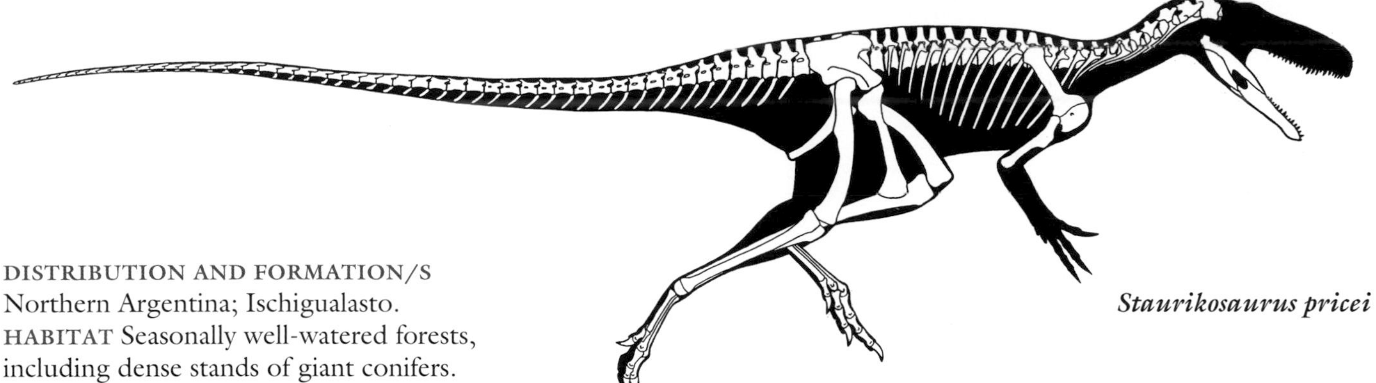
Staurikosaurus pricei

DISTRIBUTION AND FORMATION/S Northern Argentina; Ischigualasto.
HABITAT Seasonally well-watered forests, including dense stands of giant conifers.
HABITS Probably omnivorous, hunted smaller game and consumed some easily digested plant material, prey included *Panphagia* and *Pisanosaurus.* Main enemy *Herrerasaurus.*
NOTES One of the most (if not the most) primitive dinosaurs, this omnivore may not be a theropod, possibly a very basal saurischian, or a very basal sauropodomorph.

Alwalkeria maleriensis
1.5 m (5 ft) TL, 2 kg (4 lb)

FOSSIL REMAINS Minority of skull and skeleton.
ANATOMICAL CHARACTERISTICS Back teeth bladed, front teeth less so, no teeth serrated.
AGE Late Triassic, Carnian.
DISTRIBUTION AND FORMATION/S Southeast India; Lower Maleri.
HABITS Probably omnivorous, hunted smaller game and consumed some easily digested plant material.
NOTES Possible close relative of *Eoraptor*, may not be a theropod, possibly a very basal saurischian.

Chindesaurus bryansmalli
2.4 m (8 ft) TL, 15 kg (30 lb)

FOSSIL REMAINS Minority of skeleton, isolated bones.
ANATOMICAL CHARACTERISTICS Appears to be standard for group.
AGE Late Triassic, middle Norian.
DISTRIBUTION AND FORMATION/S Arizona, New Mexico? Texas?; middle Chinle, Bull Canyon? Tecovas?
HABITAT Well-watered forests, including dense stands of giant conifers.
NOTES It is uncertain whether the remains outside original Chinle specimen belong to one taxon; remains from the Tecovas may be *Caseosaurus crosbyensis.*

Staurikosaurus pricei
2.1 m (7 ft) TL, 12 kg (26 lb)

FOSSIL REMAINS Minority of skull and majority of skeleton.
ANATOMICAL CHARACTERISTICS Standard for group.
AGE Late Triassic, early Carnian.
DISTRIBUTION AND FORMATION/S Southeast Brazil; lower Santa Maria.
HABITS Prey included *Saturnalia.*

Herrerasaurus ischigualastensis
4.5 m (15 ft) TL, 200 kg (450 lb)

FOSSIL REMAINS Two complete skulls and several partial skeletons.
ANATOMICAL CHARACTERISTICS Standard for group.
AGE Late Triassic, Carnian.
DISTRIBUTION AND FORMATION/S Northern Argentina; lower Ischigualasto.
HABITAT Seasonally well-watered forests, including dense stands of giant conifers.
HABITS Prey included large herbivorous thecodonts and reptiles. Potential prey of larger predatory thecodonts.

Herrerasaurus ischigualastensis
(see also next page)

Herrerasaurus ischigualastensis

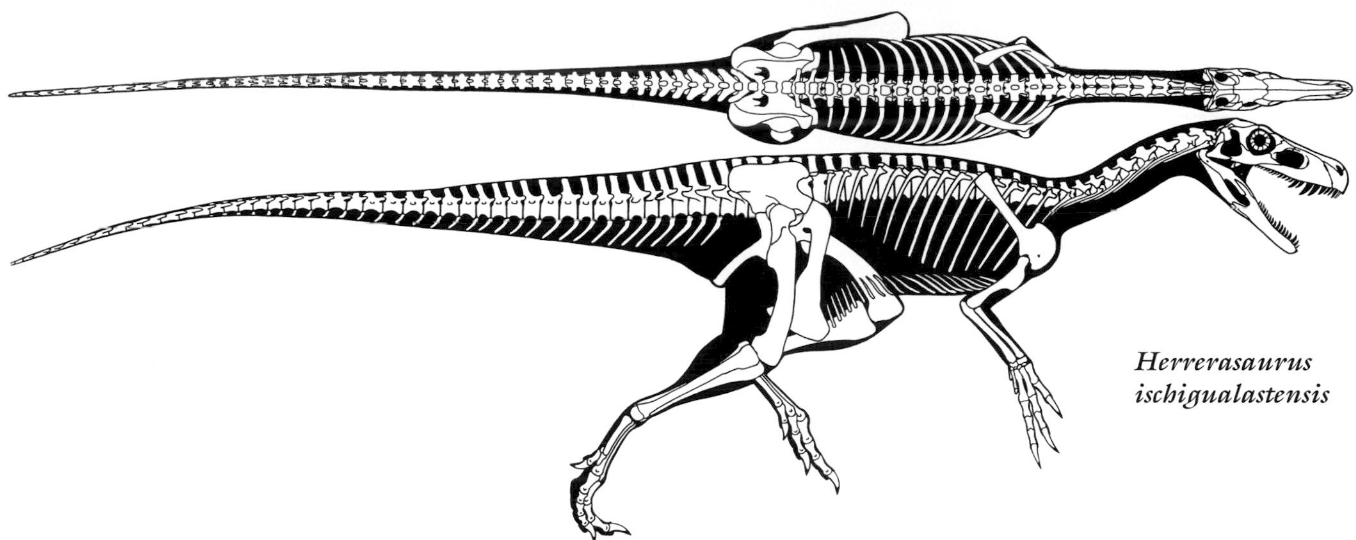
Herrerasaurus ischigualastensis

NOTES The classic archaic theropod, includes *Frenguellisaurus ischigualastensis* and *Ischisaurus cattoi*. Family Herrerasauridae may include some but not all other baso-theropods, especially *Staurikosaurus* and *Chindesaurus*.

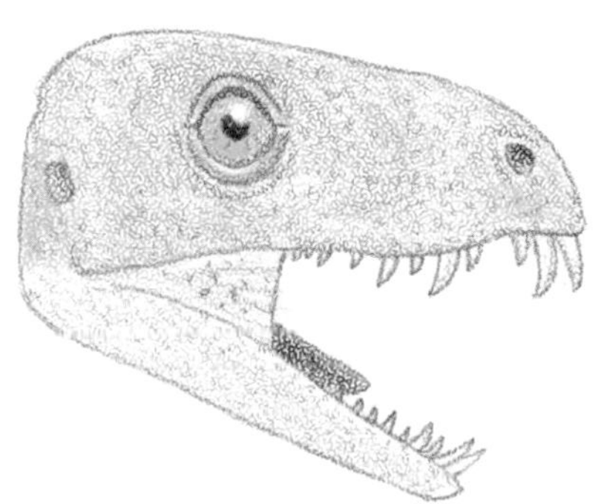
Daemonosaurus chauliodus

Eodromaeus murphi

1.8 m (6 ft) TL, 5 kg (11 lb)

FOSSIL REMAINS Majority of skull and skeleton.
ANATOMICAL CHARACTERISTICS Somewhat more robust than other baso-theropods.
AGE Late Triassic, Carnian.
DISTRIBUTION AND FORMATION/S Northern Argentina; lower Ischigualasto.
HABITAT Seasonally well-watered forests, including dense stands of giant conifers.
HABITS Prey included *Panphagia* and *Pisanosaurus*. Main enemy *Herrerasaurus*.

Daemonosaurus chauliodus

Adult size uncertain

FOSSIL REMAINS Nearly complete skull and minority of skeleton.

Eodromaeus murphi

ANATOMICAL CHARACTERISTICS Head deep, subtriangular, snout short, eyes large, upper teeth very large, those in front procumbent.
AGE Late Triassic, Rhaetian.
DISTRIBUTION AND FORMATION/S New Mexico; probably upper Chinle.
HABITS May have caught fish as well as land prey.
NOTES Ontogenetic stage of specimen is uncertain, so how much unusual form is juvenile or distinctive is also uncertain. Shared its habitat with *Coelophysis bauri.*

Tawa hallae
2.5 m (8 ft) TL, 15 kg (30 lb)

FOSSIL REMAINS Nearly complete skull and majority of a few skeletons, almost completely known.
ANATOMICAL CHARACTERISTICS Snout pointed, upper teeth large. Leg slender.
AGE Late Triassic, middle Norian.
DISTRIBUTION AND FORMATION/S New Mexico; upper Chinle.
HABITAT Seasonally well-watered forests, including dense stands of giant conifers.
HABITS Probably fastest of known baso-theropods, prey included protodinosaurs and other small animals.
NOTES Closest known baso-theropod to avepods.

AVEPODS

SMALL TO GIGANTIC, THREE-TOED PREDATORY AND HERBIVOROUS THEROPODS, MOST PREDATORS, FROM THE LATE TRIASSIC TO THE END OF THE DINOSAUR ERA (WITH BIRDS SURVIVING BEYOND), ALL CONTINENTS.

ANATOMICAL CHARACTERISTICS Highly variable. Head size and shape variable, teeth large and bladed to absent. Neck long to fairly short. Trunk short, stiff. Tail long to very short. Fused furcula often present, arm very long to severely reduced, fingers four to one, usually three, fingers long and slender to short, claws large to reduced. Pelvis large, leg long, usually three main toes, inner toe a short hallux, sometimes four load-bearing toes, or two. Skeletons pneumatic, birdlike, air-sac-ventilated respiratory system developing. Brains vary from reptilian in size and form to similar to birds.
HABITS Diets ranged from classic hunting in most to full herbivory in some specialized groups.
ENERGETICS Energy levels and food consumption probably similar to those of ratite birds except as noted.
NOTES Theropods lacking contact of metatarsal 1 with ankle, or ancestors with same that are in the clade that includes extant birds; includes Neotheropoda and more basal taxa with the feature. Distinctly birdlike from the start.

Tawa hallae

BASO-AVEPODS

SMALL TO GIGANTIC PREDATORS, FROM THE LATE TRIASSIC TO THE END OF THE DINOSAUR ERA.

ANATOMICAL CHARACTERISTICS Variable. Head size and shape variable, neck long to fairly short, tail long to very short, teeth bladed. Arm moderately long to severely reduced, four fingers. Pelvis moderately to very large. Brains reptilian. Skeletal pneumaticity partly developed, so birdlike respiratory system developing.
HABITS Pursuit and ambush predators.
NOTES The primitive avepod theropods.

COELOPHYSOIDS

SMALL TO LARGE BASO-AVEPODS, LIMITED TO THE LATE TRIASSIC AND EARLY JURASSIC, NORTHERN HEMISPHERE AND AFRICA.

ANATOMICAL CHARACTERISTICS Fairly uniform. Generally lightly built. Head long, snout pointed, narrow, indentation at front of upper jaw often present, lightly constructed paired crests over snout often present. Neck long. Trunk not deep. Tail very long, slender. Teeth bladed. Arm moderately long, fingers moderately long, claws modest in size. Pelvis moderately large.
ONTOGENY Growth rates moderate.
HABITS Although predominantly fast pursuit predators, snaggly teeth at tip of kinked upper jaw indicate these were also fishers. Crests when present too delicate for head butting; for lateral visual display within the species, may or may not have been brightly colored at least during breeding season.
NOTES The most primitive avepods, first large examples show avepod theropods reached considerable size as early as the Triassic. Group splittable into a number of subdivisions, or major divisions. Absence from at least some other continents probably reflects lack of sufficient sampling.

Dracoraptor hanigani
1.8 m (6 ft) TL, 4 kg (10 lb)

FOSSIL REMAINS Partial skull and minority of skeleton.
ANATOMICAL CHARACTERISTICS Standard for small coelophysoids.
AGE Early Jurassic, early Hettangian.
DISTRIBUTION AND FORMATION/S Wales; lower Blue Lias.
NOTES Found as drift in nearshore marine deposits.

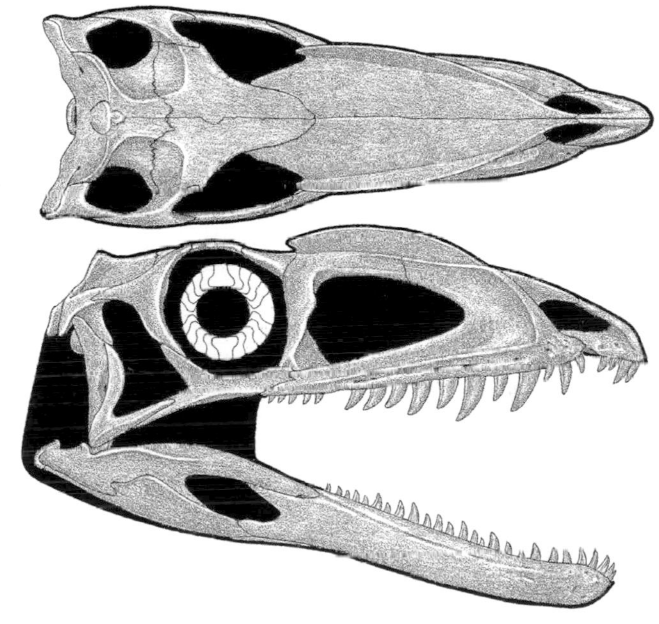

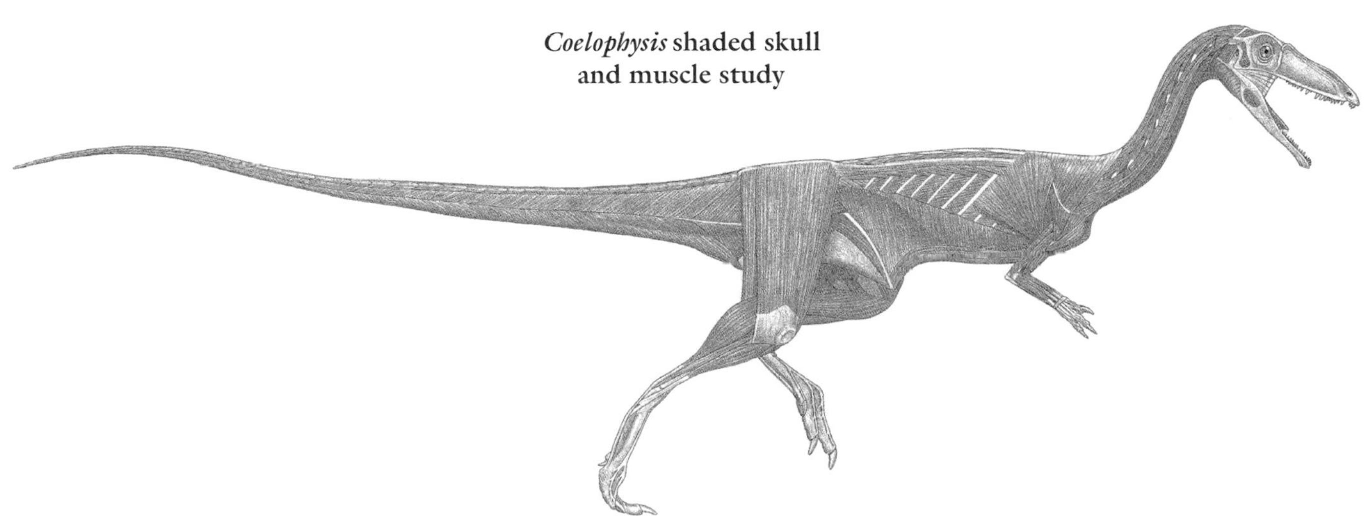

Coelophysis **shaded skull and muscle study**

Procompsognathus triassicus
1.1 m (3.5 ft) TL, 1 kg (2 lb)

FOSSIL REMAINS Poorly preserved partial skeleton with possible skull.
ANATOMICAL CHARACTERISTICS Standard for small coelophysoids.
AGE Late Triassic, middle Norian.
DISTRIBUTION AND FORMATION/S Germany; middle Lowenstein.
NOTES Remains that belong to this taxon not entirely certain. Appears to be the most primitive and smallest member of the group. Name incorrectly implies an ancestral relationship with the very different *Compsognathus*. Whether head crests were present is uncertain.

Coelophysis bauri
3 m (10 ft) TL, 25 kg (50 lb)

FOSSIL REMAINS Hundreds of skulls and skeletons, many complete, juvenile to adult, completely known.
ANATOMICAL CHARACTERISTICS Very lightly built and gracile, overall very long bodied. Head long and shallow, bite not powerful, crests absent, teeth numerous and small. Neck long and slender.
AGE Late Triassic, Rhaetian.
DISTRIBUTION AND FORMATION/S New Mexico; uppermost Chinle.
HABITS Predominantly small game hunter but may have occasionally attacked larger prosauropods and herbivorous thecodonts.
NOTES The classic early avepod theropod. In accord with a decision of the committee that handles taxonomic issues, the specimen that the taxon is based on was shifted from inadequate fossils in the Chinle to a complete specimen from the famous Ghost Ranch Quarry. How hundreds of skeletons came to be concentrated in the quarry remains unsettled. This, *Panguraptor*, *Procompsognathus*, and *Podokesaurus* may form family Coelophysidae.

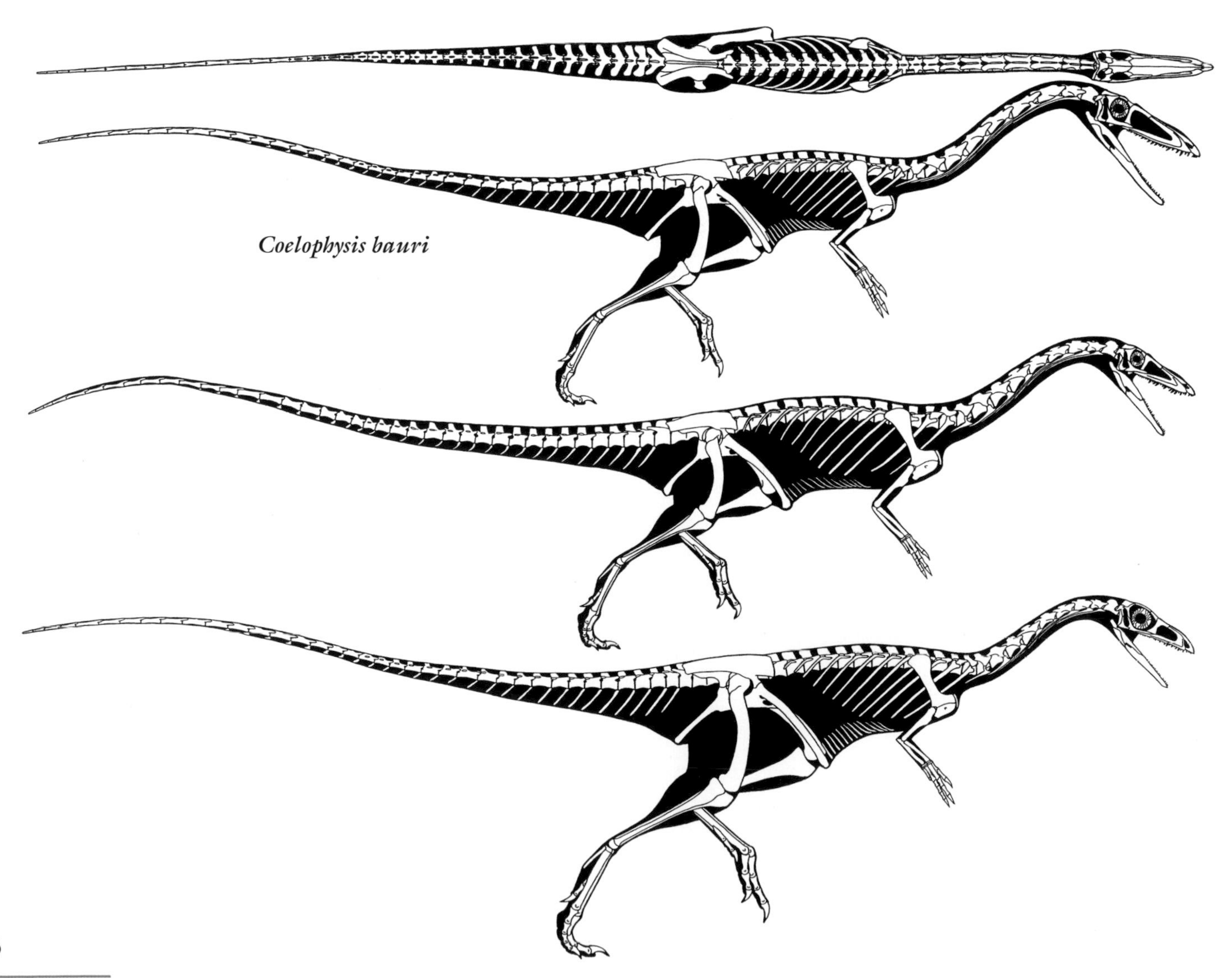

Coelophysis bauri

Coelophysis bauri

Coelophysis rhodesiensis
2.2 m (7 ft) TL, 13 kg (30 lb)

FOSSIL REMAINS Hundreds of skulls and skeletons, juvenile to adult, completely known.
ANATOMICAL CHARACTERISTICS Same as *C. bauri* except leg longer relative to body.
AGE Early Jurassic, Hettangian.
DISTRIBUTION AND FORMATION/S Zimbabwe; Forest Sandstone?
HABITAT Desert with dunes and oases.
HABITS Same as *C. bauri*, except thecodonts not present.
NOTES Originally *Syntarsus*, that name preoccupied by an insect; species is now often accepted as a species of the very similar *Coelophysis*. Whether remains from other South African formations belong to this species is uncertain.

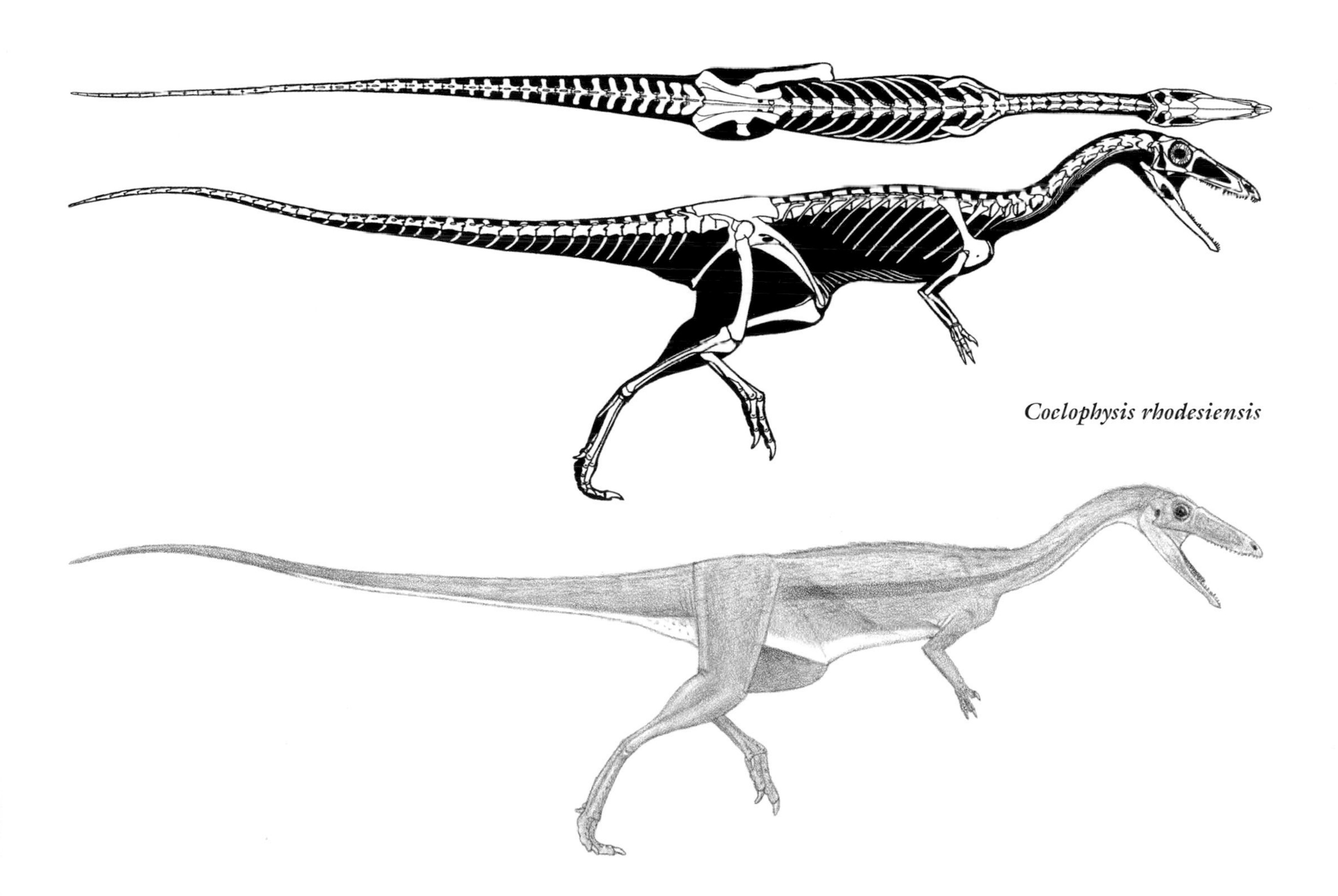
Coelophysis rhodesiensis

Coelophysis? kayentakatae
2.5 m (9 ft) TL, 30 kg (60 lb)

FOSSIL REMAINS Complete skull and minority of skeleton, other partial remains.
ANATOMICAL CHARACTERISTICS Head fairly deep, snout crests well developed, teeth fairly large and less numerous than in other *Coelophysis*.
AGE Early Jurassic, early Sinemurian.
DISTRIBUTION AND FORMATION/S Arizona; middle Kayenta.
HABITAT Semiarid.
HABITS More robust head and larger teeth indicate this species tended to hunt larger game than other *Coelophysis*. Prey included *Scutellosaurus*.
NOTES Originally placed in *Syntarsus*, may be placeable in *Coelophysis*.

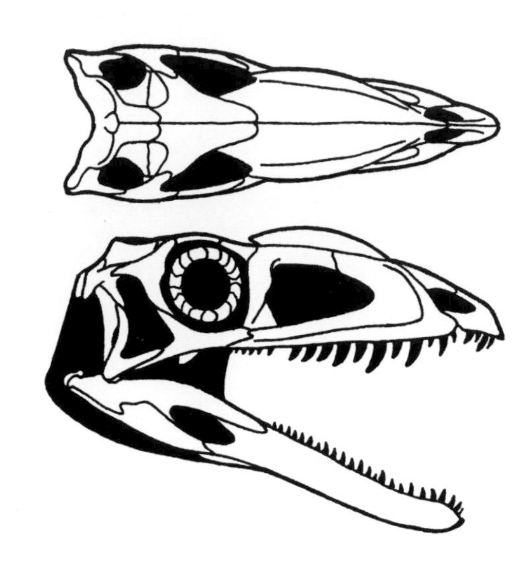

Coelophysis? kayentakatae **skull and head**

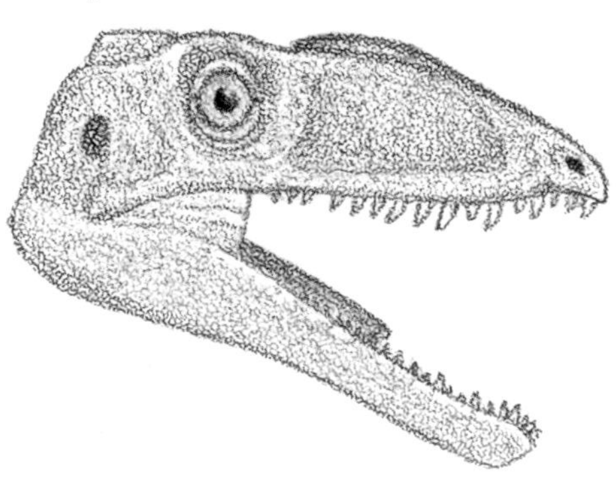

Podokesaurus (or *Coelophysis*) *holyokensis*
1 m (3 ft) TL, 1 kg (2 lb)

FOSSIL REMAINS Partial skeleton, possibly juvenile.
ANATOMICAL CHARACTERISTICS Standard for small coelophysoids.
AGE Early Jurassic, Pliensbachian or Toarcian.
DISTRIBUTION AND FORMATION/S Massachusetts; Portland?
HABITAT Semiarid rift valley with lakes.
NOTES Lost in a fire; the original location and age of this specimen are not entirely certain. Not known whether head crests were present.

Panguraptor lufengensis
Adult size uncertain

FOSSIL REMAINS Majority of skull and skeleton, probably a subadult.
ANATOMICAL CHARACTERISTICS Standard for small coelophysoids, snout crests absent.
AGE Early Jurassic, Hettangian.
DISTRIBUTION AND FORMATION/S Southwest China; lower Lower Lufeng.
HABITS Prey included *Eshanosaurus*.
NOTES Main enemy *Sinosaurus*.

Gojirasaurus quayi
6 m (20 ft) TL, 150 kg (350 lb)

FOSSIL REMAINS Small portion of skeleton.
ANATOMICAL CHARACTERISTICS Insufficient information.
AGE Late Triassic, middle Norian.
DISTRIBUTION AND FORMATION/S New Mexico; Cooper Canyon.
HABITAT Well-watered forests, including dense stands of giant conifers.
HABITS Prey included large prosauropods and thecodonts.

Panguraptor lufengensis

NOTES Remains that belong to this taxon uncertain. Not known whether head crests were present.

Lophostropheus airelensis
Adult size uncertain

FOSSIL REMAINS Partial skeleton.
ANATOMICAL CHARACTERISTICS Insufficient information.
AGE Latest Triassic and/or Early Jurassic, late Rhaetian and/or early Hettangian.
DISTRIBUTION AND FORMATION/S Northern France; Moon-Airel.

Segisaurus halli
1 m (3 ft) TL, 5 kg (10 lb)

FOSSIL REMAINS Partial skeleton, large juvenile.
ANATOMICAL CHARACTERISTICS Standard for small coelophysoids.
AGE Early Jurassic, Pliensbachian or Toarcian.
DISTRIBUTION AND FORMATION/S Arizona; Navajo Sandstone.
HABITAT Desert with dunes and oases.
HABITS Largely a small game hunter, probably small prosauropods and ornithischians also.
NOTES Not known whether head crests were present.

Liliensternus liliensterni
5.2 m (17 ft) TL, 130 kg (300 lb)

FOSSIL REMAINS Majority of skull and two skeletons.
ANATOMICAL CHARACTERISTICS Lightly built like smaller coelophysoids.
AGE Late Triassic, late Norian.
DISTRIBUTION AND FORMATION/S Central Germany; Knollenmergel.
HABITS Prey included prosauropods, herbivorous thecodonts.
NOTES Not known whether head crests were present. Prey included *Plateosaurus longiceps*.

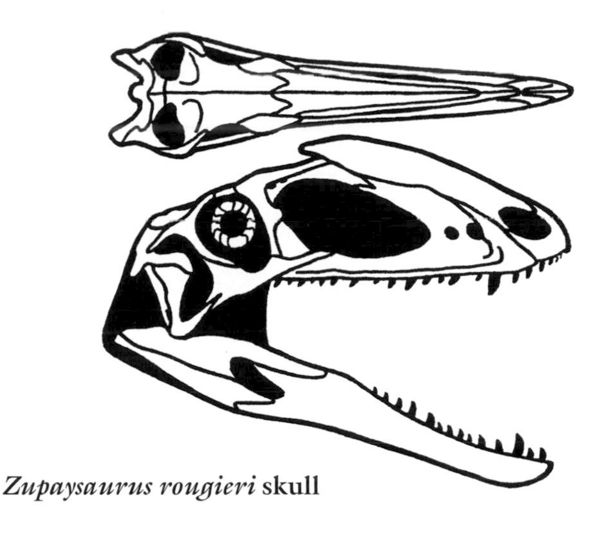

Zupaysaurus rougieri **skull**

Zupaysaurus rougieri
6 m (20 ft) TL, 250 kg (550 lb)

FOSSIL REMAINS Almost complete skull and partial skeleton.
ANATOMICAL CHARACTERISTICS Skull moderately deep, snout very large, adorned with well-developed paired crests, teeth not large.
AGE Late Triassic, Norian.
DISTRIBUTION AND FORMATION/S Northern Argentina; Los Colorados.
HABITAT Seasonally wet woodlands.
HABITS Prey included large prosauropods and thecodonts.
NOTES When first described this was considered to be the earliest tetanurian theropod, but other research indicates it is a coelophysoid related to *Dilophosaurus*.

Dracovenator regenti
6 m (20 ft) TL, 250 kg (550 lb)

FOSSIL REMAINS Two partial skulls, juvenile and adult.
ANATOMICAL CHARACTERISTICS Snout crests apparently not large, teeth large.

Liliensternus liliensterni

AGE Early Jurassic, late Hettangian or Sinemurian.
DISTRIBUTION AND FORMATION/S Southeast Africa; Upper Elliot.
HABITAT Arid.
HABITS Big game hunter, prey included *Massospondylus*, *Aardonyx*, *Pulanesaura*, *Lesothosaurus*, and *Heterodontosaurus*.

Dilophosaurus wetherilli
7 m (22 ft) TL, 400 kg (900 lb)

FOSSIL REMAINS Majority of several skulls and skeletons.
ANATOMICAL CHARACTERISTICS More robustly constructed than smaller coelophysoids. Head large, deep, snout crests large, teeth large.
AGE Early Jurassic, Hettangian or Sinemurian.
DISTRIBUTION AND FORMATION/S Arizona; lower Kayenta.
HABITAT Semiarid.
HABITS Prey included large prosauropods and early armored ornithischians.
NOTES This, *Liliensternus*, *Zupaysaurus*, and *Dracovenator* may form family Dilophosauridae, or these may form a group close to but outside coelophysoids.

Dilophosaurus wetherilli

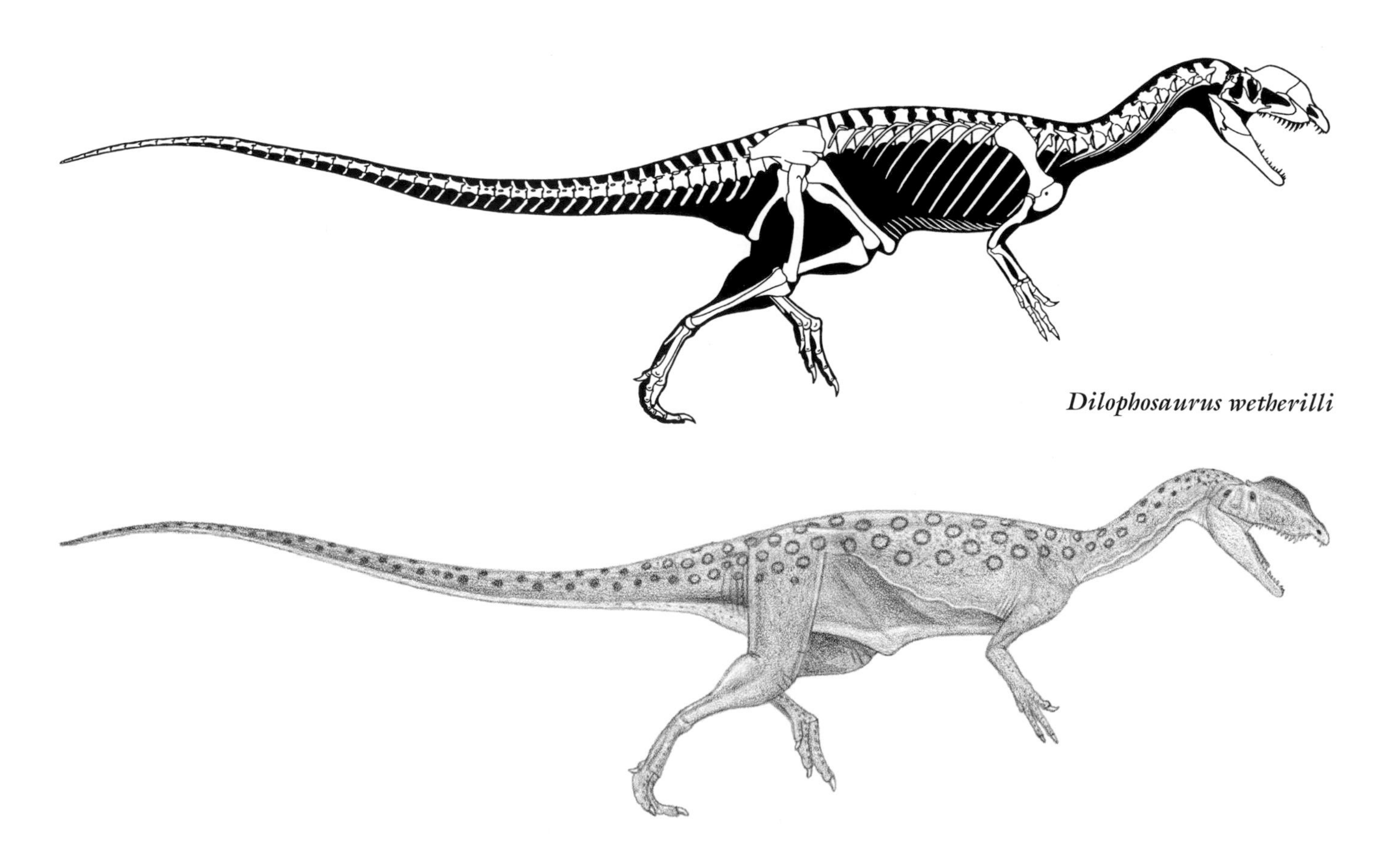

Dilophosaurus wetherilli

AVEROSTRANS

SMALL TO GIGANTIC PREDATORY AND HERBIVOROUS AVEPODS FROM THE EARLY JURASSIC TO THE END OF THE DINOSAUR ERA (WITH BIRDS SURVIVING BEYOND), ON ALL CONTINENTS.

ANATOMICAL CHARACTERISTICS Highly variable. Brains reptilian to avian. Nasal sinuses better developed. Birdlike respiratory system better developed. Pelvic plate large.

Baso-averostrans

LARGE TO GIGANTIC AVEROSTRANS OF THE EARLY JURASSIC TO THE EARLY LATE CRETACEOUS

NOTES The relationships of the following primitive and partially known averostrans are uncertain. Fragmentary Early Jurassic, South American *Tachiraptor admirabilis* may be in this grouping.

Berberosaurus liassicus
5 m (15 ft) TL, 300 kg (600 lb)

FOSSIL REMAINS Minority of skeleton.
ANATOMICAL CHARACTERISTICS Insufficient information.
AGE Early Jurassic, Pliensbachian or Toarcian.
DISTRIBUTION AND FORMATION/S Morocco; Toundoute series.
NOTES *Berberosaurus* confirms that the primitive abelisaurs were present early in the dinosaur era.

Bahariasaurus ingens and/or *Deltadromeus agilis*
11 m (35 ft) TL, 4 tonnes

FOSSIL REMAINS Minority of skeletons.
ANATOMICAL CHARACTERISTICS Shoulder girdle massively constructed. Leg long and gracile.
AGE Late Cretaceous, early Cenomanian.
DISTRIBUTION AND FORMATION/S Morocco; Bahariya.
HABITAT Coastal mangroves.
HABITS Fast-running pursuit predator.
NOTES The relationships of *Bahariasaurus* and *Deltadromeus* to other theropods and each other are uncertain; the latter may be a juvenile of the former.

Elaphrosaurs

MEDIUM-SIZED AVEROSTRANS, LIMITED TO THE LATE JURASSIC OF ASIA, AFRICA, AND NORTH AMERICA.

ANATOMICAL CHARACTERISTICS Overall build gracile. Head modest in size, lightly built, toothless, with blunt beak. Arm slender, hand reduced. Pelvis moderately large, leg long and gracile.
HABITS Possibly omnivorous, predominantly herbivorous combined with some small animals and insects. Main defense speed, also kicks from legs.
NOTES These Jurassic ostrich mimics evolved feeding and running adaptations broadly similar to those of the even faster and longer-armed Cretaceous ornithomimids. Absence from at least some other continents probably reflects lack of sufficient sampling.

Spinostropheus gauthieri
4 m (14 ft) TL, 200 kg (450 lb)

FOSSIL REMAINS Minority of skeleton.
ANATOMICAL CHARACTERISTICS Insufficient information.
AGE Late Middle or early Late Jurassic.
DISTRIBUTION AND FORMATION/S Niger; Tiouraren.
HABITAT Well-watered woodlands.
NOTES Originally thought to be from the Early Cretaceous; researchers now place the Tiouraren in the later Jurassic. Prey of *Afrovenator*.

Elaphrosaurus bambergi
6 m (20 ft) TL, 200 kg (450 lb)

FOSSIL REMAINS Majority of skeleton.
ANATOMICAL CHARACTERISTICS Insufficient information.
AGE Late Jurassic, late Kimmeridgian and/or early Tithonian.
DISTRIBUTION AND FORMATION/S Tanzania; middle Tendaguru.
HABITAT Coastal, seasonally dry with heavier vegetation farther inland.
NOTES Shared its habitat with *Dryosaurus lettowvorbecki*.

Elaphrosaurus? unnamed species
4.5 m (15 ft) TL, 100 kg (220 lb)

FOSSIL REMAINS Small portion of skeleton.
ANATOMICAL CHARACTERISTICS Insufficient information.
AGE Late Jurassic, late Oxfordian and/or Kimmeridgian.
DISTRIBUTION AND FORMATION/S Colorado; lower and middle Morrison.

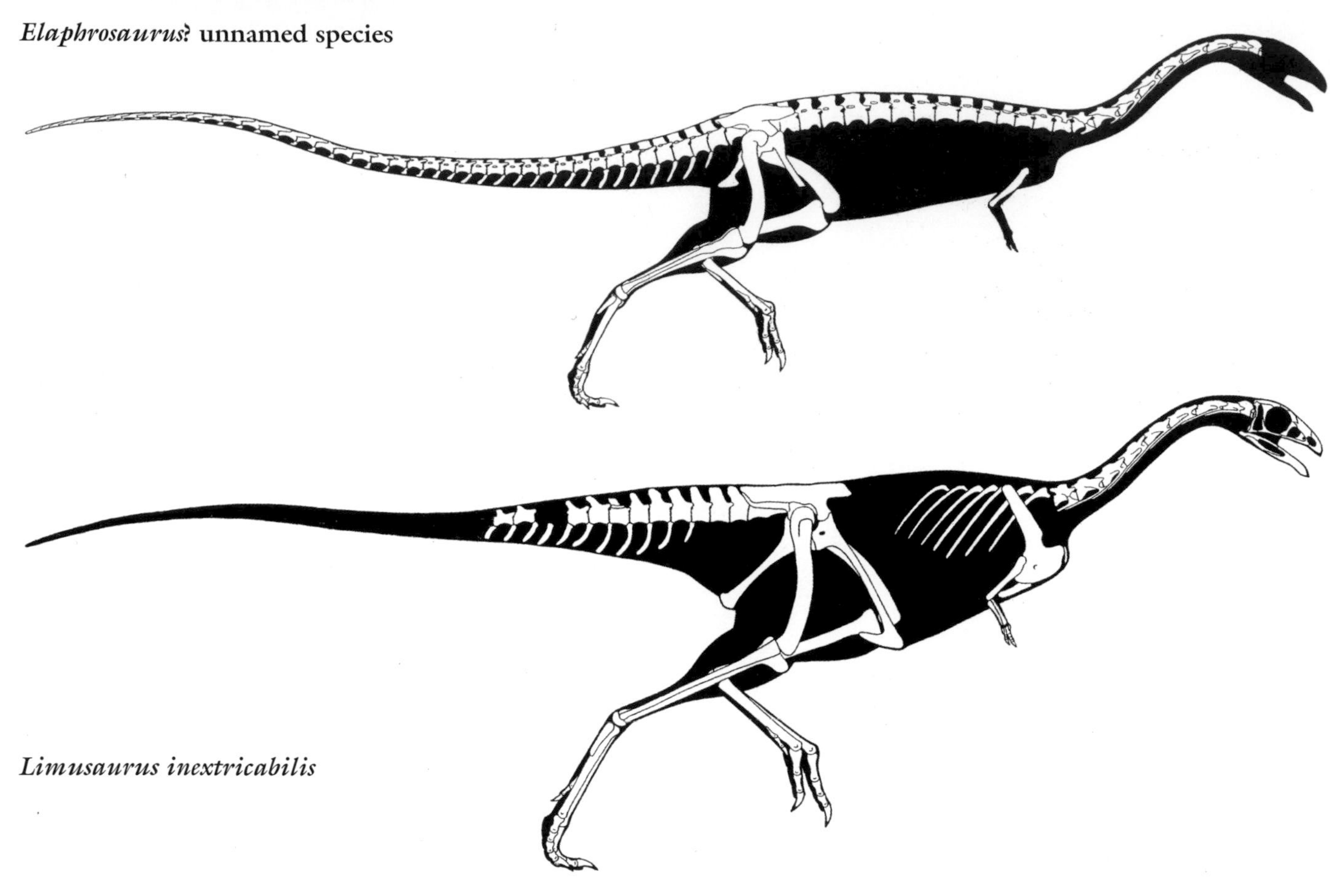

Elaphrosaurus? unnamed species

Limusaurus inextricabilis

HABITAT Short wet season, otherwise semiarid with open floodplains and riverine forests.
NOTES It is uncertain whether these remains are the same genus as *Elaphrosaurus*; they probably constitute two species over time.

Limusaurus inextricabilis
2 m (6 ft) TL, 15 kg (30 lb)

FOSSIL REMAINS Complete skull and majority of two skeletons, gastroliths present.
ANATOMICAL CHARACTERISTICS Head moderately deep. Ossified sternum present. Two functional fingers. Inner toe reduced.
AGE Late Jurassic, early Oxfordian.
DISTRIBUTION AND FORMATION/S Northwest China; upper Shishugou.
NOTES Shared its habitat with *Haplocheirus* and *Yinlong*. Prey of *Zuolong*.

Ceratosaurs

LARGE PREDATORY AVEROSTRANS FROM THE LATE JURASSIC OF THE AMERICAS, EUROPE, AND AFRICA.

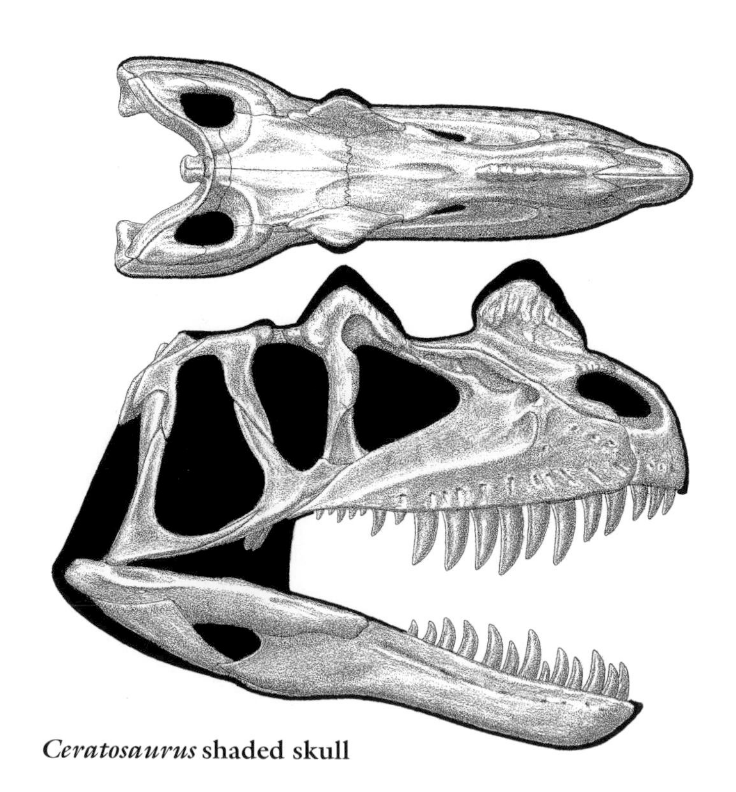

Ceratosaurus shaded skull

ANATOMICAL CHARACTERISTICS Uniform. Stoutly built. Four fingers. Brains reptilian.
NOTES Fragmentary *Genyodectes serus* from the Aptian of Argentina may indicate survival of group into Early Cretaceous. Absence from some continents may reflect lack of sufficient sampling.

Sarcosaurus woodi
3 m (10 ft) TL, 70 kg (150 lb)

FOSSIL REMAINS Minority of skeleton.
ANATOMICAL CHARACTERISTICS Insufficient information.
AGE Early Jurassic, late Sinemurian.
DISTRIBUTION AND FORMATION/S England; Lower Lias.
HABITS Prey included *Scelidosaurus.*
NOTES The relationships of this Early Jurassic theropod are uncertain.

Ceratosaurus nasicornis
6 m (20 ft) TL, 600 kg (1,300 lb)

FOSSIL REMAINS Two skulls and some skeletons including a juvenile.
ANATOMICAL CHARACTERISTICS Head large, long, rectangular, narrow; large, narrow nasal horn, orbital hornlets large, subtriangular; teeth large. Tail deep and heavy. Arm and hand short. Leg not long. Single row of small bony scales along back.
AGE Late Jurassic, late Oxfordian to early Tithonian.
DISTRIBUTION AND FORMATION/S Colorado, Utah; lower and middle Morrison.
HABITAT Short wet season, otherwise semiarid with open floodplain prairies and riverine forests.
HABITS Ambush predators. Large bladed teeth indicate that this hunted large prey including sauropods and stegosaurs by delivering slashing wounds and that the head was a much more important weapon than the small arms. Deep tail may have been used as sculling organ while swimming. Nasal horn probably for display and head butting within the species.
NOTES The species *C. magnicornis* is so similar that it appears to be a member of *C. nasicornis*, or it may represent a descendant of *C. nasicornis. Ceratosaurus* shared its habitat with the much more common, faster *Allosaurus* and similarly uncommon, stouter *Torvosaurus.*

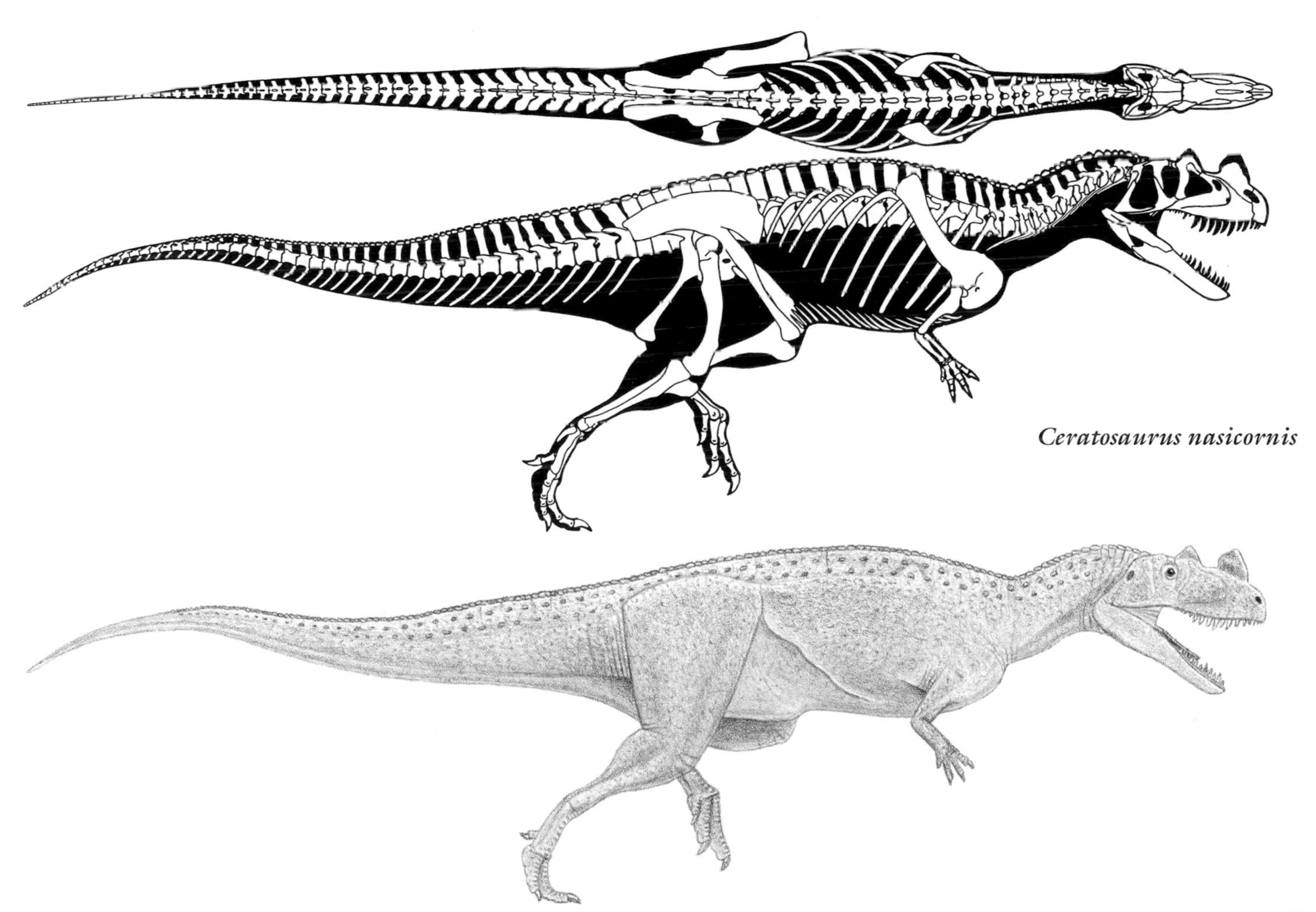

Ceratosaurus nasicornis

Ceratosaurus dentisulcatus
7 m (21 ft) TL, 700 kg (1,500 lb)

FOSSIL REMAINS Part of a skull and skeleton.
ANATOMICAL CHARACTERISTICS Head deeper, lower jaw not as curved, teeth not as proportionally large as in *C. nasicornis*.
AGE Late Jurassic, middle Tithonian.
DISTRIBUTION AND FORMATION/S Utah; upper Morrison.
HABITAT Wetter than earlier Morrison, otherwise semiarid with open floodplain prairies and riverine forests.
HABITS Similar to *C. nasicornis*.
NOTES It is uncertain whether *C. dentisulcatus* had a nasal horn or not. May have been the direct descendant of *C. nasicornis*. Shared its habitat with later species of *Allosaurus*.

Ceratosaurus? unnamed species
6 m (20 ft) TL, 600 kg (1,300 lb)

FOSSIL REMAINS Minority of skeleton.
ANATOMICAL CHARACTERISTICS Insufficient information.
AGE Late Jurassic, late Kimmeridgian or early Tithonian.
DISTRIBUTION AND FORMATION/S Portugal; Lourinha.
HABITAT Large, seasonally dry island with open woodlands.
NOTES Assignment by some researchers of this specimen to *C. dentisulcatus* is uncertain. Shared its habitat with *Allosaurus europaeus*.

Abelisauroids

SMALL TO GIGANTIC AVEROSTRANS FROM THE EARLY JURASSIC TO THE END OF THE DINOSAUR ERA, LARGELY LIMITED TO THE SOUTHERN HEMISPHERE.

ANATOMICAL CHARACTERISTICS Highly variable. Arm short, four fingers. Vertebrae often flat topped. Pelvis large. Birdlike respiratory system well developed.
NOTES Abelisaurs show that relatively archaic avepods were able to thrive in the Southern Hemisphere to the end of the dinosaur era as they evolved into specialized forms. Absence from Antarctica probably reflects lack of sufficient sampling.

ABELISAURIDS Large to gigantic abelisauroids of the Middle Jurassic to the end of the dinosaur era, largely limited to the Southern Hemisphere.

ANATOMICAL CHARACTERISTICS Fairly uniform. Head heavily constructed, short and deep, lower jaw slender, teeth short and stout. Arm reduced. Tubercle scales set amid fairly large, flat scales.
HABITATS Seasonally dry to well-watered woodlands.
HABITS Reduction of arms indicates that the stout head was the primary weapon, but how the combination of a deep short skull, slender lower jaw that indicates modest musculature, and short teeth functioned is obscure. Prey included titanosaur juveniles and adults, and ankylosaurs.
NOTES Fragmentary Late Cretaceous *Arcovenator escotae* from France indicates presence of group in Northern Hemisphere.

Eoabelisaurus mefi
6.5 m (21 ft) TL, 800 kg (1,800 lb)

FOSSIL REMAINS Minority of skull and majority of skeleton.
ANATOMICAL CHARACTERISTICS Robustly built.
AGE Middle Jurassic, Aalenian and/or Bajocian.
DISTRIBUTION AND FORMATION/S Southern Argentina; Asfalto.

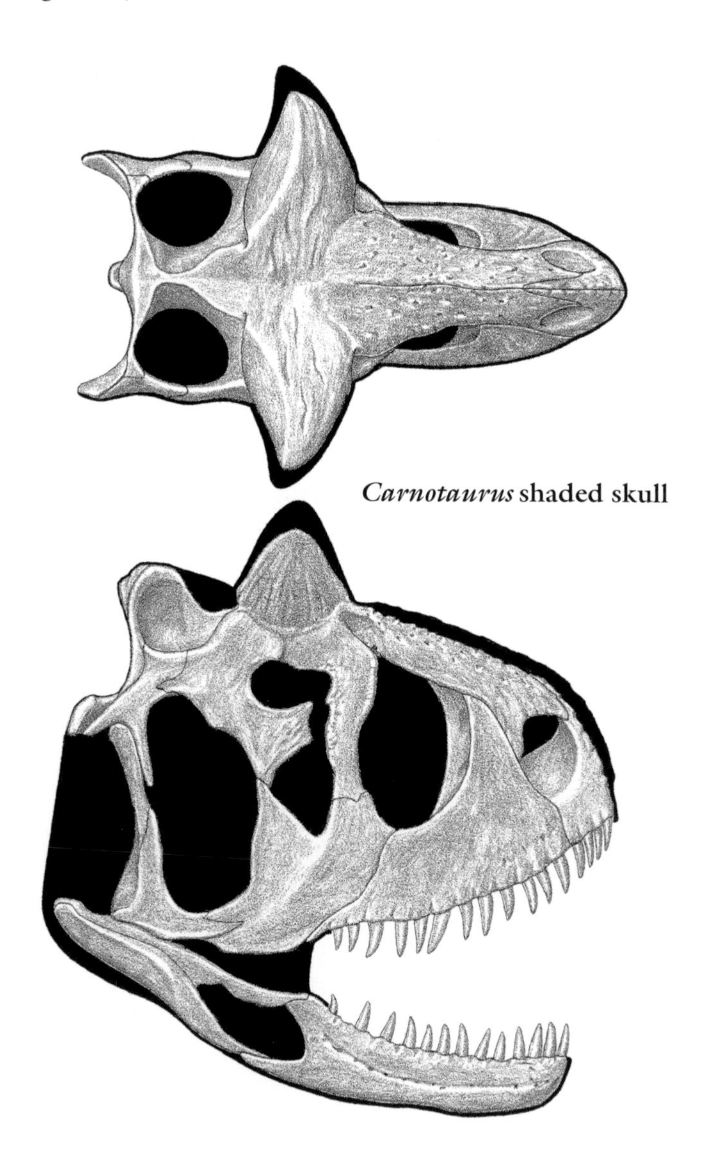

Carnotaurus shaded skull

Kryptops palaios
Adult size uncertain

FOSSIL REMAINS Minority of skull and skeleton.
ANATOMICAL CHARACTERISTICS Insufficient information.
AGE Early Cretaceous, Aptian.
DISTRIBUTION AND FORMATION/S Niger; Elrhaz, level uncertain.
HABITAT Coastal river delta.
NOTES The one specimen may be a large juvenile. Shared its habitat with *Eocarcharia*.

Rugops primus
6 m (20 ft) TL, 750 kg (1,600 lb)

FOSSIL REMAINS Partial skull.
ANATOMICAL CHARACTERISTICS Snout very deep and robust, possible low paired crests on snout.
AGE Late Cretaceous, Cenomanian.
DISTRIBUTION AND FORMATION/S Niger; Echkar.
NOTES Shared its habitat with *Carcharodontosaurus iguidensis* and a similar-sized semiterrestrial crocodilian.

Xenotarsosaurus bonapartei
6 m (20 ft) TL, 750 kg (1,700 lb)

FOSSIL REMAINS Minority of skeleton.
ANATOMICAL CHARACTERISTICS Leg long and gracile.
AGE Late Cretaceous, late Cenomanian or Turonian.
DISTRIBUTION AND FORMATION/S Southern Argentina; lower Bajo Barreal.
HABITS Pursuit predator.

Majungasaurus crenatissimus
6 m (20 ft) TL, 750 kg (1,700 lb)

FOSSIL REMAINS Nearly perfect skull and extensive skeletal material, nearly completely known.
ANATOMICAL CHARACTERISTICS Low central horn above orbits. Leg stout, not elongated.
AGE Late Cretaceous, Campanian.
DISTRIBUTION AND FORMATION/S Madagascar; Maevarano.
HABITAT Seasonally dry floodplain with coastal swamps and marshes.
HABITS Probably used horn for display and head butting within the species. Prey mainly sauropods including *Rapetosaurus*, large ornithischians apparently absent.

Ilokelesia aguadagrandensis
4 m (14 ft) TL, 200 kg (450 lb)

FOSSIL REMAINS Minority of skeleton.
ANATOMICAL CHARACTERISTICS Base of tail exceptionally broad.
AGE Late Cretaceous, late Cenomanian.

Rugops primus

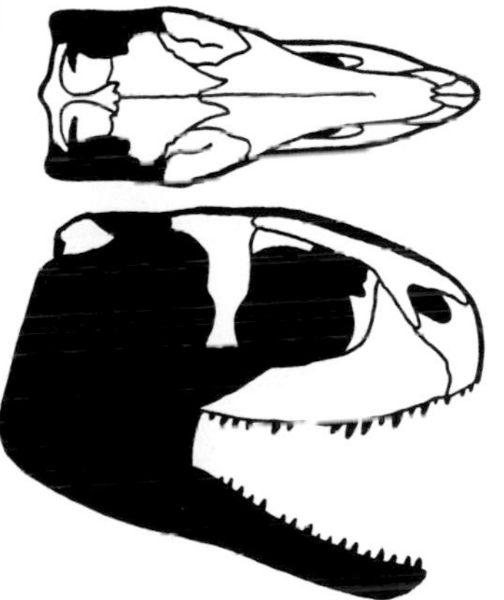

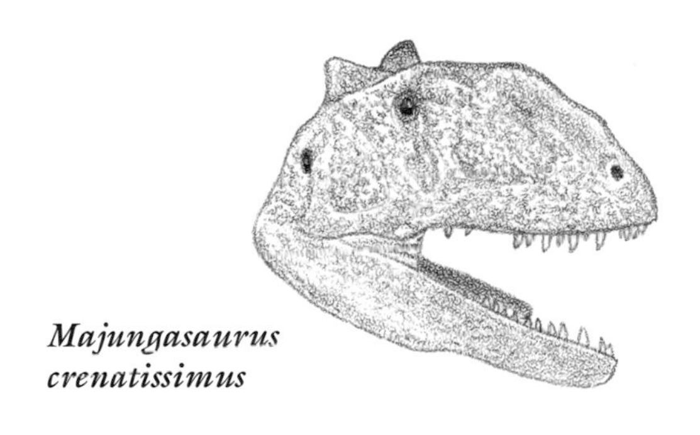

Majungasaurus crenatissimus

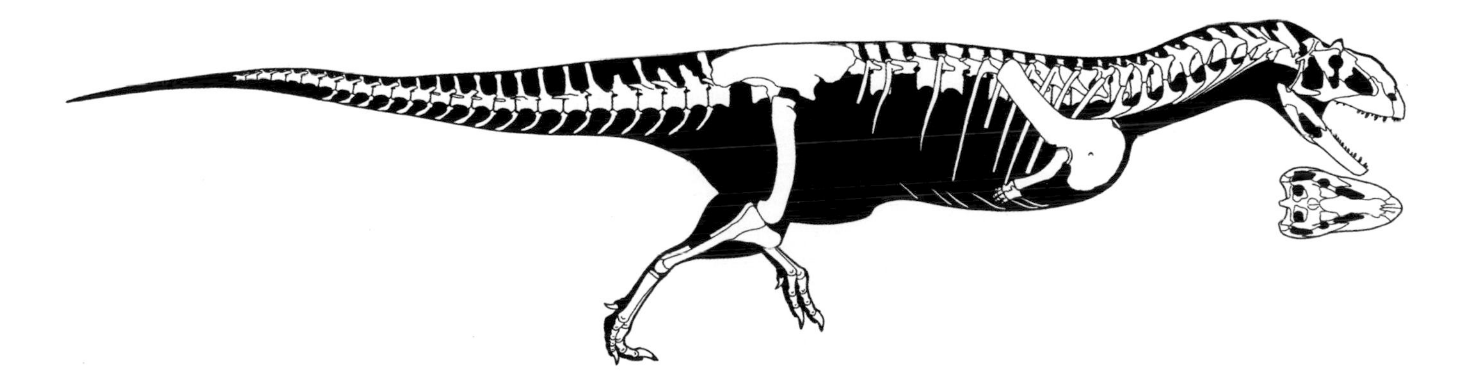

DISTRIBUTION AND FORMATION/S Western Argentina; middle Huincul.
HABITAT Short wet season, otherwise semiarid with open floodplains and riverine forests.
NOTES Shared its habitat with *Mapusaurus.*

Rajasaurus narmadensis
11 m (35 ft) TL, 4 tonnes

FOSSIL REMAINS Majority of skull and partial skeleton.
ANATOMICAL CHARACTERISTICS Back of head adorned by central crest. Leg stout.
AGE Late Cretaceous, Maastrichtian.
DISTRIBUTION AND FORMATION/S Central India; Lameta.
HABITS Probably used horn for display and head butting within the species. Prey included *Isisaurus.*
NOTES Shared its habitat with smaller, longer-legged *Rahiolisaurus*, unless latter was a juvenile of this taxon.

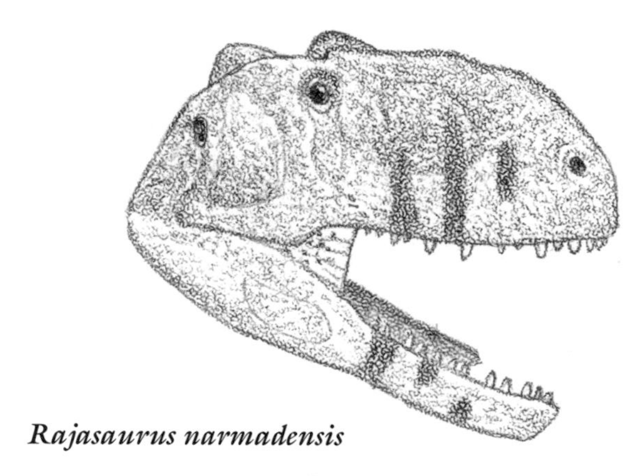

Rajasaurus narmadensis

Rahiolisaurus gujaratensis
8 m (27 ft) TL, 2 tonnes

FOSSIL REMAINS Minority of skull and skeletons.
ANATOMICAL CHARACTERISTICS Less robust than *Rajasaurus.*
AGE Late Cretaceous, Maastrichtian.
DISTRIBUTION AND FORMATION/S Central India; Lameta.

Ekrixinatosaurus novasi
6.5 m (21 ft) TL, 800 kg (1,800 lb)

FOSSIL REMAINS Minority of skull and partial skeleton.
ANATOMICAL CHARACTERISTICS Insufficient information.
AGE Late Cretaceous, early Cenomanian.
DISTRIBUTION AND FORMATION/S Western Argentina; Candeleros.
HABITAT Short wet season, otherwise semiarid with open floodplains and riverine forests.
NOTES Shared its habitat with *Giganotosaurus.*

Skorpiovenator bustingorryi
7.5 m (25 ft) TL, 1.67 tonnes

FOSSIL REMAINS Complete skull and majority of skeleton.
ANATOMICAL CHARACTERISTICS Rugose area around eye socket. Leg long.
AGE Late Cretaceous, Middle Cenomanian.

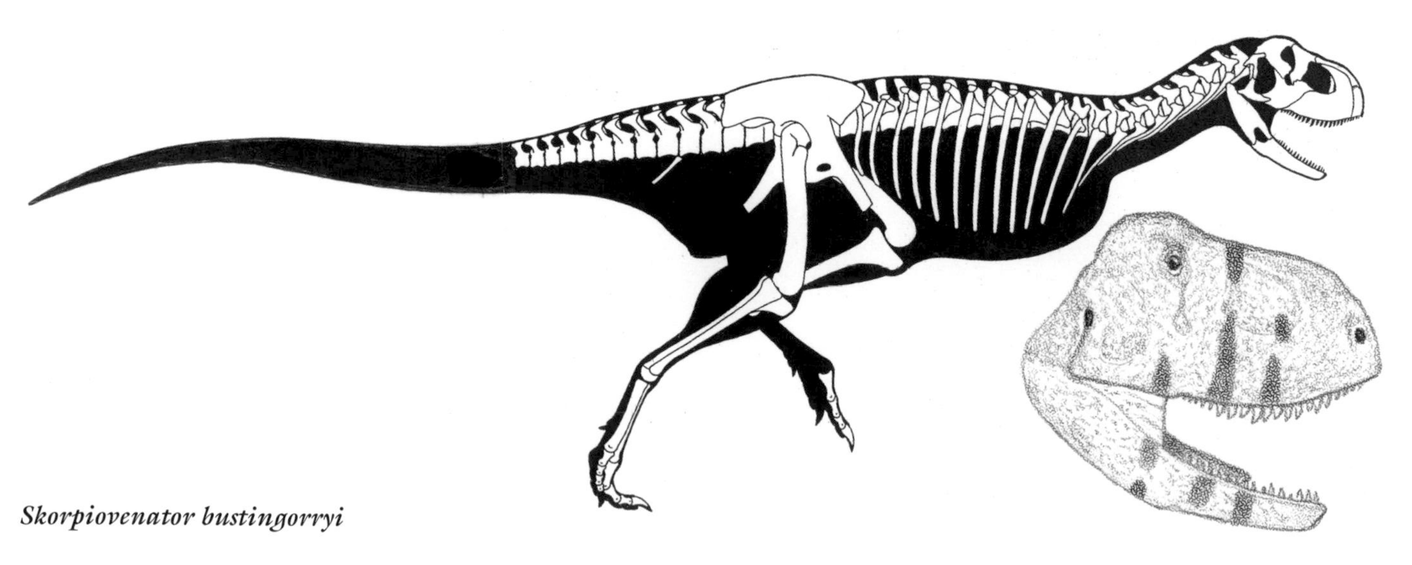

Skorpiovenator bustingorryi

DISTRIBUTION AND FORMATION/S Western Argentina; lower Huincul.
HABITAT Well-watered woodlands with short dry season.
HABITS Prey included *Cathartesaura*.

Abelisaurus comahuensis
10 m (30 ft) TL, 3 tonnes

FOSSIL REMAINS Partial skull.
ANATOMICAL CHARACTERISTICS Head unadorned.
AGE Late Cretaceous, late Santonian and/or early Campanian.
DISTRIBUTION AND FORMATION/S Western Argentina; Anacleto.
HABITS Prey included titanosaurs.

Abelisaurus (= *Aucasaurus*) *garridoi*
5.5 m (18 ft) TL, 700 kg (1,500 lb)

FOSSIL REMAINS Complete skull and nearly complete skeleton.

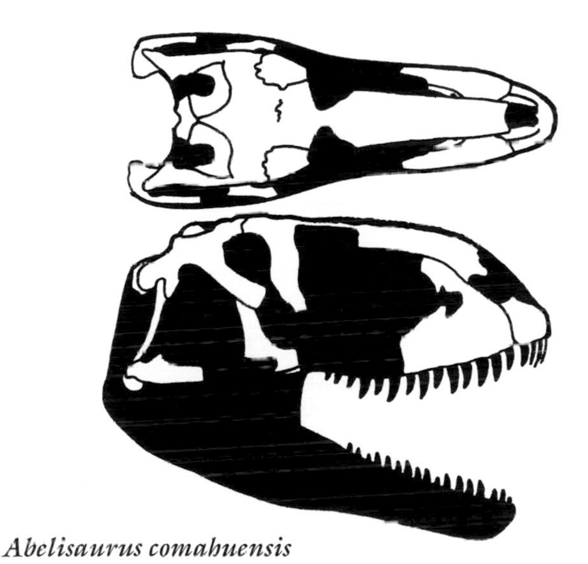

Abelisaurus comahuensis

ANATOMICAL CHARACTERISTICS Head unadorned. Lower arm and hand atrophied. Leg long and gracile, inner toe reduced, toe claws small.
AGE Late Cretaceous, late Santonian and/or early Campanian.
DISTRIBUTION AND FORMATION/S Western Argentina; Anacleto.
HABITS Pursuit predator able to chase prey at high speed. Prey included *Gasparinisaura*; shared its habitat with *Aerosteon*.
NOTES Named in a new genus, *Aucasaurus*; the only reason this does not appear to be a juvenile *Abelisaurus comahuensis* is that fusion of skeletal elements indicates it is an adult.

Pycnonemosaurus nevesi
7 m (23 ft) TL, 1.2 tonnes

FOSSIL REMAINS Minority of skeleton.
ANATOMICAL CHARACTERISTICS Insufficient information.
AGE Late Cretaceous, Campanian or Maastrichtian.
DISTRIBUTION AND FORMATION/S Southwest Brazil; Bauro Group.

Carnotaurus sastrei
7.5 m (25 ft) TL, 2 tonnes

FOSSIL REMAINS Complete skull and majority of skeleton, skin patches.
ANATOMICAL CHARACTERISTICS Head very deep, large, stout brow horns directed sideways. Lower arm and hand atrophied.
AGE Late Cretaceous, Campanian or early Maastrichtian.
DISTRIBUTION AND FORMATION/S Southern Argentina; La Colonia.

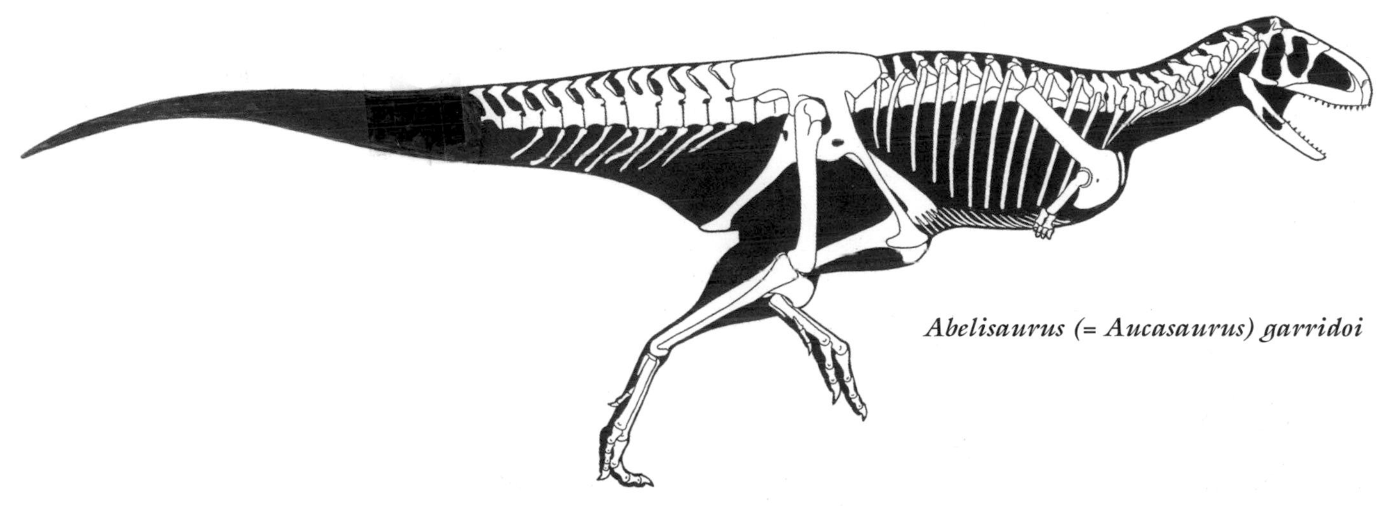

Abelisaurus (= Aucasaurus) garridoi

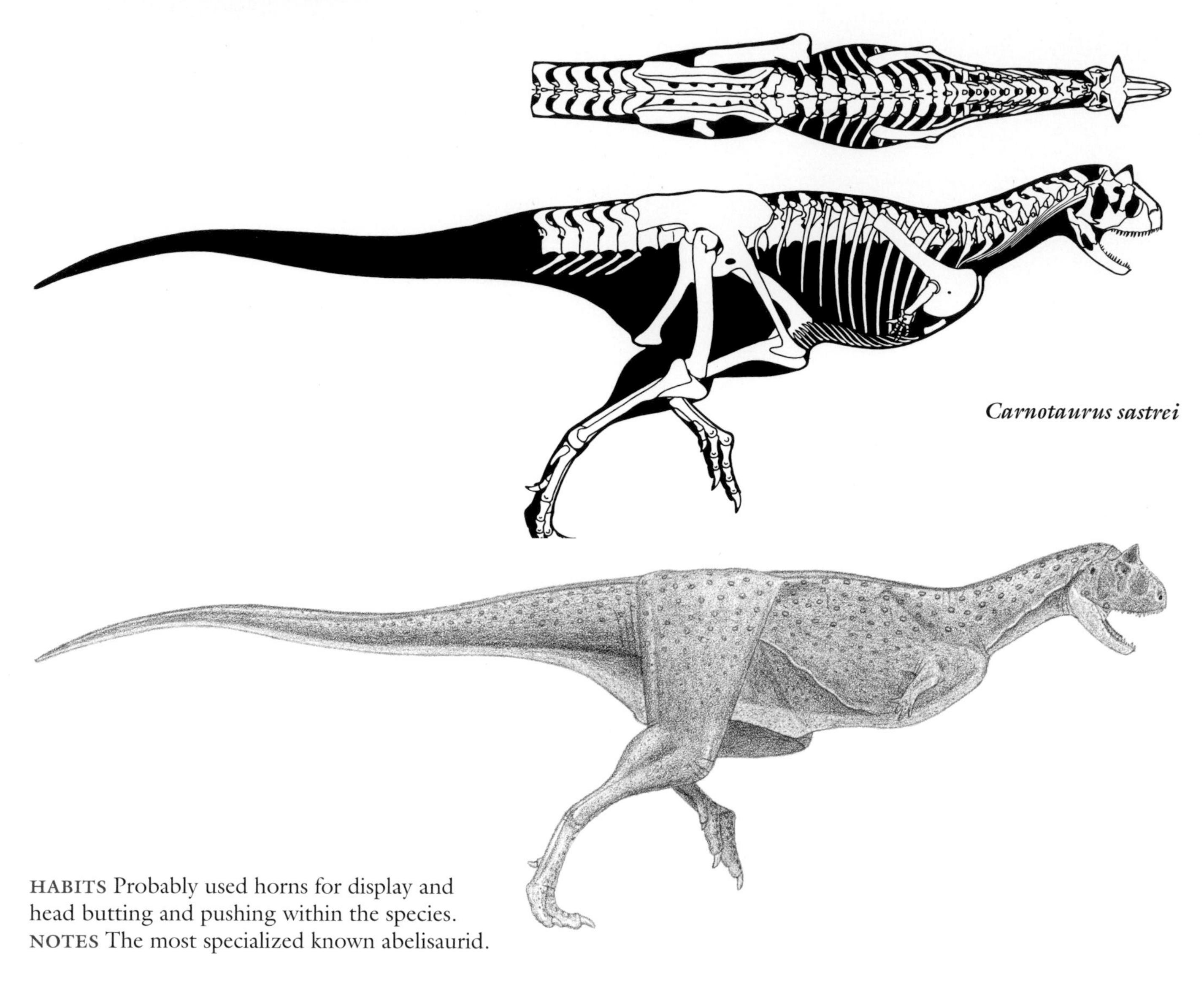

Carnotaurus sastrei

HABITS Probably used horns for display and head butting and pushing within the species.
NOTES The most specialized known abelisaurid.

NOASAURIDS Small to medium-sized abelisauroids.

ANATOMICAL CHARACTERISTICS Highly variable. Most lightly constructed of the abelisaurs. Arm better developed than in other abelisaurs.

Ligabueino andesi
0.6 m (2 ft) TL, 0.5 kg (1 lb)

FOSSIL REMAINS Minority of skeleton.
ANATOMICAL CHARACTERISTICS Standard for group.
AGE Early Cretaceous, Barremian and/or early Aptian.
DISTRIBUTION AND FORMATION/S Western Argentina; La Amarga.
HABITAT Well-watered coastal woodlands with short dry season.
HABITS Small game hunter.
NOTES If not a juvenile, this is one of the smallest theropods outside the birdlike maniraptors.

Masiakasaurus knopfleri
2 m (7 ft) TL, 20 kg (50 lb)

FOSSIL REMAINS Minority of skull and skeleton.
ANATOMICAL CHARACTERISTICS Front teeth of lower jaw form a procumbent whorl and are long and weakly serrated; back teeth are more conventional.
AGE Late Cretaceous, Campanian.
DISTRIBUTION AND FORMATION/S Madagascar; Maevarano.
HABITAT Seasonally dry floodplain with coastal swamps and marshes.
HABITS Probably hunted small prey, especially fish.

Noasaurus leali
1.5 m (5 ft) TL, 15 kg (30 lb)

FOSSIL REMAINS Minority of a skull and skeleton.
ANATOMICAL CHARACTERISTICS Standard for group.

Masiakasaurus knopfleri

AGE Late Cretaceous, probably early Maastrichtian.
DISTRIBUTION AND FORMATION/S Northern Argentina; Lecho.
HABITS Pursuit predator.
NOTES It was thought that a large claw was a sickle-toe weapon like those of dromaeosaurids, but more likely it belonged to the hand.

Genusaurus sisteronis
3 m (10 ft) TL, 35 kg (70 lb)

FOSSIL REMAINS Minority of skeleton.
ANATOMICAL CHARACTERISTICS Insufficient information.
AGE Early Cretaceous, Albian.
DISTRIBUTION AND FORMATION/S Southeast France; Bevon Beds.
HABITAT Forested coastline.
NOTES Found as drift in nearshore marine deposits. Placement in noasaurids uncertain, if correct indicates that abelisauroids migrated to the Northern Hemisphere.

TETANURANS

SMALL TO GIGANTIC PREDATORY AND HERBIVOROUS AVEROSTRANS FROM THE MIDDLE JURASSIC TO THE END OF THE DINOSAUR ERA (WITH BIRDS SURVIVING BEYOND), ON ALL CONTINENTS.

ANATOMICAL CHARACTERISTICS Highly variable. Arm very long to very reduced. Birdlike respiratory system better developed. Brains reptilian to avian.

Baso-tetanurans

SMALL TO LARGE PREDATORY AND HERBIVOROUS TETANURANS FROM THE JURASSIC.

HABITS Crests when present too delicate for head butting; for visual display, may or may not have been brightly colored at least during breeding season.
NOTES The relationships of the following primitive and usually partially known tetanurans are uncertain.

Sinosaurus sinensis (Illustration overleaf)
5.5 m (18 ft) TL, 300 kg (650 lb)

FOSSIL REMAINS Nearly complete skull and skeleton.
ANATOMICAL CHARACTERISTICS. Overall build slender. Head adorned by large paired crests. Body fairly gracile.
AGE Early Jurassic, probably Hettangian.
DISTRIBUTION AND FORMATION/S Southwest China; lower Lower Lufeng.
HABITS Long crests for lateral display. Prey included *Lufengosaurus* and *Yunnanosaurus*.
NOTES Comparable in time, size, and overall appearance to similarly crested *Dilophosaurus*, it was assumed to be a member of the same genus, but detailed anatomy indicates this is a more derived tetanuran avepod.

Cryolophosaurus ellioti
6 m (20 ft) TL, 350 kg (800 lb)

FOSSIL REMAINS Partial skull and minority of skeleton.
ANATOMICAL CHARACTERISTICS Paired crests low at front of snout, above orbits, arc toward middle and join to form large transverse crest.

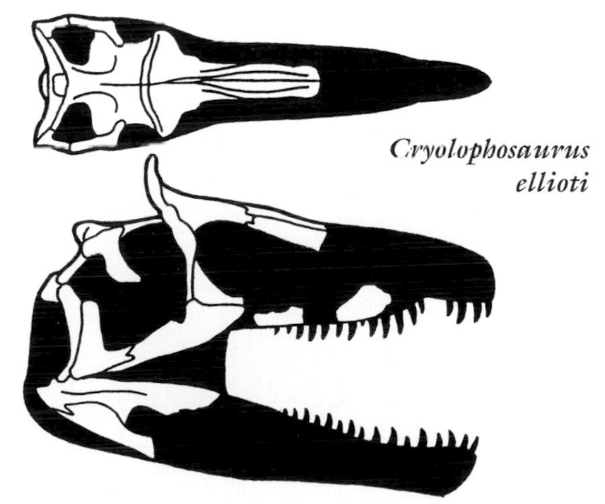

Cryolophosaurus ellioti

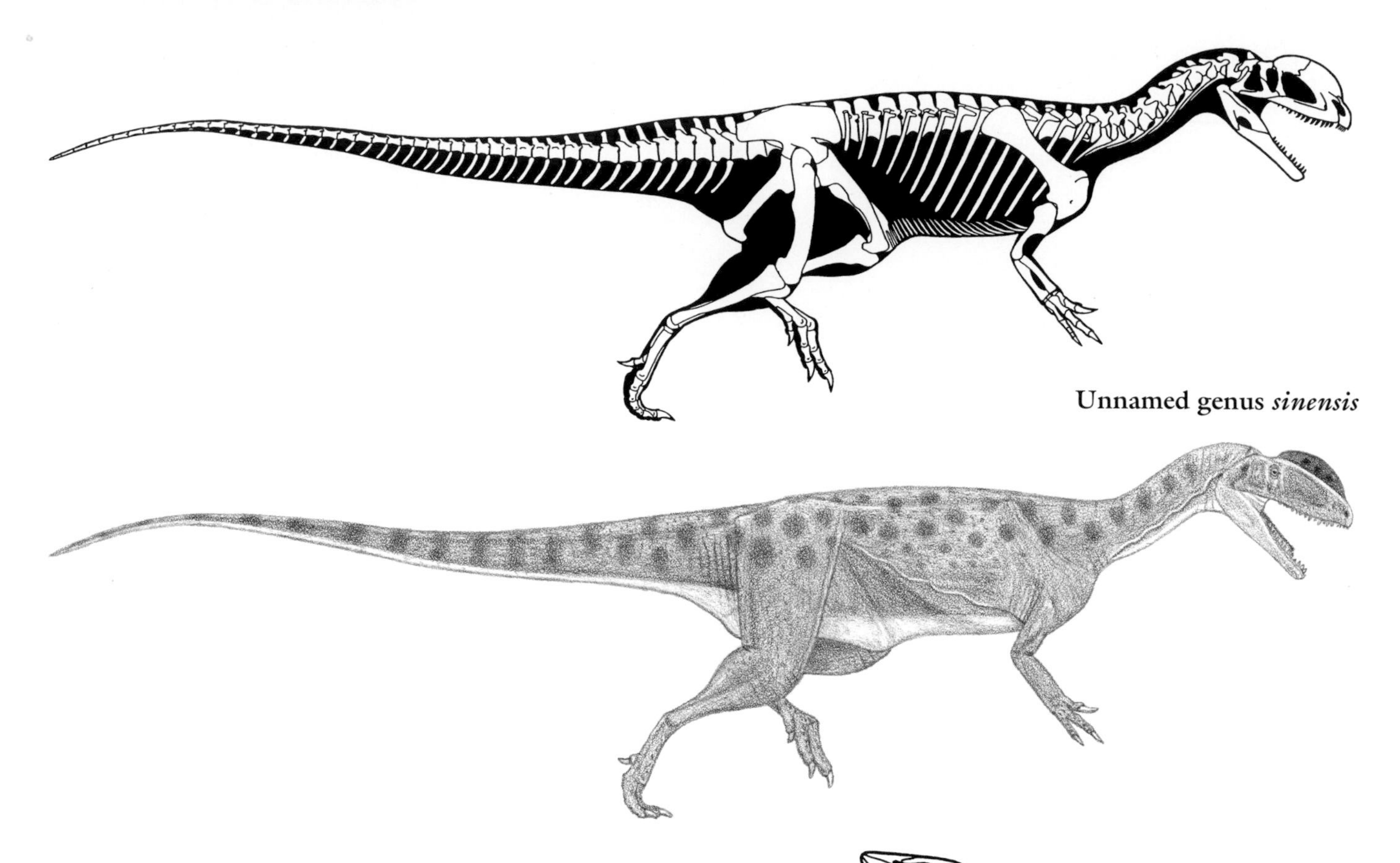
Unnamed genus *sinensis*

AGE Early Jurassic, Sinemurian or Pliensbachian.
DISTRIBUTION AND FORMATION/S Central Antarctica; Hanson.
HABITAT Polar forests with warm, daylight-dominated summers and cold, dark winters.
HABITS Broad, tall crest for frontal display. Prey included large prosauropods.
NOTES The only theropod yet named from Antarctica; this is an artifact stemming from the lack of more extensive exposed deposits and difficult research conditions.

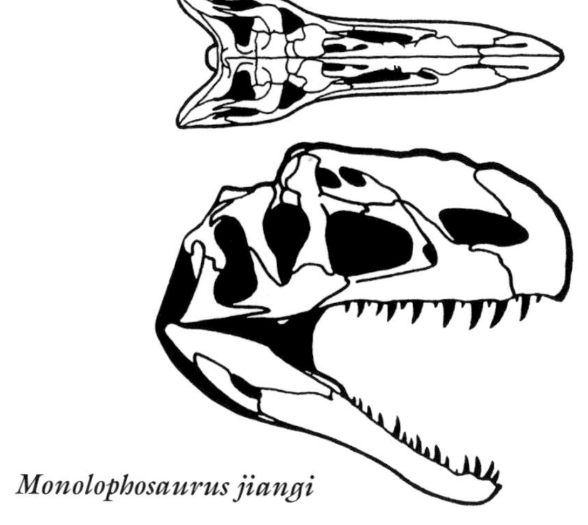
Monolophosaurus jiangi

Monolophosaurus jiangi
5.5 m (18 ft) TL, 475 kg (1,000 lb)

FOSSIL REMAINS Complete skull and majority of skeleton.
ANATOMICAL CHARACTERISTICS Overall build robust. Snout ridges united and enlarged into an enormous midline crest.
AGE Middle Jurassic, late Callovian.
DISTRIBUTION AND FORMATION/S Northwest China; lower Shishugou.
HABITS Prey included primitive sauropods such as *Bellusaurus*. Crest too delicate for head butting; for visual display within the species.
NOTES Shared its habitat with smaller *Guanlong*.

Shidaisaurus jinae
6 m (20 ft) TL, 700 kg (1,600 lb)

FOSSIL REMAINS Minority of skull and partial skeleton.
ANATOMICAL CHARACTERISTICS Neural spines form shallow sail over trunk and base of tail.
AGE Early Middle Jurassic.
DISTRIBUTION AND FORMATION/S Southwest China; Upper Lufeng.

Xuanhanosaurus qilixiaensis
4.5 m (15 ft) TL, 250 kg (500 lb)

FOSSIL REMAINS Minority of skeleton.
ANATOMICAL CHARACTERISTICS Stoutly built. Arm and hand well developed.
AGE Late Jurassic, Bathonian and/or Callovian.
DISTRIBUTION AND FORMATION/S Central China; Shaximiao.
HABITAT Heavily forested.
HABITS Arms probably important in handling prey.
NOTES Shared its habitat with *Gasosaurus* and *Yangchuanosaurus zigongensis.*

Chilesaurus diegosuarezi
2.5 m (8 ft) TL, 15 kg (35 lb)

FOSSIL REMAINS Partial skull and majority of skeletons.
ANATOMICAL CHARACTERISTICS Head small, teeth leaf shaped, lightly serrated. Neck fairly long. Body elongated, pubis retroverted. Arm robust, third digit absent. Inner toes well developed, so foot almost tetradactyl.
AGE Late Jurassic, middle Tithonian.
DISTRIBUTION AND FORMATION/S Chile; Toqui.
HABITS Herbivorous or omnivorous.
NOTES Exact relationships with other avepods uncertain. This is another example of avepods evolving herbivory, and of reenlarging the inner toe—almost as much as in therizinosaurs and similar to *Balaur*—and another example of dinosaurs retroverting the pubis, in this case to enlarge the gut for herbivory. Fragmentary Asian Early Jurassic *Eshanosaurus deguchiianus* may be a relative.

Megalosauroids

PREDATORY TETANURANS LIMITED TO THE MIDDLE AND LATE JURASSIC ON MOST CONTINENTS.

ANATOMICAL CHARACTERISTICS Head large, low.
NOTES The relationships of the following megalosauroids are uncertain. Absence from Australia and Antarctica probably reflects lack of sufficient sampling.

Eustreptospondylus oxoniensis
6 m (20 ft) TL, 500 kg (1,000 lb)

FOSSIL REMAINS Majority of skull and skeleton.
ANATOMICAL CHARACTERISTICS Lightly built. Teeth widely spaced.
AGE Middle Jurassic, late Callovian.
DISTRIBUTION AND FORMATION/S Southern England; Middle Oxford Clay.

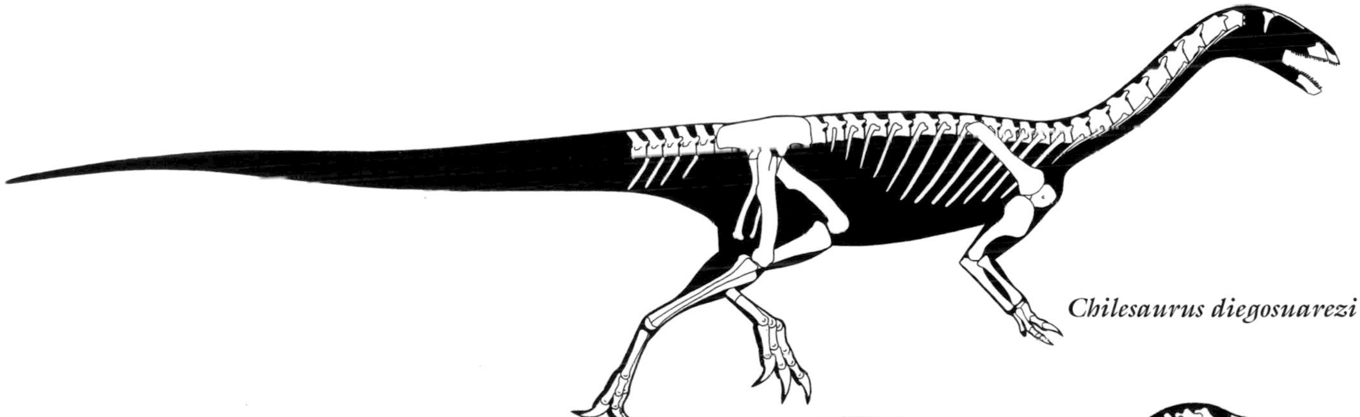

Chilesaurus diegosuarezi

Eustreptospondylus oxoniensis

Marshosaurus bicentesimus
4.5 m (15 ft) TL, 200 kg (400 lb)

FOSSIL REMAINS Minority of skeletons.
ANATOMICAL CHARACTERISTICS Insufficient information.
AGE Late Jurassic, middle Tithonian.
DISTRIBUTION AND FORMATION/S Utah; middle Morrison.
HABITAT Wetter than earlier Morrison, otherwise semiarid with open floodplain prairies and riverine forests.

Piatnitzkysaurus floresi
4.5 m (15 ft) TL, 275 kg (600 lb)

FOSSIL REMAINS Minority of skull and majority of skeleton.
ANATOMICAL CHARACTERISTICS Lightly built.
AGE Middle Jurassic.
DISTRIBUTION AND FORMATION/S Southern Argentina; Canadon Asfalto.
NOTES Shared its habitat with *Condorraptor.*

Condorraptor currumili
4.5 m (15 ft) TL, 200 kg (400 lb)

FOSSIL REMAINS Minority of skeleton.
ANATOMICAL CHARACTERISTICS Insufficient information.
AGE Middle Jurassic.
DISTRIBUTION AND FORMATION/S Southern Argentina; Canadon Asfalto.

Afrovenator abakensis
8 m (25 ft) TL, 1 tonne

FOSSIL REMAINS Majority of skull and partial skeleton.
ANATOMICAL CHARACTERISTICS Lightly built. Orbital hornlet modest, teeth large. Leg long.
AGE Late Middle or early Late Jurassic.
DISTRIBUTION AND FORMATION/S Niger; Tiouraren.
HABITAT Well-watered woodlands.
HABITS Pursuit predator. Prey included *Spinostropheus.*
NOTES Originally thought to be from the Early Cretaceous; researchers now place the Tiouraren in the later Jurassic. This, *Dubreuillosaurus*, and *Magnosaurus* may form subfamily Afrovenatorinae.

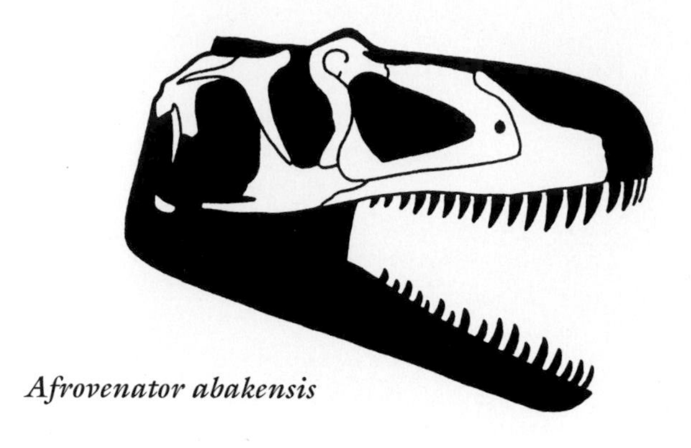
Afrovenator abakensis

Dubreuillosaurus valesdunensis
5 m (15 ft) TL, 250 kg (500 lb)

FOSSIL REMAINS Majority of skull and partial skeleton.
ANATOMICAL CHARACTERISTICS Teeth large.
AGE Middle Jurassic, middle Bathonian.
DISTRIBUTION AND FORMATION/S Northwest France; Calcaire de Caen.

Dubreuillosaurus valesdunensis

Piatnitzkysaurus floresi

HABITAT Coastal mangroves.
NOTES Not a species of *Poekilopleuron*, as originally thought.

Magnosaurus nethercombensis
4.5 m (15 ft) TL, 200 kg (400 lb)

FOSSIL REMAINS Minority of skull and skeleton.
ANATOMICAL CHARACTERISTICS Insufficient information.
AGE Middle Jurassic, Aalenian or Bajocian.
DISTRIBUTION AND FORMATION/S Southwestern England; Inferior Oolite.

MEGALOSAURIDS Very large predatory tetanurans limited to the Middle and Late Jurassic of Europe and North America.

ANATOMICAL CHARACTERISTICS Fairly uniform. Massively constructed. Teeth stout. Lower arm short and stout. Pelvis broad and shallow. Brains reptilian.
HABITAT Seasonally dry to well-watered woodlands.
HABITS Ambush predators, prey included sauropods and stegosaurs.

Duriavenator hesperis
7 m (23 ft) TL, 1 tonne

FOSSIL REMAINS Partial skull.
ANATOMICAL CHARACTERISTICS Teeth in lower jaw widely spaced.
AGE Middle Jurassic, late Bajocian.
DISTRIBUTION AND FORMATION/S Southern England; upper Inferior Oolite.

Megalosaurus bucklandi
6 m (20 ft) TL, 700 kg (1,600 lb)

FOSSIL REMAINS Lower jaw and possibly skeletal parts.
ANATOMICAL CHARACTERISTICS Standard for group.
AGE Middle Jurassic, middle Bathonian.
DISTRIBUTION AND FORMATION/S Central England; Stonesfield Slate.
NOTES Over the years *Megalosaurus* became a taxonomic grab bag into which a large number of remains from many places and times were placed. The genus and species may be limited to the original specimens, the full extent of which is not entirely certain.

Poekilopleuron? bucklandii
7 m (23 ft) TL, 1 tonne

FOSSIL REMAINS Partial skeleton.
ANATOMICAL CHARACTERISTICS Appears to be standard for group.
AGE Middle Jurassic, middle Bathonian.
DISTRIBUTION AND FORMATION/S Northwest France; Calcaire de Caen.
NOTES Because at least some bones of this and *Megalosaurus bucklandi* are very similar, it is possible that this is the same genus and even species as the latter, or some of the original material placed in the British megalosaur may belong to this theropod. Original remains destroyed in World War II.

Torvosaurus tanneri
9 m (30 ft) TL, 2 tonnes

FOSSIL REMAINS Majority of a skull and partial skeletons.
ANATOMICAL CHARACTERISTICS Standard for group.
AGE Late Jurassic, early Tithonian.
DISTRIBUTION AND FORMATION/S Colorado, Wyoming, Utah; middle Morrison.
HABITAT Short wet season, otherwise semiarid with open floodplain prairies and riverine forests.
NOTES Shared habitat with much more common *Allosaurus* and similarly uncommon *Ceratosaurus*. Remains imply this taxon or a close relative was present in the lower Morrison and/or the Portuguese Lourinha Formation.

Torvosaurus tanneri

composite megalosaurid

Spinosaurs

LARGE TO GIGANTIC FISHING AND PREDATORY TETANURANS OF THE CRETACEOUS OF AFRICA AND EURASIA.

ANATOMICAL CHARACTERISTICS Fairly uniform. Long bodied. Head very long and shallow; snout elongated, narrow, and tip hooked; tip of lower jaw expanded, teeth conical; low central crest above orbits, lower jaws could bow outward. Arm well developed, three fingers, claws large hooks. Leg short. Brains reptilian.
HABITAT Large watercourses or coastlines.
HABITS Probably able to prey on large animals, but predominantly small game hunters with specializations for fishing using crocodilian-like heads and teeth, outward-bowing pelican-like mandibles, and hooked hand claws. Head crests probably for display within the species.
NOTES Absence from at least some other continents may reflect lack of sufficient sampling.

Ichthyovenator laosensis
8.5 m (27 ft) TL, 2 tonnes

FOSSIL REMAINS Partial skeleton.
ANATOMICAL CHARACTERISTICS Vertebral spines moderately tall and broad.
AGE Early Cretaceous, late Barremian or early Cenomanian.
DISTRIBUTION AND FORMATION/S Laos; Gres Superieurs.
NOTES Shared its habitat with *Tangvayosaurus.*

Baryonyx walkeri
7.5 m (25 ft) TL, 1.2 tonnes

FOSSIL REMAINS Partial skull and skeleton.
ANATOMICAL CHARACTERISTICS Small central crest over orbits.
AGE Early Cretaceous, Barremian.
DISTRIBUTION AND FORMATION/S Southeast England; Weald Clay.

Baryonyx (= Suchomimus) tenerensis
9.5 m (30 ft) TL, 2.5 tonnes

FOSSIL REMAINS Partial skull and skeleton.
ANATOMICAL CHARACTERISTICS Small central crest over orbits. Vertebral spines moderately tall.
AGE Early Cretaceous, late Aptian.
DISTRIBUTION AND FORMATION/S Niger; upper Elrhaz.
HABITAT Coastal river delta.
NOTES Probably includes *Cristatusaurus lapparenti.*

Irritator challengeri
7.5 m (25 ft) TL, 1 tonne

FOSSIL REMAINS Majority of skull.
ANATOMICAL CHARACTERISTICS Long, low midline crest over back of head, back of head deep.
AGE Early Cretaceous, probably Albian.
DISTRIBUTION AND FORMATION/S Eastern Brazil; Santana.
NOTES Found as drift in marine deposits; a snout tip labeled *Angaturama limai* from the same formation may belong to this species or even same specimen. There is evidence of predation on a pterosaur.

Spinosaurus aegypticus
14 m (45 ft) TL, 10 tonnes

FOSSIL REMAINS Minority of skull and skeleton, additional remains problematic.

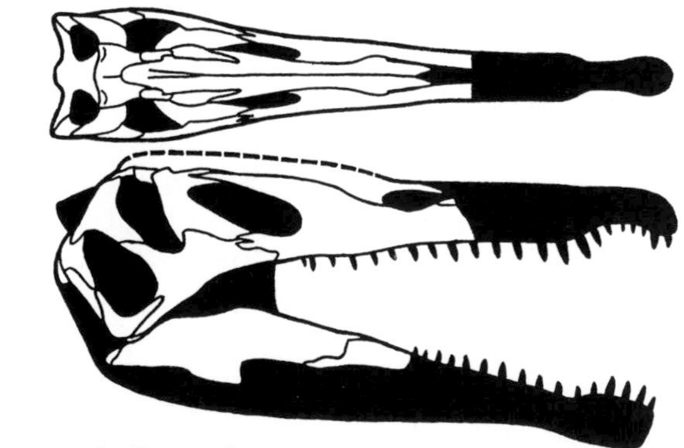

Irritator challengeri

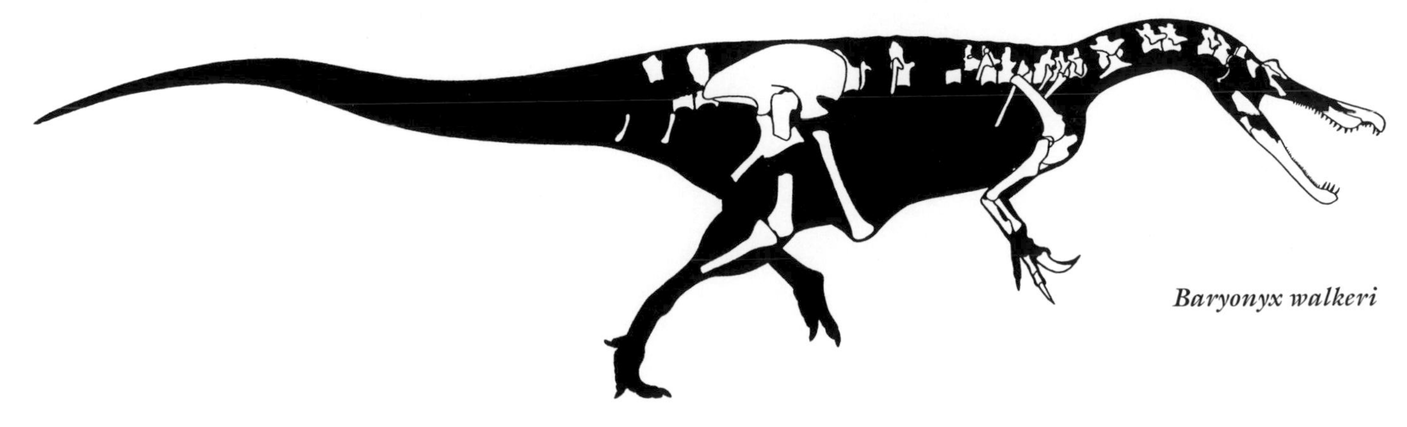

Baryonyx walkeri

Baryonyx (=Suchomimus) tenerensis

ANATOMICAL CHARACTERISTICS Very tall vertebral spines over trunk form enormous finback sail.
AGE Late Cretaceous, early Cenomanian.
DISTRIBUTION AND FORMATION/S Egypt; Bahariya.
HABITAT Coastal mangroves.
NOTES Because remains are very incomplete, weight estimate is tentative; rivals *Giganotosaurus carolinii* as the largest known theropod. Shared its habitat with the similarly large and more powerful *Carcharodontosaurus saharicus*. Original remains destroyed by Allied bombing during World War II. Restorations as an extremely short-legged semiquadruped based on errant inclusion of remains of other spinosaurs—the skeletal proportions of which are also uncertain—from other regions of northern Africa, including fragmentary but different Moroccan *Sigilmassasaurus brevicollis*, in this species are inaccurate, although the legs may well have been rather short.

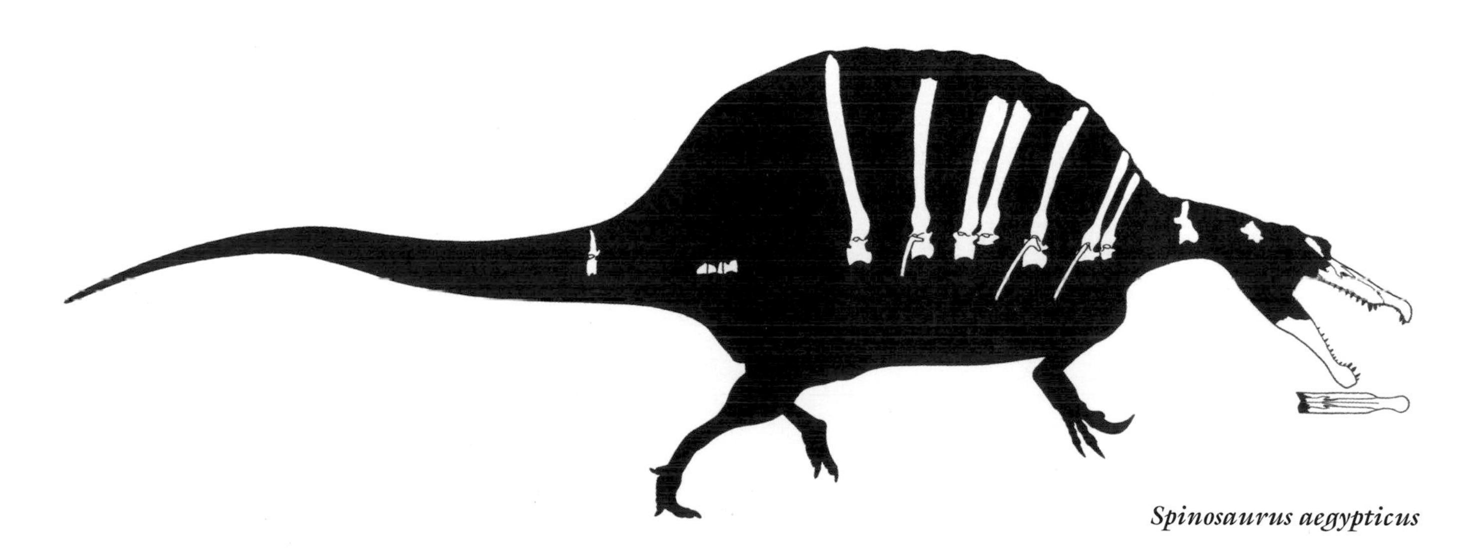
Spinosaurus aegypticus

AVETHEROPODS

SMALL TO GIGANTIC PREDATORY AND HERBIVOROUS TETANURANS FROM THE MIDDLE JURASSIC TO THE END OF THE DINOSAUR ERA (WITH BIRDS SURVIVING BEYOND), ON MOST CONTINENTS.

ANATOMICAL CHARACTERISTICS Highly variable. Extra joint in lower jaw usually better developed. Arm very long to very reduced. Birdlike respiratory system highly developed. Brains reptilian to avian.
NOTES Absence from Mesozoic Antarctica probably reflects lack of sufficient sampling.

ALLOSAUROIDS

LARGE TO GIGANTIC PREDATORY AVETHEROPODS APPROACHING 10 TONNES FROM THE MIDDLE JURASSIC TO THE EARLY LATE CRETACEOUS, ON MOST CONTINENTS.

ANATOMICAL CHARACTERISTICS Moderately variable. Conventional avetheropod form. Head only moderately robustly built and moderately muscled, not very broad, orbital hornlet modest to well developed, bladed teeth not very large. Tail long. Arm length medium to short. Leg moderately long. Brains reptilian.
HABITS Ambush and pursuit predators. Heads and arms used as weapons. Extreme size of some species indicates that adults hunted adult as well as younger sauropods, armored ornithischians, and large ornithopods using heads and long tooth rows powered by powerful neck muscles to dispatch victims with slashing bites intended to cripple prey so it could be safely consumed. Arms used to handle and control prey when necessary. Adult prey included sauropods, stegosaurs, ankylosaurs, ornithopods; juveniles focused on juveniles and smaller game.
NOTES Fragmentary remains imply presence in Australia. If megaraptorids are allosauroids rather than tyrannosaurids, as some research indicates, then this group survived until the end of the dinosaur era. The standard big theropod type. Group has been labeled carnosaurs.

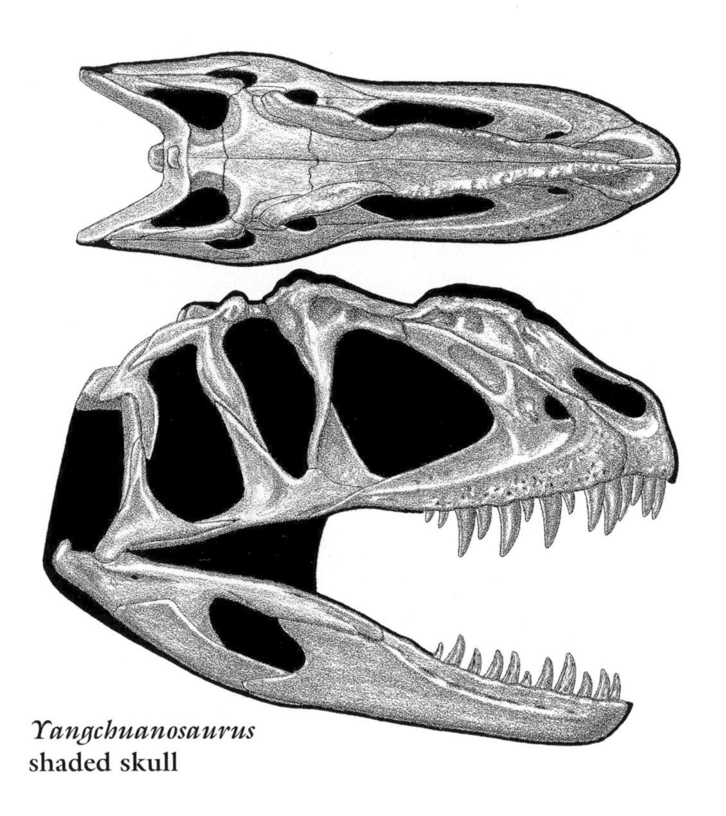

Yangchuanosaurus **shaded skull**

Allosaurus **muscle study**

Yangchuanosaurids Large to gigantic allosauroids limited to the Middle and Late Jurassic of Eurasia.

ANATOMICAL CHARACTERISTICS Uniform. Remnant of fourth finger present.
NOTES Relationships within group and naming of group are uncertain. Incomplete Asian, Middle Jurassic, juvenile *Gasosaurus constructus* and fragmentary European, Late Jurassic *Metriacanthosaurus parkeri* may belong to this group. Fragmentary *Siamotyrannus isanensis* may or may not indicate survival of the group into Early Cretaceous.

Yangchuanosaurus? zigongensis
Adult size uncertain

FOSSIL REMAINS Minority of two skeletons.
ANATOMICAL CHARACTERISTICS Insufficient information.
AGE Middle Jurassic, Bathonian or Callovian.
DISTRIBUTION AND FORMATION/S Central China; Shaximiao.
HABITAT Heavily forested.
HABITS Arms probably important in handling prey.
NOTES Originally placed in *Szechuanosaurus*, which is based on inadequate remains; the identity of these remains is uncertain. Shared its habitat with *Xuanhanosaurus* and *Gasosaurus*.

Yangchuanosaurus (= Sinraptor) dongi
8 m (26 ft) TL, 1.3 tonnes

FOSSIL REMAINS Complete skulls and majority of a skeleton.
ANATOMICAL CHARACTERISTICS Snout ridges not well developed.
AGE Late Jurassic, early Oxfordian.
DISTRIBUTION AND FORMATION/S Northwest China; upper Shishugou.
HABITS Prey included mamenchisaur sauropods and stegosaurs.
NOTES This species barely differs from *Y. shangyuensis.*

Yangchuanosaurus shangyuensis
11 m (35 ft) TL, 3 tonnes

FOSSIL REMAINS A few complete skulls and the majority of some skeletons, completely known.
ANATOMICAL CHARACTERISTICS Snout ridges well developed.
AGE Late Jurassic, probably Oxfordian.
DISTRIBUTION AND FORMATION/S Central China; Shangshaximiao.

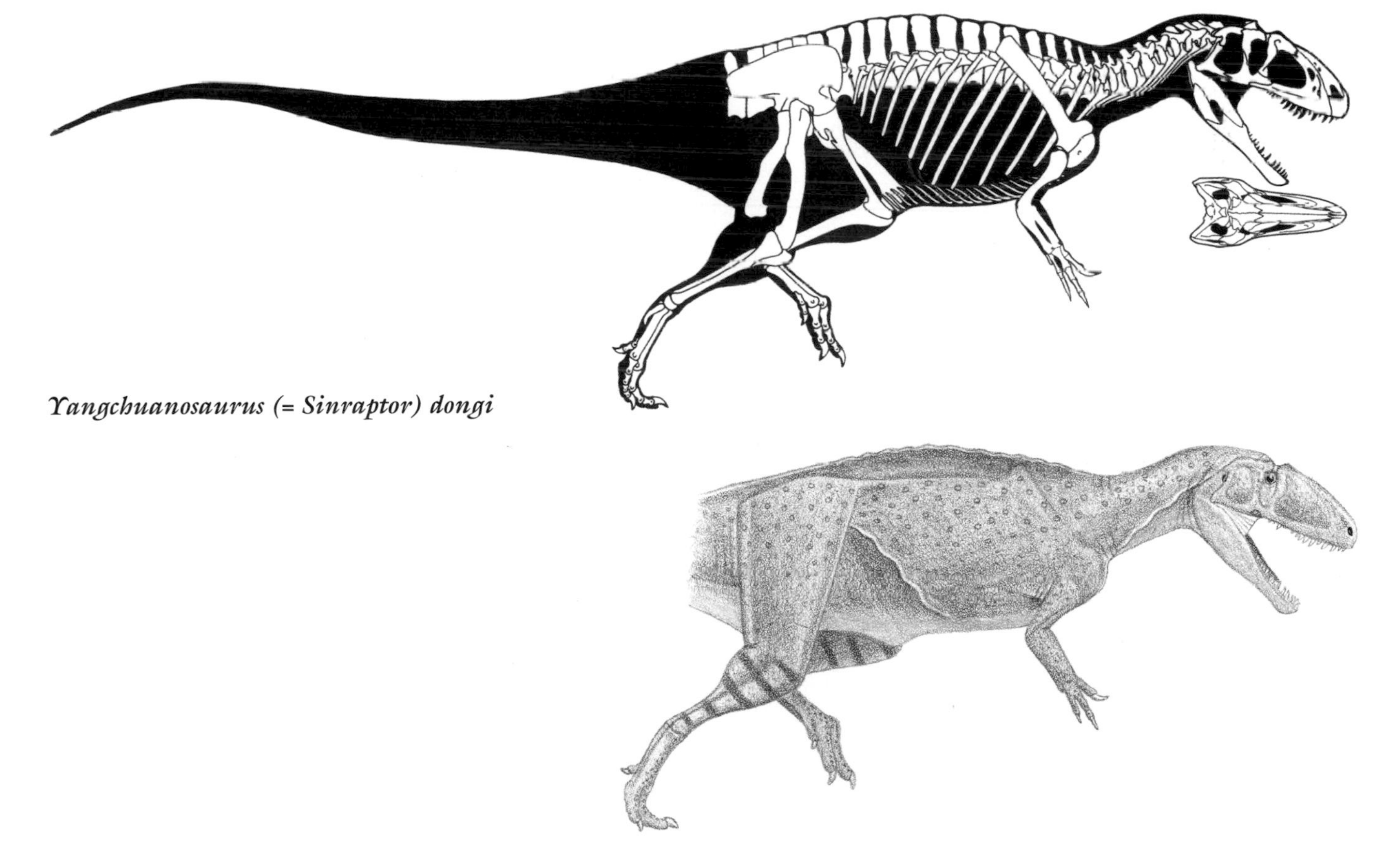

Yangchuanosaurus (= Sinraptor) dongi

Yangchuanosaurus shangyuensis

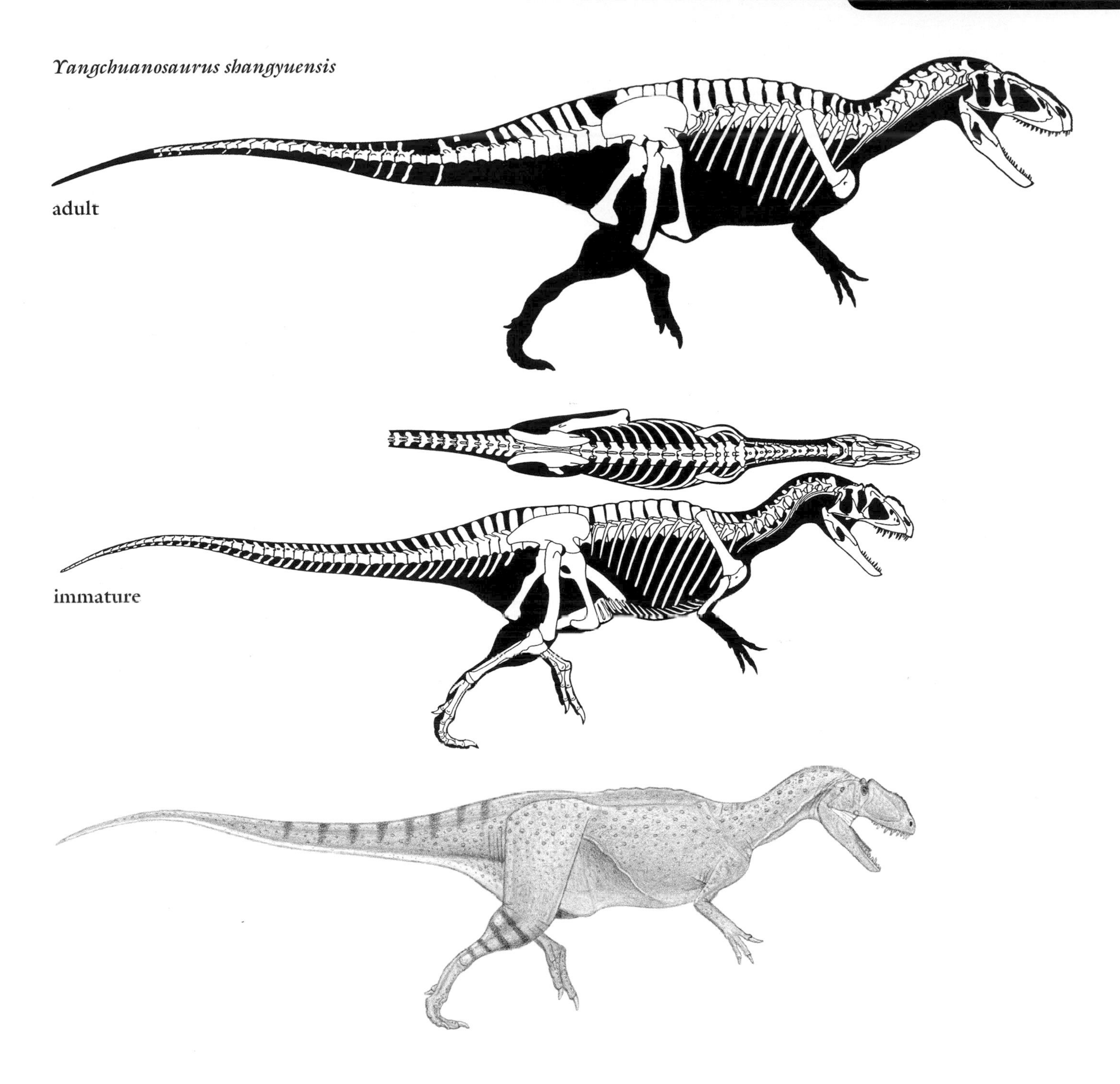

HABITAT Heavily forested.
HABITS Prey included mamenchisaur sauropods and stegosaurs.
NOTES From the same formation, very similar and progressively larger in size, *Y. hepingensis*, *Y. shangyuensis*, and *Y. magnus* appear to form a progressive growth series within a single species. It is possible that the modest-sized, fairly complete but poorly described *Leshansaurus qianweiensis* from this formation is a juvenile of this species, or it could be an afrovenator.

ALLOSAURIDS Large to gigantic allosauroids limited to the Late Jurassic, in North America, Europe, and Africa.

ANATOMICAL CHARACTERISTICS Uniform. Head not especially large, back of head more rigidly braced, triangular and sharp-tipped brow hornlets present. Tail long. Boot of pubis large. Fourth finger entirely lost.
ONTOGENY Growth rates moderately rapid, adult size reached in about two decades; life span normally not exceeding three decades.

Allosaurus fragilis
8.5 m (28 ft) TL, 1.7 tonnes

FOSSIL REMAINS At least one nearly complete skull and skeleton, possibly more skulls, skeletons, and elements.
ANATOMICAL CHARACTERISTICS Head rather short, deep, and subtriangular, orbital hornlet large, subtriangular. Arm large at least in adults.
AGE Late Jurassic, late Oxfordian and early Kimmeridgian.
DISTRIBUTION AND FORMATION/S Colorado, Utah; lower Morrison.
HABITAT Short wet season, otherwise semiarid with open floodplain prairies and riverine forests.
HABITS Normally hunted smaller individual camarasaurs, diplodocines, and apatosaurs as well as stegosaurs and camptosaurs.
NOTES Despite *Allosaurus* being one of the best-known dinosaurs, its taxonomy has not been well studied, and the remains that the genus *Allosaurus* and its species are based on are not adequate, so taxonomic designations are uncertain. All Morrison *Allosaurus* have usually been lumped into this species, but there is considerable diversity among the specimens, especially in the length/height ratio of the skull, and it is improbable that any one species spanned the 7 million years or more of the Morrison. A complete skull and skeleton from lower in the Morrison may be a juvenile of *A. fragilis*, or a new species with smaller arms. *A. lucasi* is based on inadequate remains.

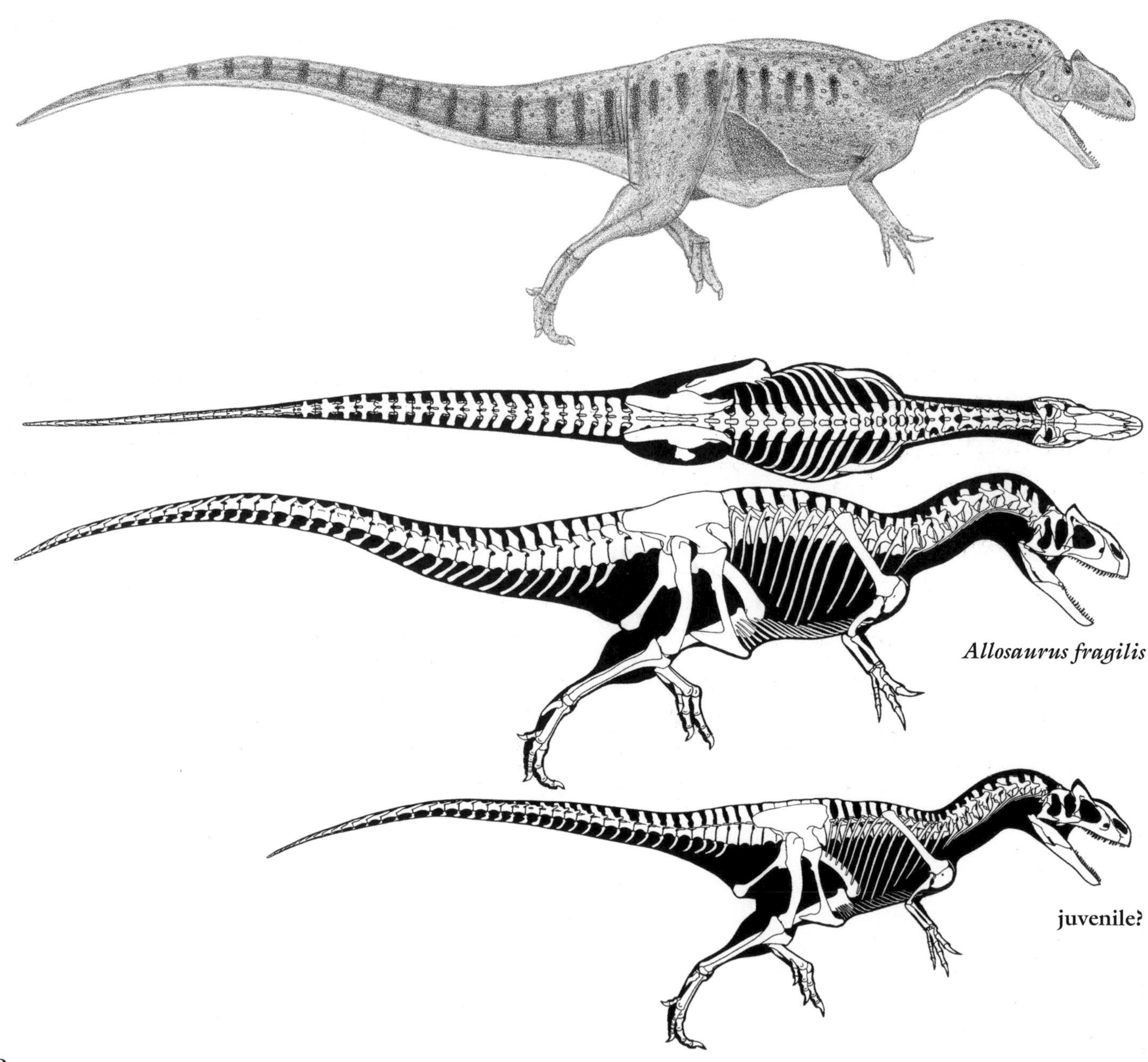

Allosaurus unnamed species

8.5 m (28 ft) TL, 1.7 tonnes

FOSSIL REMAINS A large number of complete and partial skulls and skeletons.
ANATOMICAL CHARACTERISTICS Skull long, shallow, and subrectangular, orbital hornlet large, subtriangular. Arm large.
AGE Late Jurassic, late Kimmeridgian to middle Tithonian.
DISTRIBUTION AND FORMATION/S Utah, Wyoming, Colorado; middle Morrison.
HABITAT Short wet season, otherwise semiarid with open floodplain prairies and riverine forests.
HABITS Normally hunted smaller individual camarasaurs, diplodocines, and apatosaurines as well as stegosaurs and camptosaurs.
NOTES This has been placed in *A. atrox*, which is based on inadequate remains. By far the most common theropod in the Morrison, some *Allosaurus* species shared their habitats with *Ceratosaurus* and *Torvosaurus*. There may be more than one *Allosaurus* species in the middle Morrison. The classic nontyrannosaur large theropod.

Allosaurus (or *Saurophaganax*) *maximus*

10.5 m (35 ft) TL, 3 tonnes

FOSSIL REMAINS Minority of the skeleton.
ANATOMICAL CHARACTERISTICS Insufficient information.
AGE Late Jurassic, middle Tithonian.
DISTRIBUTION AND FORMATION/S Oklahoma; upper Morrison.
HABITAT Wetter than earlier Morrison, otherwise semiarid with open floodplain prairies and riverine forests.
HABITS Able to hunt larger sauropods.
NOTES Not enough of the skeleton is known to decide whether it is a giant *Allosaurus* or a distinct genus, as some details imply; may be the descendant of one of the earlier Morrison *Allosaurus* species.

Allosaurus europaeus?

7 m (23 ft) TL, 1 tonne

FOSSIL REMAINS Partial skull and minority of skeleton.
ANATOMICAL CHARACTERISTICS Orbital hornlet large, subtriangular.
AGE Late Jurassic, late Kimmeridgian or early Tithonian.
DISTRIBUTION AND FORMATION/S Portugal; Lourinha.

Allosaurus europaeus?

Allosaurus unnamed species

HABITAT Large, seasonally dry island with open woodlands.
NOTES At this time the European archipelago was very close to North America, and whether this is distinct from all known Morrison *Allosaurus* species is uncertain.

Lourinhanosaurus (or *Allosaurus*) *antunesi*
Adult size uncertain

FOSSIL REMAINS Minority of skeleton, possibly juvenile.
ANATOMICAL CHARACTERISTICS Insufficient information.
AGE Late Jurassic, late Kimmeridgian or Tithonian.
DISTRIBUTION AND FORMATION/S Portugal; Amoreira-Porto Novo.
HABITAT Large, seasonally dry island with open woodlands.

CARCHARODONTOSAURIDS Large to gigantic allosauroids of the Cretaceous of the Western Hemisphere, Eurasia, and Africa.

ANATOMICAL CHARACTERISTICS Fairly variable. Boot of pubis further enlarged. Arm reduced.
HABITS Arms used less when hunting than in other allosauroids.
NOTES Fragmentary Early Cretaceous *Datanglong guangxiensis* may indicate group's presence in Asia. Absence from additional continents may reflect lack of sufficient sampling.

Concavenator corcovatus
5 m (16 ft) TL, 320 kg (700 lb)

FOSSIL REMAINS Nearly complete skull and skeleton.
ANATOMICAL CHARACTERISTICS Vertebral spines immediately in front of and behind pelvis tall, forming double sail back, especially in front of hips. Bumps on trailing edge of upper arm indicate large quills.
AGE Early Cretaceous, late Barremian.
DISTRIBUTION AND FORMATION/S Spain; Calizas de la Huerguina.
NOTES Possible quill nodes imply presence of feathery structures in allosauroids.

Acrocanthosaurus atokensis
11 m (35 ft) TL, 4.4 tonnes

FOSSIL REMAINS Complete skull and majority of skeletons.
ANATOMICAL CHARACTERISTICS Back of lower jaw deep. Tall vertebral spines from neck to tail form a low sail.

Concavenator corcovatus

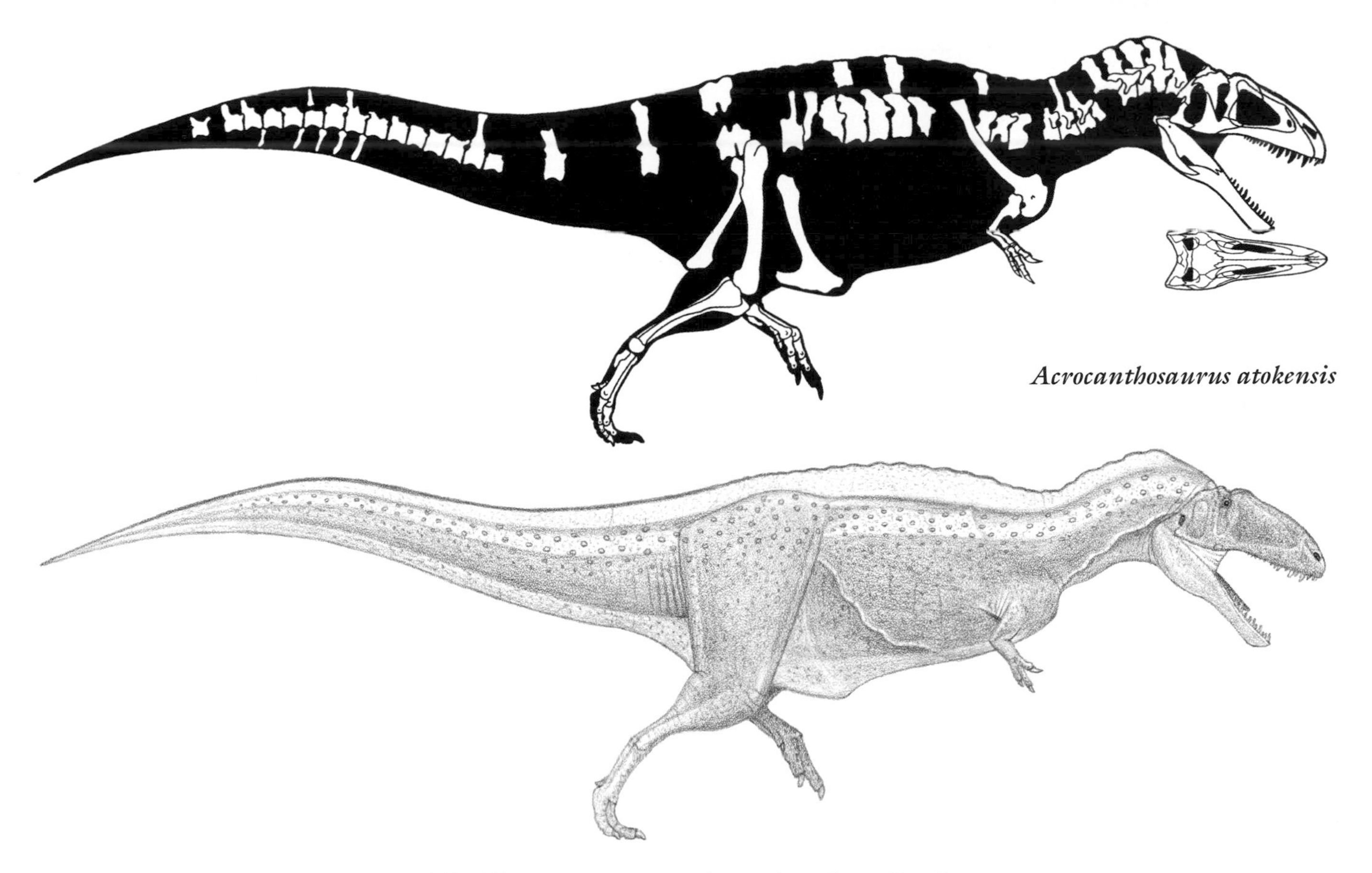
Acrocanthosaurus atokensis

AGE Early Cretaceous, Aptian to middle Albian.
DISTRIBUTION AND FORMATION/S Oklahoma, Texas; Antlers, Twin Mountains.
HABITAT Floodplain with coastal swamps and marshes.
HABITS Prey included *Sauroposeidon*.
NOTES Researchers disagree as to whether this is an allosaurid, a carcharodontosaurid, or its own group.

Eocarcharia dinops
Adult size uncertain

FOSSIL REMAINS Minority of skull.
ANATOMICAL CHARACTERISTICS Insufficient information.
AGE Early Cretaceous, Aptian.
DISTRIBUTION AND FORMATION/S Niger; Elrhaz, level uncertain.
NOTES The one specimen may be a large juvenile. Shared its habitat with *Kryptops*.

Tyrannotitan chubutensis
13 m (42 ft) TL, 7 tonnes

FOSSIL REMAINS Minority of skull and skeleton.
ANATOMICAL CHARACTERISTICS Vertebral spines over tail rather tall.
AGE Early Cretaceous, Aptian.
DISTRIBUTION AND FORMATION/S Southern Argentina; Cerro Barcino.
NOTES Among the largest avepods, prey included *Chubutisaurus*.

Carcharodontosaurus saharicus
12 m (40 ft) TL, 6 tonnes

FOSSIL REMAINS Partial skull and parts of skeletons.
ANATOMICAL CHARACTERISTICS Standard for group.
AGE Late Cretaceous, early Cenomanian.
DISTRIBUTION AND FORMATION/S Egypt, possibly Morocco and other parts of North Africa; Bahariya, upper Kem Kem Beds, etc.
HABITAT Coastal mangroves.

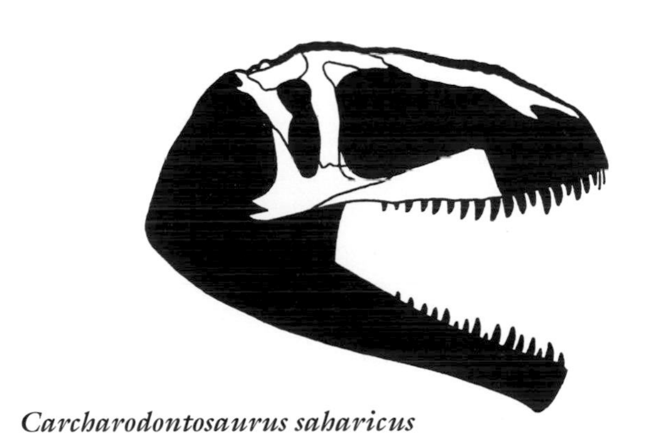
Carcharodontosaurus saharicus

HABITS Prey included *Paralititan*.
NOTES Whether specimens from a large number of formations actually belong to this species is problematic. Shared its habitat with the even more gigantic but less powerful *Spinosaurus aegypticus*.

Carcharodontosaurus iguidensis
10 m (34 ft) TL, 4 tonnes

FOSSIL REMAINS Minority of several skulls and small portion of skeleton.
ANATOMICAL CHARACTERISTICS Standard for group.
AGE Late Cretaceous, early Cenomanian.
DISTRIBUTION AND FORMATION/S Niger; Echkar.
NOTES Was placed in *C. saharicus*. Shared its habitat with *Rugops primus* and a large semiterrestrial crocodilian.

Giganotosaurus (or *Carcharodontosaurus*) *carolinii*
13–14 m (42–45 ft) TL, 7–8 tonnes

FOSSIL REMAINS Majority of skull and skeleton.
ANATOMICAL CHARACTERISTICS Standard for group.
AGE Late Cretaceous, early Cenomanian.
DISTRIBUTION AND FORMATION/S Western Argentina; Candeleros.
HABITAT Short wet season, otherwise semiarid with open floodplains and riverine forests.
HABITS Prey of this giant included the whale-sized titanosaur *Andesaurus*, among other sauropods.
NOTES The incomplete skulls of carcharodontosaurs have been restored with too great a length. Shared its habitat with *Ekrixinatosaurus*. Rivals *Spinosaurus aegypticus* as the largest known theropod.

Mapusaurus roseae
11.5 m (38 ft) TL, 5 tonnes

FOSSIL REMAINS Large number of skull and skeletal bones.
ANATOMICAL CHARACTERISTICS Standard for group.
AGE Late Cretaceous, middle Cenomanian.
DISTRIBUTION AND FORMATION/S Western Argentina; lower Huincul.
HABITAT Short wet season, otherwise semiarid with open floodplains and riverine forests.
NOTES Shared its habitat with *Ilokelesia*.

Shaochilong maortuensis
Adult size uncertain

FOSSIL REMAINS Partial skeleton.
ANATOMICAL CHARACTERISTICS Insufficient information.
AGE Late Cretaceous, Turonian.
DISTRIBUTION AND FORMATION/S Northern China; Ulansuhai.
NOTES Shared its habitat with *Chilantaisaurus*.

NEOVENATORIDS Medium-sized to gigantic allosauroids of the Cretaceous of Eurasia.

ANATOMICAL CHARACTERISTICS Fairly variable. Arm well developed.

Neovenator salerii
7 m (23 ft) TL, 1 tonne

FOSSIL REMAINS Minority of skull and skeleton.
ANATOMICAL CHARACTERISTICS Lightly built, leg long. Head narrow.
AGE Early Cretaceous, Barremian.
DISTRIBUTION AND FORMATION/S Isle of Wight, England; Wessex.
HABITS Prey included armored ankylosaurs, sauropods.
NOTES That researchers have disagreed whether this is a basal tyrannosauroid or an allosauroid reinforces the close relationship of the two groups. Shared its habitat with smaller *Eotyrannus* and *Aristosuchus*.

Chilantaisaurus tashuikouensis
11 m (35 ft) TL, 4 tonnes

FOSSIL REMAINS Partial skeleton.
ANATOMICAL CHARACTERISTICS Skeleton heavily constructed. Arm well developed.
AGE Late Cretaceous, Turonian.
DISTRIBUTION AND FORMATION/S Northern China; Ulansuhai.
HABITS Prey included *Gobisaurus*.
NOTES Shared its habitat with *Shaochilong*, the last of the known allosauroids, unless megaraptorids are allosauroids.

Giganotosaurus (or *Carcharodontosaurus*) *carolinii*

COELUROSAURS

SMALL TO GIGANTIC PREDATORY AND HERBIVOROUS AVETHEROPODS OF THE MIDDLE JURASSIC TO THE END OF THE DINOSAUR ERA (WITH BIRDS SURVIVING BEYOND), ON MOST CONTINENTS.

ANATOMICAL CHARACTERISTICS Highly variable. Tail long to very short. Arm from longer than leg to severely reduced. Leg extremely gracile to robust, toes four to three. Feather fibers often preserved.
HABITS Extremely variable, from big game predators to fully herbivorous.

TYRANNOSAUROIDS

SMALL TO GIGANTIC PREDATORY AVETHEROPODS OF THE MIDDLE JURASSIC TO THE END OF THE DINOSAUR ERA, ALL CONTINENTS EXCEPT ANTARCTICA.

ANATOMICAL CHARACTERISTICS In most regards form conventional for avepod theropods. Front teeth of upper jaw D-shaped in cross section. Arm long to severely reduced. Leg long. Brains reptilian.
HABITS Pursuit and ambush predators. Dispatched victims with powerful, deep, punch-like bites rather than slashing, wounding bites.
NOTES Proving to be long-lived and widely distributed. Although tyrannosauroids are not descended directly from allosauroids, the two groups may share a close common ancestor.

BASO-TYRANNOSAUROIDS Small to gigantic tyrannosauroids of the Middle Jurassic to the end of the dinosaur era.

ANATOMICAL CHARACTERISTICS Fairly variable. Arm not reduced. Leg not as gracile as those of tyrannosaurs.
HABITS Arms as well as head used to handle and wound prey.

Proceratosaurus bradleyi
3–4 m (10–13 ft) TL, 50–100 kg (100–200 lb)

FOSSIL REMAINS Majority of skull.
ANATOMICAL CHARACTERISTICS Head subrectangular, snout fairly deep and adorned with nasal hornlet or crest, back of head rigidly built, teeth fairly large.
AGE Middle Jurassic, middle Bathonian.
DISTRIBUTION AND FORMATION/S Central England; Forest Marble.

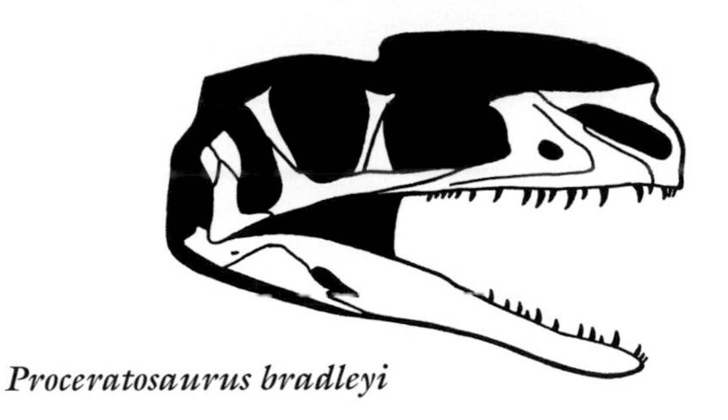

Proceratosaurus bradleyi

HABITS Crest too delicate for head butting; probably for display within the species.
NOTES Name incorrectly implies an ancestral relationship with the very different *Ceratosaurus*. This, *Guanlong*, *Sinotyrannus*, and *Juratyrant* may form family Proceratosauridae.

Guanlong wucaii (Illustration overleaf)
3.5 m (11 ft) TL, 125 kg (250 lb)

FOSSIL REMAINS Nearly complete skull and partial skeleton.
ANATOMICAL CHARACTERISTICS Snout ridges united and enlarged into an enormous midline crest with a backward projection.
AGE Middle Jurassic, late Callovian.
DISTRIBUTION AND FORMATION/S Northwest China; lower Shishugou.
HABITS Crest too delicate for head butting; probably for display within the species.
NOTES Shared its habitat with *Monolophosaurus*.

Sinotyrannus kazuoensis
9 m (30 ft) TL, 2.5 tonnes

FOSSIL REMAINS Partial skull.
AGE Early Cretaceous, early or middle Aptian.
DISTRIBUTION AND FORMATION/S Northeast China; Jiufotang.
HABITAT Well-watered forests and lakes, winters chilly with some snow.

Juratyrant langhami
5 m (16 ft) TL, 300 kg (600 lb)

FOSSIL REMAINS Partial skeleton.
ANATOMICAL CHARACTERISTICS Lightly built.
AGE Late Jurassic, early Tithonian.
DISTRIBUTION AND FORMATION/S Southern England; Kimmeridge Clay.
NOTES Originally placed in the smaller baso-tyrannosaurid *Stokesosaurus*, the original species of which, *S. clevelandi* from the Late Jurassic of western North America, is based on one bone.

Guanlong wucaii

Aviatyrannis jurassica

1 m (3 ft) TL, 4 kg (10 lb)

FOSSIL REMAINS Small portion of skeleton, possibly juvenile.
ANATOMICAL CHARACTERISTICS Insufficient information.
AGE Late Jurassic, Kimmeridgian.
DISTRIBUTION AND FORMATION/S Portugal; Camadas de Alcobaca.
HABITAT Large, seasonally dry island with open woodlands.
NOTES Shared its habitat with *Lourinhasaurus.*

Dilong paradoxus

1.3 m (4 ft) TL, 6 kg (13 lb)

FOSSIL REMAINS A few nearly complete skulls and partial skeletons, external fibers.
ANATOMICAL CHARACTERISTICS Head long and low, low Y-shaped crest on snout. Hand fairly long. Leg very long. Full extent of protofeather covering uncertain.
AGE Early Cretaceous, Barremian.
DISTRIBUTION AND FORMATION/S Northeast China; lower Yixian.
HABITAT Well-watered forests and lakes, winters chilly with some snow.
HABITS Shallow, broad crest probably used in a frontal display, rather than the side display used by dinosaurs with deeper crests.

Yutyrannus huali

7.5 m (25 ft) TL, 1.1 tonnes

FOSSIL REMAINS Two nearly complete skulls and skeletons, external fibers.
ANATOMICAL CHARACTERISTICS Fairly robustly built.

Dilong paradoxus

Yutyrannus huali

Snout fairly deep. Arm long. External fibers apparently covered much of body including upper foot.
AGE Early Cretaceous, early Aptian.
DISTRIBUTION AND FORMATION/S Northeast China; Yixian.
HABITAT Well-watered forests and lakes, winters chilly with some snow.
HABITS Prey included *Jianchangosaurus*, *Beipiaosaurus*, *Dongbeititan*, and *Hongshanosaurus*.

Eotyrannus lengi

3 m (10 ft) TL, 70 kg (150 lb)

FOSSIL REMAINS Minority of skull and skeleton.
ANATOMICAL CHARACTERISTICS Skull strongly built, front of head deep. Skeleton lightly built. Arm long. Leg long and gracile.
AGE Early Cretaceous, Barremian.
DISTRIBUTION AND FORMATION/S Isle of Wight, England; Wessex.

Bagaraatan ostromi

Adult size uncertain

FOSSIL REMAINS Minority of skeleton, immature.
ANATOMICAL CHARACTERISTICS Lightly constructed. Tail stiffened.
AGE Late Cretaceous, late Campanian and/or early Maastrichtian.
DISTRIBUTION AND FORMATION/S Mongolia; Nemegt.
HABITAT Well-watered woodland with seasonal rain, winters cold.

Fukuiraptor kitadaniensis

5 m (16 ft) TL, 300 kg (600 lb)

FOSSIL REMAINS Partial skeleton.
ANATOMICAL CHARACTERISTICS Lightly built, leg long.
AGE Early Cretaceous, Albian.
DISTRIBUTION AND FORMATION/S Main island Japan; Kitadani.

Australovenator wintonensis

6 m (20 ft) TL, 500 kg (1,000 lb)

FOSSIL REMAINS Minority of skull and skeleton.
ANATOMICAL CHARACTERISTICS Lightly built, leg long.
AGE Early Cretaceous, latest Albian.
DISTRIBUTION AND FORMATION/S Northeast Australia; Winton.
HABITAT Well-watered, cold winters with heavy snows.

Megaraptor namunhuaiquii
8 m (25 ft) TL, 1 tonne

FOSSIL REMAINS Minority of a few skeletons.
ANATOMICAL CHARACTERISTICS Hand claws slender.
AGE Late Cretaceous, late Turonian.
DISTRIBUTION AND FORMATION/S Western Argentina; Portezuelo.
HABITAT Well-watered woodlands with short dry season.
HABITS Prey included *Macrogryphosaurus.*
NOTES Incorrectly thought to be the biggest dromaeosaurid; others consider this a spinosaur or an allosauroid. This, *Fukuiraptor*, *Australovenator*, *Aerosteon*, and *Orkoraptor* may form family Megaraptoridae.

Aerosteon riocoloradensis
6 m (20 ft) TL, 500 kg (1,000 lb)

FOSSIL REMAINS Minority of skull and partial skeleton.
ANATOMICAL CHARACTERISTICS Lightly built, leg long.
AGE Late Cretaceous, late Santonian and/or early Campanian.
DISTRIBUTION AND FORMATION/S Western Argentina; Anacleto.
HABITS Prey included *Gasparinisaura.*
NOTES Shared its habitat with *Abelisaurus.*

Orkoraptor burkei
6 m (20 ft) TL, 500 kg (1,000 lb)

FOSSIL REMAINS Minority of skull and skeleton.
ANATOMICAL CHARACTERISTICS Insufficient information.
AGE Late Cretaceous, early Maastrichtian.
DISTRIBUTION AND FORMATION/S Southern Argentina; Pari Aike.
HABITS Prey included *Talenkauen.*
NOTES If megaraptorids were allosauroids, as some researchers indicate, then *Orkoraptor* indicates that allosauroids lasted until close to and probably up to the end of the dinosaur era.

Santanaraptor placidus
1.5 m (5 ft) TL, 15 kg (30 lb)

FOSSIL REMAINS Minority of skeleton.
ANATOMICAL CHARACTERISTICS Insufficient information.
AGE Early Cretaceous, probably Albian.
DISTRIBUTION AND FORMATION/S Eastern Brazil; Santana.
NOTES Found as drift in marine deposits.

Xiongguanlong baimoensis
5 m (15 ft) TL, 200 kg (450 lb)

FOSSIL REMAINS Majority of a distorted skull and minority of skeleton.
ANATOMICAL CHARACTERISTICS Head and especially snout long, low.
AGE Early Cretaceous, probably Aptian or Albian.
DISTRIBUTION AND FORMATION/S Central China; lower Xinminpu.
NOTES Shows that some tyrannosauroids were fairly large in the Mid-Cretaceous. Prey included *Beishanlong.*

Dryptosaurus aquilunguis
7.5 m (25 ft) TL, 1.5 tonnes

FOSSIL REMAINS Minority of skeleton.
ANATOMICAL CHARACTERISTICS Arm and finger claws large.
AGE Late Cretaceous, late Campanian or early Maastrichtian.
DISTRIBUTION AND FORMATION/S New Jersey; Marshalltown.
HABITS Arms used as weapons. Prey included hadrosaurs.
NOTES Found as drift in marine deposits.

Labocania anomala
7 m (23 ft) TL, 1.5 tonnes

FOSSIL REMAINS Small portion of skull and skeleton.
ANATOMICAL CHARACTERISTICS Massively constructed.
AGE Late Cretaceous, probably Campanian.
DISTRIBUTION AND FORMATION/S Baja California, Mexico; La Bocana Roja.
HABITS Ambush big game hunter.

TYRANNOSAURIDS Large to gigantic tyrannosauroids, limited to the later Late Cretaceous of North America and Asia.

ANATOMICAL CHARACTERISTICS Fairly uniform, but juveniles and smaller species gracile, large adults more robust. Head large and long, robustly constructed, stout bars in the temporal region invade the side openings and further strengthen the skull, skulls of juveniles and smaller species very long, shallow, and graceful, those of adults deeper and shorter snouted, midline ridge on snout rugose, probably bore low ridge boss, small brow hornlets or bosses over orbits, back half of skull a broad box accommodating exceptionally powerful jaw muscles, eyes face partly forward, and some degree of stereo vision possible, front of snout broader and more rounded than usual, supporting a U-shaped arc of teeth that are D-shaped in cross section, teeth stouter and more conical than in general, lower jaw deep, especially back half. Neck strongly constructed, powerfully muscled. Trunk short and deep. Tail shorter and lighter than standard in other large theropods. Arm severely reduced in size, outer finger severely reduced to only two developed

fingers, yet hands still functional. Pelvis very large and leg very long, so leg muscles exceptionally well developed, foot very long and strongly compressed from side to side. Reduction of tail and arm in favor of enlarged and elongated leg indicates greater speed potential than in other giant theropods. Scales small and pebbly, may have been mixed with fibers. Skeletons of juveniles very gracile, becoming increasingly robust as size increases, but basic characteristics unaltered. Brains larger than usual in large theropods, olfactory bulbs especially large.
ONTOGENY Growth rates moderately rapid, adult size reached in about two decades; life span normally not exceeding three decades. Some small species that have been named are the juveniles of giant taxa; whether any species were small as adults is uncertain.
HABITS Pursuit and perhaps ambush predators; able to chase running prey at unusually high speeds. Head the primary if not sole weapon. Long snouts of juveniles imply they were independent hunters. Smaller individuals probably hunted swift ornithomimids and ornithopods as well as protoceratopsians, pachycephalosaurs, juvenile hadrosaurs, and ceratopsians. Giant adults preyed on hadrosaurs and ankylosaurs in all known habitats, as well as ceratopsians and titanosaur sauropods where available, using their massive heads and strong teeth to dispatch victims with powerful, deep, punch-like bites rather than slashing, wounding bites aimed with forward vision, powered by very strong jaw and neck muscles, and intended to cripple prey so it could be safely consumed. Function of arms poorly understood: they appear too short and small to be useful in handling prey; may have provided grip for males while mating. Head bosses presumably for head butting during intraspecific contests.

NOTES Overall the most advanced and sophisticated of large theropods. Large numbers of hunting juveniles may have swamped their habitats, suppressing the populations of smaller theropods such as dromaeosaurids and troodontids.

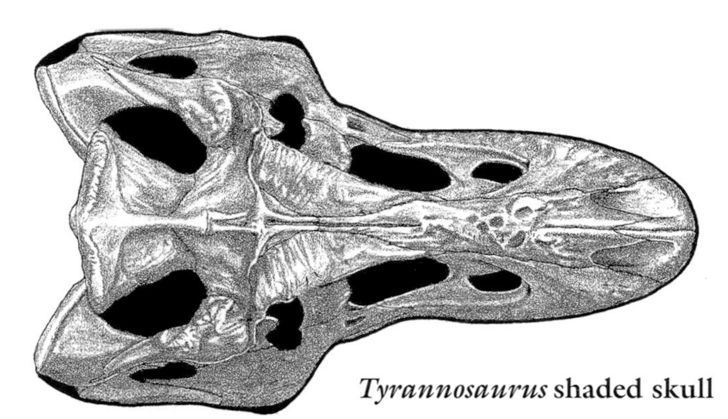

Tyrannosaurus shaded skull

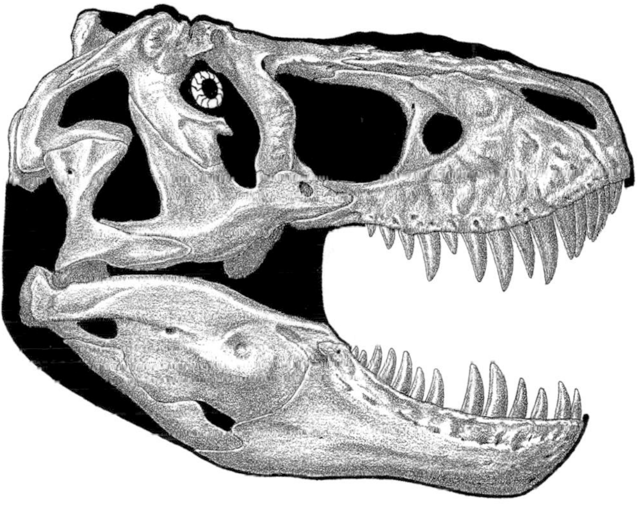

Tyrannosaurus muscle study

Alectrosaurus olseni
Adult size uncertain

FOSSIL REMAINS Partial skull, skeleton, possibly immature.
ANATOMICAL CHARACTERISTICS Typically gracile for smaller tyrannosaurids.
AGE Late Late Cretaceous.
DISTRIBUTION AND FORMATION/S China; Iren Dabasu.
HABITAT Seasonally wet-dry woodlands.
HABITS Assuming the known specimens are adults, pursued similar-sized dinosaurs including the fastest species.
NOTES Prey included *Archaeornithomimus.*

Alioramus remotus
Adult size uncertain

FOSSIL REMAINS Skull, some parts of skeleton, possibly immature.
ANATOMICAL CHARACTERISTICS Typically gracile for smaller tyrannosaurids. Crenulated midline crest on snout.
AGE Late Late Cretaceous.
DISTRIBUTION AND FORMATION/S Mongolia; Nogoon Tsav.
NOTES This and *Qianzhousaurus* form subfamily Alioramini.

Alioramus altai
Adult size uncertain

FOSSIL REMAINS Immature skulls, some parts of skeleton.
ANATOMICAL CHARACTERISTICS Skull unusually long and low even for a tyrannosaur of this size. Crenulated midline crest on snout.
AGE Late Cretaceous, early Maastrichtian.
DISTRIBUTION AND FORMATION/S Mongolia: Nemegt.
NOTES Thought to be somewhat different in time from *A. remotus*; if not, may be the same species. Competed with juvenile *T. bataar.*

Qianzhousaurus sinensis
Adult size uncertain

FOSSIL REMAINS Nearly complete skull and minority of skeleton.
ANATOMICAL CHARACTERISTICS Snout unusually slender even for a tyrannosaur of this size.
AGE Late Cretaceous, early Maastrichtian.
DISTRIBUTION AND FORMATION/S Northern China; Yuanpu.
HABITS Prey included *Nanshiungosaurus.*

Appalachiosaurus montgomeriensis
Adult size uncertain

FOSSIL REMAINS Partial skull, skeleton.
ANATOMICAL CHARACTERISTICS Typically gracile for smaller tyrannosaurids.
AGE Late Cretaceous, early Campanian.
DISTRIBUTION AND FORMATION/S Alabama; Dermopolis.

Albertosaurus (Gorgosaurus) libratus
8 m (27 ft) TL, 2.5 tonnes

FOSSIL REMAINS A number of skulls and skeletons from juvenile to adult, small skin patches, completely known.
ANATOMICAL CHARACTERISTICS A standard giant tyrannosaur. Brow hornlets fairly prominent. Skeleton not heavily built.

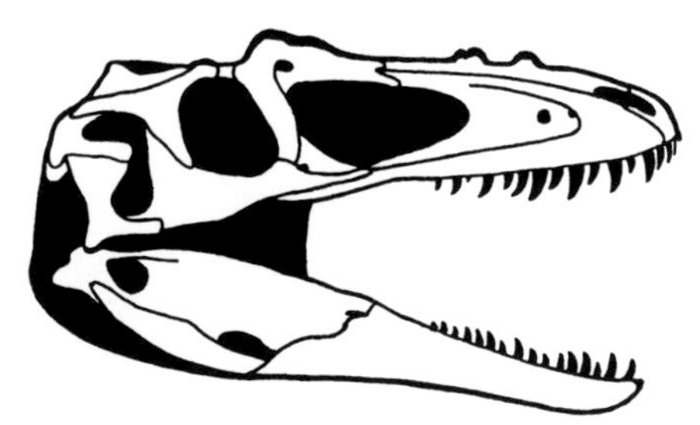

Alioramus remotus

Qianzhousaurus sinensis

AGE Late Cretaceous, late Campanian.
DISTRIBUTION AND FORMATION/S Alberta, Montana?; at least middle Dinosaur Park, possibly Judith River and Upper Two Medicine.
HABITAT Well-watered, forested floodplain with coastal swamps and marshes, cool winters, uplands drier.
HABITS Relatively gracile build indicates adults specialized in hunting unarmed hadrosaurs, although ceratopsians and ankylosaurs were probably occasional victims.
NOTES A separate genus according to some; very similar to *Albertosaurus sarcophagus*. Whether *A. libratus* lived through the entire time span of the Dinosaur Park Formation is uncertain.

Albertosaurus (Gorgosaurus) libratus

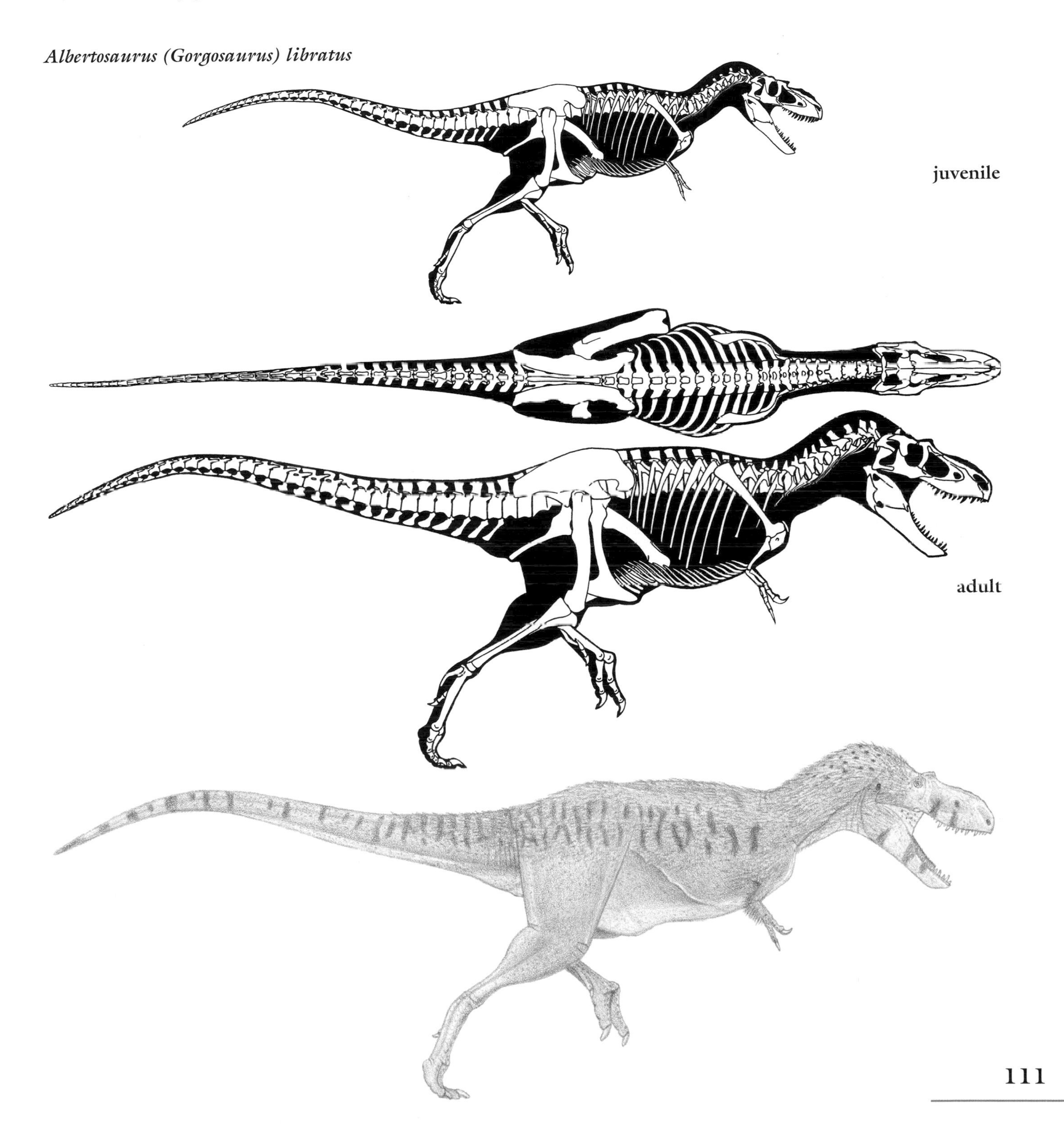

Albertosaurus (Albertosaurus) sarcophagus
8 m (27 ft) TL, 2.5 tonnes

FOSSIL REMAINS Some skulls and partial skeletons, well known.
ANATOMICAL CHARACTERISTICS Very similar to *A. libratus*, which may have been its ancestor. Leg may have been somewhat longer.
AGE Late Cretaceous, early Maastrichtian.
DISTRIBUTION AND FORMATION/S Alberta, Montana; lower to middle Horseshoe Canyon.
HABITAT Well-watered, forested floodplain with coastal swamps and marshes, cool winters.
HABITS Relatively gracile build indicates this species also preyed mainly on hadrosaurs.
NOTES Includes *A. arctunguis*, may be a direct descendant of *A. libratus*.

Daspletosaurus unnamed species
9 m (30 ft) TL, 2.5 tonnes

FOSSIL REMAINS Skulls and partial remains.
ANATOMICAL CHARACTERISTICS Similar to *D. torosus*.
AGE Late Cretaceous, middle and/or late Campanian.
DISTRIBUTION AND FORMATION/S Montana; Upper Two Medicine.
HABITAT Seasonally dry upland woodlands.
NOTES Not yet described; separation from *D. torosus* not documented. *Daspletosaurus* may be a member of *Tyrannosaurus*.

Daspletosaurus torosus
9 m (30 ft) TL, 2.5 tonnes

FOSSIL REMAINS Complete skulls and majority of skeleton, other remains including juveniles.
ANATOMICAL CHARACTERISTICS Skull broad, strongly constructed. Orbital hornlets reduced, teeth robust. Skeleton robustly built. Leg shorter than usual for group.
AGE Late Cretaceous, middle Campanian.
DISTRIBUTION AND FORMATION/S Alberta; upper Oldman.
HABITAT Well-watered, forested floodplain with coastal swamps and marshes, cool winters.

Albertosaurus (Albertosaurus) sarcophagus

HABITS Stout build indicates this species was specialized to cope with horned ceratopsids, and armored ankylosaurs when available, expanding the resouce base it could prey on, although more vulnerable hadrosaurs were probably still common prey.

Daspletosaurus unnamed species
9 m (30 ft) TL, 2.5 tonnes

FOSSIL REMAINS Some skulls and skeletons of varying completeness including juveniles.
ANATOMICAL CHARACTERISTICS Skull broad, strongly constructed. Orbital hornlets reduced, teeth robust. Skeleton robustly built, leg shorter than usual for group.
AGE Late Cretaceous, late Campanian.
DISTRIBUTION AND FORMATION/S Alberta; Dinosaur Park.
HABITAT Well-watered, forested floodplain with coastal swamps and marshes, cool winters.
HABITS Similar to *D. torosus.*
NOTES May be multiple species. Shared its habitat with the equal-sized but more lightly built and somewhat more common *Albertosaurus libratus*, which probably specialized in hunting hadrosaurs.

Raptorex? kriegstenis (Illustration overleaf)
Adult size uncertain

FOSSIL REMAINS Nearly complete juvenile skull and skeleton.
ANATOMICAL CHARACTERISTICS Standard for juvenile tyrannosaurid.
AGE Late Late Cretaceous.
DISTRIBUTION AND FORMATION/S Northern China or Mongolia; formation uncertain.
NOTES Formation found is uncertain—original premise that it is from Early Cretaceous Yixian is incorrect. Probably a juvenile of a new taxon, possibly of *T. bataar.*

Lythronax argestes
5 m (15 ft) TL, 500 kg (1,000 lb)

FOSSIL REMAINS Majority of skull and minority of skeleton.
ANATOMICAL CHARACTERISTICS Standard for robust tyrannosaurid.
AGE Late Cretaceous, middle Campanian.
DISTRIBUTION AND FORMATION/S Southern Utah; Wahweap.
NOTES May be a member of *Tyrannosaurus.*

Lythronax argestes

Daspletosaurus torosus

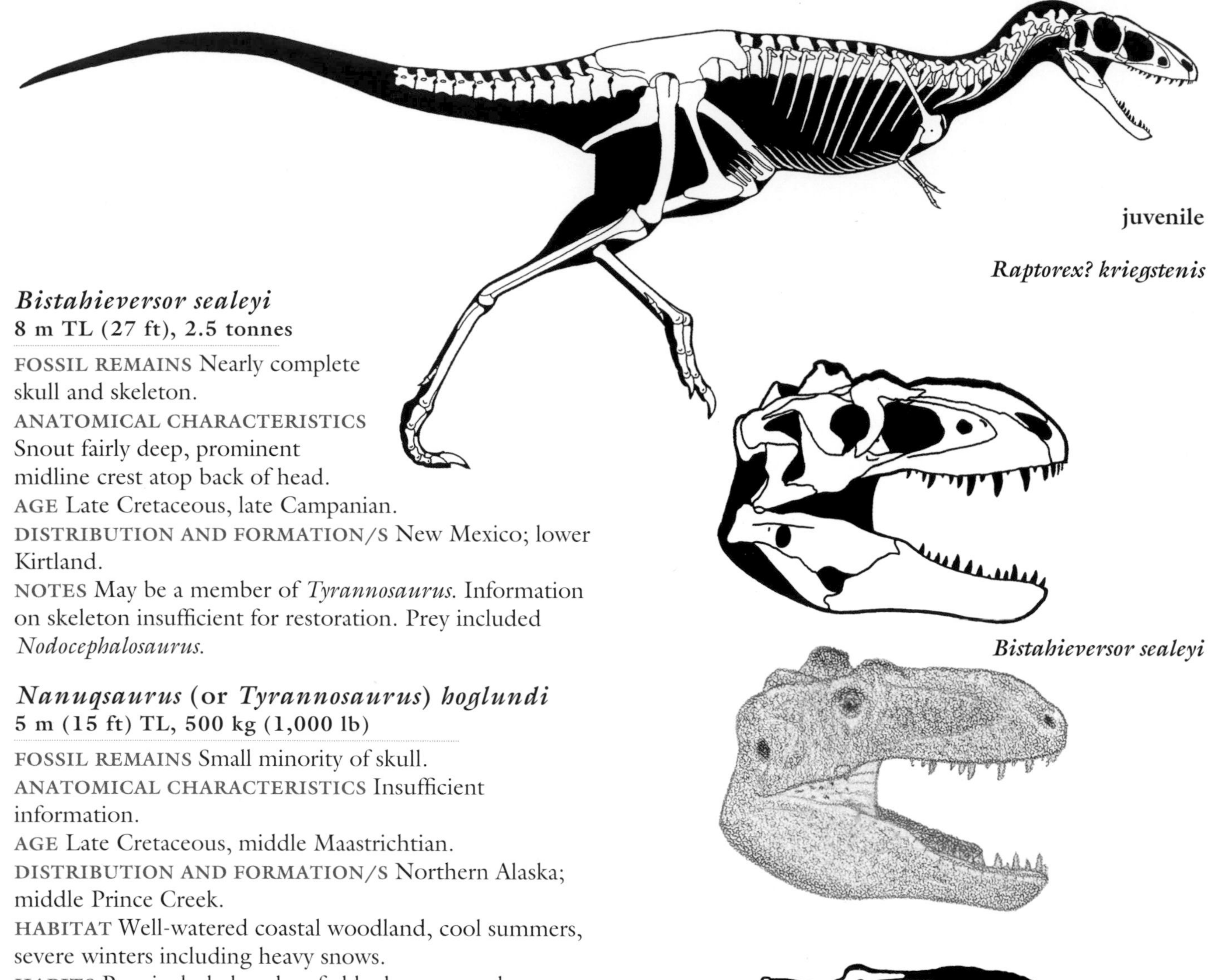

juvenile

Raptorex? kriegstenis

Bistahieversor sealeyi

Bistahieversor sealeyi
8 m TL (27 ft), 2.5 tonnes

FOSSIL REMAINS Nearly complete skull and skeleton.
ANATOMICAL CHARACTERISTICS Snout fairly deep, prominent midline crest atop back of head.
AGE Late Cretaceous, late Campanian.
DISTRIBUTION AND FORMATION/S New Mexico; lower Kirtland.
NOTES May be a member of *Tyrannosaurus.* Information on skeleton insufficient for restoration. Prey included *Nodocephalosaurus.*

Nanuqsaurus (or *Tyrannosaurus*) *hoglundi*
5 m (15 ft) TL, 500 kg (1,000 lb)

FOSSIL REMAINS Small minority of skull.
ANATOMICAL CHARACTERISTICS Insufficient information.
AGE Late Cretaceous, middle Maastrichtian.
DISTRIBUTION AND FORMATION/S Northern Alaska; middle Prince Creek.
HABITAT Well-watered coastal woodland, cool summers, severe winters including heavy snows.
HABITS Prey included undwarfed hadrosaurs and ceratopsids.
NOTES Based on inadequate remains, this is included because if remains are adult, indicates presence of a dwarf derived tyrannosaurid in a harsh Arctic environment. Probably heavily feathered, especially in winter. Main competitor large troodonts. May be a member of *Tyrannosaurus.*

Tyrannosaurus (Teratophoneus) curriei

Tyrannosaurus (*Teratophoneus*) *curriei*
8 m TL (27 ft), 2.5 tonnes

FOSSIL REMAINS Majority of skull and minority of skeleton.
ANATOMICAL CHARACTERISTICS Back of skull especially broad to accommodate oversized jaw and neck muscles, eyes face more strongly forward, increasing overlap of fields of vision.
AGE Late Cretaceous, late Campanian.
DISTRIBUTION AND FORMATION/S Utah; middle Kaiparowits.
HABIT Tremendous strength of head indicates specialization for hunting horned ceratopsids including *Nasutoceratops.*
NOTES Extreme broadening of back of head similar to *Tyrannosaurus rex*, may be ancestral to latter.

Tyrannosaurus (Tarbosaurus) bataar
9.5 m (31 ft) TL, 4 tonnes

FOSSIL REMAINS A number of skulls and skeletons from juvenile to adult, completely known. Small skin patches.
ANATOMICAL CHARACTERISTICS Skull very large, but even largest examples are not unusually broad. Bosses above orbits strongly suppressed, teeth not exceptionally large and robust. Skeleton moderately robust.
AGE Late Cretaceous, late Campanian and/or early Maastrichtian.
DISTRIBUTION AND FORMATION/S Mongolia and northern China; Nemegt, Nemegt Svita, Yuanpu, Quiba, etc.
HABITAT Well-watered woodland with seasonal rain, winters cold.
HABITS Lacking large horned ceratopsids in its habitat, not as powerful as North American *Tyrannosaurus*. Adult prey consisted primarily of *Saurolophus*, titanosaurids, and ankylosaurids.
NOTES Juveniles competed with *Alioramus altai*. Very fragmentary *Zhuchengtyrannus magnus* indicates that a tyrannosaurine larger than *T. bataar* lived in the region a little earlier.

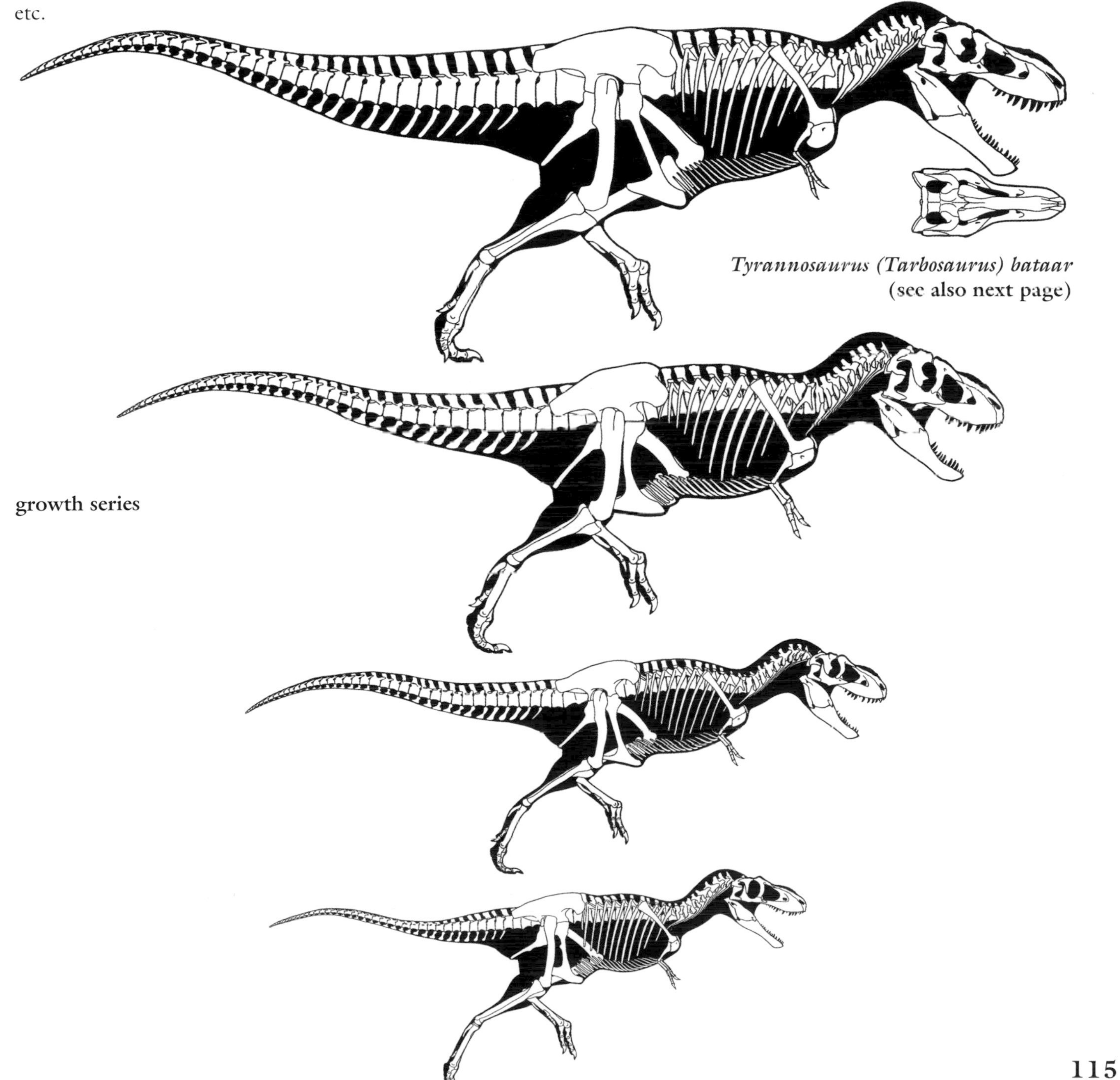

Tyrannosaurus (Tarbosaurus) bataar
(see also next page)

growth series

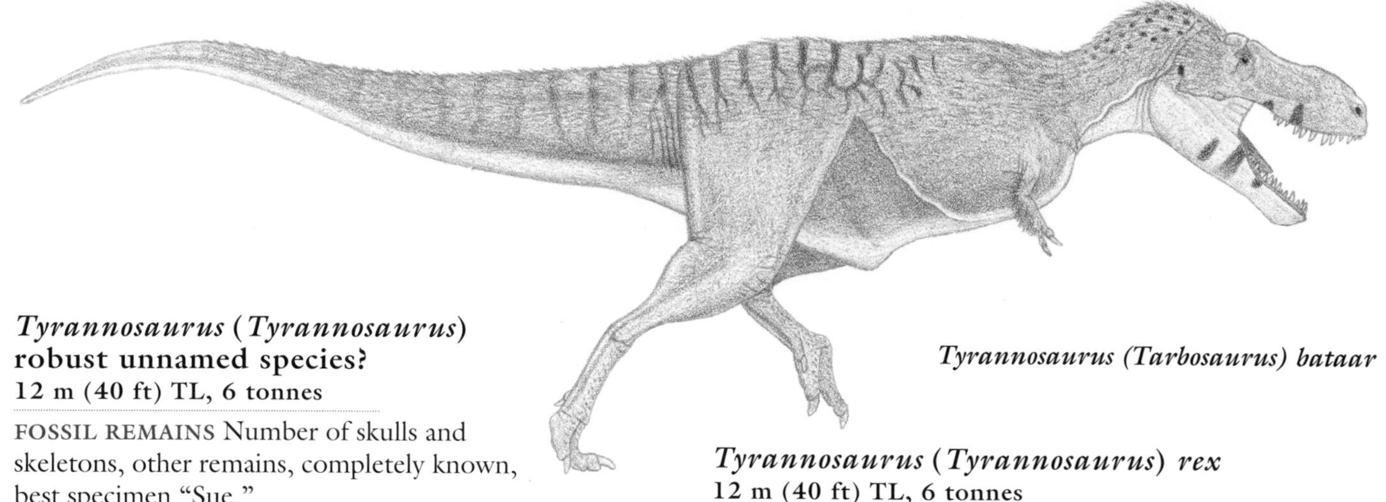

Tyrannosaurus (Tarbosaurus) bataar

Tyrannosaurus (*Tyrannosaurus*) robust unnamed species?

12 m (40 ft) TL, 6 tonnes

FOSSIL REMAINS Number of skulls and skeletons, other remains, completely known, best specimen "Sue."

ANATOMICAL CHARACTERISTICS Overall build robust. Otherwise similar to *T. rex*.

AGE Late Cretaceous, late Maastrichtian.

DISTRIBUTION AND FORMATION/S Montana, Dakotas, Wyoming, Colorado; lower Lance, lower Hell Creek.

HABITAT Well-watered coastal woodlands, climate cooler than in latest Maastrichtian, possibly chilly in winter.

HABITS Same as for *T. rex*, primary prey *Triceratops horridus*.

NOTES The most derived of the uniform tyrannosaurids are being badly oversplit at the genus level. Those from *Daspletosaurus* on up are the subfamily Tyrannosaurinae. This species may be ancestral to either or both of the later *Tyrannosaurus* species.

Tyrannosaurus (*Tyrannosaurus*) *rex*

12 m (40 ft) TL, 6 tonnes

FOSSIL REMAINS Partial skull and skeleton, a few elements.

ANATOMICAL CHARACTERISTICS Overall build robust. Skull much more heavily constructed and stouter than those of other tyrannosaurids, back of skull especially broad to accommodate oversized jaw and neck muscles at all ages; no other land predator with as powerful a bite. Eyes face more strongly forward, increasing overlap of fields of vision. Snout also broad. Lower jaws very deep. Brow bosses robust but not prominent, teeth unusually large and conical. Neck very stout. Head relatively small, teeth bladed, arm and hand unusually large, and leg extremely long in half-sized potential juveniles.

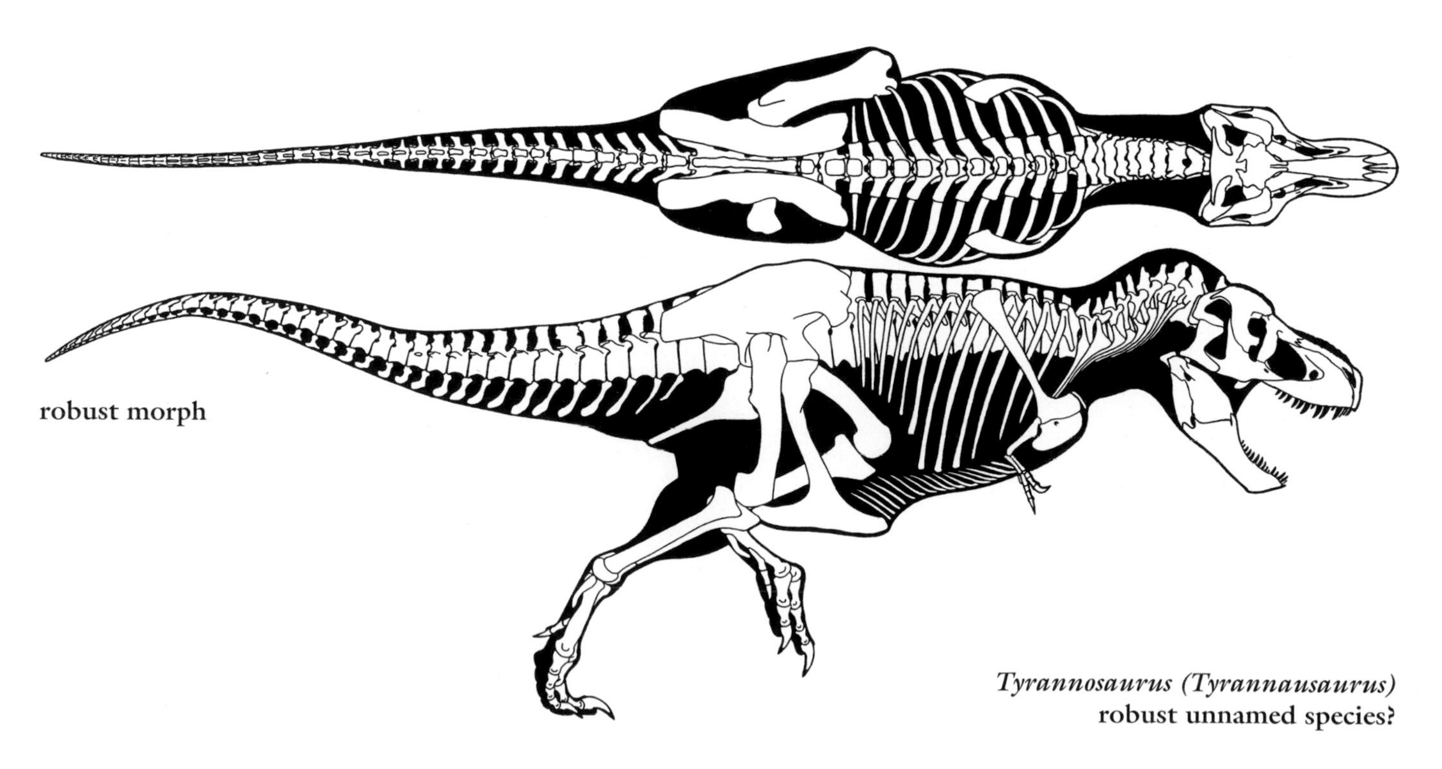

robust morph

Tyrannosaurus (Tyrannausaurus) **robust unnamed species?**

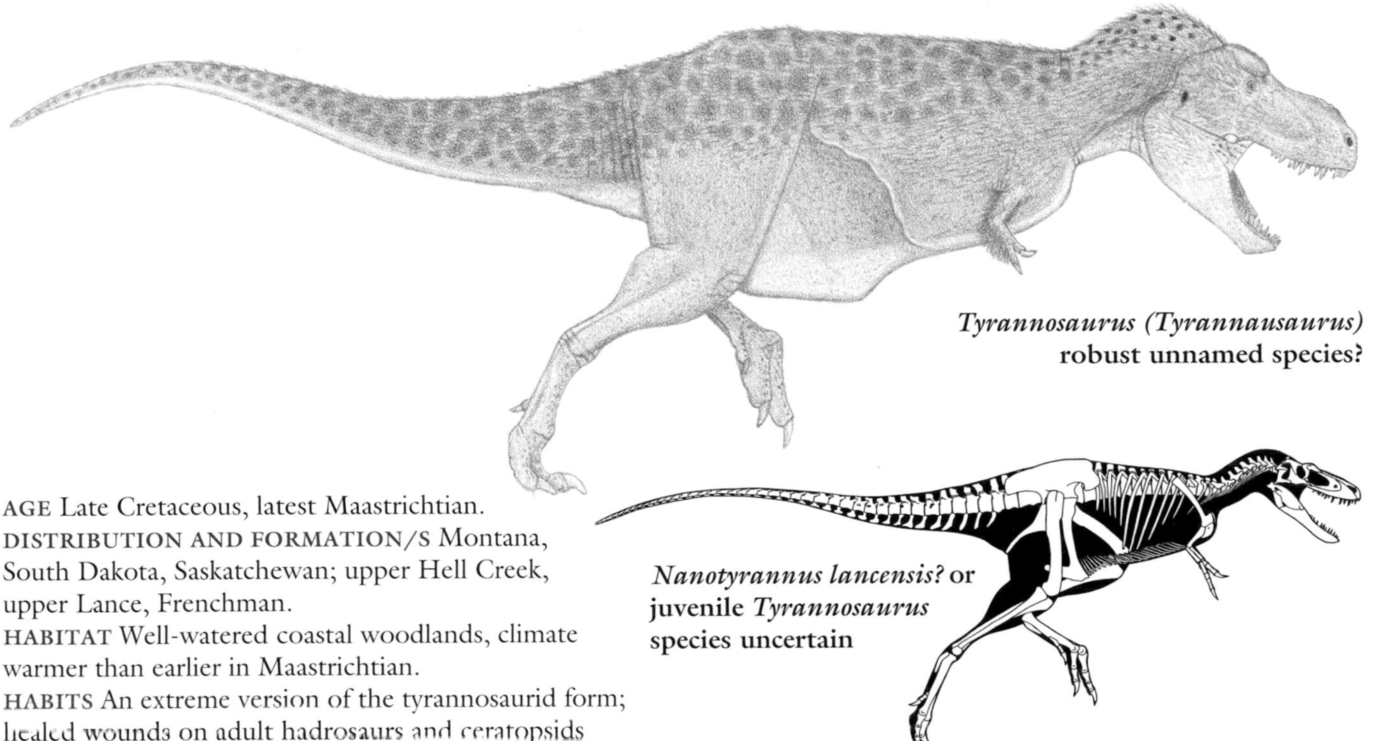

Tyrannosaurus (Tyrannausaurus) **robust unnamed species?**

Nanotyrannus lancensis? **or juvenile** ***Tyrannosaurus*** **species uncertain**

AGE Late Cretaceous, latest Maastrichtian.
DISTRIBUTION AND FORMATION/S Montana, South Dakota, Saskatchewan; upper Hell Creek, upper Lance, Frenchman.
HABITAT Well-watered coastal woodlands, climate warmer than earlier in Maastrichtian.
HABITS An extreme version of the tyrannosaurid form; healed wounds on adult hadrosaurs and ceratopsids indicate that adults hunted similarly elephant-sized prey on a regular basis, using the tremendously powerful head and teeth to lethally wound victims. Such firepower and size was more than was needed to hunt less-dangerous juveniles or just adult hadrosaurs, which made up a minority of the herbivore population dominated by dangerous horned *Triceratops prorsus*.
NOTES Although *Tyrannosaurus* skeletons were once rare, their high financial value has encouraged the discovery of a number of specimens, all of which have been placed in *T. rex*. That is problematic. Despite the popularity of *Tyrannosaurus*, the detailed taxonomy is only now being intensively researched. The unusually great variation in the robustness of the specimens, and the pattern of the distribution of varying robustness not being uniform throughout the stratigraphic range, as should be true if the robusts and graciles were females and males, indicate that multiple species evolved over 1.5 million years. If correct, then the main competitor of robust *T. rex* was the much more common gracile *Tyrannosaurus* species, with *T. rex* more specialized in preying on *Triceratops*. Assignment of specimens from locations well to the south of Montana and South Dakota to *T. rex* are highly problematic. Very gracile, juvenile remains with heads very broad and back sections sometimes labeled *Nanotyrannus lancensis* may be juveniles of the more massive adults, or one or more separate taxa whose adults were smaller than those of the giant *Tyrannosaurus*—more specimens are needed to resolve the situation. If nanotyrannos are juveniles of massive *Tyrannosaurus*, then no other avepod is known to have undergone such an extreme change in form with growth, including a shift from bladed to massive conical

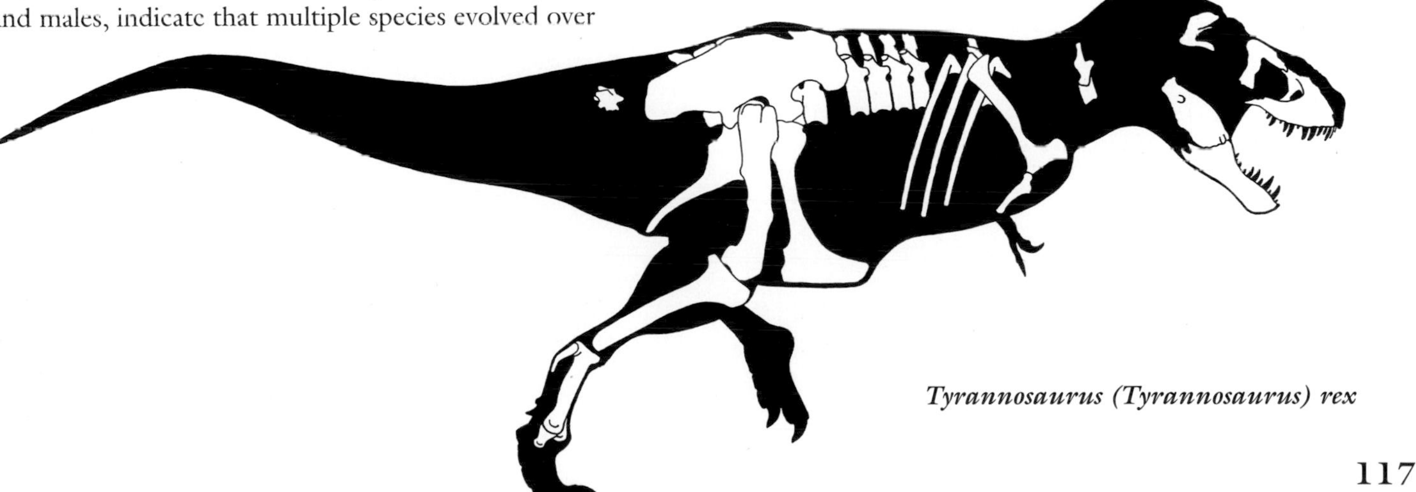

Tyrannosaurus (Tyrannosaurus) rex

Tyrannosaurus

teeth; if real, the drastic transformation with growth is probably a result of the radical shift from hunting fast prey such as ornithomimids to elephant-sized, fast-charging adult horned dinosaurs. If the nanotyrannos are instead distinct taxa, then the situation was similar to that in Asia, where gracile alioramins and stout tarbosaurs shared the same habitats. Absence of *T. rex* from other latest Maastrichtian formations of the region probably reflects lack of sufficient sampling.

Tyrannosaurus (*Tyrannosaurus*) gracile unnamed species?

12 m (40 ft) TL, 6 tonnes

FOSSIL REMAINS Number of skulls and skeletons, other remains, completely known, best specimens "Stan" and "Wankel."
ANATOMICAL CHARACTERISTICS Overall build more gracile. Otherwise similar to *T. rex*.
AGE Late Cretaceous, latest Maastrichtian.
DISTRIBUTION AND FORMATION/S Montana, Dakotas, Wyoming, Colorado, Alberta; upper Lance, upper Hell Creek, Denver, Laramie, Frenchman, Scollard.
HABITAT Well-watered coastal woodlands, climate warmer than earlier in Maastrichtian.
HABITS Same as for *T. rex*, primary prey *Triceratops prorsus*, although better adapted for hunting hadrosaurs than more robust *T. rex*.
NOTES Much more common than main competitor *T. rex*; two species of giant *Tyrannosaurus* may have been present in the same region at the final extinction.

NEOCOELUROSAURS

SMALL TO GIGANTIC PREDATORY AND HERBIVOROUS COELUROSAURS OF THE LATE JURASSIC TO THE END OF THE DINOSAUR ERA (WITH BIRDS SURVIVING BEYOND), MOST CONTINENTS.

ANATOMICAL CHARACTERISTICS Highly variable. Head from large to small, orbital hornlet absent, toothed to toothless and beaked, when teeth present serrations tend to be reduced in some manner or absent. Neck moderately to very long. Tail very long to very short. Shoulder girdle usually like that in birds, with horizontal scapula blade and vertical, anterior-facing coracoid, arm very long to short, wrist usually with a large, half-moon-shaped carpal block that allowed arm to be folded like bird's, hand usually long, fingers three to one. Brains enlarged, semiavian in form.
ONTOGENY Growth rates apparently moderate.

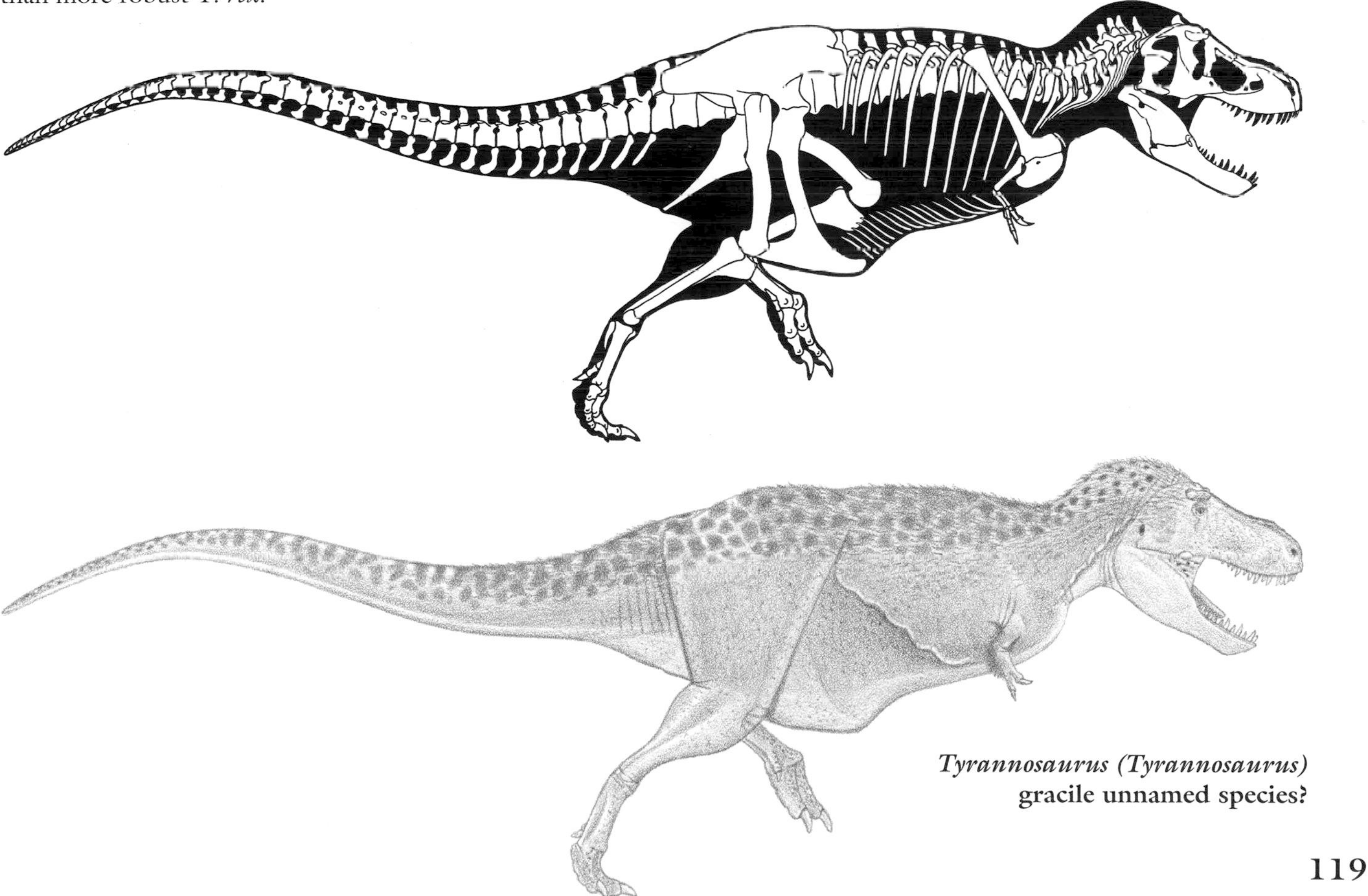

Tyrannosaurus (Tyrannosaurus) **gracile unnamed species?**

HABITS Reproduction generally similar to that of ratites and tinamous; in at least some cases males incubated the eggs and were probably polygamous; egg hatching in a given clutch not synchronous.
NOTES Coelurosaurs more derived than tyrannosauroids; some examples such as *Bicentenaria*, *Sciurumimus*, and *Zuolong* may be basal to this group.

BASO-NEOCOELUROSAURS Small neocoelurosaurs of the Late Jurassic to the Late Cretaceous.

ANATOMICAL CHARACTERISTICS Uniform. Heads small to moderately large, elongated, crests absent. Skeletons gracile. Tail long. Arm and hand well developed. Leg and foot long and slender.
HABITS Fast small game hunters, also fishers, similar in function to earlier coelophysids.
NOTES The classic small coelurosaurs were a widespread element in dinosaur faunas, the generalized small canids of their time; group splittable into subdivisions.

Bicentenaria argentina
3 m (10 ft) TL, 60 kg (130 lb)

FOSSIL REMAINS Partial skull and skeletal elements.
ANATOMICAL CHARACTERISTICS Insufficient information.
AGE Late Cretaceous, early Cenomanian.
DISTRIBUTION AND FORMATION/S Western Argentina; upper Candeleros.
HABITAT Short wet season, otherwise semiarid with open floodplains and riverine forests.
NOTES Enemies included juvenile *Ekrixinatosaurus* and *Giganotosaurus*.

Sciurumimus albersdoerferi
Adult size uncertain

FOSSIL REMAINS Complete juvenile skull and skeleton.
ANATOMICAL CHARACTERISTICS Insufficient information due to juvenile status of specimen.
AGE Late Jurassic, late Kimmeridgian.
DISTRIBUTION AND FORMATION/S Southern Germany; upper Rogling level.

Zuolong salleei

Zuolong salleei
3 m (10 ft) TL, 60 kg (130 lb)

FOSSIL REMAINS Majority of skull and minority of skeleton.
ANATOMICAL CHARACTERISTICS Standard for group.
AGE Late Jurassic, early Oxfordian.
DISTRIBUTION AND FORMATION/S Northwest China; upper Shishugou.
HABITAT Well-watered woodlands with short dry season.
NOTES Enemies included juvenile *Yangchuanosaurus dongi*, prey included *Limusaurus*, *Haplocheirus*, and *Yinlong*.

Aorun zhaoi
Adult size uncertain

FOSSIL REMAINS Majority of juvenile skull and minority of skeleton.
ANATOMICAL CHARACTERISTICS Standard for group.
AGE Late Jurassic, late Callovian.
DISTRIBUTION AND FORMATION/S Northwest China; lower Shishugou.
HABITAT Well-watered woodlands with short dry season.

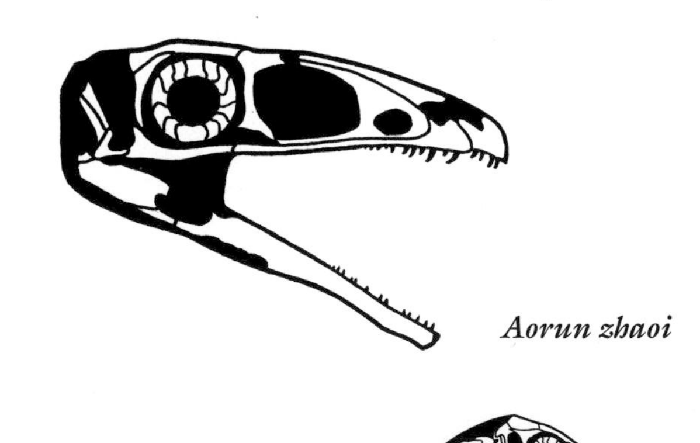

Aorun zhaoi

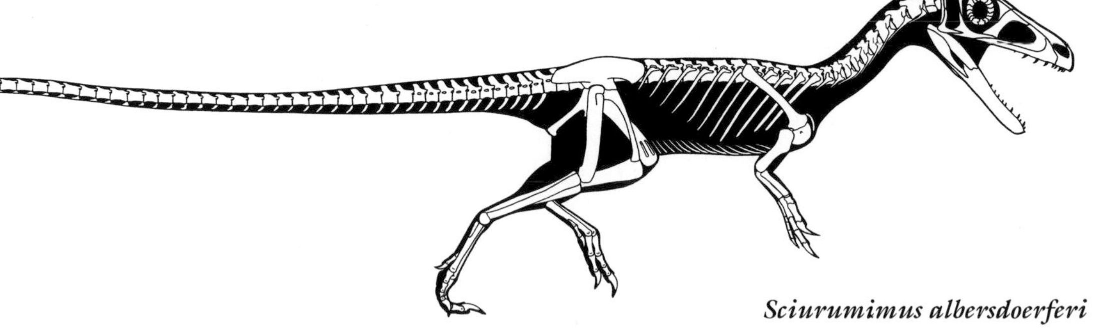

Sciurumimus albersdoerferi

Ornitholestes hermanni
2 m (7 ft) TL, 13 kg (30 lb)

FOSSIL REMAINS Nearly complete skull and majority of skeleton.
ANATOMICAL CHARACTERISTICS Head subrectangular, rather small relative to body, teeth on lower jaw restricted to front end. Leg moderately long.
AGE Late Jurassic, late Oxfordian.
DISTRIBUTION AND FORMATION/S Wyoming; lower Morrison.
HABITAT Short wet season, otherwise semiarid with open floodplain prairies and riverine forests.
HABITS Probably ambushed and chased small game as well as fished.
NOTES A classic coelurosaur. Shared its habitat with *Coelurus* and *Tanycolagreus.*

Coelurus fragilis
2.5 m (8 ft) TL, 15 kg (33 lb)

FOSSIL REMAINS Majority of skeleton.
ANATOMICAL CHARACTERISTICS Lightly built. Fingers long and slender.
AGE Late Jurassic, late Oxfordian.
DISTRIBUTION AND FORMATION/S Wyoming; lower Morrison.
HABITAT Short wet season, otherwise semiarid with open floodplain prairies and riverine forests.
HABITS Able to pursue faster prey than *Ornitholestes.*
NOTES Remains imply close relatives higher in the Morrison. This and *Tanycolagreus* may form the family Coeluridae.

Tanycolagreus topwilsoni
4 m (13 ft) TL, 120 kg (250 lb)

FOSSIL REMAINS Much of the skull and majority of the skeleton.
ANATOMICAL CHARACTERISTICS Head large, long, subrectangular. Leg long and gracile.
AGE Late Jurassic, late Oxfordian.
DISTRIBUTION AND FORMATION/S Wyoming; lower Morrison.
HABITAT Short wet season, otherwise semiarid with open floodplain prairies and riverine forests.
HABITS Prey included fairly large game.

Nedcolbertia justinhofmanni
Adult size uncertain

FOSSIL REMAINS Minority of several skeletons, immature.
ANATOMICAL CHARACTERISTICS Leg long and gracile.
AGE Early Cretaceous, probably Barremian.
DISTRIBUTION AND FORMATION/S Utah; Lower Cedar Mountain.
HABITAT Short wet season, otherwise semiarid with floodplain prairies and open woodlands, and riverine forests.

Ornitholestes hermanni

Tugulusaurus facilis
2 m (7 ft) TL, 13 kg (30 lb)

FOSSIL REMAINS Minority of skeleton.
ANATOMICAL CHARACTERISTICS Insufficient information.
AGE Early Cretaceous.
DISTRIBUTION AND FORMATION/S Northwest China; Lianmuging.

Aniksosaurus darwini
2.5 m (9 ft) TL, 30 kg (65 lb)

FOSSIL REMAINS Several partial skeletons.
ANATOMICAL CHARACTERISTICS Robustly built. Posterior pelvis broad.
AGE Late Cretaceous, late Cenomanian or Turonian.
DISTRIBUTION AND FORMATION/S Southern Argentina; lower Bajo Barreal.

Scipionyx samniticus
Adult size uncertain

FOSSIL REMAINS Complete skull and almost complete skeleton, juvenile, some internal organs preserved.
ANATOMICAL CHARACTERISTICS Proportions characteristic for juvenile.
AGE Early Cretaceous, early Albian.
DISTRIBUTION AND FORMATION/S Central Italy; unnamed formation.
HABITS Juveniles probably hunted small vertebrates, insects.
NOTES May be a compsognathid.

COMPSOGNATHIDS Small predatory neocoelurosaurs limited to the Late Jurassic and Early Cretaceous of Eurasia and South America.

ANATOMICAL CHARACTERISTICS Uniform. In most regards standard for small neocoelurosaurs. Neck moderately long. Tail very long. Hand strongly asymmetrical because thumb and claw unusually stout and outer finger slender. Boot on pubis large; leg moderately long.
HABITS Ambushed and chased small game, also fish in some cases. Thumb an important weapon for hunting and/or combat within species.
NOTES A common and widely distributed group of small neocoelurosaurs, the foxes of their time.

Compsognathus longipes
1.25 m (4 ft) TL, 2.5 kg (5.5 lb)

FOSSIL REMAINS Two nearly complete skulls and skeletons.
ANATOMICAL CHARACTERISTICS Snout subtriangular, teeth small.
AGE Late Jurassic, late Kimmeridgian.
DISTRIBUTION AND FORMATION/S Southern Germany, southern France; Solnhofen.
HABITAT Found as drift in lagoonal deposits near probably arid, brush-covered islands.
HABITS Prey included *Archaeopteryx.*
NOTES The second dinosaur known from a largely complete skull and skeleton.

Juravenator starki
Adult size uncertain

FOSSIL REMAINS Nearly complete juvenile skull and skeleton, small skin patches.
ANATOMICAL CHARACTERISTICS Skull subrectangular, snout fairly deep, indentation in snout, teeth large. Skin with small scales on leg and tail, rest of body covering uncertain.
AGE Late Jurassic, late Kimmeridgian.
DISTRIBUTION AND FORMATION/S Southern Germany; Solnhofen.
HABITAT Found as drift in lagoonal deposits near probably arid, brush-covered islands.
HABITS Large teeth indicate it hunted fairly large animals; kink in upper jaw indicates it also fished. Prey included *Archaeopteryx.*
NOTES Shared its habitat with *Compsognathus.*

Scipionyx samniticus juvenile

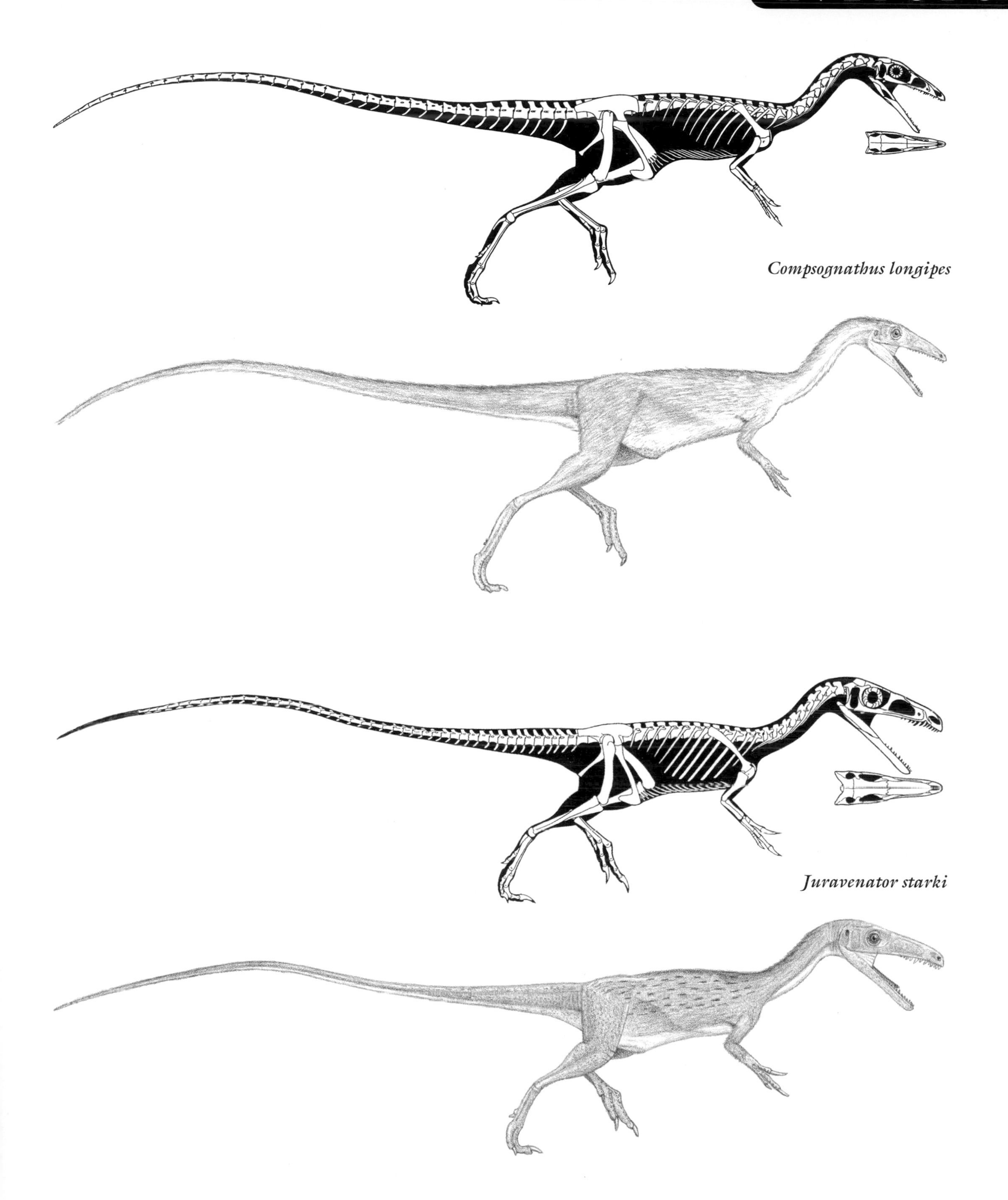
Compsognathus longipes
Juravenator starki

Sinosauropteryx prima
1 m (3 ft) TL, 1 kg (2.2 lb)

FOSSIL REMAINS A few complete skulls and skeletons, extensive external fibers, eggs.
ANATOMICAL CHARACTERISTICS Snout subtriangular, teeth small. Arm shorter and thumb and claw stouter than in other compsognathids. Simple protofeathers on most of head except front of snout and on most of body except hands and feet; protofeathers atop head and body and dark bands on tail dark brown or reddish brown, lighter bands in between. Elongated eggs 4 cm (1.5 in) long.
AGE Early Cretaceous, early Aptian.
DISTRIBUTION AND FORMATION/S Northeast China; Yixian.
HABITAT Well-watered forests and lakes, winters chilly with some snow.
HABITS Eggs formed and laid in pairs.
NOTES Shared its habitat with larger and more powerful *Huaxiagnathus*, *Sinocalliopteryx*, *Sinornithosaurus millenii*, *Tianyuraptor*, and *Zhenyuanlong*.

Sinosauropteryx? unnamed species
1 m (3 ft) TL, 1 kg (2.5 lb)

FOSSIL REMAINS Nearly complete skull and skeleton, some external fibers.
ANATOMICAL CHARACTERISTICS Head subtriangular, teeth large. Tail rather short. Arm and hand rather small. Leg long. Simple protofeathers over most of body including tuft at end of tail.
AGE Early Cretaceous, early Aptian.
DISTRIBUTION AND FORMATION/S Northeast China; Yixian.
HABITAT Well-watered forests and lakes, winters chilly with some snow.
HABITS Fast pursuit predator.
NOTES Incorrectly placed in *S. prima*.

Sinocalliopteryx gigas
2.3 m (7.5 ft) TL, 20 kg (40 lb)

FOSSIL REMAINS Complete skull and skeleton, extensive external fibers.

Sinosauropteryx prima and *Confuciusornis sanctus*

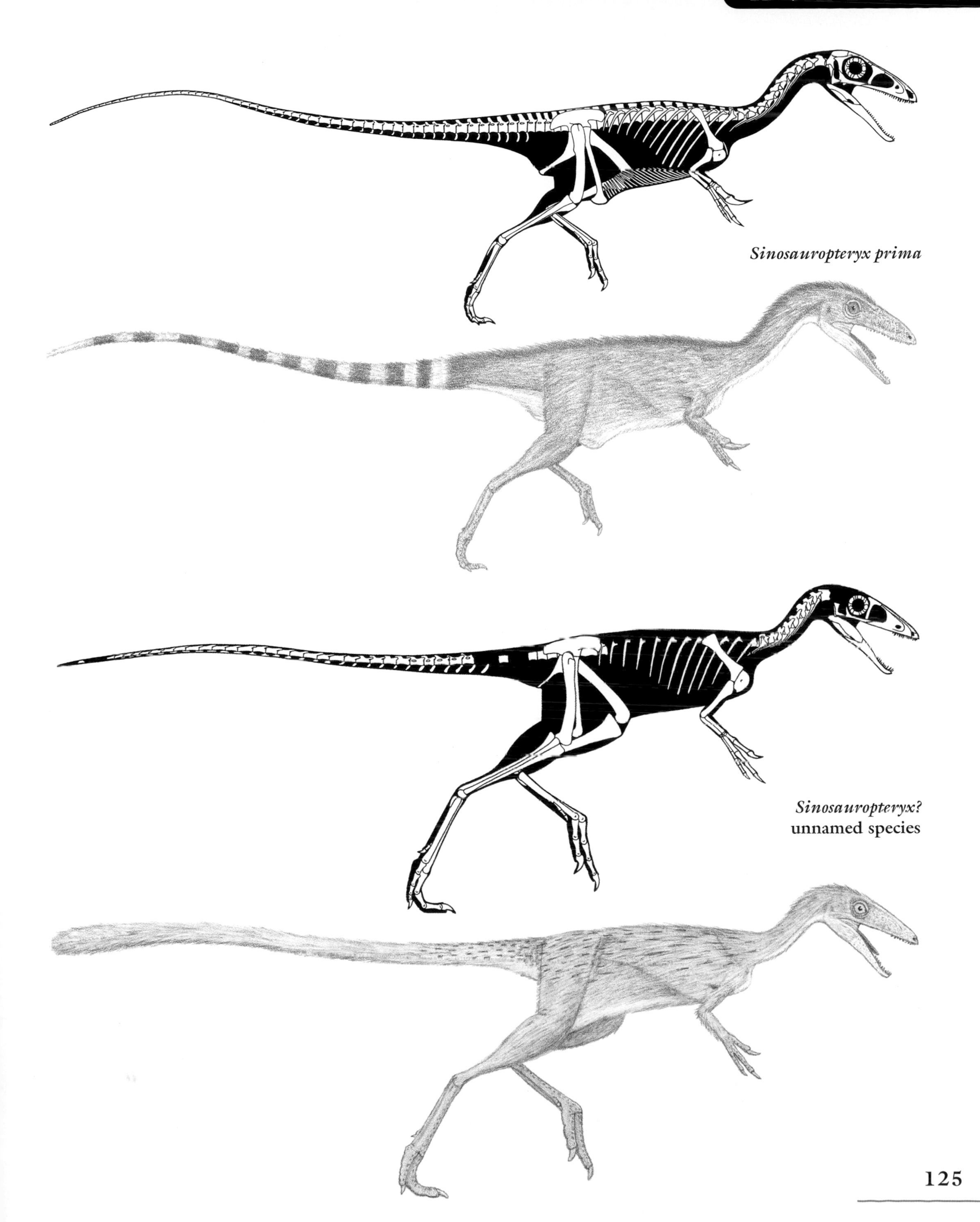

Sinosauropteryx prima

Sinosauropteryx?
unnamed species

ANATOMICAL CHARACTERISTICS Head subtriangular, small paired crestlets atop snout, teeth fairly large. Tail rather short. Leg long. Simple protofeathers over most of body including upper feet, especially long at hips, tail base, and thigh, forming tuft at end of tail.
AGE Early Cretaceous, early Aptian.
DISTRIBUTION AND FORMATION/S Northeast China; Yixian.
HABITAT Well-watered forests and lakes, winters chilly with some snow.
HABITS Fast pursuit predator that hunted larger prey than smaller *Huaxiagnathus*. Foot feathers probably for display.

Huaxiagnathus orientalis
1.7 m (5.5 ft) TL, 5 kg (12 lb)

FOSSIL REMAINS Nearly complete skull and skeleton.
ANATOMICAL CHARACTERISTICS Head subrectangular, front fairly deep, teeth not very large.
AGE Early Cretaceous, early Aptian.
DISTRIBUTION AND FORMATION/S Northeast China; Yixian.
HABITAT Well-watered forests and lakes, winters chilly with some snow.
HABITS Hunted larger prey than smaller *Sinosauropteryx*, including the latter.

Aristosuchus pusillus
2 m (6 ft) TL, 7 kg (15 lb)

FOSSIL REMAINS Minority of skeleton.
ANATOMICAL CHARACTERISTICS Insufficient information.
AGE Early Cretaceous, Barremian.
DISTRIBUTION AND FORMATION/S Isle of Wight, England; Wessex.

Mirischia asymmetrica
2 m (6 ft) TL, 7 kg (15 lb)

FOSSIL REMAINS Minority of skeleton. Some internal organs preserved.
ANATOMICAL CHARACTERISTICS Standard for group.
AGE Early Cretaceous, probably Albian.
DISTRIBUTION AND FORMATION/S Eastern Brazil; Santana.
NOTES Found as drift in marine deposits.

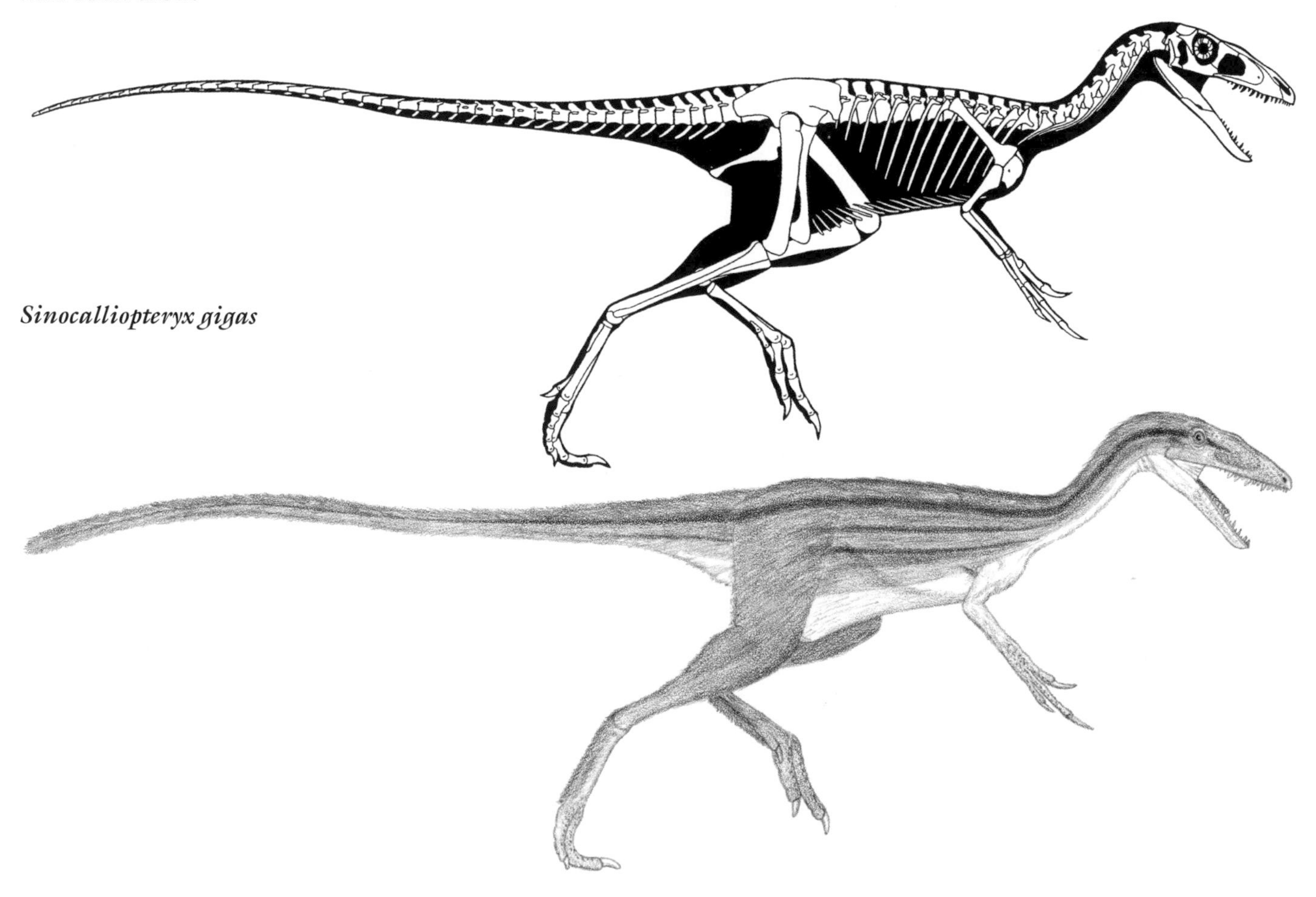

Sinocalliopteryx gigas

Huaxiagnathus orientalis

ORNITHOMIMOSAURS

SMALL TO GIGANTIC NONPREDATORY NEOCOELUROSAURS OF THE CRETACEOUS, NORTHERN HEMISPHERE AND AFRICA.

ANATOMICAL CHARACTERISTICS Usually uniform. Usually not heavily built. Head small, shallow, and narrow, teeth reduced or absent and shallow, blunt beak present, eyes face partly forward, and some degree of stereo vision possible, extra joint in lower jaw absent. Neck long, at least fairly slender. Arm and hand long and slender. Leg long, toes short, claws not sharp. Brains semiavian in structure and size, olfactory bulbs reduced. At least some examples had large feathers on arms. Gizzard stones sometimes present.
HABITAT Well-watered areas.
HABITS Small slender skulls, unhooked beaks, and lightly constructed necks bar these from being predators. Possibly omnivorous, combining some small animals and insects with plant material gathered with assistance of long arms and hands. Main defense speed, also kicks from powerful legs and slashing with large hand claws.

Gallimimus shaded skull

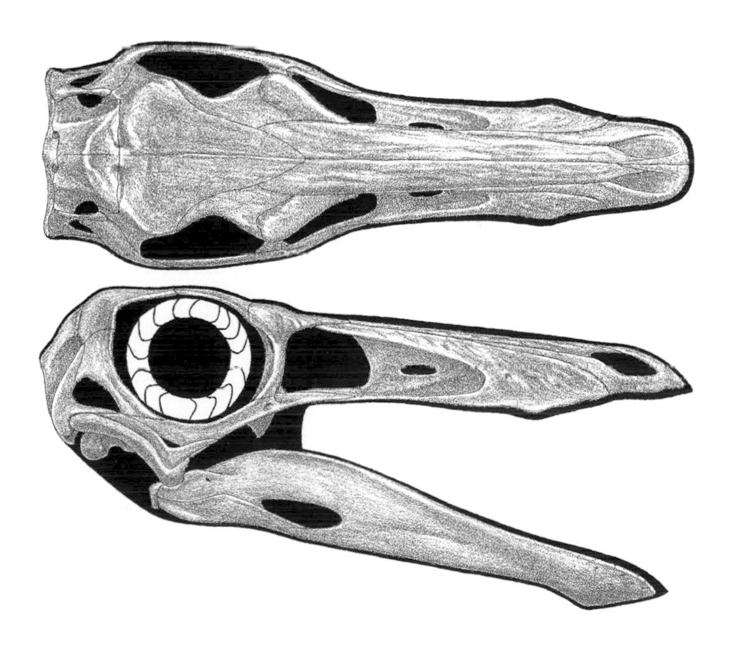

Gallimimus muscle study

NOTES The dinosaurs most similar to ostriches and other big ratites. Fragmentary remains imply possible presence in Australia. Fragmentary Siberian *Lepidocheirosaurus natatilis* may indicate presence of group in Late Jurassic. A frequent and sometimes common element of Cretaceous faunas.

BASO-ORNITHOMIMOSAURS Small to large ornithomimosaurs of the Cretaceous of Eurasia.

ANATOMICAL CHARACTERISTICS Not as gracile as ornithomimids, pelvis usually not as large, foot not as compressed, hallux usually still present. May be splittable into a larger number of divisions.

Nqwebasaurus thwazi
1 m (3 ft) TL, 1 kg (2.5 lb)

FOSSIL REMAINS Minority of skull and majority of skeleton.
ANATOMICAL CHARACTERISTICS Hand moderately long, thumb enlarged. Boot on pubis small, leg very long and gracile.
AGE Late Jurassic or Early Cretaceous.
DISTRIBUTION AND FORMATION/S Southern South Africa; Upper Kirkwood.
NOTES Indicates that at least early ornithomimosaurs were present in Southern Hemisphere, and perhaps in the Late Jurassic.

Pelecanimimus polyodon
2.5 m (8 ft) TL, 30 kg (60 lb)

FOSSIL REMAINS Complete skull and front part of skeleton, some soft tissues.
ANATOMICAL CHARACTERISTICS Snout long and tapering, small hornlets above orbits, hundreds of tiny teeth concentrated in front of jaws. Fingers subequal in length, claws nearly straight. Small soft crest at back of head, throat pouch, no feathers preserved on limited areas of smooth, unscaly skin.
AGE Early Cretaceous, late Barremian.
DISTRIBUTION AND FORMATION/S Central Spain; Calizas de la Huergina.
HABITS Teeth may have been for cutting plants and/or filtering small organisms, throat pouch may have been for containing fish. Hornlets and crest for display within the species.
NOTES Found as drift in marine deposits.

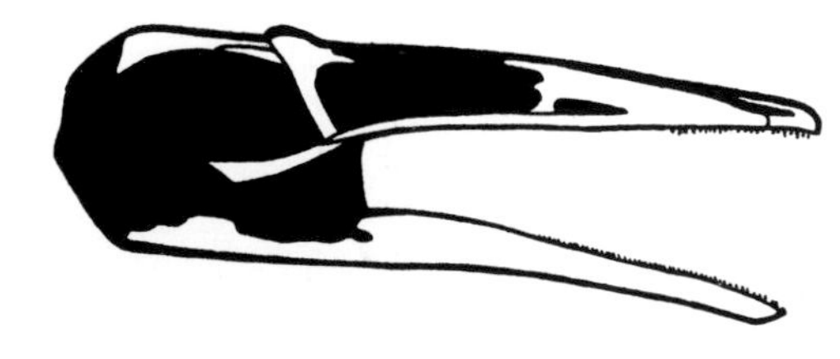

Pelecanimimus polyodon

Hexing qingyi
Adult size uncertain

FOSSIL REMAINS Complete skull and minority of skeleton, possibly immature.
ANATOMICAL CHARACTERISTICS A few small teeth at front end of lower jaw. Thumb much shorter than other fingers, claws fairly short and curved.

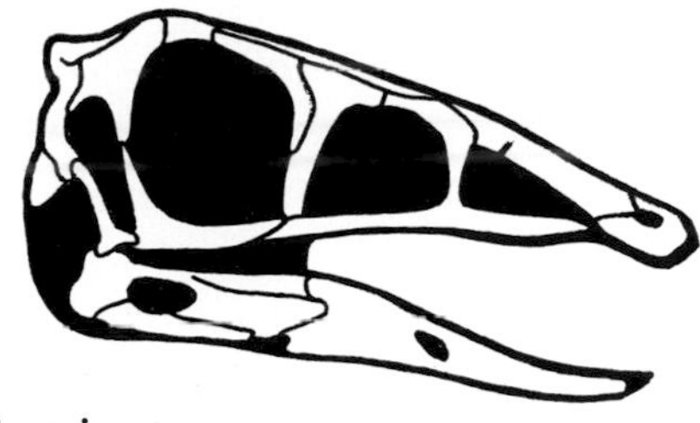

Hexing qingyi

AGE Early Cretaceous, Valanginian or early Barremian.
DISTRIBUTION AND FORMATION/S Northeast China; lowest Yixian.
HABITAT Well-watered forests and lakes, winters chilly with some snow.

Shenzhousaurus orientalis
1.6 m (5 ft) TL, 10 kg (20 lb)

FOSSIL REMAINS Complete skull and majority of skeleton.
ANATOMICAL CHARACTERISTICS A few small, conical teeth at front end of lower jaw. Thumb not as long as other fingers, claws nearly straight.
AGE Early Cretaceous, Barremian.
DISTRIBUTION AND FORMATION/S Northeast China; lower Yixian.
HABITAT Well-watered forests and lakes, winters chilly with some snow.

Harpymimus okladnikovi
3 m (10 ft) TL, 50 kg (110 lb)

FOSSIL REMAINS Nearly complete skull and majority of skeleton.
ANATOMICAL CHARACTERISTICS A few small teeth at tip of lower jaw. Thumb not as long as other fingers, claws gently curved.

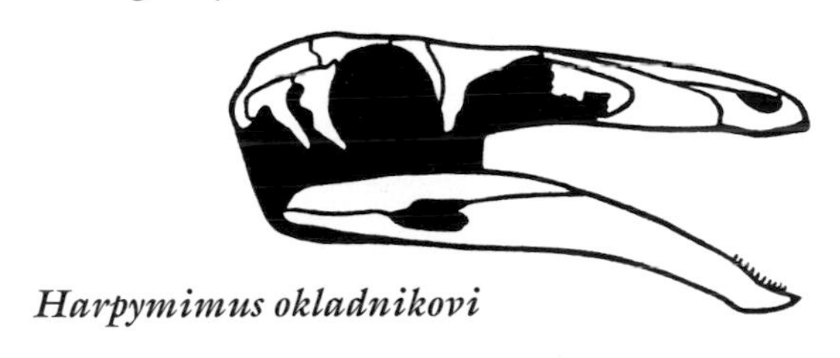

Harpymimus okladnikovi

AGE Early Cretaceous, late Albian.
DISTRIBUTION AND FORMATION/S Mongolia; Shinekhudag Svita.

Beishanlong grandis
7 m (23 ft) TL, 550 kg (1,200 lb)

FOSSIL REMAINS Minority of skeletons.
ANATOMICAL CHARACTERISTICS Fairly robustly built.
AGE Early Cretaceous, probably Aptian or Albian.
DISTRIBUTION AND FORMATION/S Central China; lower Xinminpu.
NOTES Prey of *Xiongguanlong*.

Garudimimus brevipes
2.5 m (8 ft) TL, 30 kg (60 lb)

FOSSIL REMAINS Complete skull and majority of skeleton.
ANATOMICAL CHARACTERISTICS Toothless and beaked.
AGE Early Late Cretaceous.
DISTRIBUTION AND FORMATION/S Mongolia; Bayanshiree.
NOTES Shared its habitat with *Achillobator*.

Deinocheirus mirificus
11.5 m (38 ft) TL, 5 tonnes

FOSSIL REMAINS Complete skull, majority of a skeleton and minority of two others.
ANATOMICAL CHARACTERISTICS Skull not as small relative to skeleton as in other ornithomimosaurs, very slender and narrow with very long beak flaring out to a small duck bill, eye sockets not large, lower jaw very deep, jaws weakly muscled, teeth absent. Neck fairly stout. Trunk vertebrae articulated in a very strong arc up from hip and down to shoulders, vertebral spines form tall sail just in front of hips. Tail ends with a small pygostyle. Hips broader than in other ornithomimosaurs. Arm 2.5 m (9 ft) long, rather slender, fingers subequal in length, claws blunt-tipped hooks. Hip very large and deep, leg robust but not massive, feet moderately long, toes short and ending with blunt, hoof-like claws.

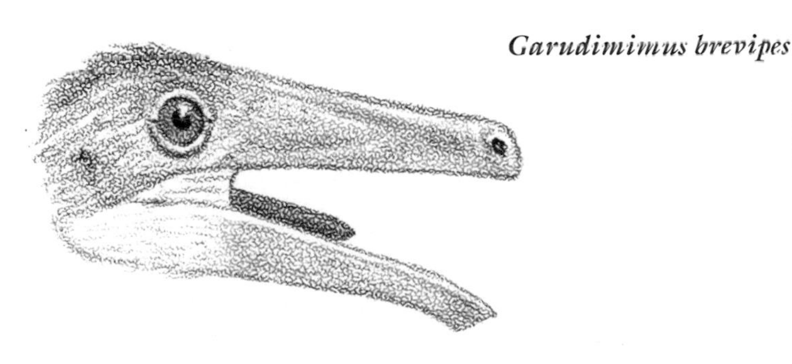

Garudimimus brevipes

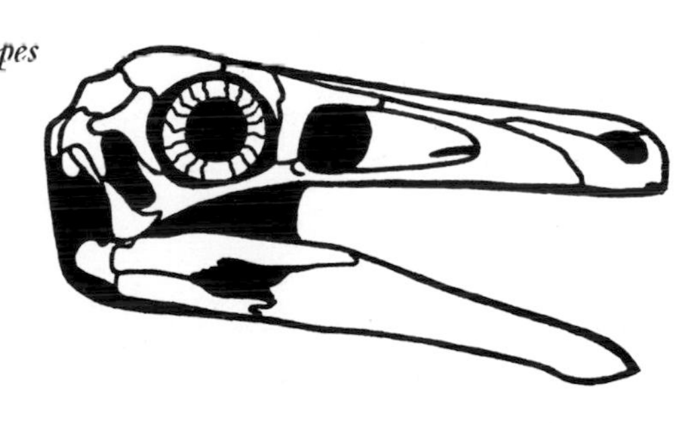

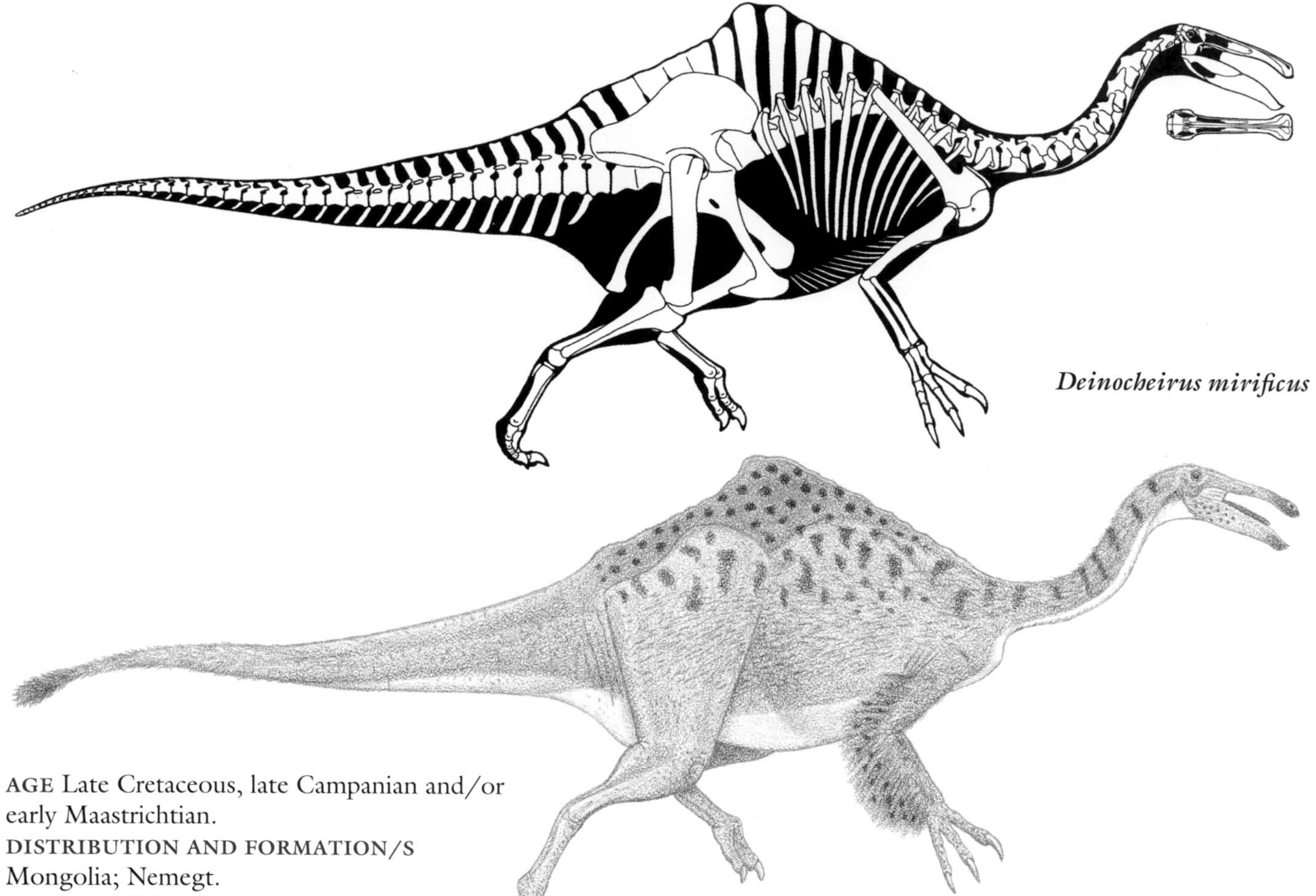

Deinocheirus mirificus

AGE Late Cretaceous, late Campanian and/or early Maastrichtian.
DISTRIBUTION AND FORMATION/S Mongolia; Nemegt.
HABITAT Well-watered woodland with seasonal rain, winters cold.
HABITS Probably an omnivore that fed on softer vegetation and aquatic creatures as indicated by fish remains in apparent stomach contents. Deep jaws indicate strong tongue and possible suction action. Arms could be used for gathering vegetation, possibly digging. Main enemy *Tyrannosaurus bataar*, better able to defend itself against predators with large, clawed arms than smaller, faster ornithomimosaurs.
NOTES Full form only recently realized from more complete remains long after discovery of isolated gigantic arms in the 1960s. This, *Beishanlong*, and *Garudimimus* may form family Deinocheiridae. Shared its habitat with *Therizinosaurus*, main enemy *T. bataar*.

ORNITHOMIMIDS Medium-sized ornithomimosaurs of the Cretaceous, limited to the Northern Hemisphere.

ANATOMICAL CHARACTERISTICS Highly uniform. Gracile build. Toothless and beaked. Fingers subequal in length, claws at least fairly long and not strongly curved. Trunk compact. Tail shorter and lighter than standard for theropods. Pelvis very large and leg very long, so leg muscles exceptionally well developed, foot very long and strongly compressed from side to side, hallux completely lost, so speed potential very high. At least some examples had large feathers on arm.
HABITS Main defense very high speed.
NOTES Prey of dromaeosaurids, troodontids, and juvenile tyrannosaurids when present.

Kinnareemimus khonkaensis
Size not available

FOSSIL REMAINS Minority of skeleton.
ANATOMICAL CHARACTERISTICS Insufficient information.
AGE Early Cretaceous, Valanginian or Hauterivian.
DISTRIBUTION AND FORMATION/S Thailand; Sao Khua.
NOTES Prey of *Siamotyrannus*.

Archaeornithomimus asiaticus
Adult size uncertain

FOSSIL REMAINS Minority of skeleton.
ANATOMICAL CHARACTERISTICS Insufficient information.

AGE Late Cretaceous.
DISTRIBUTION AND FORMATION/S China; Iren Dabasu.
HABITAT Seasonally wet-dry woodlands.
NOTES Prey of *Alectrosaurus*.

Sinornithomimus dongi
2.5 m (8 ft) TL, 45 kg (100 lb)

FOSSIL REMAINS Over a dozen skulls and skeletons, many complete, juvenile to adult, completely known.
ANATOMICAL CHARACTERISTICS Skull somewhat shorter, and skeleton not quite as gracile, as in most later ornithomimids.
AGE Late Cretaceous, Toronian.
DISTRIBUTION AND FORMATION/S Northern China; Ulansuhai.
NOTES Shared its habitat with *Chilantaisaurus* and *Shaochilong*.

Ansermimus planinychus
3 m (10 ft) TL, 50 kg (110 lb)

FOSSIL REMAINS Minority of skeleton.
ANATOMICAL CHARACTERISTICS Hand moderately elongated.
AGE Late Cretaceous, early Maastrichtian.
DISTRIBUTION AND FORMATION/S Mongolia; Nemegt Svita.
HABITAT Well-watered woodland with seasonal rain.

Gallimimus bullatus (Illustration overleaf)
6 m (20 ft) TL, 450 kg (1,000 lb)

FOSSIL REMAINS Several complete skulls and skeletons, juveniles to adult, completely known.
ANATOMICAL CHARACTERISTICS Beak elongated. Shorter armed and legged than other advanced ornithomimids, and presumably not quite as swift.
AGE Late Cretaceous, late Campanian and/or early Maastrichtian.
DISTRIBUTION AND FORMATION/S Mongolia; Nemegt.
HABITAT Well-watered woodland with seasonal rain, winters cold.
NOTES Prey of juvenile *Alioramus altai*, *T. bataar*, and adult *Saurornithoides junior*.

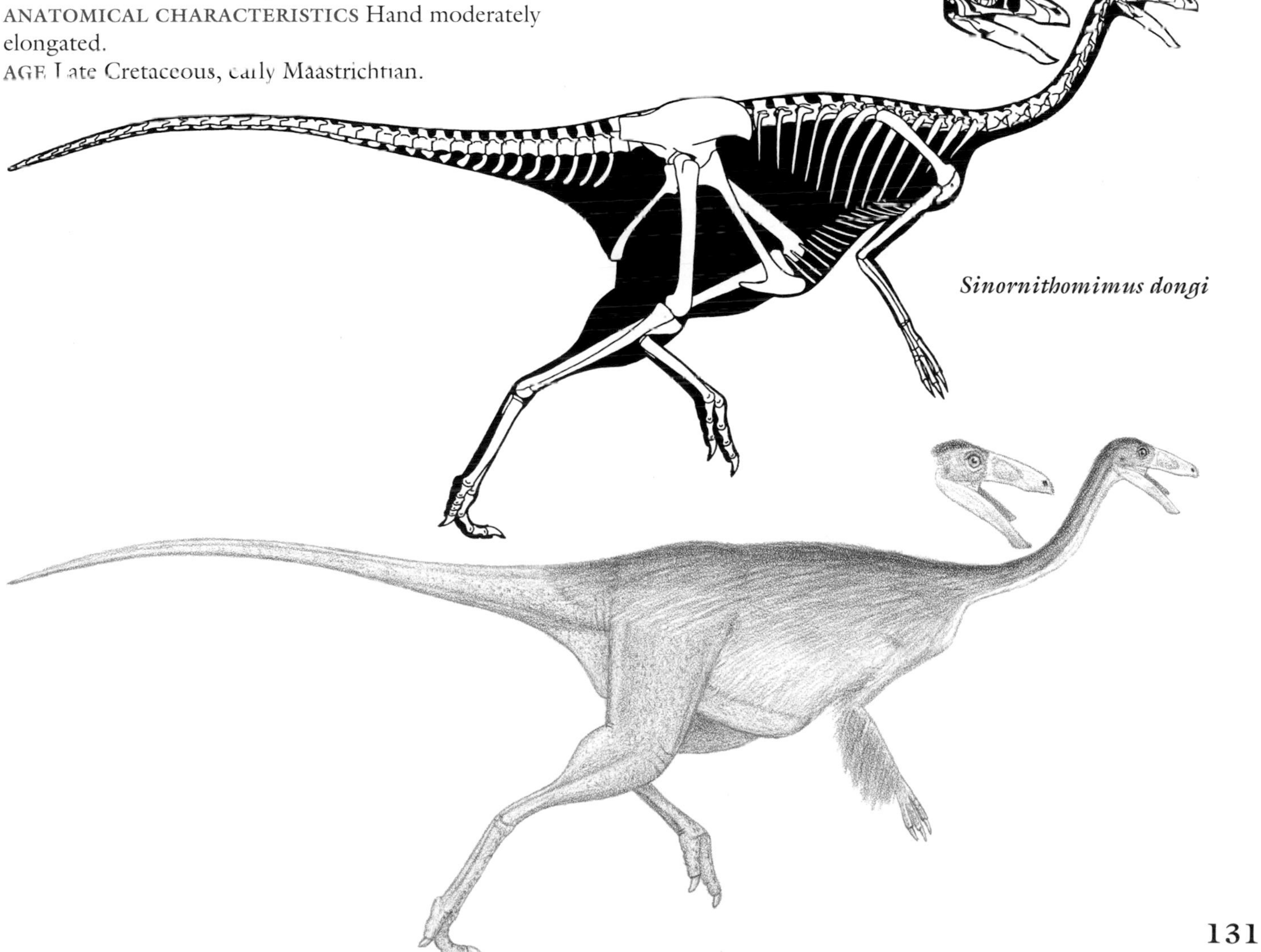

Sinornithomimus dongi

Struthiomimus altus
4 m (13 ft) TL, 150 kg (330 lb)

FOSSIL REMAINS Nearly complete and partial skeletons.
ANATOMICAL CHARACTERISTICS Leg long.
AGE Late Cretaceous, late Campanian.
DISTRIBUTION AND FORMATION/S Alberta; Dinosaur Park, level uncertain.
HABITAT Well-watered, forested floodplain with coastal swamps and marshes, cool winters.
NOTES The adequacy of the remains the genus *Struthiomimus* is based on is problematic. Prey of juvenile *Albertosaurus libratus* and adult *Stenonychosaurus.*

Struthiomimus edmontonicus
3.8 m (12 ft) TL, 170 kg (370 lb)

FOSSIL REMAINS Several complete skulls and skeletons.
ANATOMICAL CHARACTERISTICS Skull gracile. Fingers nearly equal in length, claws long, nearly straight, and delicate. Leg very long. Short and medium-length feathers covering most of body and tail, large feathers on arm, no feathers on leg from thigh on down.

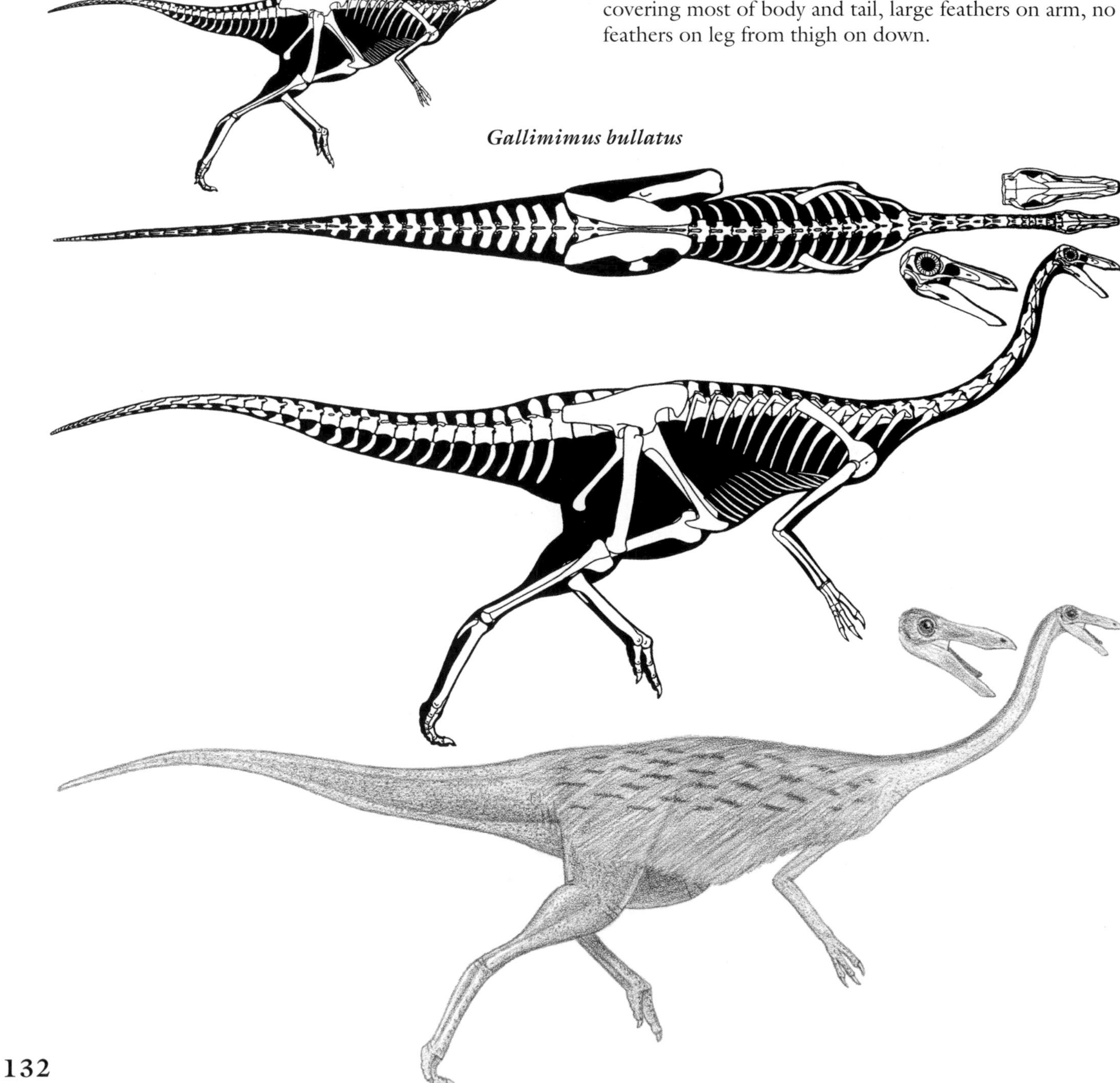

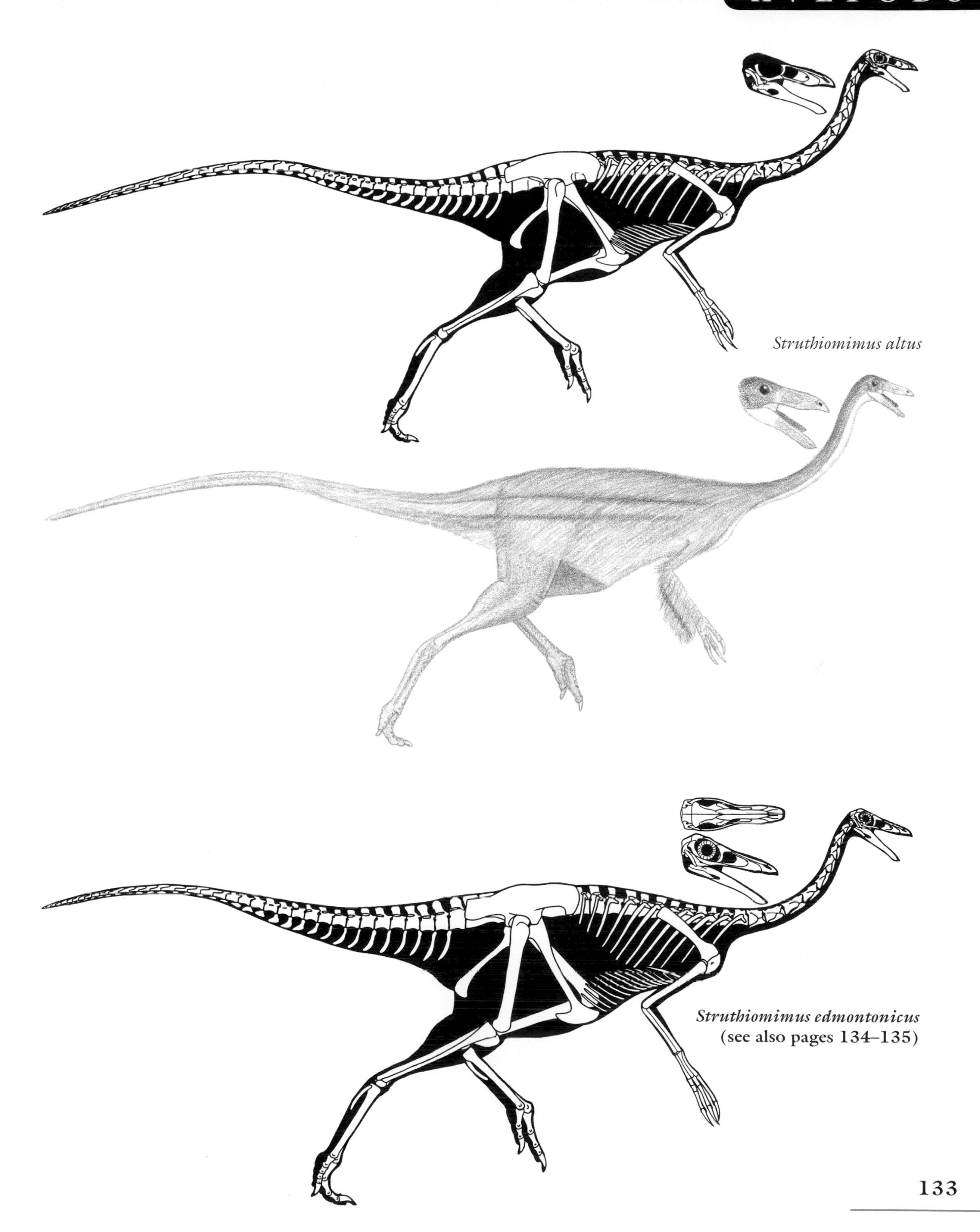

Struthiomimus altus

Struthiomimus edmontonicus
(see also pages 134–135)

AGE Late Cretaceous, latest Campanian and/or early Maastrichtian.
DISTRIBUTION AND FORMATION/S Alberta; lower Horseshoe Canyon.
HABITAT Well-watered, forested floodplain with coastal swamps and marshes, cool winters.
NOTES Probably includes *Dromicieomimus brevitertius.* Ostrichlike plumage previously predicted by this researcher. Prey of juvenile *Albertosaurus sarcophagus.*

Struthiomimus? sedens?
4.8 m (16 ft) TL, 350 kg (700 lb)

FOSSIL REMAINS Partial skeletons.
ANATOMICAL CHARACTERISTICS Leg long.
AGE Late Cretaceous, late Maastrichtian.
DISTRIBUTION AND FORMATION/S Colorado, Wyoming, South Dakota; Denver, Hell Creek, Ferris, levels uncertain.
HABITAT Well-watered coastal woodlands.
NOTES May include *Ornithomimus velox*, which is based on entirely inadequate remains; the remains that *S. sedens* is based on are problematic. May be more than one stratigraphic species. Main enemy gracile/juvenile tyrannosaurids.

Struthiomimus edmontonicus

Struthiomimus edmontonicus

MANIRAPTORS

SMALL TO GIGANTIC PREDATORY AND HERBIVOROUS NEOCOELUROSAURS OF THE MIDDLE JURASSIC TO THE END OF THE DINOSAUR ERA (WITH BIRDS SURVIVING BEYOND), MOST CONTINENTS.

ANATOMICAL CHARACTERISTICS Highly variable.

Alvarezsaurs

SMALL MANIRAPTORS FROM THE LATE JURASSIC TO THE END OF THE DINOSAUR ERA.

ANATOMICAL CHARACTERISTICS Head lightly built, long, shallow, snout semitubular, teeth increased in number and reduced in size and serrations. Neck slender. Tail moderately long. Thumb robust, three to one functional fingers.
HABITS Fed on termite and other insect colonies, using massive hand claws to break into hardened soil or wood nests and tubular snout to gather up insects, possibly with an elongated tongue. Main defense high speed.
NOTES Originally thought to be aveairfoilans very close to birds, but lack of flight adaptations and other features indicates they are outside that group and may be closely related to ornithomimosaurs. Skulls and legs developed very birdlike features.

Haplocheirids Small alvarezsaurs from the Late Jurassic of Asia.

ANATOMICAL CHARACTERISTICS Snout not as tubular as in alvarezsaurids, postorbital bar complete, teeth bladed and serrated. Arm and hand moderately long, thumb not massive, three functional fingers present. Pubis vertical, booted. Foot moderately elongated. Not as small as alvarezsaurids.
NOTES Limited distribution may be an artifact of sampling.

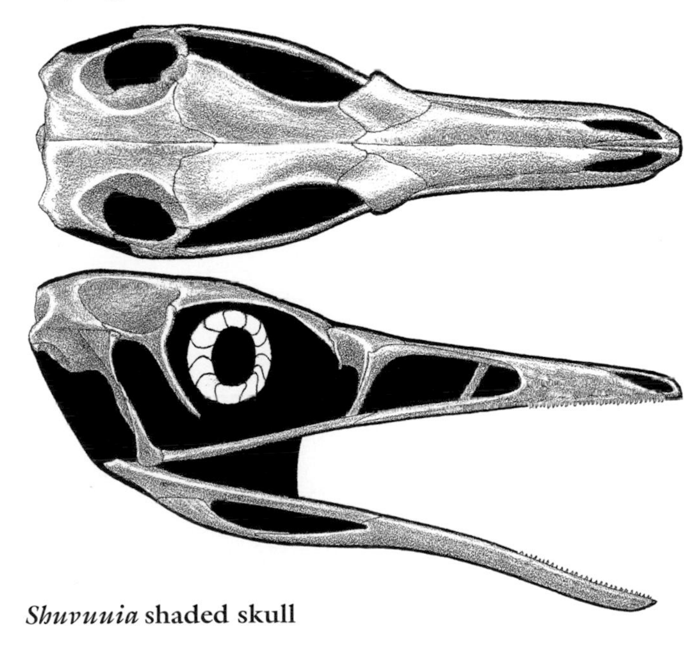
Shuvuuia **shaded skull**

Haplocheirus sollers
2 m (6 ft) TL, 18 kg (40 lb)

ANATOMICAL CHARACTERISTICS As for group.
FOSSIL REMAINS Nearly complete skull and skeleton.
AGE Late Jurassic, early Oxfordian.
DISTRIBUTION AND FORMATION/S Northwest China; upper Shishugou.
NOTES Shows that alvarezsaurs were present as early as the Jurassic. Prey of *Zuolong*.

ALVAREZSAURIDS Small alvarezsaurs from the Early Cretaceous to the end of the dinosaur era.

ANATOMICAL CHARACTERISTICS Snout semitubular, teeth very numerous and small, postorbital bar incomplete, as in birds. Neck slender. Arm very short and stout, powerfully muscled, hand reduced to one massive functional finger and robust claw. Pubis retroverted, unbooted, leg and foot very long and slender.
HABITS Main defense very high speed.
NOTES The number of Late Cretaceous genera may be excessive.

Alvarezsaurus calvoi
1 m (3.3 ft) TL, 3 kg (6.5 lb)

FOSSIL REMAINS Minority of skeleton.
ANATOMICAL CHARACTERISTICS Foot not strongly compressed from side to side.
AGE Late Cretaceous, Santonian.
DISTRIBUTION AND FORMATION/S Western Argentina; Bajo de la Carpa.
NOTES *Achillesaurus manazzonei* probably the adult of this species.

Patagonykus puertai
1 m (3.3 ft) TL, 3.5 kg (8 lb)

FOSSIL REMAINS Minority of skeleton.
ANATOMICAL CHARACTERISTICS Pubis not strongly retroverted.
AGE Late Cretaceous, Turonian or Coniacian.
HABITAT Well-watered woodlands with short dry season.
DISTRIBUTION AND FORMATION/S Western Argentina; Rio Neuquen.

Linhenykus monodactylus
0.5 m (1.6 ft) TL, 0.5 kg (1.1 lb)

FOSSIL REMAINS Minority of skeleton.
ANATOMICAL CHARACTERISTICS Standard for group.
AGE Late Cretaceous, early Campanian.
DISTRIBUTION AND FORMATION/S Mongolia; Wulansuhai.
HABITAT Semidesert with some dunes and oases.
NOTES Shared its habitat with *Wulatelong*. Prey of *Linhevenator*.

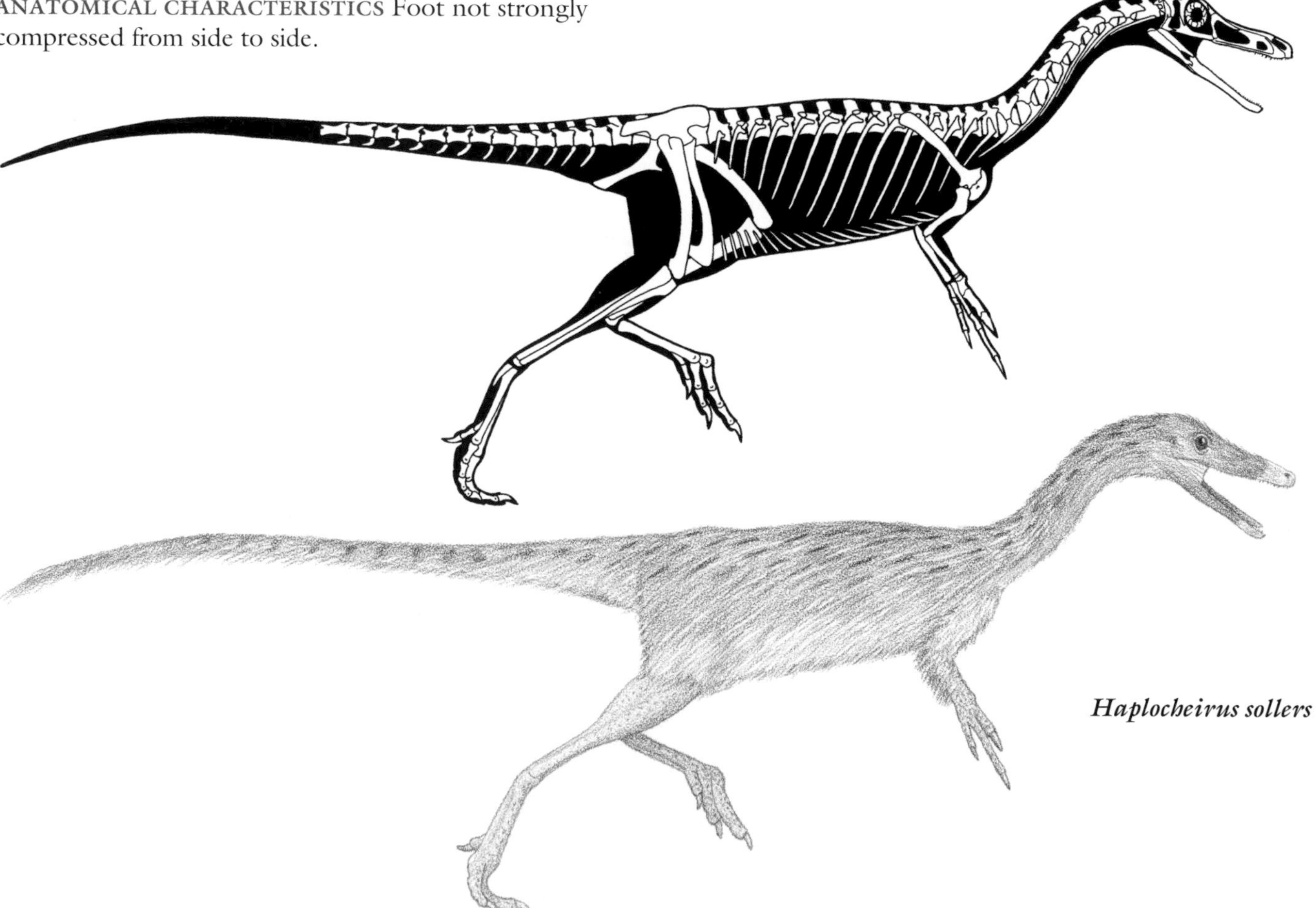

Haplocheirus sollers

Albertonykus borealis

1.1 m (3.5 ft) TL, 5 kg (12 lb)

FOSSIL REMAINS Minority of skeleton.
ANATOMICAL CHARACTERISTICS Insufficient information.
AGE Late Cretaceous, middle Maastrichtian.
DISTRIBUTION AND FORMATION/S Alberta; upper Horseshoe Canyon.
HABITAT Well-watered, forested floodplain with coastal swamps and marshes, cool winters.

Ceratonykus oculatus

0.6 m (2 ft) TL, 1 kg (2 lb)

FOSSIL REMAINS Partial skull and minority of skeleton.
ANATOMICAL CHARACTERISTICS Foot strongly compressed from side to side.
AGE Late Cretaceous, Santonian or Campanian.
DISTRIBUTION AND FORMATION/S Mongolia; Barun Goyot.
HABITAT Semidesert with some dunes and oases.
NOTES Shared its habitat with *Tylocephale* and *Bagaceratops*.

Kol ghuva

1.8 m (6 ft) TL, 20 kg (40 lb)

FOSSIL REMAINS Minority of skeleton.
ANATOMICAL CHARACTERISTICS Insufficient information.
AGE Late Cretaceous, Campanian.
DISTRIBUTION AND FORMATION/S Mongolia; Djadokhta.
HABITAT Desert with dunes and oases.
NOTES Prey of *Tsaagan*, *Velociraptor*, *Gobivenator*, and *Saurornithoides mongoliensis*.

Xixianykus zhangi

Adult size uncertain

FOSSIL REMAINS Minority of possibly juvenile skeleton.
ANATOMICAL CHARACTERISTICS Insufficient information.
AGE Late Cretaceous, late Coniacian or Santonian.
DISTRIBUTION AND FORMATION/S Eastern China; Majiacun.

Parvicursor remotus

0.4 m (1.3 ft) TL, 0.2 kg (0.4 lb)

FOSSIL REMAINS Partial skeleton.
ANATOMICAL CHARACTERISTICS Pubis strongly retroverted, foot strongly compressed from side to side.
AGE Early Late Cretaceous.
DISTRIBUTION AND FORMATION/S Mongolia; Bayenshiree Svita.

Shuvuuia deserti

1 m (3.3 ft) TL, 3.5 kg (8 lb)

FOSSIL REMAINS Two nearly complete skulls and several partial skeletons, external fibers.
ANATOMICAL CHARACTERISTICS Pubis strongly retroverted, foot strongly compressed from side to side. Short, hollow fibers on head and body.
AGE Late Cretaceous, Campanian.
DISTRIBUTION AND FORMATION/S Mongolia; Djadokhta.
HABITAT Desert with dunes and oases.
NOTES Main enemy *Velociraptor*. Shared its habitat with *Kol* and *Mononychus*.

Shuvuuia deserti

Mononykus olecranus

1 m (3.3 ft) TL, 3.5 kg (8 lb)

FOSSIL REMAINS Partial skeletons.

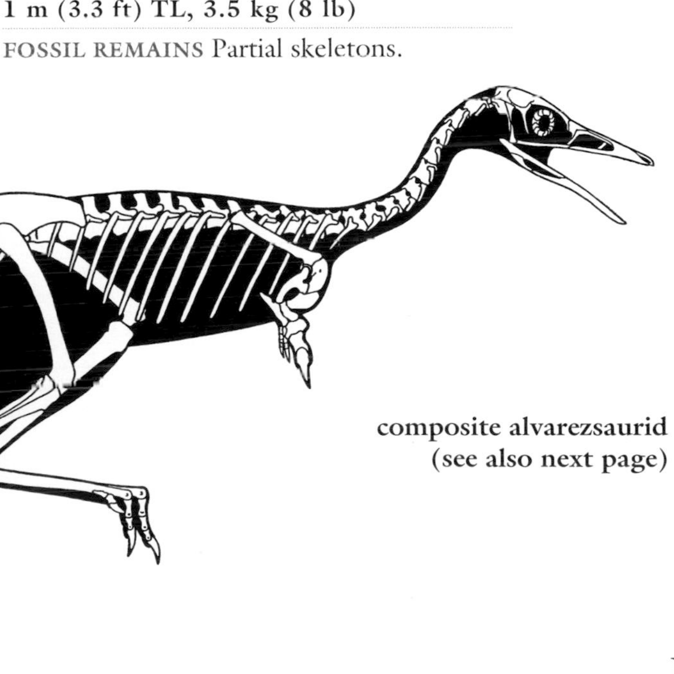

composite alvarezsaurid (see also next page)

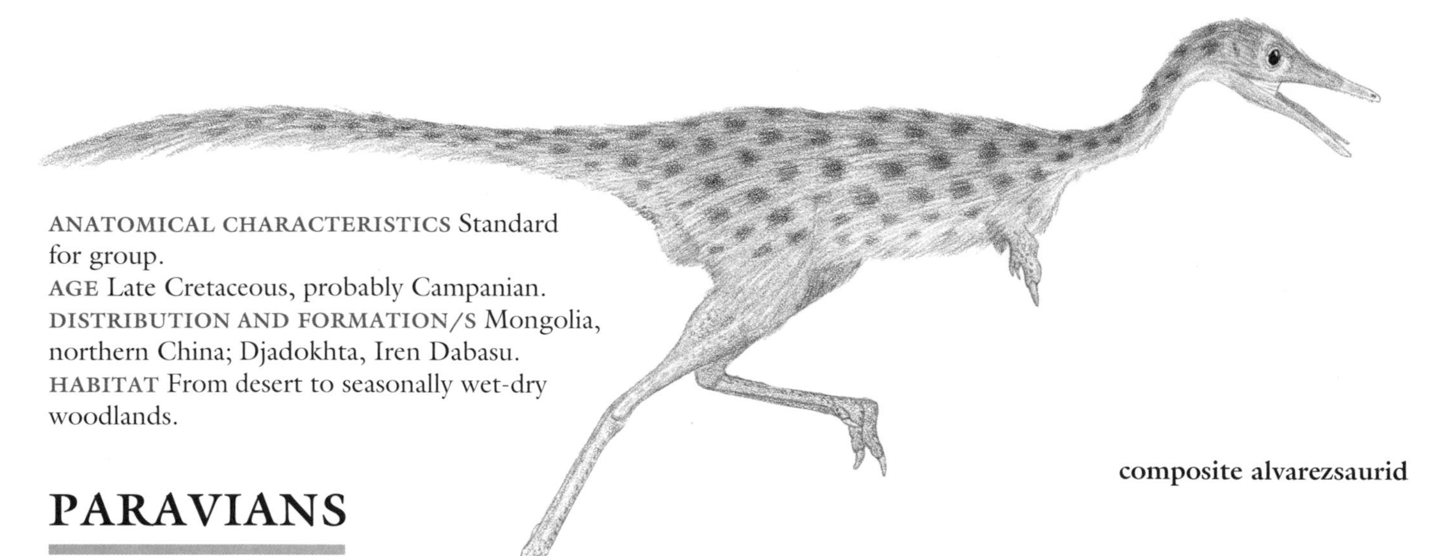

ANATOMICAL CHARACTERISTICS Standard for group.
AGE Late Cretaceous, probably Campanian.
DISTRIBUTION AND FORMATION/S Mongolia, northern China; Djadokhta, Iren Dabasu.
HABITAT From desert to seasonally wet-dry woodlands.

composite alvarezsaurid

PARAVIANS

SMALL TO GIGANTIC PREDATORY AND HERBIVOROUS MANIRAPTORS OF THE MIDDLE JURASSIC TO THE END OF THE DINOSAUR ERA (WITH BIRDS SURVIVING BEYOND), MOST CONTINENTS.

ANATOMICAL CHARACTERISTICS Highly variable. Wrist mobility increased.

Scansoriopterygids

SMALL PARAVIANS OF THE MIDDLE/LATE JURASSIC OF ASIA.

ANATOMICAL CHARACTERISTICS Head short and broad, lower jaw shallow, a few procumbent, pointed teeth at front of jaws. Neck medium length. Body unusually shallow because pubis quite short. Tail not very long. Small ossified sternal plate present in at least some examples. Arm long, hand strongly asymmetrical because outer finger hyperelongated, extra lateral elongated strut in at least some examples, arm and hand appear to support wing membrane, finger claws large. Pelvis shallow, at least in juveniles, leg not elongated, hallux partly reversed.
HABITS Probably strongly arboreal, trunk hugging facilitated by flattened body. Some form of flight, at least gliding or possibly marginally powered, apparently present. Probably insectivorous.
NOTES Relationships with other neocoelurosaurs uncertain. Appears to have been a brief evolutionary experiment in dinosaur flight that lost out to the birdlike aveairfoilans, the earliest known examples of which—*Anchiornis*, *Aurornis*, *Eosinopteryx*, and *Xiaotingia*—lived in the same habitat.

Scansoriopteryx heilmanni
0.5 m (1 ft) TL, 0.25 kg (0.5 lb)

FOSSIL REMAINS Probably minority of an adult, and majority of two juvenile skeletons, all with complete or partial skulls, some feather fibers.
ANATOMICAL CHARACTERISTICS Shallow midline crest on snout, teeth small. Tail not abbreviated, but its total length and the length of adults are uncertain. Arm and hand very elongated, extra strut present.

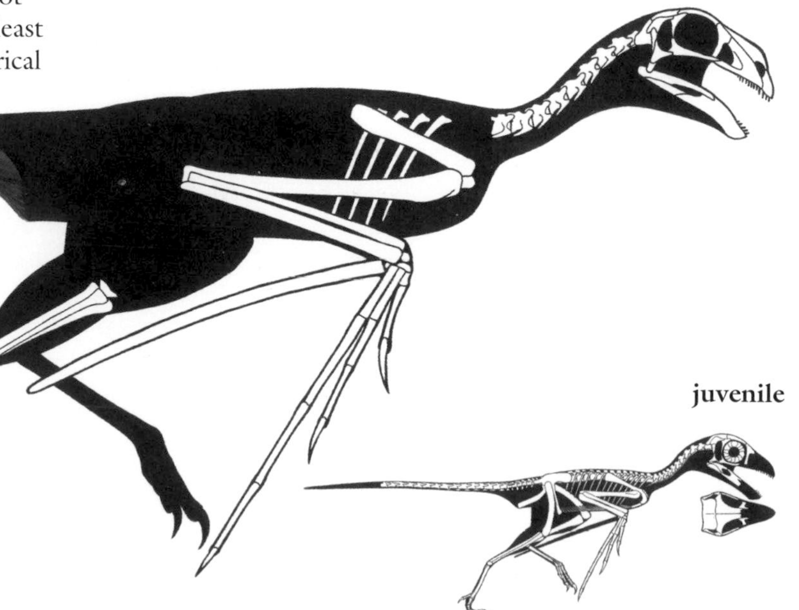

Scansoriopteryx heilmanni

juvenile

AGE Middle or Late Jurassic, Callovian or Oxfordian.
DISTRIBUTION AND FORMATION/S Northeast China; Tiaojishan.
NOTES The name *Scansoriopteryx* appears to have edged out *Epidendrosaurus ningchengensis* in the race for priority. *Yi qi*—the shortest dinosaur name—is probably the adult form of this species. Lack of preserved extra hand strut in juveniles may be due to lack of sufficient growth and/or ossification. Larger wing suggests better flight performance than in *Epidexipteryx*.

Epidexipteryx hui

0.3 m (1 ft) TL, 0.22 kg (0.5 lb)

FOSSIL REMAINS Complete skull and majority of skeleton, some feather fibers.
ANATOMICAL CHARACTERISTICS Teeth of lower jaw large and procumbent. Tail abbreviated. Small ossified sternal plates present. Hand strongly elongated, presence of extra strut uncertain. Arm feathers apparently short, four very long banded feathers trail from tail, simpler feathers cover much of body.
AGE Middle or Late Jurassic, Callovian or Oxfordian.
DISTRIBUTION AND FORMATION/S Northeast China; Tiaojishan.
HABITAT Well-watered forests and lakes.
HABITS Long tail feathers for display within the species.
NOTES May compete with *Scansoriopteryx* and *Aurornis* for title of smallest known dinosaur—all three contenders lived in the same habitat.

AVEAIRFOILANS

SMALL TO GIGANTIC PREDATORY AND HERBIVOROUS PARAVIANS OF THE LATE JURASSIC TO THE END OF THE DINOSAUR ERA (WITH BIRDS SURVIVING BEYOND), MOST CONTINENTS.

ANATOMICAL CHARACTERISTICS Highly variable. Head toothed to toothless and beaked; when teeth are present, serrations tend to be reduced in some manner or absent. Tail very long to very short. Shoulder girdle usually like that in birds, with horizontal scapula blade and vertical, anterior-facing coracoid, furcula often large, arm very long to short, wrist usually with a large, half-moon-shaped carpal block that allowed arm to be folded like bird's, hand usually long. Brains enlarged, semiavian in form. Pennaceous feathers often present. Overall appearance very birdlike.
ONTOGENY Growth rates apparently moderate.
HABITS Reproduction generally similar to that of ratites and tinamous; in at least some cases males incubated the eggs and were probably polygamous; egg hatching in a given clutch not synchronous.
NOTES Paravians with feather wings or ancestors with same that are in the clade that includes extant birds. Prone to evolving and especially losing flight, perhaps repeating cycle in some cases.

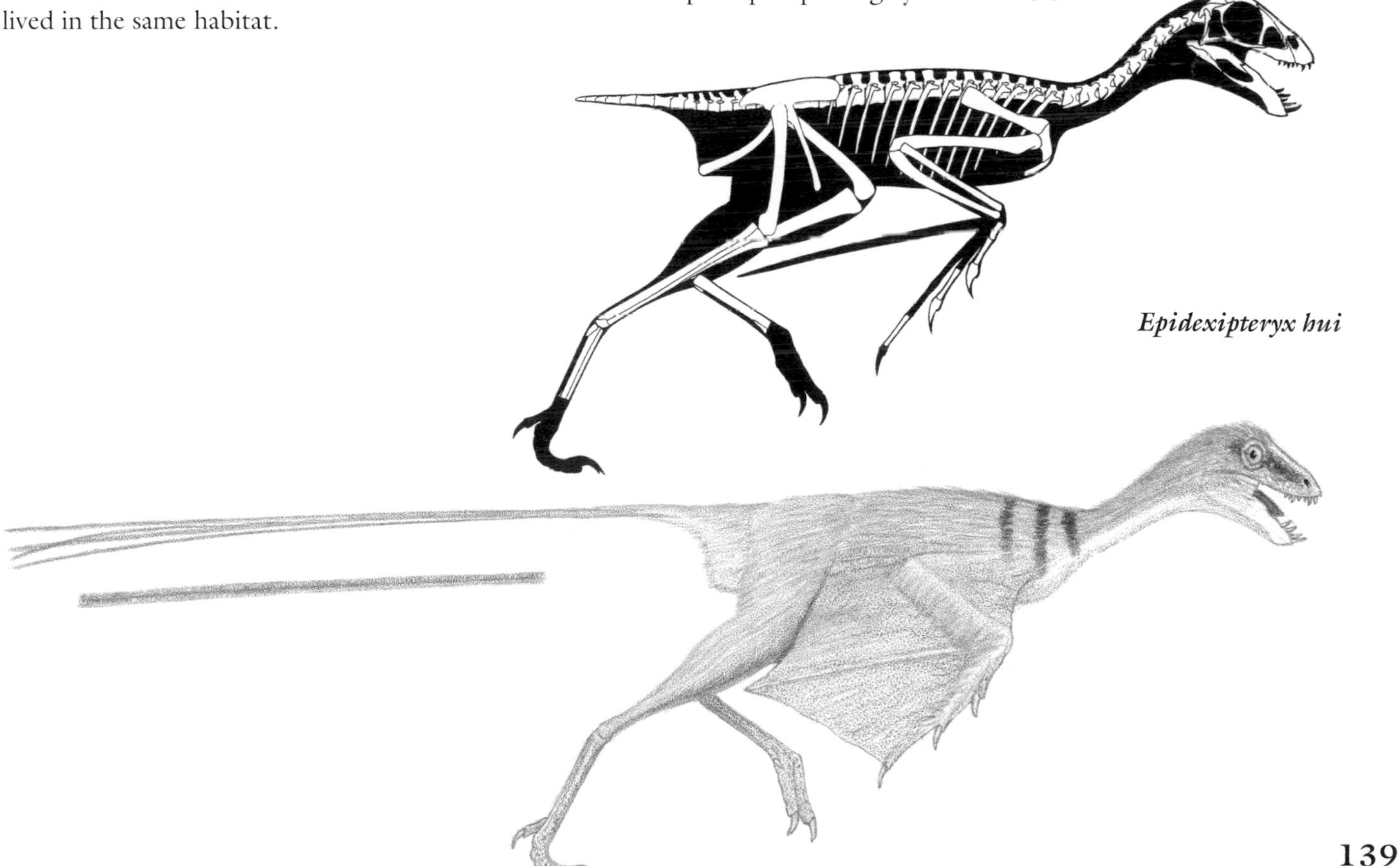

Epidexipteryx hui

AVEAIRFOILAN MISCELLANEA

Yixianosaurus longimanus
1 m (3 ft) TL, 1 kg (2.2 lb)

FOSSIL REMAINS Arms.
ANATOMICAL CHARACTERISTICS Hand elongated, finger claws large and strongly hooked.
AGE Early Cretaceous, Barremian.
DISTRIBUTION AND FORMATION/S Northeast China; Yixian.
HABITAT Well-watered forests and lakes, winters chilly with some snow.
HABITS Well-developed arms suitable for handling prey and climbing.
NOTES Relationships very uncertain, may not be a aveairfoilan.

Balaur bondoc
2.5 m (9 ft) TL, 15 kg (30 lb)

FOSSIL REMAINS Partial skeleton.
ANATOMICAL CHARACTERISTICS Robustly built. Upper hand elements fused, outer finger reduced. Foot short, broad, inner toe large so foot effectively tetradactyl, inner two toes hyperextendable, with large claws.
AGE Late Cretaceous, middle Maastrichtian.
DISTRIBUTION AND FORMATION/S Romania; Sebes.
HABITAT Forested island.
HABITS Possibly herbivorous.
NOTES Originally thought to be a predatory dromaeosaur, more probably a secondarily flightless, herbivorous, island-adapted near-bird.

DEINONYCHOSAURS

SMALL TO MEDIUM-SIZED PREDATORY AND OMNIVOROUS AVEAIRFOILANS OF THE LATE JURASSIC TO THE END OF THE DINOSAUR ERA, ON MOST CONTINENTS.

ANATOMICAL CHARACTERISTICS Fairly variable. Eyes face partly forward and some degree of stereo vision possible, tooth serrations reduced or absent. Tail slender, base very flexible, especially upward. Arm and hand well developed, sometimes very long, finger claws large hooks. Second toe hyperextendable and/or claw enlarged.
HABITS Very agile, sophisticated predators and omnivores, prey varying from insects and small game to big game. Climbing ability generally good, especially in smaller species, longer-armed species, and juveniles, hyperextendable toe probably used as hook and spike during climbing in species living in areas with trees. Two-toed trackways confirm that hyperextendable claw was normally carried clear of ground; relative scarcity of such trackways indicates most deinonychosaurs did not spend much time patrolling shorelines.
NOTES It is not known whether deinonychosaurs were a distinct group or a stage within aviremigians, membership also uncertain. Presence of large sternal plates, ossified sternal ribs, and ossified uncinate processes in most flightless deinonychosaurs indicates they were secondarily flightless.

DEINONYCHOSAUR MISCELLANEA

NOTE Neither the placement of these aveairfoilans in the deinonychosaurs nor their placement within the group is certain.

Pedopenna daohugouensis
1 m (3 ft) TL, 1 kg (2.2 lb)

FOSSIL REMAINS Lower leg and foot, some feathers.
ANATOMICAL CHARACTERISTICS Claw on second toe markedly larger than others. Large and symmetrical pennaceous feathers on upper feet.
AGE Middle or Late Jurassic, Callovian or Oxfordian.
DISTRIBUTION AND FORMATION/S Northeast China; Tiaojishan.
HABITAT Well-watered forests and lakes, winters chilly with some snow.
HABITS Symmetry of foot feathers indicates they were for display rather than aerodynamic purposes.
NOTES May not be a deinonychosaur, or may be an archaeopterygian.

Richardoestesia* (or *Ricardoestesia*) *gilmorei
2 m (3.5 ft) TL, 10 kg (20 lb)

FOSSIL REMAINS Minority of skull.
ANATOMICAL CHARACTERISTICS Lower jaw very slender.
AGE Late Cretaceous, late Campanian.
DISTRIBUTION AND FORMATION/S Alberta; Dinosaur Park, level uncertain.
HABITAT Well-watered, forested floodplain with coastal swamps and marshes, cool winters.
HABITS Hunted small game, possibly fished.
NOTES May be a dromaeosaurid or troodontid. Remains imply type was common in other late Late Cretaceous habitats.

ARCHAEOPTERYGIANS

SMALL-BODIED, LARGE-ARMED PREDATORY DEINONYCHOSAURS LIMITED TO THE LATER JURASSIC OF EURASIA.

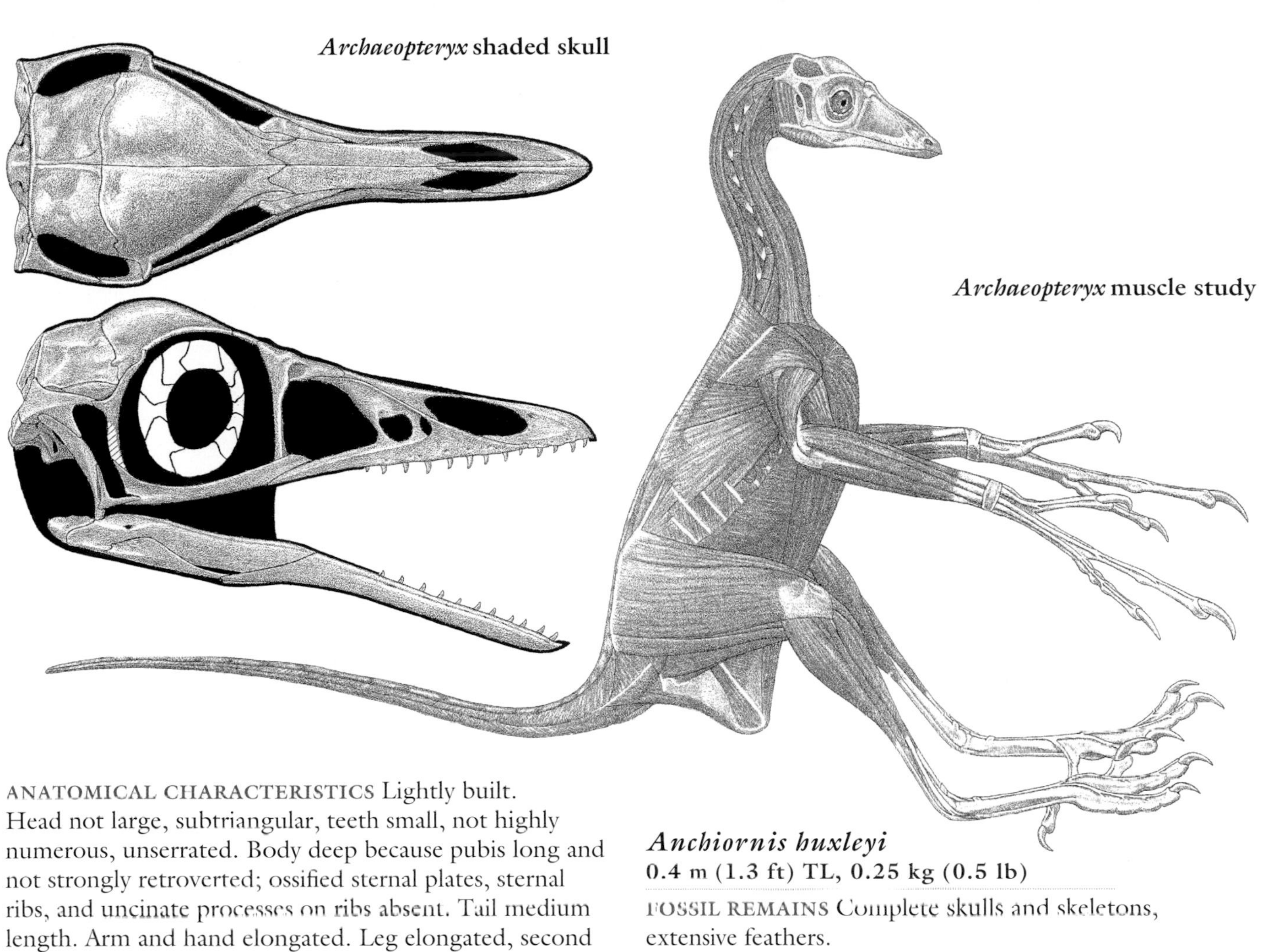

Archaeopteryx shaded skull

Archaeopteryx muscle study

ANATOMICAL CHARACTERISTICS Lightly built. Head not large, subtriangular, teeth small, not highly numerous, unserrated. Body deep because pubis long and not strongly retroverted; ossified sternal plates, sternal ribs, and uncinate processes on ribs absent. Tail medium length. Arm and hand elongated. Leg elongated, second toe slender and claw not markedly enlarged.
HABITS Diet included insects, small game, fish. Flight performance minimal to significant.
NOTES Extent of group and relationships of these early aveairfoilans are uncertain.

Anchiornis huxleyi

0.4 m (1.3 ft) TL, 0.25 kg (0.5 lb)

FOSSIL REMAINS Complete skulls and skeletons, extensive feathers.
ANATOMICAL CHARACTERISTICS Head short, subtriangular. Arm not as long as leg. Toe claws not

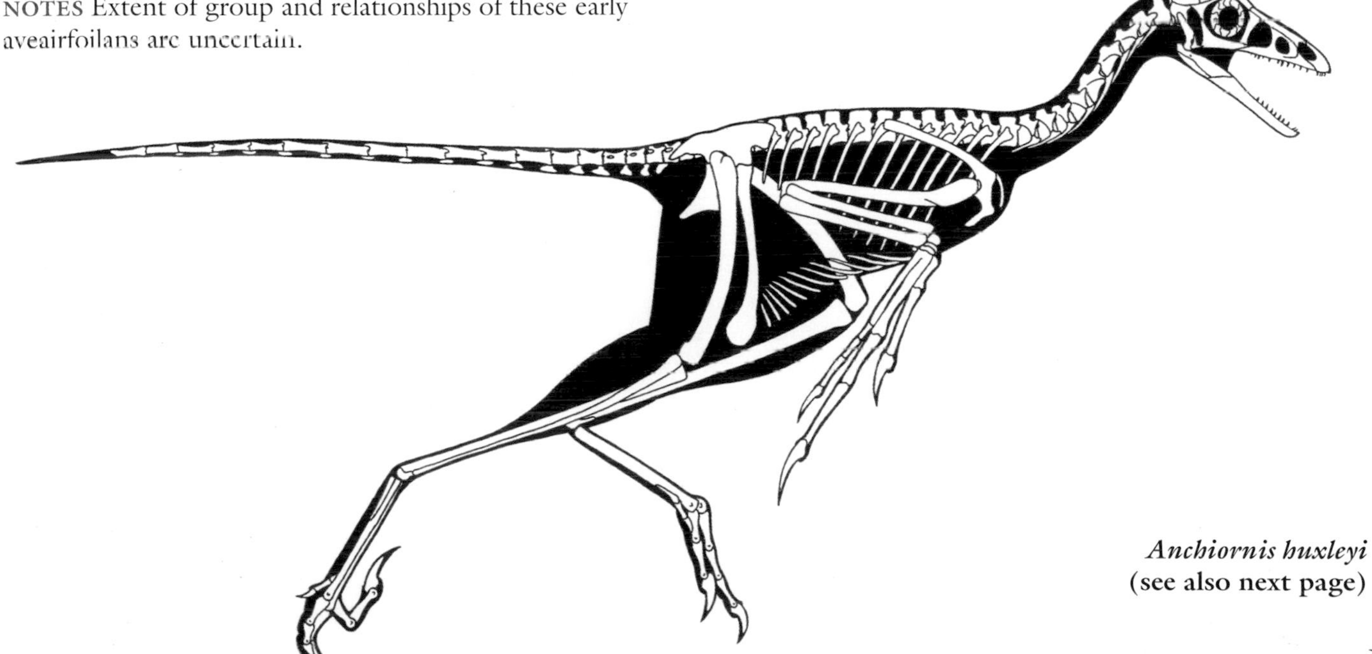

Anchiornis huxleyi
(see also next page)

Anchiornis huxleyi

strongly curved. Well-developed head feather crest, arm primary feathers symmetrical, moderately long on arm, same on leg, short feathers on toes, most feathers dark gray or black, head feathers speckled reddish brown, head crest partly brown or reddish brown, broad whitish bands on arm and leg feathers interrupted by narrow dark irregular bands, primary feather tips black.
AGE Middle or Late Jurassic, Callovian or Oxfordian.
DISTRIBUTION AND FORMATION/S Northeast China; Tiaojishan.
HABITAT Well-watered forests and lakes, winters chilly with some snow.
HABITS Lack of strongly curved toe claws indicates was not highly arboreal. Arm wing too small and primary feathers too symmetrical for flight, at most a parachuting ability was present, potentially secondarily flightless.
NOTES May be a troodont. Shared its habitat with *Scansoriopteryx*, *Epidexipteryx*, *Pedopenna*, *Aurornis*, *Eosinopteryx*, and *Xiaotingia*.

Aurornis xui

0.5 m (1.5 ft) TL, 0.25 kg (0.5 lb)

FOSSIL REMAINS Complete skull and skeleton, poorly preserved feathers.
ANATOMICAL CHARACTERISTICS Head elongated. Arm not as long as leg. Toe claws not strongly curved. Preservation of feathers insufficient for assessment.
AGE Middle or Late Jurassic, Callovian or Oxfordian.
DISTRIBUTION AND FORMATION/S Northeast China; Tiaojishan.

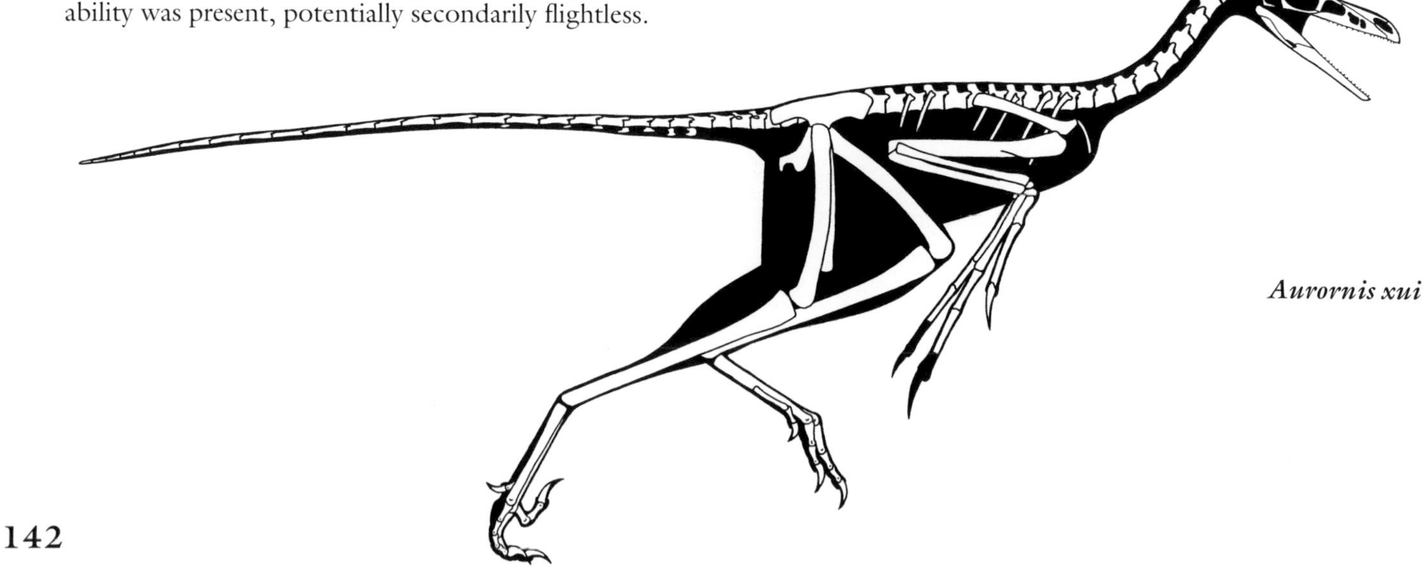

Aurornis xui

Anchiornis huxleyi

HABITAT Well-watered forests and lakes, winters chilly with some snow.
HABITS Not highly arboreal. Arms too short for significant flight, potentially secondarily flightless.
NOTES May be a troodont. Competes with *Epidexipteryx* and *Scansoriopteryx* for title of smallest known dinosaur.

Eosinopteryx brevipenna
0.3 m (1 ft) TL, 0.1 kg (0.2 lb)

FOSSIL REMAINS Complete skull and skeleton, extensive feathers.
ANATOMICAL CHARACTERISTICS Head short, subrectangular. Arm not as long as leg. Toe claws not strongly curved. Primary feathers on arm symmetrical, moderately long, pennaceous feathers apparently absent on leg and tail.
AGE Middle or Late Jurassic, Callovian or Oxfordian.
DISTRIBUTION AND FORMATION/S Northeast China; Tiaojishan.
HABITAT Well-watered forests and lakes, winters chilly with some snow.
HABITS Not highly arboreal. Arm wing too small and primary feathers too symmetrical for flight, absence of tail feathers indicates same, at most a parachuting ability was present, potentially secondarily flightless.
NOTES May be a troodont.

Xiaotingia zhengi
0.6 m (2 ft) TL, 0.6 kg (1.4 lb)

FOSSIL REMAINS Majority of skull and skeleton, poorly preserved feathers.
ANATOMICAL CHARACTERISTICS Head short, subtriangular, teeth blunt. Arm not as long as leg. Toe claws strongly curved. Long pennaceous feathers on leg, otherwise preservation of feathers insufficient for assessment.
AGE Middle or Late Jurassic, Callovian or Oxfordian.
DISTRIBUTION AND FORMATION/S Northeast China; Tiaojishan.
HABITS Strongly arboreal. Arm wing too small for significant flight, at most a parachuting ability was present, potentially secondarily flightless.
HABITAT Well-watered forests and lakes, winters chilly with some snow.
NOTES May be a dromaeosaur or troodont, or the closest known relative of *Archaeopteryx.*

Archaeopteryx lithographica
0.5 m (1.7 ft) TL, 0.7 m (2.3 ft) wingspan, w0.5 kg (1.1 lb)

FOSSIL REMAINS Several complete and partial skulls and skeletons, extensive feathers, completely known.
ANATOMICAL CHARACTERISTICS Head subtriangular, snout pointed, teeth subconical, unserrated. Body shallower because pubis more retroverted. Tail modest in length. Arm longer and more strongly built than leg. Hallux fairly large and semireversed, toe claws varying in curvature. Most of body covered by short feathers, arm and hand supporting well-developed, broad chord wings made of asymmetrical feathers. Lower leg supporting a modest-sized feather airfoil, tail supporting a long set of feather vanes forming an airfoil.

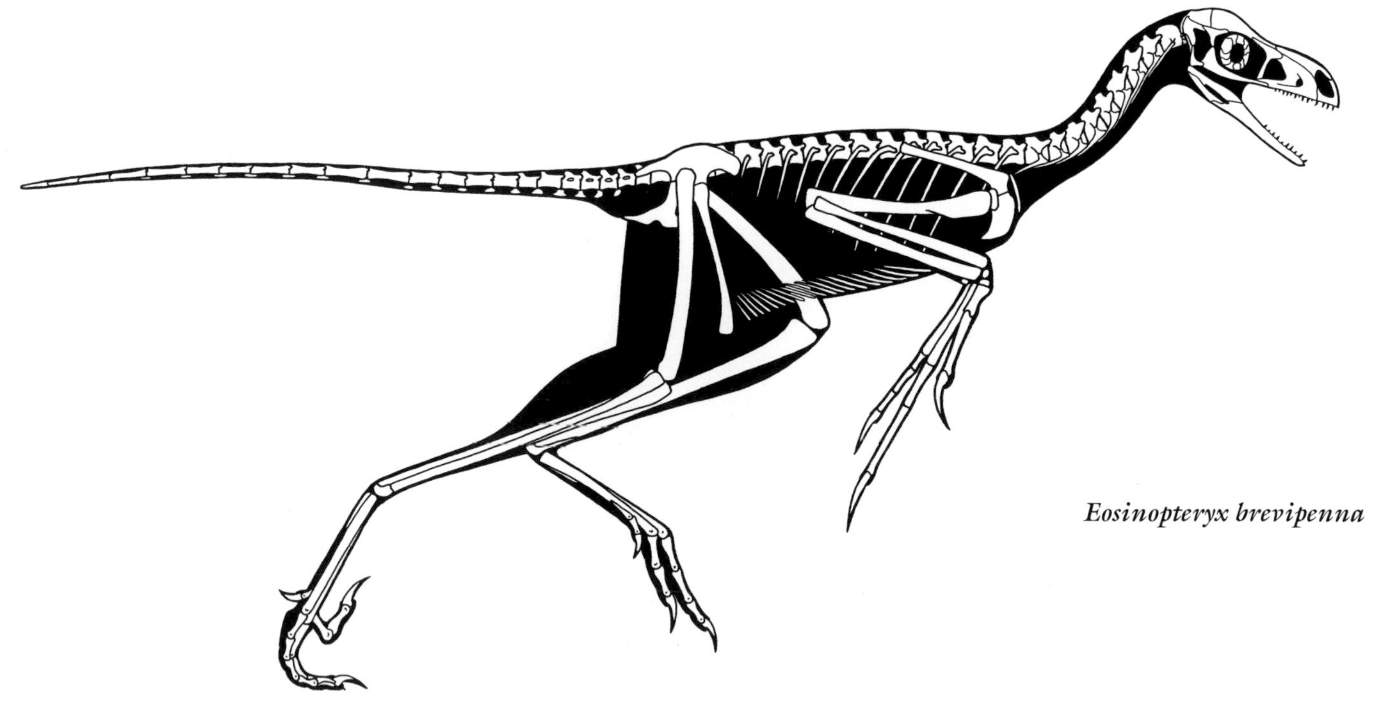

Eosinopteryx brevipenna

Archaeopteryx lithographica

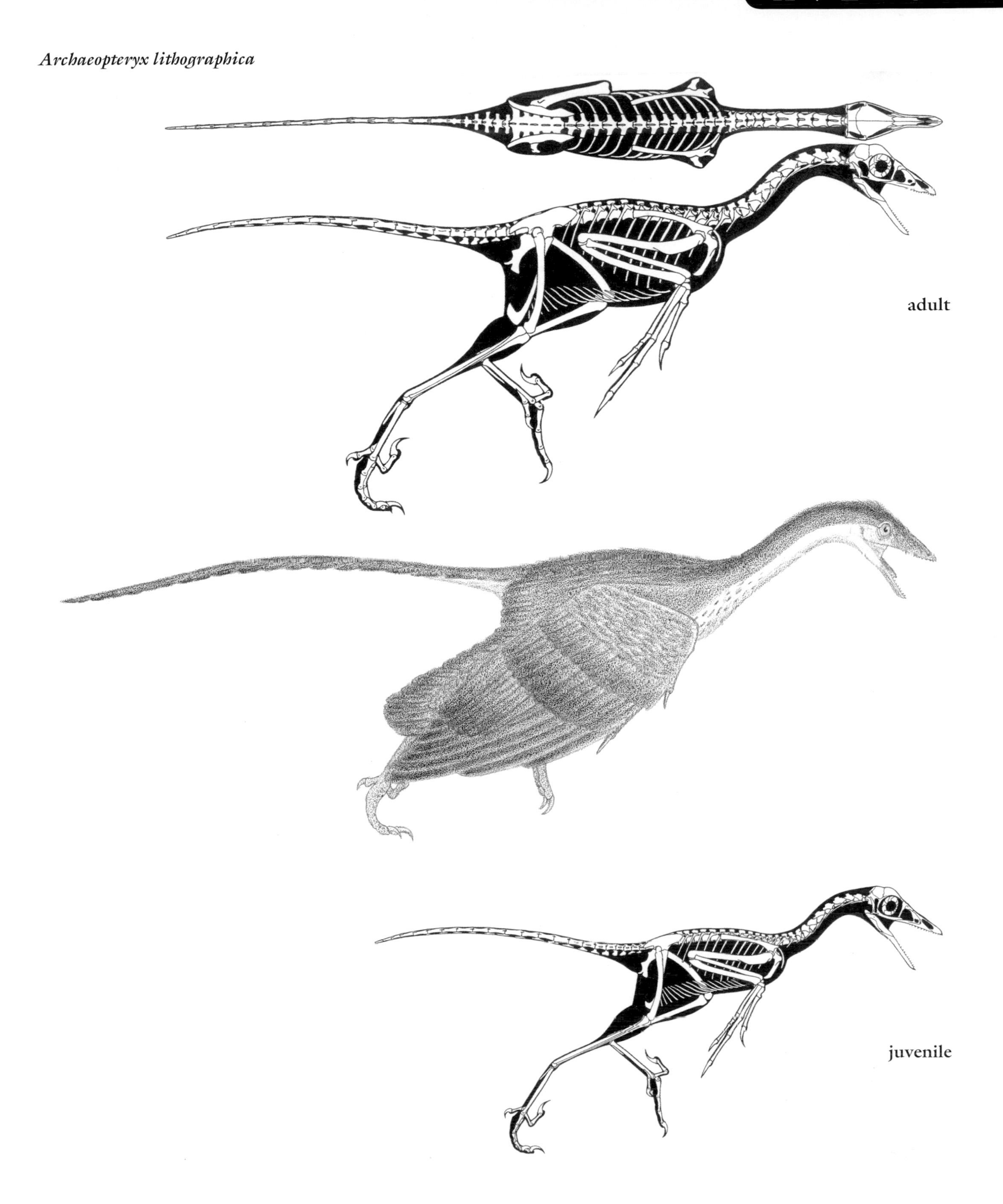

AGE Late Jurassic, late Kimmeridgian.
DISTRIBUTION AND FORMATION/S Southern Germany; lower and upper Solnhofen.
HABITAT Found as drift in lagoonal deposits near probably arid, brush- and mangrove-covered islands then immediately off the northeast coast of North America.
HABITS Probably semiarboreal. Capable of low-grade powered flight and gliding probably a little inferior to that of *Sapeornis*. Legs could not splay out nearly flat so feathers probably used as auxiliary rudders and air brakes. May have been able to swim with wings. Defense included climbing and flight, and claws.
NOTES It is uncertain whether there is one or more species within *Archaeopteryx*, which includes *Wellnhoferia*. Long known as the first bird, it is now known to be one among an array of later Jurassic dinobirds. Some researchers contend all specimens are juveniles and that maximum mass was over 25 percent heavier. This and *Xiaotingia* may form family Archaeopterygidae. Prey of *Compsognathus* and *Juravenator*.

DROMAEOSAURS

SMALL TO MEDIUM-SIZED FLYING AND FLIGHTLESS PREDATORY DEINONYCHOSAURS OF THE CRETACEOUS, ON MOST CONTINENTS.

ANATOMICAL CHARACTERISTICS Fairly variable. Teeth bladed, serrations limited to back edge. Arm large to very large. Tail long, ensheathed in very long and slender ossified tendons. Large ossified sternal plates, sternal ribs, and uncinate processes present. Large sickle claw on robust hyperextendable toe. Olfactory bulbs enlarged.

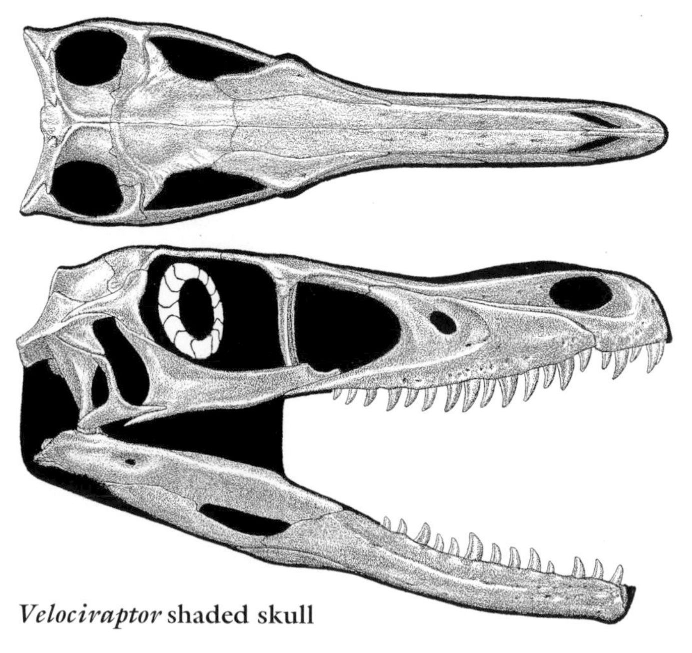

***Velociraptor* shaded skull**

HABITS Archpredators equipped to ambush, pursue, and dispatch relatively large prey using sickle claw as a primary weapon, as well as smaller game. Sickle claw also facilitated climbing taller prey. Leaping performance when arboreal or attacking prey excellent. Flight performance from well developed to none; the retention of the pterosaur-like ossified tendons in flightless dromaeosaurids is further evidence that they were secondarily flightless. Juveniles of large-bodied species with longer arms may have possessed some flight ability.
NOTES Teeth imply that small members of the group may have evolved by the Late Jurassic. Fragmentary remains indicate presence in Australia.

DROMAEOSAURIDS Small to medium-sized flying and flightless predatory deinonychosaurs of the Cretaceous, on most continents.

Microraptorines Small flying dromaeosaurids of the Cretaceous of the Northern Hemisphere.

ANATOMICAL CHARACTERISTICS Highly uniform. Lightly built. Postorbital bar probably incomplete, as in birds, frontmost teeth not serrated. Body shallow because pubis strongly retroverted. Furcula and sternal plates large, arm and hand very large, longer and stronger than leg, outer, upper hand bone curved, central finger stiffened and base flattened to better support fully developed, broad chord wings made of asymmetrical feathers. Hip socket more upwardly directed than usual, leg very long, supported well-developed second wing made of asymmetrical feathers that extended onto upper foot, head of femur more spherical than in other theropods, sickle claw well developed, other toe claws strongly curved. Part of head and most of body covered by short, simple feathers, tail supported a long set of feather vanes forming an airfoil.
HABITAT Well-watered forests and lakes.
HABITS Large stiff foot feathers not well suited for running, and strongly curved toe claws indicate strong arboreality. Better development of sternum, ribs, uncinates, more streamlined body, modified upper hand and central finger, larger outer arm wing, extra leg wing, and pterosaur-like tail indicate sinornithosaurs were better-powered fliers than *Archaeopteryx* and *Sapeornis*. Leg appears to have been more splayable sideways than normal in theropods because hip socket faces more upward, femoral head is more spherical, and legs are often splayed sideways in articulated specimens (unlike most articulated theropods, including *Archaeopteryx*, which are usually preserved on their sides), but hind wings were not flappable and possibly provided extra wing area during glides or soaring, and air brakes when landing or ambushing prey from the air.

NOTES Similar limb design indicates all microraptorines had forewings and hind wings, preserved only in *Sinornithosaurus zhaoianus* (wing feathers are missing from a number of other Yixian bird species). The limitation of these primitive dromaeosaurids to the Northern Hemisphere may reflect lack of sufficient sampling.

Sinornithosaurus (= *Graciliraptor*) *lujiatunensis*

1 m (3 ft) TL, 1.5 kg (3.5 lb)

FOSSIL REMAINS Minority of skull and skeleton.
ANATOMICAL CHARACTERISTICS Standard for group.
AGE Early Cretaceous, Barremian.
DISTRIBUTION AND FORMATION/S Northeast China; lowest Yixian.
HABITAT Well-watered forests and lakes, winters chilly with some snow.
NOTES May be the direct ancestor of *S. millenii*.

Sinornithosaurus millenii

1.2 m (4 ft) TL, 3 kg (7 lb)

FOSSIL REMAINS Nearly complete skull and majority of skeleton, poorly preserved feathers.
ANATOMICAL CHARACTERISTICS Head large, long and shallow, all teeth serrated. Sternal plates not fused together.
AGE Early Cretaceous, early Aptian.
DISTRIBUTION AND FORMATION/S Northeast China; Yixian.
HABITAT Well-watered forests and lakes, winters chilly with some snow.
HABITS Prey included *Caudipteryx* and *Psittacosaurus*.
NOTES *S. haoina* may be an immature *S. millenii*. Poor preservation of feathers has caused many to problematically presume it was not winged despite flattening of central finger. Shared its habitat with *S. yangi*, *Tianyuraptor*, *Zhenyuanlong*, and *Sinosauropteryx*.

Sinornithosaurus (= *Changyuraptor*) *yangi*

1 m (3.4 ft) TL, 2 kg (4 lb)

FOSSIL REMAINS Complete but poorly preserved skull and skeleton, some feathers.
ANATOMICAL CHARACTERISTICS Leg feathers preserved, tail feathers exceptionally long.
AGE Early Cretaceous, early Aptian.
DISTRIBUTION AND FORMATION/S Northeast China; Yixian.

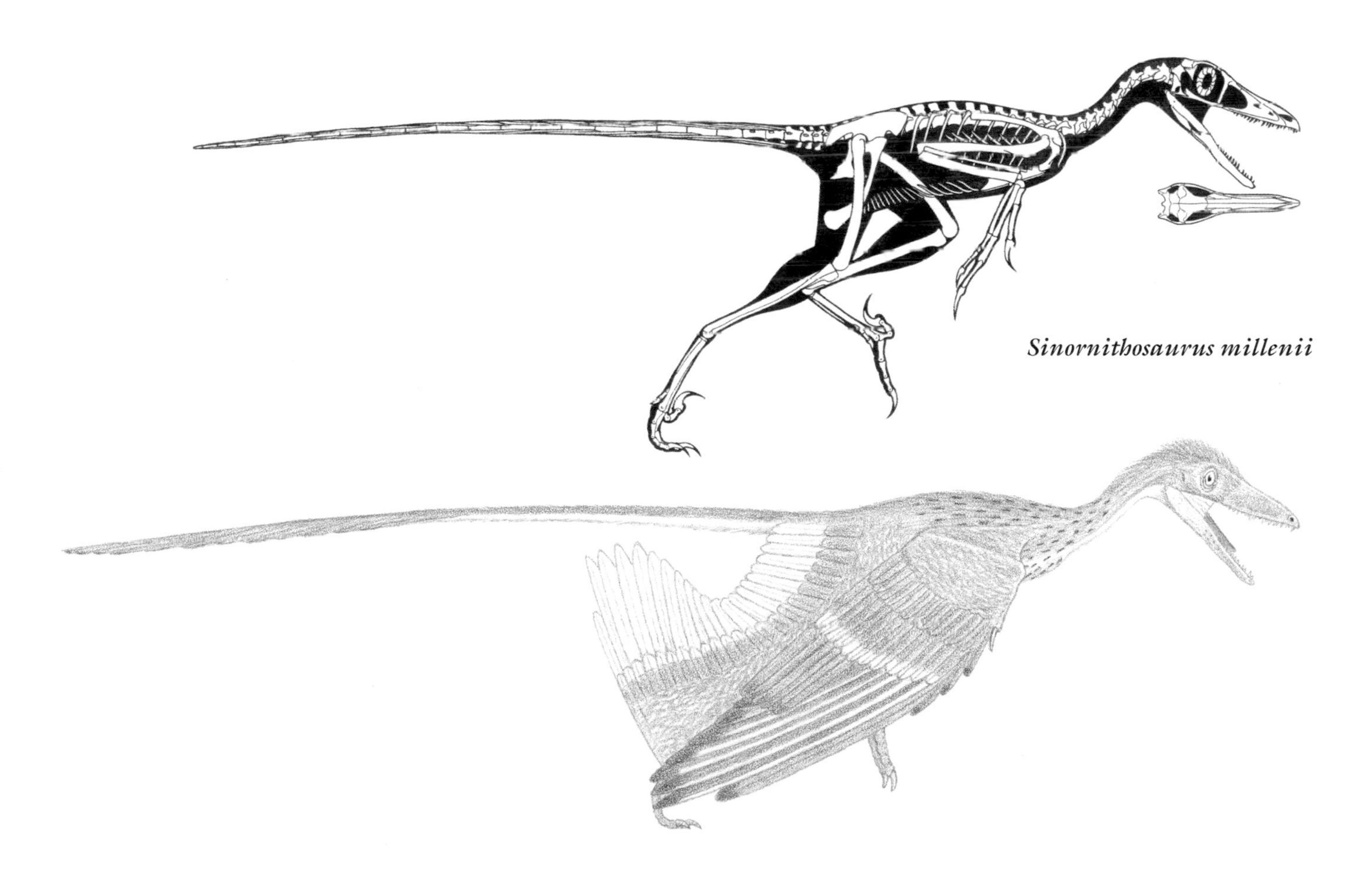

Sinornithosaurus millenii

HABITAT Well-watered forests and lakes, winters chilly with some snow.
HABITS Similar to *S. millenii.*
NOTES Poor preservation questions separation from *S. millenii.* Shows that large microraptorines were winged fliers.

Sinornithosaurus (or *Microraptor*) *zhaoianus*
0.7 m (2.5 ft) TL, 0.75 m (2.6 ft) wingspan, 0.6 kg (1.3 lb)

FOSSIL REMAINS A number of complete and partial skulls and skeletons, extensive well-preserved feathers.
ANATOMICAL CHARACTERISTICS Head not proportionally large, subtriangular, teeth less bladelike and less serrated than in *S. millenii.* Neck rather short. Sternals fused into a single plate. Leg feathers preserved, overall feathers black with iridescence.
AGE Early Cretaceous, early or middle Aptian.
DISTRIBUTION AND FORMATION/S Northeast China; Jiufotang.
HABITAT Well-watered forests and lakes, winters chilly with some snow.
HABITS Strongly curved toe claws indicate strong arboreality. Both lesser size and less bladed and serrated teeth indicate that this attacked smaller prey than *S. millenii*; gut contents include small fish, birds, mammals.
NOTES The identical *Microraptor gui* and *Cryptovolans pauli* very probably belong to *S.* (*Microraptor*) *zhaoianus*, which differs only in modest details and size from earlier *S. millenii.*

Sinornithosaurus (or *Shanag*) *ashile*
1.5 m (3 ft) TL, 5 kg (10 lb)

FOSSIL REMAINS Minority of skull.
ANATOMICAL CHARACTERISTICS Insufficient information.
AGE Early Cretaceous.
DISTRIBUTION AND FORMATION/S Mongolia; Oosh beds.
HABITS As perhaps the largest known microraptorine, probably hunted the largest prey.
NOTES Too little is known to distinguish this from *Sinornithosaurus.*

Hesperonychus elizabethae
1 m (3 ft) TL, 1.5 kg (3.5 lb)

FOSSIL REMAINS Minority of a few skeletons.
ANATOMICAL CHARACTERISTICS Well-preserved hip socket directed more upward than in nonmicroraptorine dinosaurs, indicating leg could be splayed more strongly sideways.
AGE Late Cretaceous, late Campanian.

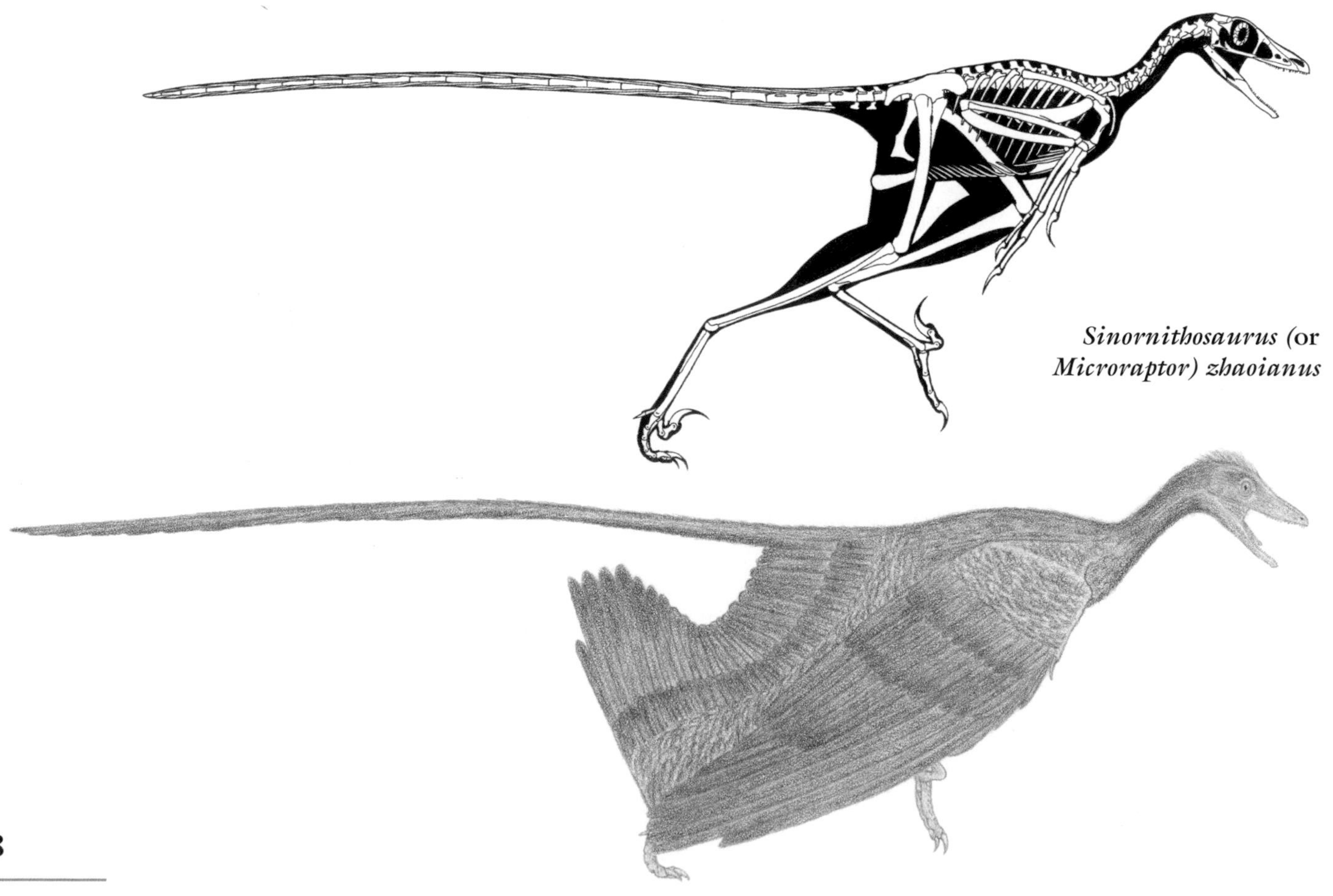

Sinornithosaurus (or *Microraptor*) *zhaoianus*

DISTRIBUTION AND FORMATION/S Alberta; at least middle and upper Dinosaur Park.
HABITAT Well-watered, forested floodplain with coastal swamps and marshes, cool winters.
NOTES *Hesperonychus* indicates that microraptorines survived into the late Late Cretaceous. Appears to have been fairly common, was the prey of *Dromaeosaurus*, *Saurornitholestes*, and *Saurornithoides*.

Dromaeosaurid miscellanea
NOTES The relationships of these dromaeosaurids are uncertain, as is that of fragmentary Early Cretaceous, North American *Yurgovuchia doellingi*.

Tianyuraptor ostromi
2.5 m (8 ft) TL, 10 kg (22 lb)

FOSSIL REMAINS Complete immature skull and skeleton.
ANATOMICAL CHARACTERISTICS Head lightly built, snout pointed. Body shallow because pubis short. Arm not elongated. Leg fairly elongated.
AGE Early Cretaceous, early Aptian.
DISTRIBUTION AND FORMATION/S Northeast China; Yixian.
HABITAT Well-watered forests and lakes, winters chilly with some snow.
HABITS Largely terrestrial predator. Prey included *Caudipteryx* and *Psittacosaurus*.
NOTES May be a close relative of microraptorines. Shared its habitat with *Huaxiagnathus*, *Sinocalliopteryx*, and *Sinornithosaurus millenii*; main enemy *Zhenyuanlong*.

Zhenyuanlong suni
2.3 m (7 ft) TL, 8 kg (18 lb)

FOSSIL REMAINS Complete skull and nearly complete skeleton, extensive feathers.
ANATOMICAL CHARACTERISTICS Head robustly built, subrectangular. Body shallow because pubis short. Arm not elongated. Leg fairly elongated, sickle claw moderate size. Most of body covered by short feathers, arm and hand supported well-developed, modest-sized winglets, no pennaceous feathers on leg, tail supported a long set of feather vanes.

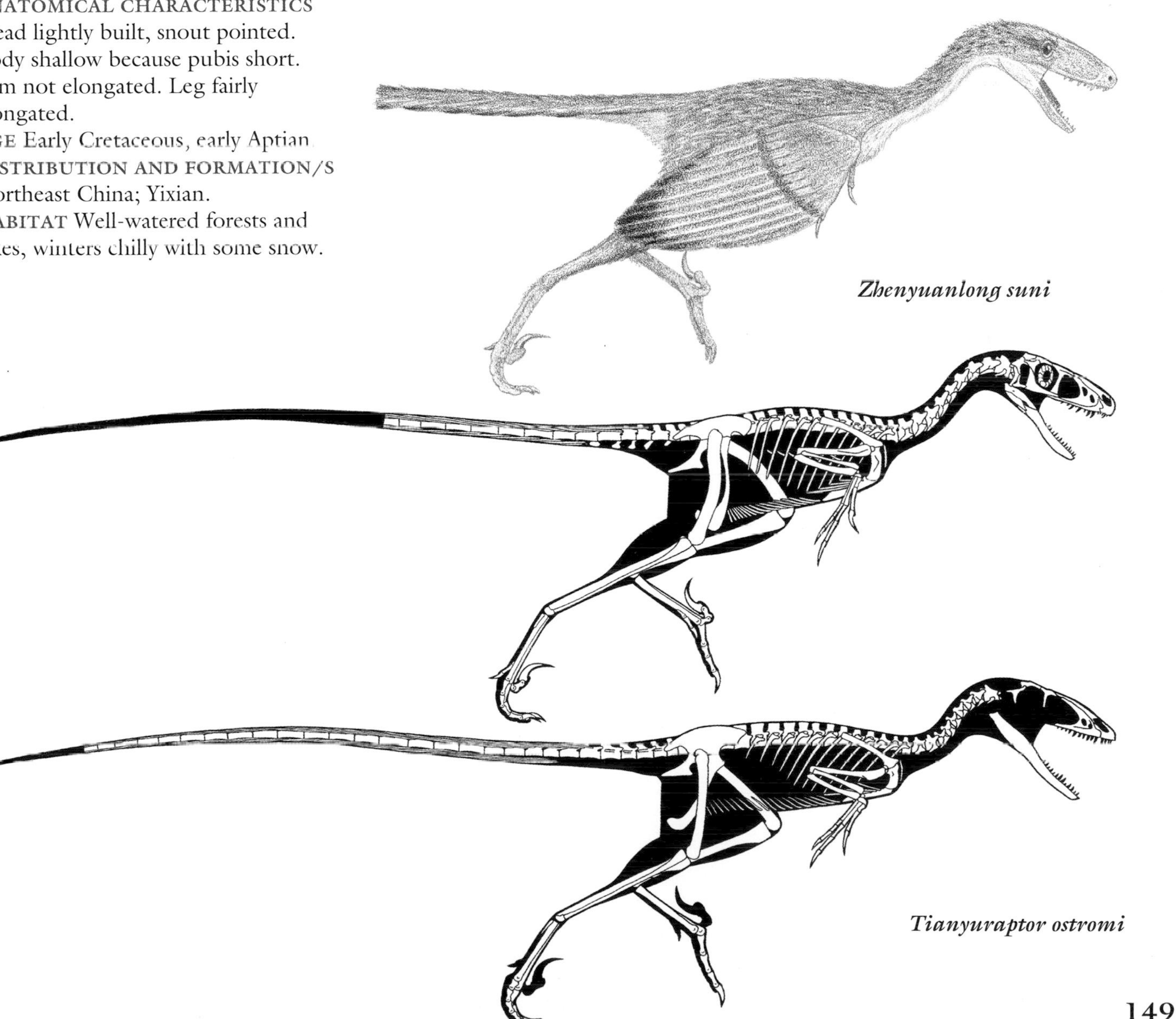

Zhenyuanlong suni

Tianyuraptor ostromi

AGE Early Cretaceous, early Aptian.
DISTRIBUTION AND FORMATION/S Northeast China; Yixian.
HABITAT Well-watered forests and lakes, winters chilly with some snow.
HABITS Largely terrestrial predator, modest-sized sickle claw indicates it did not hunt particularly large prey. Winglets too small for significant flight, primary function probably display.
NOTES May be a close relative of microraptorines but is not especially closely related to *Tianyuraptor*.

Mahakala omnogovae
0.5 m (1.7 ft) TL, 0.4 (1 lb)

FOSSIL REMAINS Minority of skull and partial skeleton.
ANATOMICAL CHARACTERISTICS Insufficient information.
AGE Late Cretaceous, Campanian.
DISTRIBUTION AND FORMATION/S Mongolia; Djadokhta.
HABITAT Desert with dunes and oases.
HABITS Hunted small game and insects.
NOTES A basal-looking dromaeosaurid despite late appearance, perhaps because of a long evolution away from a flying ancestry.

Luanchuanraptor henanensis
1.1 m (3.5 ft) TL, 2.5 kg (6 lb)

FOSSIL REMAINS Minority of skull and skeleton.
ANATOMICAL CHARACTERISTICS Insufficient information.
AGE Late Cretaceous.
DISTRIBUTION AND FORMATION/S Central China; Qiupa.
NOTES May be an Asian unenlagiine.

Unenlagiines Small to medium-sized flying and flightless dromaeosaurids limited to the Late Cretaceous of the Southern Hemisphere.

ANATOMICAL CHARACTERISTICS Variable. Pubis vertical.
HABITS Appear to have been more prone to fishing.
NOTES Unenlagiines indicate that dromaeosaurids experienced a radiation of distinctive forms in the Southern Hemisphere that included fliers; losses and/or independent evolution of flight probably occurred.

Rahonavis ostromi
0.7 m (2.2 ft) TL, 1 kg (2 lb)

FOSSIL REMAINS Partial skeleton.
ANATOMICAL CHARACTERISTICS Arm very large. Pubis vertical. Large sickle claw on hyperextendable toe. Quill nodes on trailing edge of upper arm indicate large flight feathers.
AGE Late Cretaceous, Campanian.
DISTRIBUTION AND FORMATION/S Madagascar; Maevarano.
HABITAT Seasonally dry floodplain with coastal swamps and marshes.
HABITS Diet may have included aquatic and/or terrestrial small prey, sickle claw possibly used to help dispatch larger prey. Capable of powered flight superior to that of *Archaeopteryx* and *Sapeornis*. Good climber and leaper. Defense included climbing and flight as well as sickle claws.
NOTES Not yet known whether the head was similar to the elongated form seen in larger unenlagiines.

Buitreraptor gonzalezorum
1.5 m (5 ft) TL, 3 kg (7 lb)

FOSSIL REMAINS Majority of skull and skeleton.
ANATOMICAL CHARACTERISTICS Head very long, shallow and narrow, especially snout and lower jaw, teeth small, numerous, nonserrated. Arm long, but hand rather short. Leg long and gracile. Sickle claw not large.
AGE Late Cretaceous, early Cenomanian.
DISTRIBUTION AND FORMATION/S Western Argentina; Candeleros.
HABITAT Short wet season, otherwise semiarid with open floodplains and riverine forests.
HABITS Hunted small game, probably fished. Main defense high speed or sickle claw.

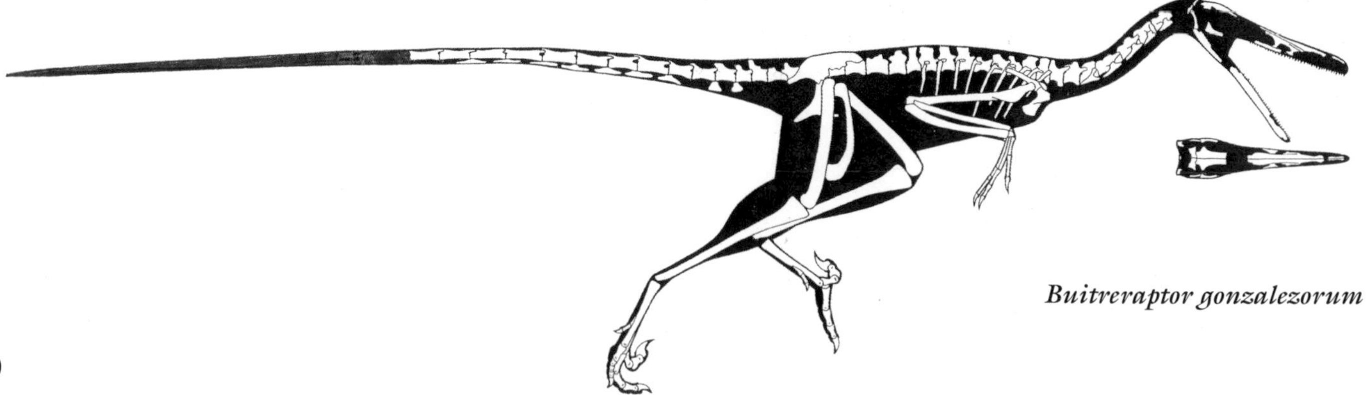

Buitreraptor gonzalezorum

Unenlagia comahuensis
3.5 m (12 ft) TL, 75 kg (170 lb)

FOSSIL REMAINS Minority of skeleton.
ANATOMICAL CHARACTERISTICS Sickle claw medium sized.
AGE Late Cretaceous, late Turonian.
DISTRIBUTION AND FORMATION/S Western Argentina; Portezuelo.
HABITAT Well-watered woodlands with short dry season.
HABITS Probably able to dispatch fairly large prey, including *Patagonykus*, as well as *Macrogryphosaurus*. May have fished like other unenlagiines.
NOTES Probably includes *U. paynemili* and *Neuquenraptor argentinus*. Shared its habitat with *Megaraptor*.

Austroraptor cabazai
6 m (20 ft) TL, 300 kg (700 lb)

FOSSIL REMAINS Majority of skull and minority of skeleton.
ANATOMICAL CHARACTERISTICS Head very long, shallow, especially snout and lower jaw, teeth small, numerous, conical. Upper arm fairly short (rest unknown).
AGE Late Cretaceous, late Campanian.
DISTRIBUTION AND FORMATION/S Central Argentina; Allen.
HABITAT Semiarid coastline.
HABITS Probably fished, terrestrial prey also plausible.
NOTES Head form parallels that of spinosaurs.

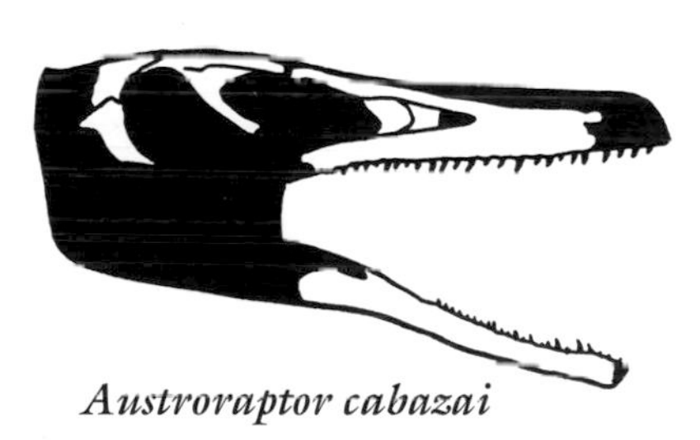
Austroraptor cabazai

Dromaeosaurines Small to large dromaeosaurids of the Cretaceous of the Northern Hemisphere.

ANATOMICAL CHARACTERISTICS Fairly variable. Robustly built. Teeth large, frontmost D-shaped in cross section.
HABITS Strong skulls and large, strong teeth indicate that dromaeosaurines used their heads to wound prey more than other dromaeosaurids.

Utahraptor ostrommaysorum
5.5 m (18 ft) TL, 300 kg (600 lb)

FOSSIL REMAINS Numerous skeletal parts, juvenile to adult.
ANATOMICAL CHARACTERISTICS Very robustly built. Sickle toe claw large.
AGE Early Cretaceous, probably Barremian.
DISTRIBUTION AND FORMATION/S Utah; Lower Cedar Mountain.
HABITAT Short wet season, otherwise semiarid with floodplain prairies and open woodlands, and riverine forests.
HABITS Not especially fast, an ambush predator that preyed on large dinosaurs including *Iguanacolossus*.

Achillobator giganticus
5 m (16 ft) TL, 250 kg (500 lb)

FOSSIL REMAINS Minority of skull and skeleton.
ANATOMICAL CHARACTERISTICS Head fairly deep. Pubis vertical, sickle claw large.
AGE Early Late Cretaceous.
DISTRIBUTION AND FORMATION/S Mongolia; Bayanshiree.
HABITS Preyed on large dinosaurs.

Atrociraptor marshalli
2 m (6 ft) TL, 15 kg (30 lb)

FOSSIL REMAINS Partial skull and small portion of skeleton.
ANATOMICAL CHARACTERISTICS Head deep, teeth stout.
AGE Late Cretaceous, latest Campanian and/or early Maastrichtian.
DISTRIBUTION AND FORMATION/S Alberta, Montana; lower Horseshoe Canyon.
HABITAT Well-watered, forested floodplain with coastal swamps and marshes, cool winters.
HABITS Able to attack relatively large prey; used strong head and teeth to wound prey more than usual for velociraptorines.

Dromaeosaurus albertensis (Illustration overleaf)
2 m (7 ft) TL, 15 kg (30 lb)

FOSSIL REMAINS Majority of skull, skeletal fragments.
ANATOMICAL CHARACTERISTICS Head broad and robust. Teeth large, stout, front tooth D-shaped in cross section.
AGE Late Cretaceous, late Campanian.
DISTRIBUTION AND FORMATION/S Alberta; Dinosaur Park, level uncertain.
HABITAT Well-watered, forested floodplain with coastal swamps and marshes, cool winters.
HABITS Able to attack relatively large prey.
NOTES Not common in its habitat.

Dakotaraptor steini
6 m (20 ft) TL, 350 kg (700 lb)

FOSSIL REMAINS Minority of skeleton.
ANATOMICAL CHARACTERISTICS Lightly built. Leg

Dromaeosaurus albertensis

long. Quill nodes on trailing edge of upper arm indicate large feather array.
AGE Late Cretaceous, latest Maastrichtian.
DISTRIBUTION AND FORMATION/S South Dakota; upper Hell Creek.
HABITAT Well-watered coastal woodlands, climate warmer than earlier in Maastrichtian.
NOTES Largest known dromaeosaurid. Main competition juvenile *Tyrannosaurus*.

Velociraptorines Small to medium-sized dromaeosaurids of the Cretaceous of the Northern Hemisphere.

ANATOMICAL CHARACTERISTICS Fairly uniform. Snout long. Skeleton lightly built.
NOTES Very fragmentary, latest Cretaceous, North American *Acheroraptor temertyorum* probably belongs to this group.

Bambiraptor feinbergi
1.3 m (4 ft) TL, 5 kg (11 lb)

FOSSIL REMAINS Almost complete skull and skeleton, less complete skeleton.
ANATOMICAL CHARACTERISTICS Lightly built. Head subrectangular. Arm and hand quite long. Pubis moderately retroverted, leg long, sickle claw large.
DISTRIBUTION AND FORMATION/S Montana; Upper Two Medicine.
AGE Late Cretaceous, middle and/or late Campanian.
HABITAT Seasonally dry upland woodlands.
HABITS Probably a generalist able to use head, arms, and sickle claw to handle and wound prey of various sizes, including small ornithopods and protoceratopsids. Long arms indicate good climbing ability and may be compatible with limited flight ability, especially in juveniles.

Bambiraptor feinbergi

Adasaurus mongoliensis
2 m (7 ft) TL, 15 kg (30 lb)

FOSSIL REMAINS Partial skull and skeleton.
ANATOMICAL CHARACTERISTICS Somewhat robustly built. Pubis moderately retroverted, sickle claw not large.
AGE Late Cretaceous, late Campanian and/or early Maastrichtian.
DISTRIBUTION AND FORMATION/S Mongolia; Nemegt.
HABITAT Well-watered woodland with seasonal rain, winters cold.
HABITS Did not use sickle claw as much as other dromaeosaurids.

Deinonychus antirrhopus
3.3 m (11 ft) TL, 60 kg (130 lb)

FOSSIL REMAINS Majority of several skulls and partial skeletons.
ANATOMICAL CHARACTERISTICS Arm fairly long. Head lightly built, very large and long, subtriangular, snout arched in one specimen, may be depressed in another. Pubis moderately retroverted, leg moderately long, sickle claw large.
AGE Early Cretaceous, middle Albian.
DISTRIBUTION AND FORMATION/S Montana; upper Cloverly.
HABITAT Short wet season, otherwise semiarid with floodplain prairies and open woodlands, and riverine forests.
HABITS Probably a generalist that ambushed and pursued small to big game. The most common predator in its habitat; the most abundant prey was *Tenontosaurus tilletti*. Juveniles longer armed than adults and may have had limited flight ability.
NOTES One of the classic dromaeosaurids, the primary basis of the *Jurassic Park* "raptors." If snouts were arched in some individuals and depressed in others, may represent sexes. Remains from lower in the Cloverly Formation usually placed in this species are probably one or more different taxa.

Tsaagan mangas
2 m (7 ft) TL, 18 kg (40 lb)

FOSSIL REMAINS Two nearly complete skulls, one with majority of skeleton.
ANATOMICAL CHARACTERISTICS Head lightly built, snout not as shallow and depressed as in *Velociraptor mongoliensis*.
AGE Late Cretaceous, Campanian.
DISTRIBUTION AND FORMATION/S Mongolia, northern China; Djadokhta.

Tsaagan mangas

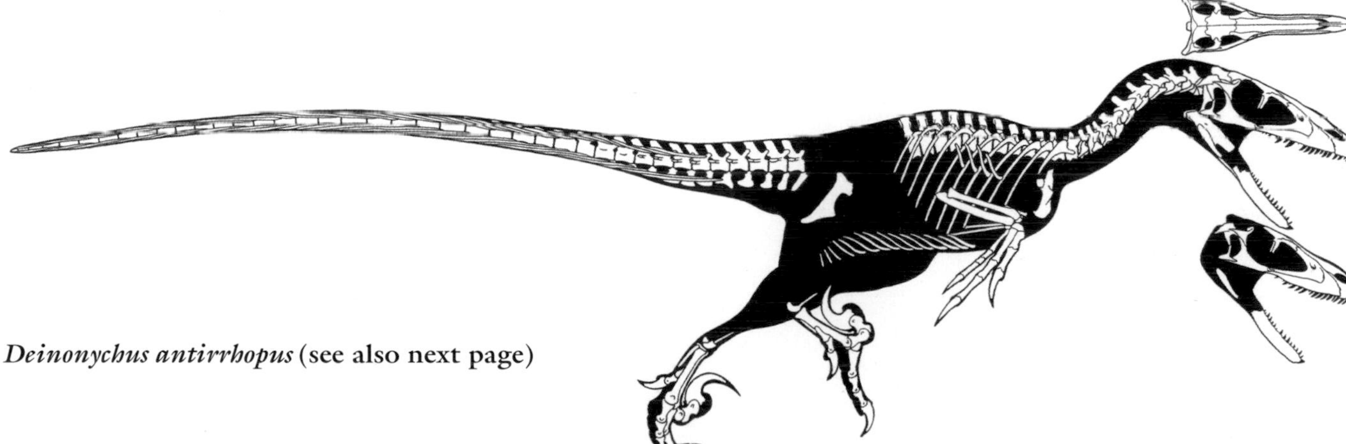

Deinonychus antirrhopus (see also next page)

Deinonychus antirrhopus

HABITAT Desert with dunes and oases.
HABITS Similar to *Velociraptor*.
NOTES Very probably includes *Linheraptor exquisitus*; it is unlikely that the Djadokhta habitat had more than two similar velociraptorine species. Differing degrees of depression of snout may represent sexes.

Velociraptor mongoliensis
2.5 m (8 ft) TL, 25 kg (55 lb)

FOSSIL REMAINS A number of complete and partial skulls and skeletons, juvenile to adult, completely known.
ANATOMICAL CHARACTERISTICS Head lightly built, long, snout shallow, strongly depressed in juveniles and less so in adults. Arm fairly long. Pubis strongly retroverted. Sickle claw large. Quill nodes on trailing edge of upper arm indicate large feather array.
AGE Late Cretaceous, Campanian.
DISTRIBUTION AND FORMATION/S Mongolia, northern China; Djadokhta, Bayan Mandahu?
HABITAT Desert with dunes and oases.
HABITS Probably a generalist that ambushed and pursued small to big game. Famous fighting pair preserves a *Velociraptor* and *Protoceratops* locked in combat. Other prey included *Citipati osmolskae* and *Oviraptor*.
NOTES The other classic dromaeosaurid. Probably includes the contemporary *V. osmolskae*. The most common predator in its habitat, which it shared with *Tsaagan*, *Gobivenator*, *Saurornithoides mongoliensis*, and *Byronosaurus*—the diversity of small predatory avepods in this barren formation is remarkable.

Saurornitholestes langstoni?
1.3 m (4 ft) TL, 5 kg (10 lb)

FOSSIL REMAINS Minority of skulls and skeletons.
ANATOMICAL CHARACTERISTICS Snout not shallow.
AGE Late Cretaceous, late Campanian.
DISTRIBUTION AND FORMATION/S Alberta, possibly

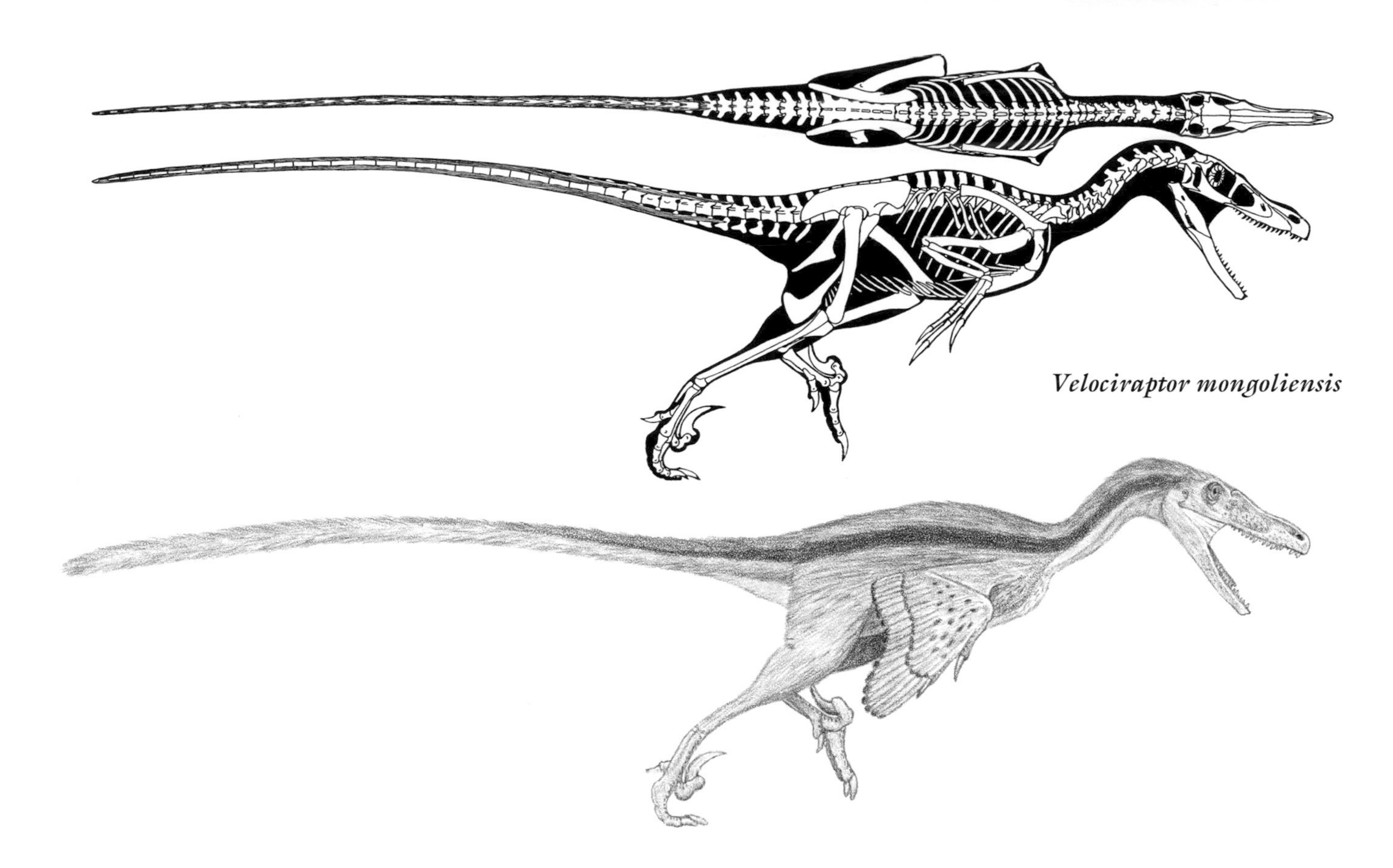

Velociraptor mongoliensis

Montana; at least lower and middle Dinosaur Park, possibly Upper Two Medicine.
HABITAT Well-watered, forested floodplain with coastal swamps and marshes and drier upland woodlands.
NOTES Based on questionably adequate remains. The most common small predator in its habitat, which it shared with *Stenonychosaurus inequalis*.

TROODONTS

SMALL TO MEDIUM-SIZED OMNIVOROUS DEINONYCHOSAURS FROM THE LATE JURASSIC UNTIL THE END OF THE DINOSAUR ERA OF THE NORTHERN HEMISPHERE.

ANATOMICAL CHARACTERISTICS Fairly variable. Lightly built. Eyes face strongly forward and often very large, teeth numerous, small, especially at front of upper jaw. Ossified sternal ribs and uncinates not present. Tail not as long as in dromaeosaurids. Ossified sternum absent. Arm not elongated. Pubis vertical or slightly retroverted, leg long and gracile, sickle claw not greatly enlarged. Eggs moderately elongated, tapering.
HABITAT Highly variable, from deserts to polar forests.
HABITS Running performance very high, leaping and climbing ability poor compared to that of other deinonychosaurs. Pursuit predators that focused on smaller game but could use sickle claws to dispatch larger prey. Probably omnivorous in that they also consumed significant plant material. Examples with very large eyes possibly more nocturnal than other dinosaurs. Eggs laid in pairs subvertically in rings, probably by more than one female in each nest, partly exposed so they could be brooded and incubated by adults sitting in center. Juveniles not highly developed so may have received care in or near nest.
NOTES Teeth imply that small troodontids evolved by the Late Jurassic, and some researchers consider *Anchiornis*, *Aurornis*, *Eosinopteryx*, and *Xiaotingia* of that age to be the earliest known members of this group. This group may be splittable into a number of subdivisions.

Sinornithoides youngi (Illustration overleaf)

1.1 m (3.5 ft) TL, 2.5 (5.5 lb)

FOSSIL REMAINS Partial skull and complete skeleton, not fully mature.
ANATOMICAL CHARACTERISTICS Standard for group.
AGE Early Cretaceous.
DISTRIBUTION AND FORMATION/S Northern China; Ejinhoro.
NOTES Skeleton found curled in sleeping position similar to *Mei*. Prey included *Psittacosaurus neimongoliensis*.

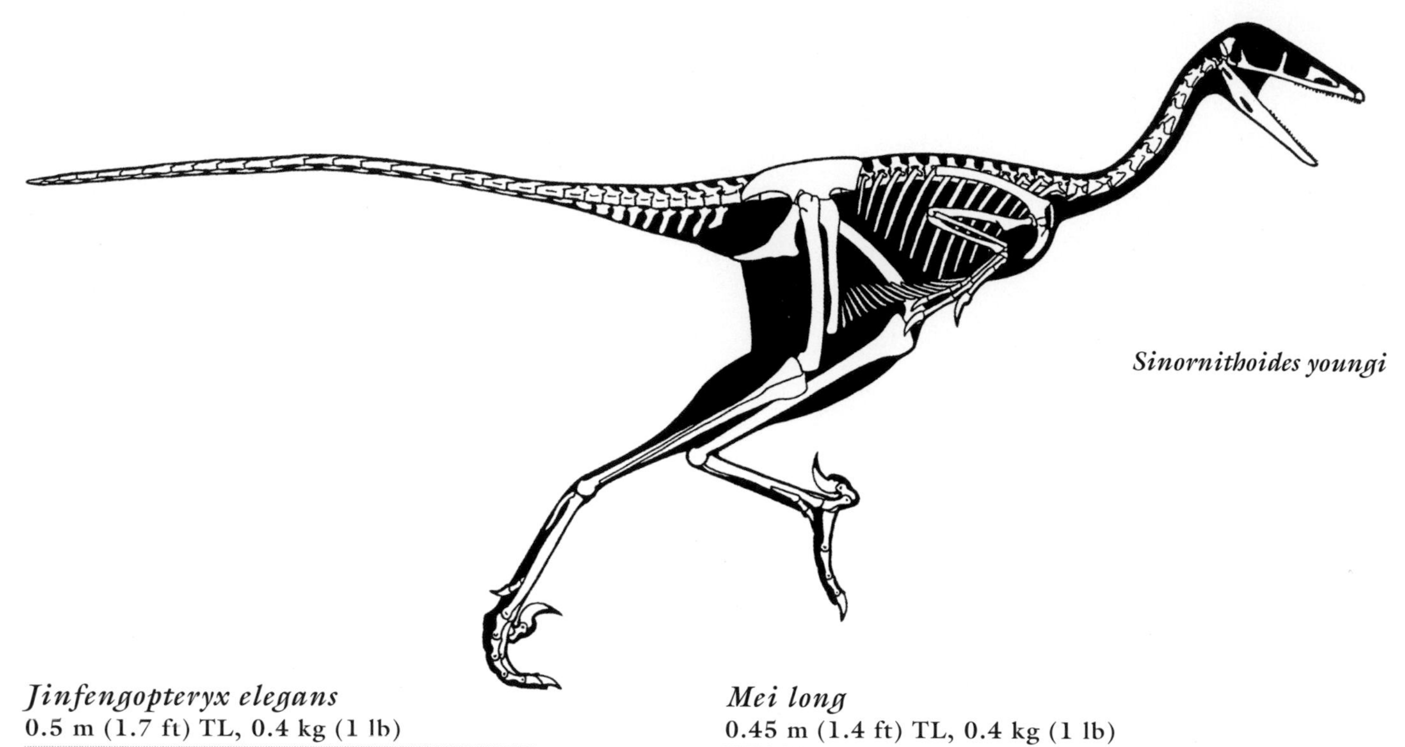

Sinornithoides youngi

Jinfengopteryx elegans

0.5 m (1.7 ft) TL, 0.4 kg (1 lb)

FOSSIL REMAINS Complete skull and skeleton, feathers.
ANATOMICAL CHARACTERISTICS Head lightly built, short, subtriangular. Well-developed pennaceous feathers line entire tail.
AGE Late Late Jurassic or early Early Cretaceous.
DISTRIBUTION AND FORMATION/S Northeast China; Qiaotou.
HABITAT Well-watered forests and lakes, winters chilly with some snow.
HABITS Diet mainly small game and insects. Roundish objects in belly region may be large seeds or nuts.
NOTES May be a juvenile. Originally thought to be a bird close to *Archaeopteryx*; the earliest certain troodontid known from skeletal remains.

Mei long

0.45 m (1.4 ft) TL, 0.4 kg (1 lb)

FOSSIL REMAINS Several nearly complete skulls and skeletons.
ANATOMICAL CHARACTERISTICS Head lightly built, short, subtriangular, postorbital bar incomplete, as in birds.
AGE Early Cretaceous, Barremian.
DISTRIBUTION AND FORMATION/S Northeast China; lower Yixian.
HABITAT Well-watered forests and lakes, winters chilly with some snow.
HABITS Diet mainly small game and insects.
NOTES Shared its habitat with *Sinovenator* and *Sinusonasus.*

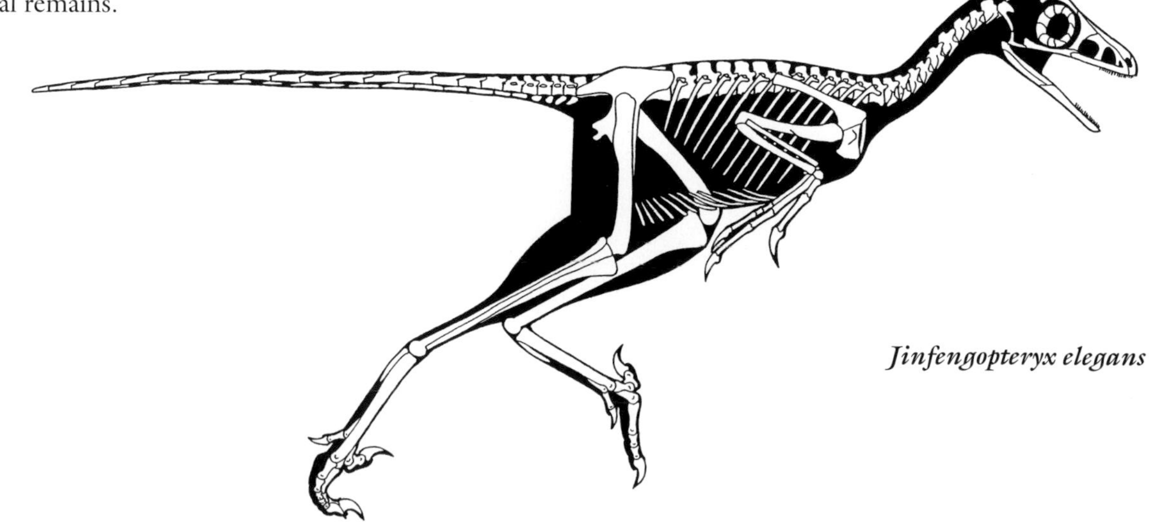

Jinfengopteryx elegans

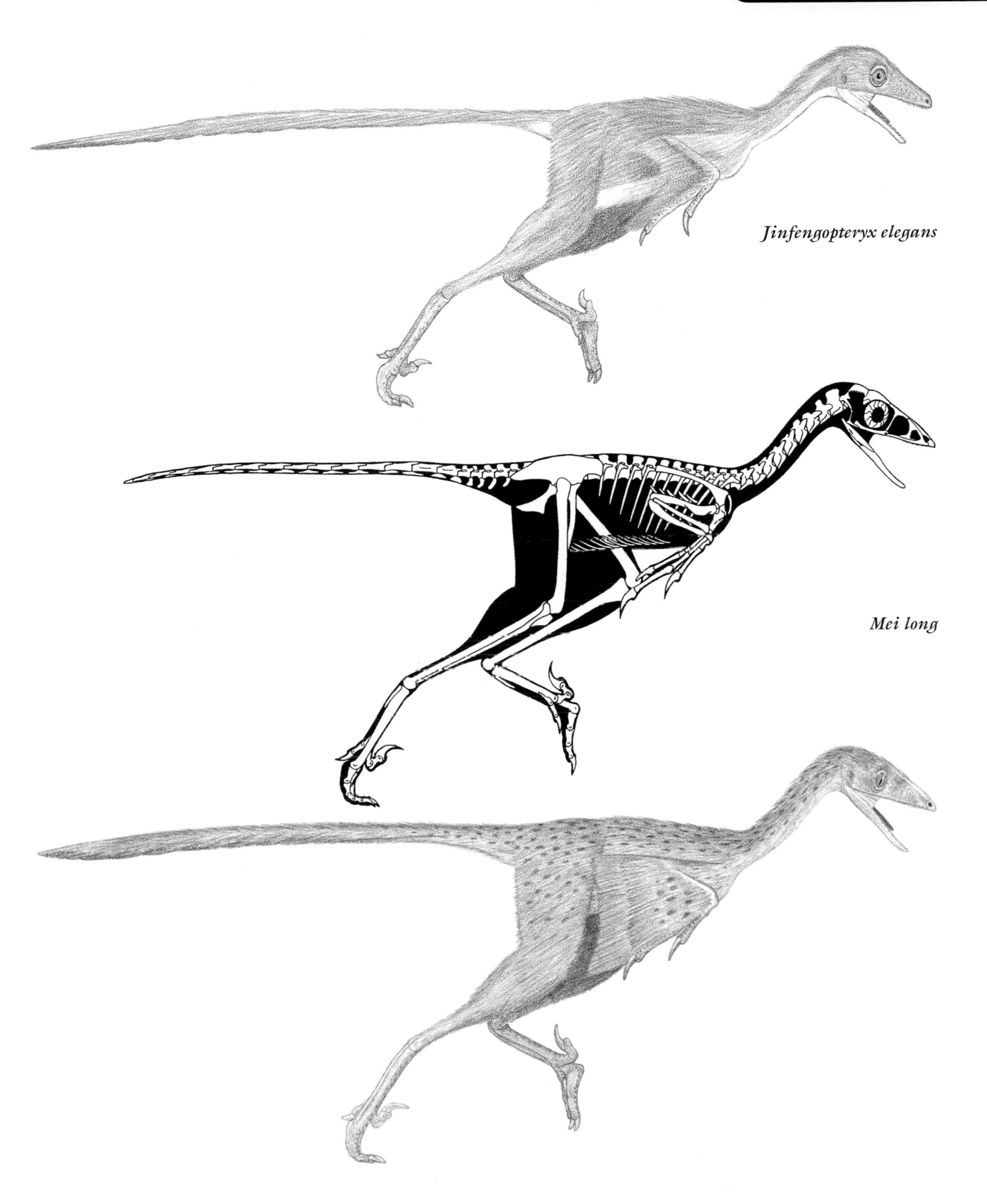

Jinfengopteryx elegans

Mei long

Sinovenator changii
1 m (3.3 ft) TL, 2.5 kg (5.5 lb)

FOSSIL REMAINS Partial skull and majority of skeleton.
ANATOMICAL CHARACTERISTICS Head short and subtriangular, serrations absent in front teeth, small and limited to back edge on rest.
AGE Early Cretaceous, Barremian.
DISTRIBUTION AND FORMATION/S Northeast China; lower Yixian.
HABITAT Well-watered forests and lakes, winters chilly with some snow.

Sinusonasus magnodens
1 m (3.3 ft) TL, 2.5 kg (5.5 lb)

FOSSIL REMAINS Partial skull and majority of skeleton.
ANATOMICAL CHARACTERISTICS Head long and shallow, serrations absent in front teeth, small and limited to back edge on rest, teeth relatively large.
AGE Early Cretaceous, Barremian.
DISTRIBUTION AND FORMATION/S Northeast China; Yixian.
HABITAT Well-watered forests and lakes, winters chilly with some snow.
HABITS Attacked bigger game than *Sinovenator*, smaller game than *Sinornithosaurus*.

Xixiasaurus henanensis
1.5 (5 ft) TL, 8 kg (18 lb)

FOSSIL REMAINS Partial skull and minority of skeleton.
ANATOMICAL CHARACTERISTICS Teeth fairly large, sharp, unserrated.
AGE Late Cretaceous, probably Campanian.
DISTRIBUTION AND FORMATION/S Eastern China; Majiacun.

Byronosaurus jaffei
2 m (7.5 ft) TL, 20 kg (40 lb)

FOSSIL REMAINS Partial skull and minority of skeleton.
ANATOMICAL CHARACTERISTICS Snout very long and very shallow, a little depressed, eyes not especially large, teeth sharp, unserrated.
AGE Late Cretaceous, Campanian.
DISTRIBUTION AND FORMATION/S Mongolia; Djadokhta.

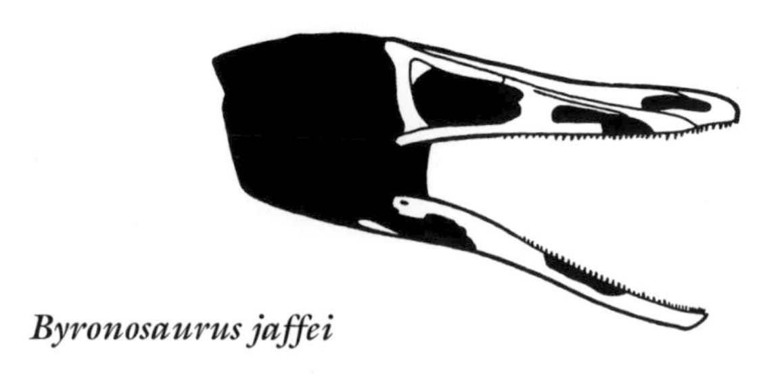

Byronosaurus jaffei

HABITAT Desert with dunes and oases.
HABITS Hunted small game, possibly fished, probably less nocturnal than most troodonts.
NOTES Shared its habitat with *Gobivenator*, *Saurornithoides*, *Tsaagan*, and *Velociraptor*.

Talos sampsoni
2 m (7.5 ft) TL, 20 kg (40 lb)

FOSSIL REMAINS Minority of skeleton.
ANATOMICAL CHARACTERISTICS Leg very slender.
AGE Late Cretaceous, late Campanian.
DISTRIBUTION AND FORMATION/S Utah; middle Kaiparowits.

Linhevenator tani
2.1 m (7 ft) TL, 25 kg (55 lb)

FOSSIL REMAINS Poorly preserved partial skull and skeleton.
ANATOMICAL CHARACTERISTICS Arm short and robust. Sickle claw large.
AGE Late Cretaceous, early Campanian.
DISTRIBUTION AND FORMATION/S Mongolia; Wulansuhai.
HABITS Strong arm and large sickle-claw foot weapon indicate it hunted bigger game than other troodonts. Prey included *Wulatelong* and *Linhenykus*.

Gobivenator mongoliensis
1.7 m (5.5 ft) TL, 9 kg (20 lb)

FOSSIL REMAINS Majority of skull and skeleton.
ANATOMICAL CHARACTERISTICS Head shallow, snout semitubular. Gracile skeleton standard for group.
AGE Late Cretaceous, Campanian.
DISTRIBUTION AND FORMATION/S Mongolia; Djadokhta.
HABITAT Desert with dunes and oases.
HABITS Hunted both small and large game, possibly fished. Possibly more nocturnal than most theropods.
NOTES The first nearly completely known troodont.

Stenonychosaurus? unnamed species
2.5 m (8 ft) TL, 35 kg (70 lb)

FOSSIL REMAINS Skull and skeletal parts. Complete nests.
ANATOMICAL CHARACTERISTICS Head fairly robustly built, shallow, snout semitubular, teeth with large denticles. Elongated eggs 18 cm (7 in) long.
AGE Late Cretaceous, middle and/or late Campanian.
DISTRIBUTION AND FORMATION/S Montana; Upper Two Medicine.
HABITAT Seasonally dry upland woodlands.
HABITS Hunted both small and large game, possibly fished. Prey included *Orodromeus*.

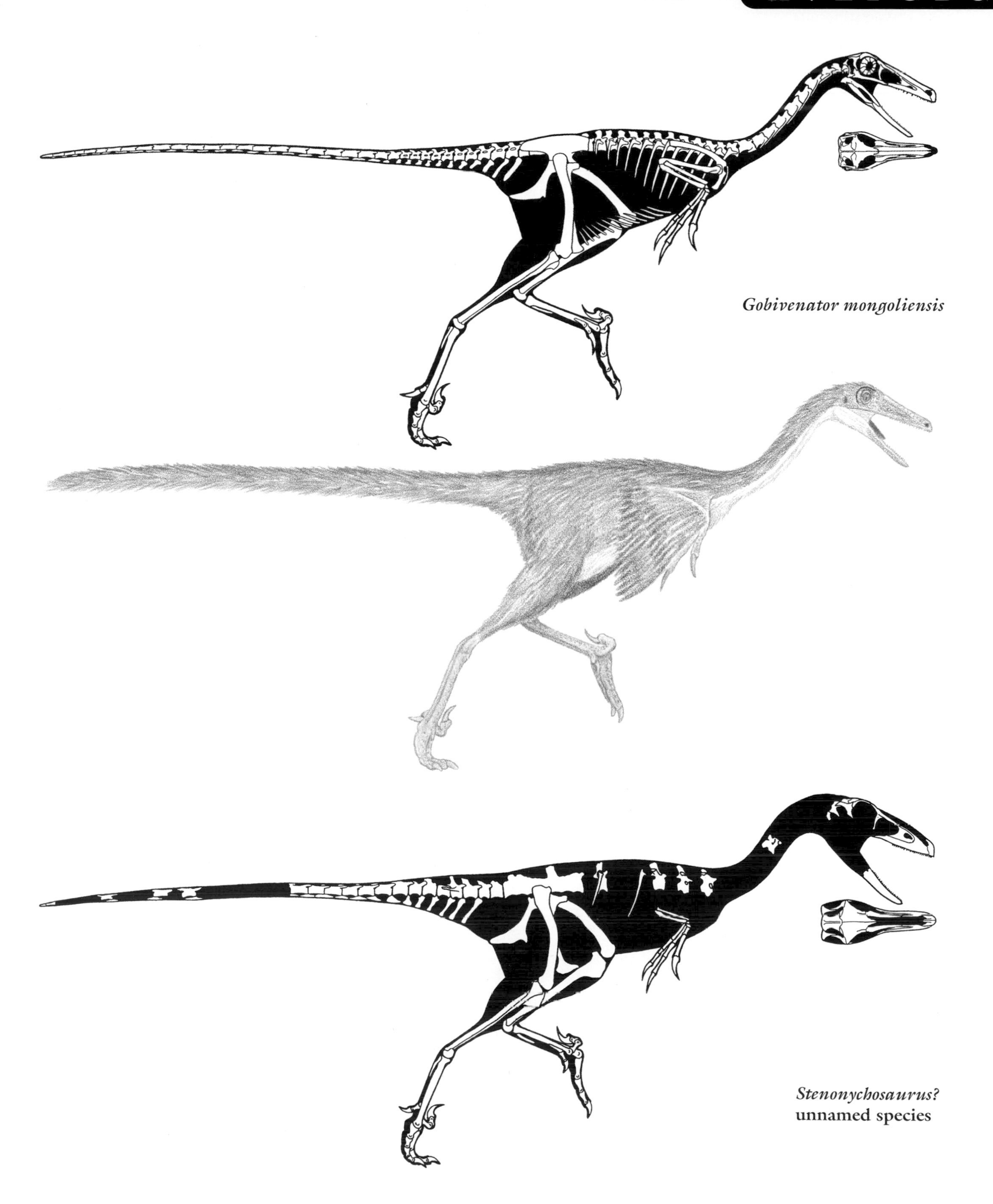

Gobivenator mongoliensis

Stenonychosaurus?
unnamed species

NOTES Usually placed in *Troodon formosus*, which is based on inadequate remains; it is uncertain whether this is same genus or species as *S. inequalis*.

Stenonychosaurus? inequalis?
2.5 m (8 ft) TL, 35 kg (70 lb)

FOSSIL REMAINS Skull and skeletal parts.
ANATOMICAL CHARACTERISTICS Head fairly robustly built, shallow, snout semitubular, teeth with large denticles.
AGE Late Cretaceous, late Campanian.
DISTRIBUTION AND FORMATION/S Alberta; probably upper Dinosaur Park.
HABITAT Well-watered, forested floodplain with coastal swamps and marshes, cool winters.
HABITS Hunted both small and large game, possibly fished.
NOTES Based on questionably adequate remains, may be the same genus as *Saurornithoides*. Shared its habitat with *Saurornitholestes*.

Stenonychosaurus? inequalis?

Saurornithoides mongoliensis
Adult size uncertain

FOSSIL REMAINS Majority of skull and minority of skeleton.
ANATOMICAL CHARACTERISTICS Head fairly robustly built, shallow, snout semitubular, teeth with large denticles.
AGE Late Cretaceous, Campanian.
DISTRIBUTION AND FORMATION/S Mongolia; Djadokhta.
HABITAT Desert with dunes and oases.
HABITS Similar to *Stenonychosaurus*. Prey included *Citipati osmolskae* and *Oviraptor*.
NOTES May have been smaller than *S. junior*. *Saurornithoides* may be same genus as *Stenonychosaurus*. Shared its habitat with *Tsaagan*, *Velociraptor*, *Byronosaurus*, and *Gobivenator*.

Saurornithoides* (= *Zanabazar*?) *junior
2.3 m (7.5 ft) TL, 25 kg (55 lb)

FOSSIL REMAINS Majority of a skull and minority of the skeleton.
ANATOMICAL CHARACTERISTICS Head fairly robustly built, shallow, snout semitubular, teeth with large denticles.
AGE Late Cretaceous, late Campanian and/or early Maastrichtian.
DISTRIBUTION AND FORMATION/S Mongolia; Nemegt.

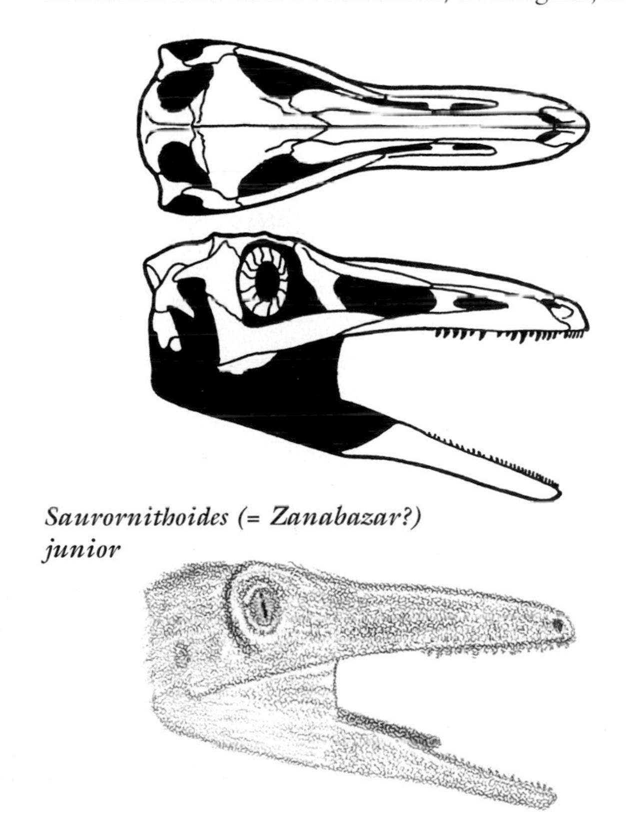
Saurornithoides (= Zanabazar?) junior

HABITAT Well-watered woodland with seasonal rain, winters cold.
HABITS Similar to *S. mongoliensis*. Prey included *Gallimimus*.
NOTES Available evidence insufficient to make separate genus from *Saurornithoides*.

THERIZINOSAURIFORMS

SMALL TO GIGANTIC FLYING AND FLIGHTLESS HERBIVOROUS AVEAIRFOILANS OF THE CRETACEOUS OF THE NORTHERN HEMISPHERE.

ANATOMICAL CHARACTERISTICS Variable. Head somewhat elongated, blunt upper beak, extra joint in lower jaw absent, teeth small, blunt, leaf shaped, not serrated. Tail from very long to very short. Arm long, lunate carpal from well to poorly developed. Foot not narrow, three to four load-bearing toes. Gastroliths often present.
HABITS Herbivores.
NOTES Jeholornids, therizinosaurians, and avians and their common ancestor, operative only if three groups form a clade that excludes all other dinosaurs except oviraptorosaurs.

JEHOLORNIDS Small flying aveairfoilans of the Early Cretaceous of Asia.

ANATOMICAL CHARACTERISTICS Beak well developed, lower jaw teeth restricted to front. Tail long and slender. Furcula and sternal plates large, arm longer and much more strongly built than leg, outer upper hand bone curved, outer and central finger stiffened and base flattened to better support fully developed, long wings made of asymmetrical feathers. Pubis moderately retroverted. Most of body covered by short feathers, base of tail supported feather fan, tail ended with palmlike feather frond, at least in some cases.
HABITS Powered flight fairly well developed. Feather fan at tail base aerodynamic, feather frond at tail tip predominantly for display. Seeds found in guts of some specimens.
NOTES Although not the ancestors of therizinosauroids or necessarily their closest relatives, jeholornids may represent the long-headed, long-tailed basal birds that therizinosaurs may have evolved from.

Jeholornis curvipes
0.7 m (2.5 ft) TL, 0.75 kg (1.5 lb)

FOSSIL REMAINS Complete but poorly preserved skulls with skeletons.

ANATOMICAL CHARACTERISTICS As for group.
AGE Early Cretaceous, early Aptian.
DISTRIBUTION AND FORMATION/S Northeast China; Yixian.
HABITAT Well-watered forests and lakes, winters chilly with some snow.
NOTES Prey of *Sinornithosaurus millenii.* May be direct ancestor of *J. prima*.

Jeholornis prima

0.7 m (2.5 ft) TL, 0.75 kg (1.5 lb)

FOSSIL REMAINS Skulls and skeletons, extensive feathers.
ANATOMICAL CHARACTERISTICS As for group.
AGE Early Cretaceous, early or middle Aptian.
DISTRIBUTION AND FORMATION/S Northeast China; Jiufotang.
HABITAT Well-watered forests and lakes, winters chilly with some snow.
NOTES Probably includes *J. palmapenis.* Shared its habitat with *Sapeornis chaoyangensis.* Prey of *Sinornithosaurus zhaoianus.*

Therizinosaurians

SMALL TO GIGANTIC HERBIVOROUS FLIGHTLESS AVEAIRFOILANS OF THE CRETACEOUS OF THE NORTHERN HEMISPHERE.

ANATOMICAL CHARACTERISTICS Variable. Head small, cheeks probably present, jaw gap limited. Neck long and slender. Trunk tilted upward from retroverted and therefore horizontal pelvis and tail, belly large. Tail from very long to very short. Furcula and sternals not large, arm long but shorter and weaker than leg, lunate carpal from well to poorly developed, finger claws large. Foot not narrow, three to four load-bearing toes, toe claws enlarged. Feathers simple.
HABITS Predominantly browsing herbivores, may have picked up small animals on occasion. Too slow to readily escape predatory theropods, main defense long arms and hand claws as well as kicks from clawed feet.
ENERGETICS Energy levels and food consumption probably low for dinosaurs.

Jeholornis prima

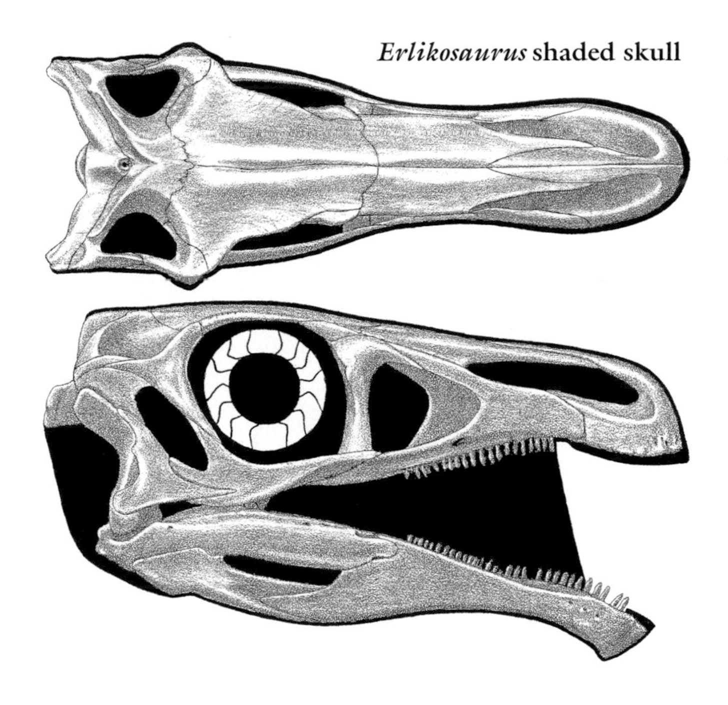

Erlikosaurus shaded skull

NOTES The most herbivorous of the theropods, therizinosaurs are so unusual in their form that before sufficient remains were found it was uncertain they were avepods, relationships to prosauropods once being an alternative. The redevelopment of a complete inner toe within the group is an especially unusual evolutionary reversal, partly paralleled in *Chilesaurus* and *Balaur*. Relationships within avepods remain uncertain. May not be aveairfoilans, may not have descended from fliers, may have descended from very early fliers near base of aveairfoilans. If secondarily flightless aveairfoilans, may have descended from long-skulled herbivorous early birds more than once, from fliers long tailed and then short tailed. Or tail reduction occurred independently in therizinosaurs if they never had flying ancestors as per conventional theory. The modest extent of initial reversals from therizinosaur-like fliers like jeholornids to the first flightless therizinosaurs, the presence of a number of flight-related features in early therizinosaurs, and their progressive reduction as the group evolved toward massive ground herbivores are strongly compatible with an origin among fliers. Lack of pennaceous feathers is paralleled in birds with a long history of flightlessness such as ratites.

Baso-therizinosaurians

MEDIUM-SIZED THERIZINOSAURIANS LIMITED TO THE EARLY CRETACEOUS.

ANATOMICAL CHARACTERISTICS Trunk only modestly tilted up, gastralia flexible. Lunate carpal well developed, finger claws hooked. Front of pelvis not strongly flared out sideways, pubis not retroverted, leg fairly long, hind foot still tridactyl with inner toe still a short hallux, claws not greatly enlarged.
HABITS Better runners than more advanced therizinosaurs.
NOTES Long tailed.

Falcarius utahensis
4 m (12 ft) TL, 100 kg (220 lb)

FOSSIL REMAINS Minority of skull and almost complete skeletal remains known from dozens of partial specimens, juvenile to adult.
ANATOMICAL CHARACTERISTICS Tail very long. Shoulder girdle birdlike.
AGE Early Cretaceous, probably early Barremian.
DISTRIBUTION AND FORMATION/S Utah; Lower Cedar Mountain.

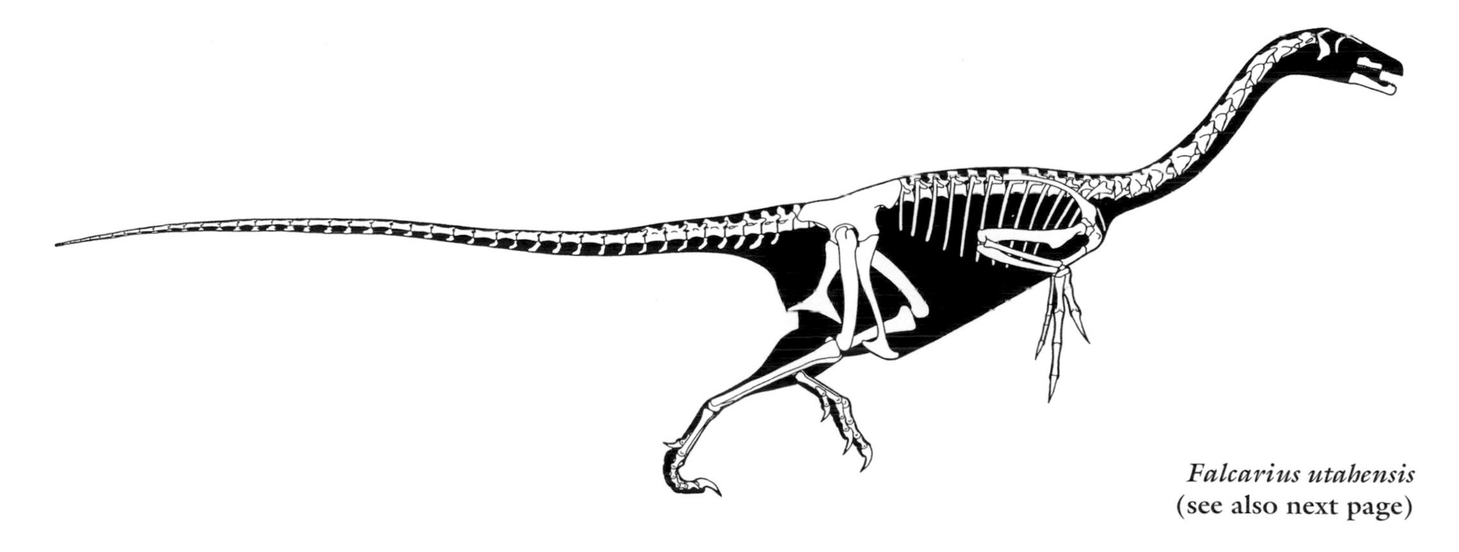

Falcarius utahensis
(see also next page)

Falcarius utahensis

HABITAT Short wet season, otherwise semiarid with floodplain prairies and open woodlands, and riverine forests.
NOTES Prey of *Utahraptor*.

Jianchangosaurus yixianensis
Total length uncertain, 20 kg (45 lb)

FOSSIL REMAINS Complete skull and majority of skeleton, some feathers.
ANATOMICAL CHARACTERISTICS Tip of lower jaw downturned. Shoulder girdle not birdlike. Tail not abbreviated but length uncertain. Leg long.
AGE Early Cretaceous, early Aptian.
DISTRIBUTION AND FORMATION/S Northeast China; Yixian.
HABITAT Well-watered forests and lakes, winters chilly with some snow.
HABITS Best runner among therizinosaurs.
NOTES Shared its habitat with *Beipiaosaurus*. Prey of *Tianyuraptor*, *Zhenyuanlong*, and *Yutyrannus*.

Therizinosauroids

SMALL TO LARGE THERIZINOSAURIANS OF THE CRETACEOUS.

ANATOMICAL CHARACTERISTICS Fairly variable. Tip of lower jaw downturned. Skeleton robustly built. Tail short. Shoulder girdle not birdlike, arm moderately long, lunate carpal block less well developed, fingers not very long, finger claws large hooks. Front of pelvis enlarged and flared sideways, and pubis retroverted to support bigger belly, foot short and broad with four complete toes, toe claws not very enlarged. Known eggs subspherical.
HABITS Buried nests and lack of evidence of brooding indicate little or no parental care. Well-developed hatchlings probably able to leave nest immediately on hatching.

Beipiaosaurus inexpectus
1.8 m (6 ft) TL, 40 kg (90 lb)

FOSSIL REMAINS Skull, two partial skeletons, feathers.

Jianchangosaurus yixianensis

Beipiaosaurus inexpectus

ANATOMICAL CHARACTERISTICS Head shallow, sharply tapering toward front. Vertebrae at tip of tail fused into a small pygostyle. Arrays of long, tapering, band-like feathers atop and beneath back of head; simple feathers on much of body.
AGE Early Cretaceous, early Aptian.
DISTRIBUTION AND FORMATION/S Northeast China; Yixian.
HABITAT Well-watered forests and lakes, winters chilly with some snow.
HABITS Feather arrays for display.
NOTES It is not entirely certain the skull and front half of the skeleton belong to this particular species.

Alxasaurus elesitaiensis (Illustration overleaf)
4 m (13 ft) TL, 400 kg (900 lb)

FOSSIL REMAINS Majority of several skeletons.
ANATOMICAL CHARACTERISTICS Standard for group.
AGE Early Cretaceous, probably Albian.
DISTRIBUTION AND FORMATION/S Northern China; Bayin-Gobi.

Unnamed genus *bohlini*
6 m (20 ft) TL, 1.3 tonnes

FOSSIL REMAINS Partial skeleton.
ANATOMICAL CHARACTERISTICS Insufficient information.

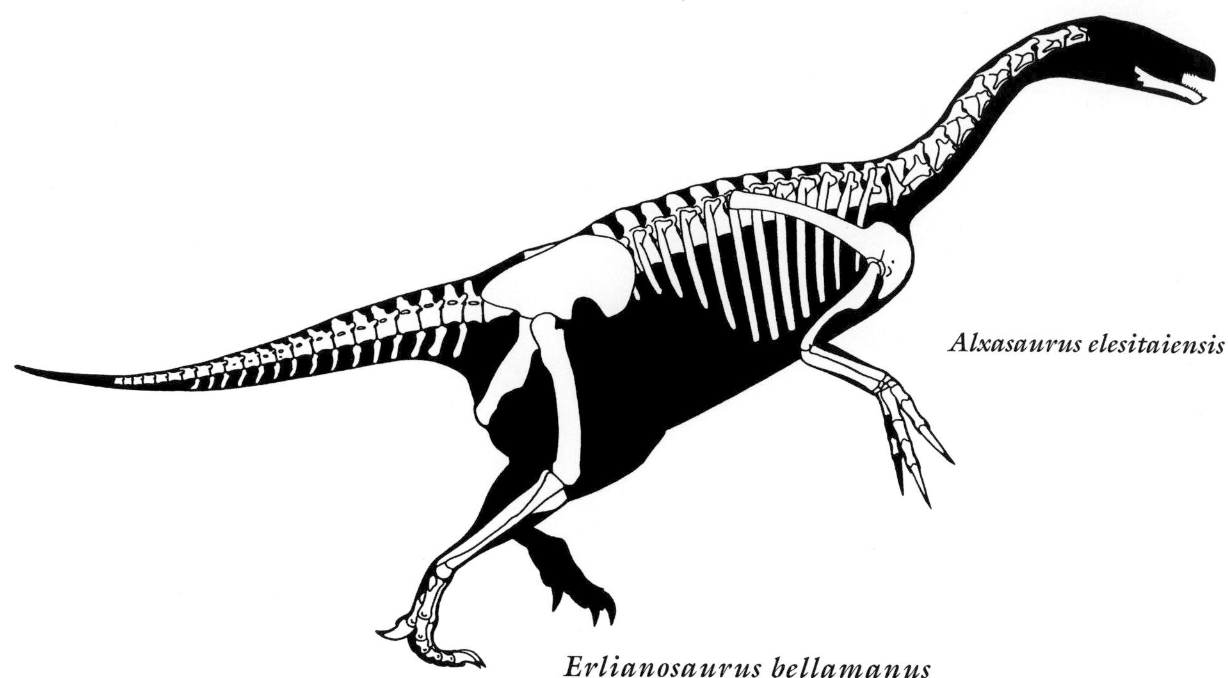

AGE Early Cretaceous, Albian.
DISTRIBUTION AND FORMATION/S Northern China; Xinminbo Group.
NOTES Whether this is a therizinosauroid or therizinosaurid is uncertain. Originally placed in the much later and different *Nanshiungosaurus*.

Suzhousaurus megatheroides
6 m (20 ft) TL, 1.3 tonnes

FOSSIL REMAINS Partial skeleton.
ANATOMICAL CHARACTERISTICS Standard for group.
AGE Early Cretaceous, Aptian or Albian.
DISTRIBUTION AND FORMATION/S Northern China; lower Xinminpu.
NOTES Whether this is a therizinosauroid or therizinosaurid is uncertain.

THERIZINOSAURIDS Medium-sized to gigantic therizinosauroids of the Late Cretaceous.

ANATOMICAL CHARACTERISTICS Uniform. Tip of lower jaw downturned. Skeleton more robustly built. Trunk more strongly tilted up. Tail short. Shoulder girdle not birdlike, lunate carpal block more poorly developed, fingers not very long but bearing very large claws. Front of pelvis further enlarged and flared sideways to support bigger belly, foot broad with four toes, toe claws large.
HABITS Strong upward tilt of body indicates these were high-level browsers.
NOTES The dinosaur group most similar to the recent giant ground sloths.

Erlianosaurus bellamanus
4 m (13 ft) TL, 400 kg (900 lb)

FOSSIL REMAINS Partial skeleton.
ANATOMICAL CHARACTERISTICS Standard for group.
AGE Late Cretaceous, Santonian.
DISTRIBUTION AND FORMATION/S Northern China; Iren Dabasu.
HABITAT Seasonally wet-dry woodlands.
NOTES Whether this is an a therizinosauroid or therizinosaurid is uncertain. Shared its habitat with *Neimongosaurus*.

Neimongosaurus yangi
3 m (10 ft) TL, 150 kg (350 lb)

FOSSIL REMAINS Minority of skull and skeleton.
ANATOMICAL CHARACTERISTICS Insufficient information.
AGE Late Cretaceous, probably Campanian.
DISTRIBUTION AND FORMATION/S Northern China; Iren Dabasu.
HABITAT Seasonally wet-dry woodlands.
NOTES Whether this is an a therizinosauroid or therizinosaurid is uncertain. Shared its habitat with *Erlianosaurus*.

Nanshiungosaurus brevispinus
5 m (16 ft) TL, 600 kg (1,300 lb)

FOSSIL REMAINS Partial skeleton.
ANATOMICAL CHARACTERISTICS Standard for group.
AGE Late Cretaceous, Campanian.
DISTRIBUTION AND FORMATION/S Northern China; Yuanpu.
NOTES Prey of *T. bataar* and *Qianzhousaurus*.

Nothronychus graffami
4.2 m (14 ft) TL, 800 kg (1,800 lb)

FOSSIL REMAINS Nearly complete skeleton.
ANATOMICAL CHARACTERISTICS Gastralia inflexible. Finger claws strongly hooked, toe claws not enlarged.
AGE Late Cretaceous, early Turonian.
DISTRIBUTION AND FORMATION/S Utah; Tropic Shale.
HABITAT Coastal swamps and marshes.
NOTES Found as drift in marine sediments. Inflexibility of gastralia probably caused by lack of major changes in volume of belly in a nongorging, constantly feeding herbivore. May be direct ancestor of *N. mckinleyi.*

Nothronychus mckinleyi
5 m (17 ft) TL, 1.2 tonnes

FOSSIL REMAINS Minority of skeleton.
ANATOMICAL CHARACTERISTICS Same as for *N. graffami.*
AGE Late Cretaceous, middle Turonian.
DISTRIBUTION AND FORMATION/S New Mexico; Moreno Hill.
HABITAT Coastal swamps and marshes.
NOTES Shared its habitat with *Zuniceratops.*

Segnosaurus galbinensis
6 m (20 ft) TL, 1.3 tonnes

FOSSIL REMAINS Minority of skull and skeleton.
ANATOMICAL CHARACTERISTICS Cheeks not as extensive as those of *Erlikosaurus.* Front of pelvis greatly enlarged, toe claws enlarged.
AGE Early Late Cretaceous.
DISTRIBUTION AND FORMATION/S Mongolia; Bayenshiree Svita.
HABITS Probably used large clawed feet for defense as well as hands.
NOTES *Enigmosaurus mongoliensis* may be the same as this species or *Erlikosaurus.*

Erlikosaurus andrewsi (Illustration overleaf)
4.5 m (15 ft) TL, 500 kg (1,100 lb)

FOSSIL REMAINS Complete skull.
ANATOMICAL CHARACTERISTICS Teeth smaller and more numerous than in *Segnosaurus*, cheeks well developed. Toe claws enlarged.
AGE Early Late Cretaceous.
DISTRIBUTION AND FORMATION/S Mongolia; Bayenshiree Svita.

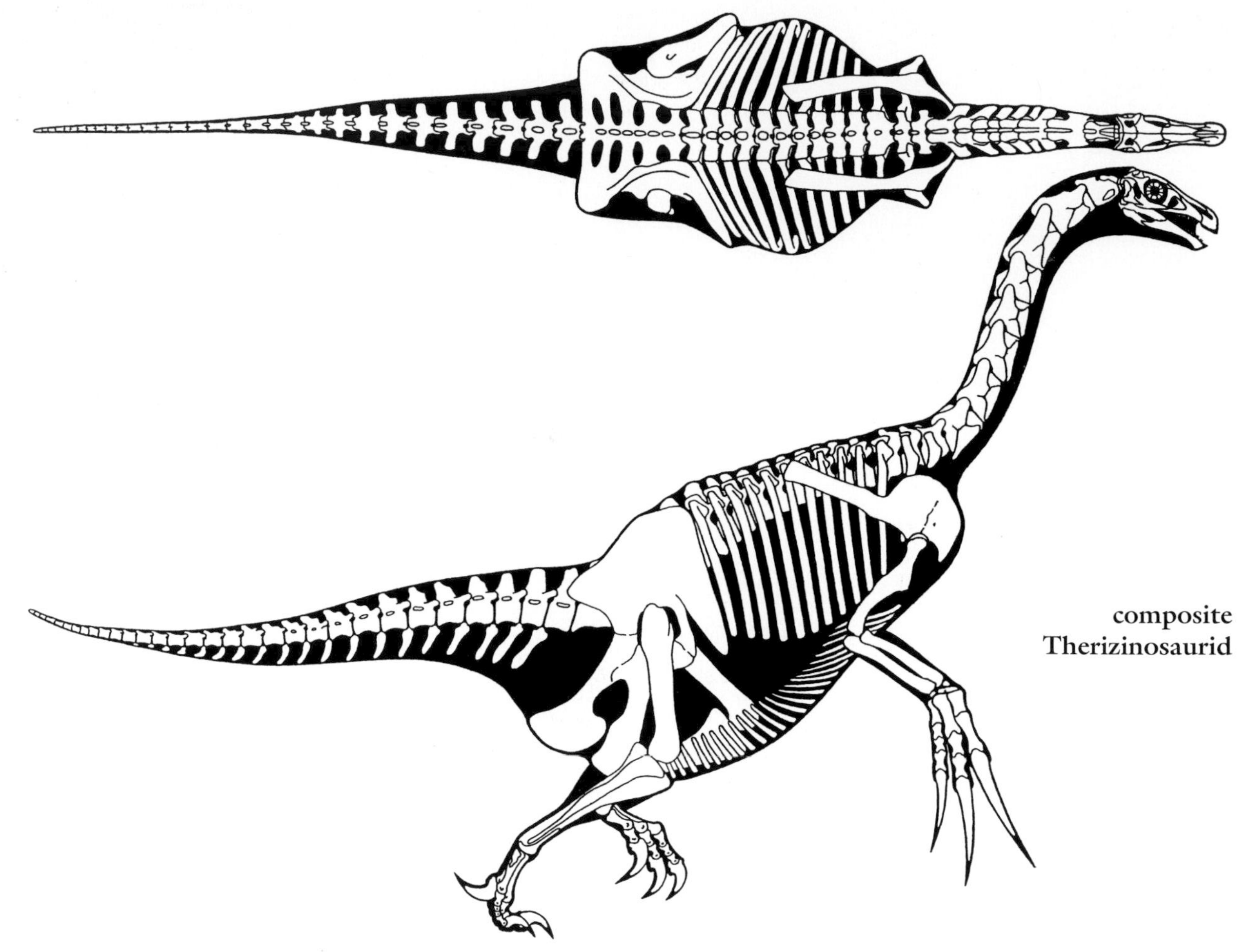

composite Therizinosaurid

Erlikosaurus andrewsi

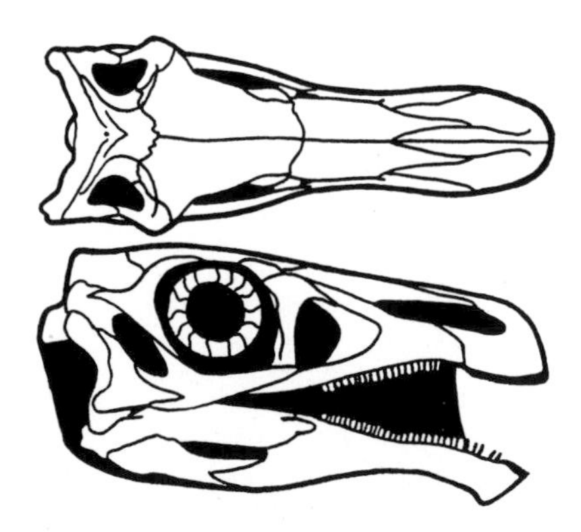

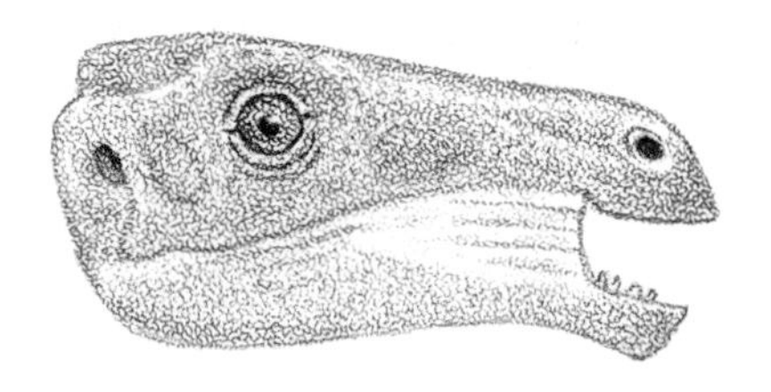

HABITS Probably used large clawed feet for defense as well as hands.
NOTES May include *Enigmosaurus mongoliensis.*

Therizinosaurus cheloniformis
10 m (33 ft) TL, 5+ tonnes

FOSSIL REMAINS Arms and some claws, parts of the hind limb.
ANATOMICAL CHARACTERISTICS Arm up to 3.5 m (11 ft) long, bearing very long, saber-shaped claws that were 0.7 m (over 2 ft) in length without their original horn sheaths.
AGE Late Cretaceous, late Campanian and/or early Maastrichtian.
DISTRIBUTION AND FORMATION/S Mongolia; Nemegt.
HABITAT Well-watered woodland with seasonal rain, winters cold.
NOTES Largest known maniraptor, another example of a very large, high-browsing theropod like *Gigantoraptor*, and the *Deinocheirus* that shared its habitat. Main enemy *T. bataar.*

***Therizinosaurus cheloniformis* opposite**

Oviraptorosauriforms

SMALL TO LARGE FLYING AND FLIGHTLESS HERBIVOROUS OR OMNIVOROUS AVEAIRFOILANS OF THE CRETACEOUS OF THE NORTHERN HEMISPHERE.

ANATOMICAL CHARACTERISTICS Fairly variable. Head not large, short and deep, sides of back of head made of slender struts, many bones including lower jaws fused together and extra joint absent, jaw joint highly mobile to allow chewing motions, teeth reduced or absent. Neck fairly long. Trunk short. Tail short. Arm short to very long, fingers three to two. Leg short to very long.
HABITS Omnivorous or herbivorous, picked up small animals at least on occasion.
NOTES Omnivoropterygids, oviraptorosaurs, and avians and their common ancestor, operative only if three groups form a clade that excludes all other dinosaurs.

Omnivoropterygids Small flying oviraptorosaurs limited to the Early Cretaceous of Asia.

ANATOMICAL CHARACTERISTICS Lower jaw shallow, a few procumbent, small, pointed teeth at front of upper jaw. No uncinate processes on ribs. Vertebrae at tip of very short tail fused into a pygostyle. Sternal plates and ossified sternal ribs may be absent, very long arm and hand indicate very large wings, outer finger severely reduced so there are only two fully functional fingers. Pubis a little retroverted, pelvis broad, leg short and not as strong as arm, toes long, hallux reversed. Wings very large, tail feather fan present, ankle feathers in some specimens.
HABITS Capable of low-grade powered, gliding, and maybe soaring flight probably a bit better than *Archaeopteryx*. Good climbers. Defense included climbing and flight.
HABITAT Well-watered forests and lakes, winters chilly with some snow.
NOTES Although not the ancestors of oviraptorosaurs, omnivoropterygids represent the parrot-headed, short-tailed basal birds that oviraptorosaurs may have evolved from. Irregular presence of ankle feathers may indicate display rather than aerodynamic function. Specific diets difficult to determine because of the unusual configuration of head and jaws.

Sapeornis unnamed species?
Adult size uncertain

FOSSIL REMAINS Complete but distorted skull with skeleton, extensive feathers.
AGE Early Cretaceous, early Aptian.
DISTRIBUTION AND FORMATION/S Northeast China; Yixian.
NOTES Probably a species different from, and may be ancestral to, *S. chaoyangensis*. Prey of *Sinornithosaurus millenii*.

Citipati shaded skull

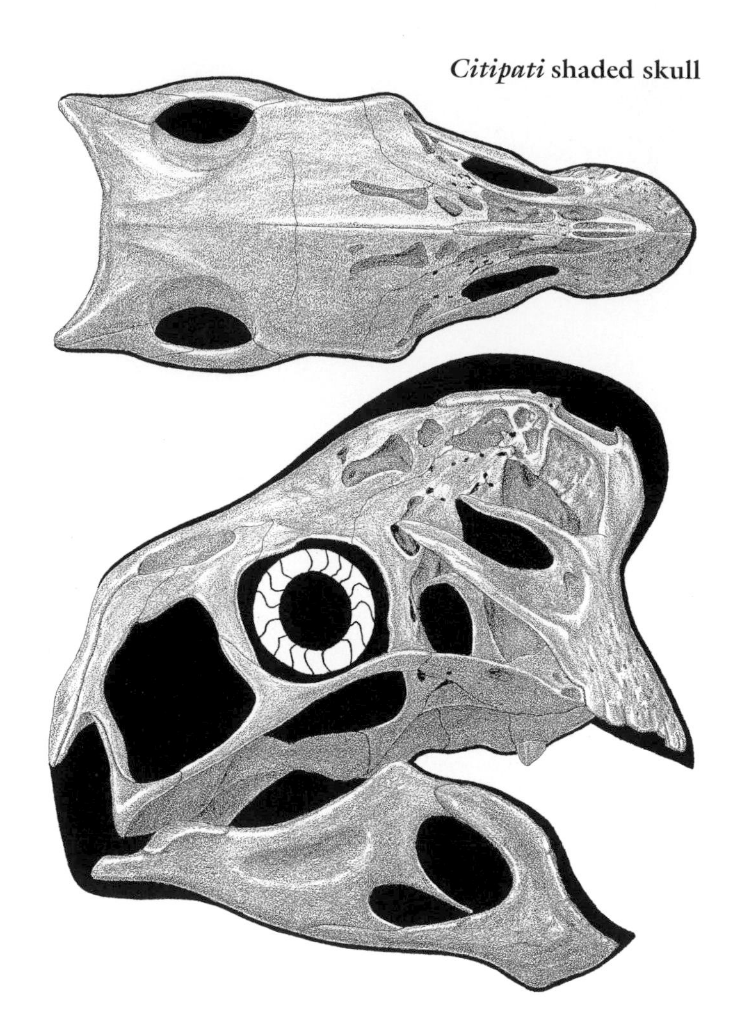

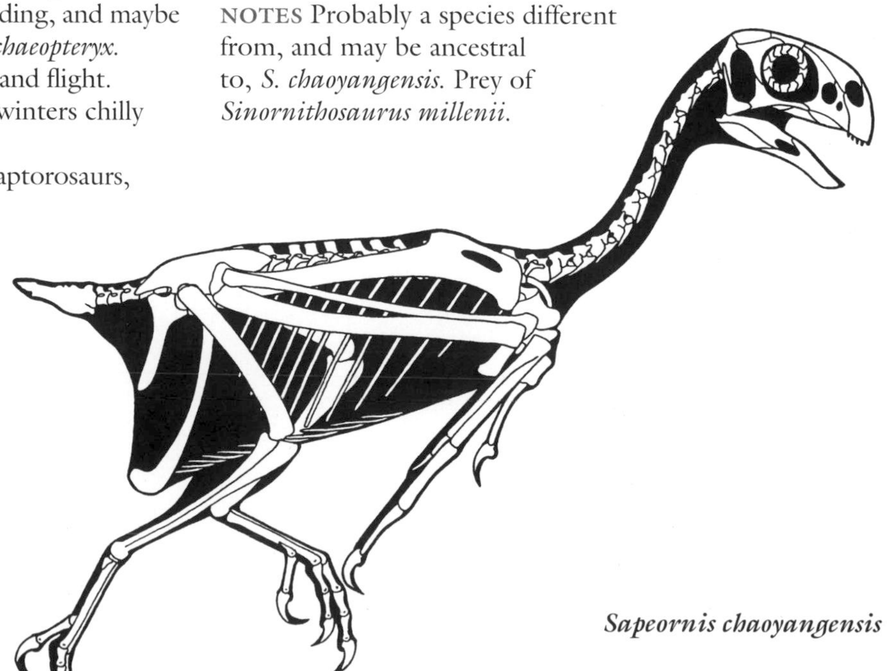

Sapeornis chaoyangensis

Sapeornis chaoyangensis

Sapeornis chaoyangensis
0.4 m (1.3 ft) TL, 1 kg (2 lb)

FOSSIL REMAINS Several complete skulls and majority of skeletons, feathers, gizzard stones.
AGE Early Cretaceous, early or middle Aptian.
DISTRIBUTION AND FORMATION/S Northeast China; Jiufotang.
NOTES A much more common component of the fauna than the preceding species, probably includes *Omnivoropteryx sinousaorum*, *Didactylornis jii*, and a number of other species. Prey of *Sinornithosaurus zhaoianus*.

OVIRAPTOROSAURS

SMALL TO LARGE HERBIVOROUS OR OMNIVOROUS AVEAIRFOILANS OF THE CRETACEOUS OF THE NORTHERN HEMISPHERE.

ANATOMICAL CHARACTERISTICS Fairly variable. Arm short to moderately long, fingers three to two. Pubis not retroverted, leg moderately to very long.
HABITS Defense included biting with beaks, slashing with hand claws, and evasion.
NOTES Presence of large sternal plates, ossified sternal ribs, ossified uncinate processes, short tail in most examples, and reduction of outer finger in some examples strongly imply that the flightless oviraptorosaurs were the secondarily flightless relatives of the flying omnivoropterygids; this requires significant but not massive reversals of some sections of the skeleton to a nonavian condition, and protarchaeopterygids, avimimids, caudipterygids, and caenagnathoids may have evolved collectively or independently from earlier fliers. Alternatively, omnivoropterygids were the flying descendants of oviraptorosaurs. In conventional theory the two groups were not closely related, in which case the heads and hands evolved in a remarkably convergent manner despite the lack of a common flight heritage. Fragmentary remains may record presence in Australia.

PROTARCHAEOPTERYGIDS Small oviraptorosaurs limited to the Early Cretaceous of Asia.

ANATOMICAL CHARACTERISTICS Highly uniform. Skull not as deep as in other oviraptorosaurs, subrectangular, roof of mouth projecting below rim of upper jaws, lower jaw shallow, frontmost teeth enlarged and well worn, rest of teeth small, blunt, and unserrated, teeth absent from tip of lower jaw. Skeleton lightly built. Large sternal plates present, arm long, three finger claws are large hooks. Leg long.
HABITS Divergence in tooth size and form is much greater than in other theropods. Incisor-like front teeth are reminiscent of rodents and imply gnawing on some form of hard plant material. Both climbing and running performance appear to be high, main defense climbing, high speed, and biting.
NOTES That these basal oviraptorosaurs are more birdlike than more-derived oviraptorosaurs is compatible with the group being secondarily flightless.

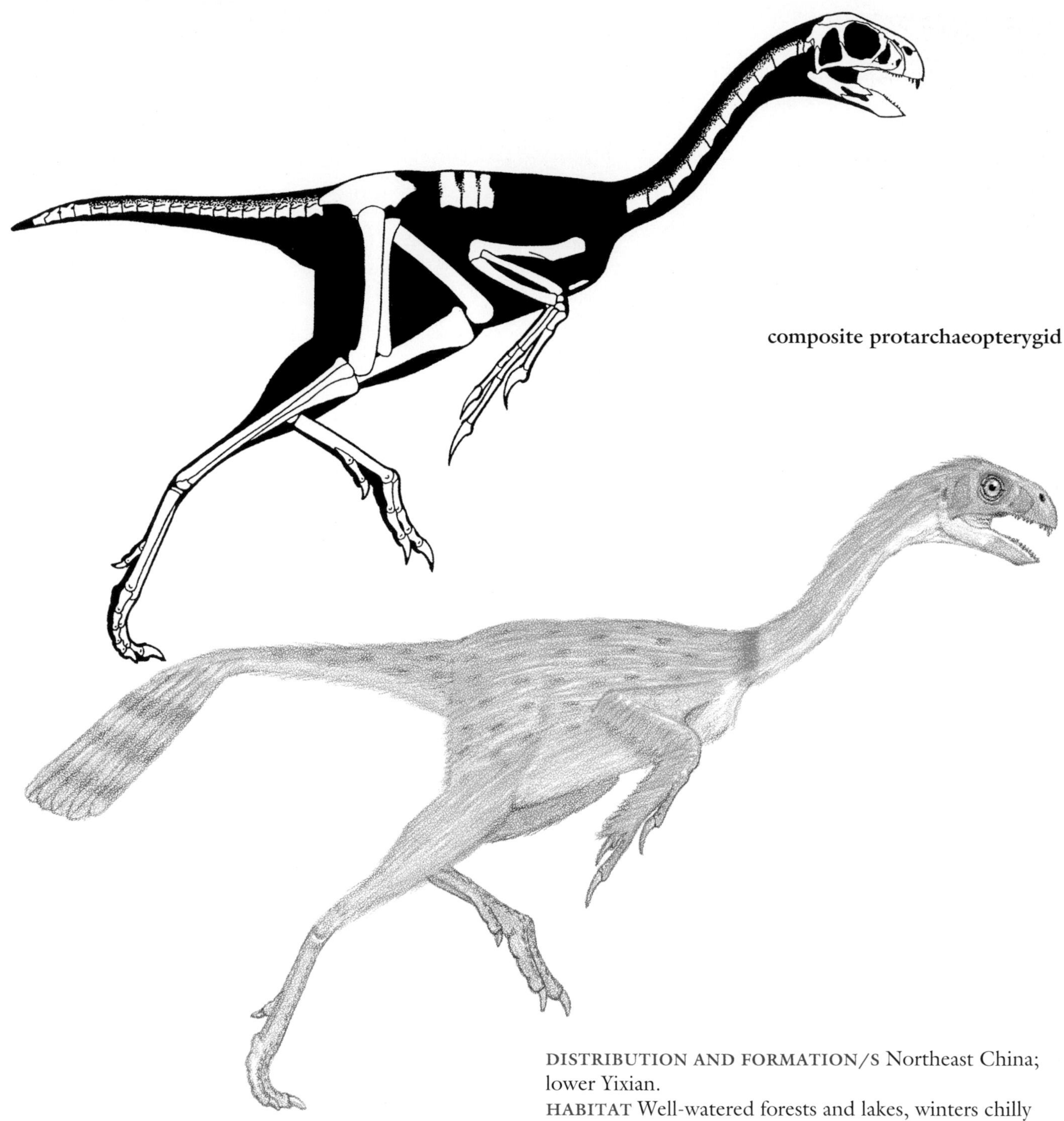

Protarchaeopteryx (or *Incisivosaurus*) *gauthieri*

0.8 m (2.7 ft) TL, 2 kg (5 lb)

FOSSIL REMAINS Almost complete skull and small portion of skeleton.

ANATOMICAL CHARACTERISTICS Standard for group, number of teeth differs from *P. robusta*.

AGE Early Cretaceous, Barremian.

DISTRIBUTION AND FORMATION/S Northeast China; lower Yixian.

HABITAT Well-watered forests and lakes, winters chilly with some snow.

NOTES Originally the genus *Incisivosaurus*, appears to be a species of *Protarchaeopteryx*, may be the ancestor of *P. robusta*.

Protarchaeopteryx robusta

0.7 m (2.3 ft) TL, 1.6 kg (3.5 lb)

FOSSIL REMAINS Majority of badly damaged skull and skeleton, some feathers.

Caudipteryx zoui

ANATOMICAL CHARACTERISTICS Arm and tail feathers fairly long, vanes asymmetrical.
AGE Early Cretaceous, early Aptian.
DISTRIBUTION AND FORMATION/S Northeast China; Yixian.
HABITAT Well-watered forests and lakes, winters chilly with some snow.
HABITS Arms not long enough and arm feathers too symmetrical for flight, some parachute capability possible.
NOTES Misnamed as a closely related predecessor to the much earlier deinonychosaur *Archaeopteryx.* May be descendant of earlier *P. gauthieri*. Shared its habitat with *Caudipteryx zoui*. Prey of an array of compsognathids and deinonychosaurs.

CAUDIPTERYGIDS Small oviraptorosaurs limited to the Early Cretaceous of Asia.

ANATOMICAL CHARACTERISTICS Head small, subtriangular, lower jaw shallow, a few procumbent, small, pointed teeth at front of upper jaw. Skeleton lightly built. Trunk short, uncinate processes on ribs. Ossified sternal plates and sternal ribs present, arm short, outer finger severely reduced so there are only two fully functional fingers, claws not large. Pelvis very large, leg very long and gracile, leg muscles exceptionally well developed, semireversed hallux small, so speed potential very high.
HABITS Climbing ability low or nonexistent, main defense high speed.
NOTES That these basal oviraptorosaurs are more birdlike in certain regards than more-derived oviraptorosaurs is compatible with the group being secondarily flightless.

Caudipteryx zoui
0.65 m (2 ft) TL, 2.2 kg (5 lb)

FOSSIL REMAINS A number of complete skulls and skeletons, extensive feathers, bundles of gizzard stones.
ANATOMICAL CHARACTERISTICS Pygostyle not present. Well-developed feather fan on hand, possibly split fan at end of tail, latter showing pigment banding, large pennaceous feathers symmetrical, simpler feathers covering much of body.
AGE Early Cretaceous, early Aptian.

Caudipteryx zoui

DISTRIBUTION AND FORMATION/S Northeast China; Yixian.
HABITAT Well-watered forests and lakes, winters chilly with some snow.
HABITS Presence of some small, sharp teeth imply *Caudipteryx* may have caught small animals, but gizzard stones verify diet of plants that required grinding. Small hand and tail feather fans probably for display within the species.
NOTES It is probable that *C. dongi* belongs to this species, which may be the direct ancestor of *C. yixianensis.* Prey of an array of compsognathids, deinonychosaurs, and juvenile *Yutyrannus.*

Caudipteryx (= *Similicaudipteryx*) *yixianensis*
1 m (3 ft) TL, 7 kg (15 lb)

FOSSIL REMAINS Majority of poorly preserved skeleton.
ANATOMICAL CHARACTERISTICS Vertebrae at tip of tail fused into a small pygostyle.
AGE Early Cretaceous, early or middle Aptian.
DISTRIBUTION AND FORMATION/S Northeast China; Jiufotang.
HABITAT Well-watered forests and lakes, winters chilly with some snow.

Avimimids Small oviraptorosaurs limited to the late Late Cretaceous of Asia.

ANATOMICAL CHARACTERISTICS Head apparently short and deep, postorbital bar incomplete as in birds, small teeth at front of upper jaw. Arm short, upper hand fused together. Pubis procumbent, pelvis large and broad, leg very long and its muscles exceptionally well developed, foot very long and strongly compressed from side to side, hallux absent and toes short, so speed potential very high.
HABITS Broad hips indicate large belly for processing plant material. Main defense high speed.

Avimimus portentosus
1.2 m (3.5 ft) TL, 12 kg (25 lb)

FOSSIL REMAINS Partial skull and skeletons.
AGE Late Cretaceous, Santonian.
DISTRIBUTION AND FORMATION/S Northern China; Iren Dabasu.
HABITAT Seasonally wet-dry woodlands.
NOTES Shared its habitat with *Gigantoraptor.*

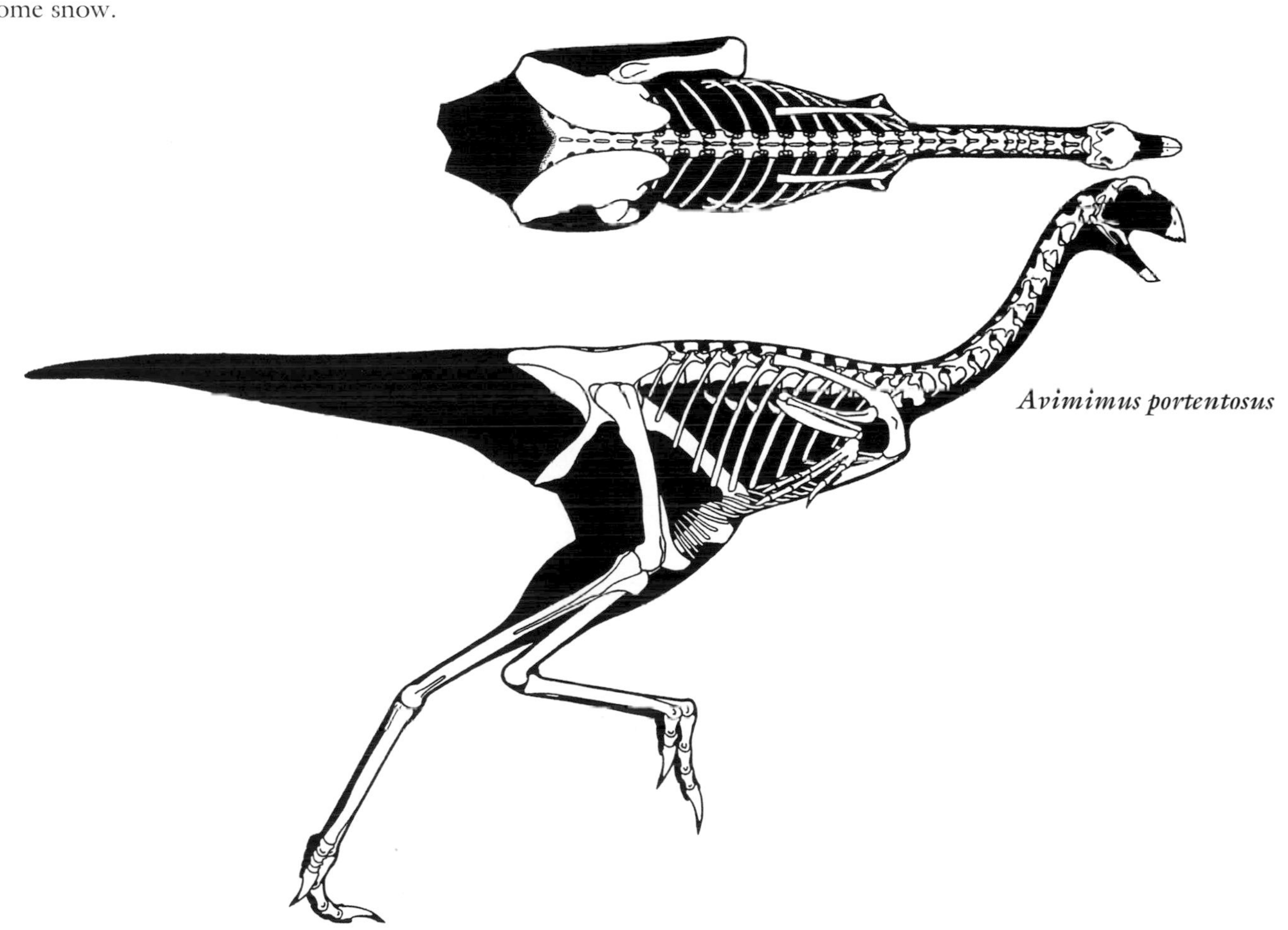

Avimimus portentosus

Caenagnathoids Small to large oviraptorosaurs of the Cretaceous of the Northern Hemisphere.

ANATOMICAL CHARACTERISTICS Most or all adults with emu-like head crests, teeth absent, and blunt beak present. Uncinate processes on ribs. Ossified sternal plates and sternal ribs present. Three finger claws well developed. Olfactory bulbs reduced. Eggs highly elongated.
HABITS Pneumatic head crests too delicate for butting, probably for visual display within species. Defense included running, climbing, hand claws, and biting. Eggs formed and laid in pairs in flat, two-layered rings, partly exposed, probably by more than one female in each nest, brooded and incubated by adult sitting in empty center of nest with feathered arms and tail draped over eggs.

CAENAGNATHIDS Small to large oviraptorosaurs of the Cretaceous of the Northern Hemisphere.

ANATOMICAL CHARACTERISTICS Fairly uniform. Lower jaw not very deep. Arm and hand long. Leg fairly long.

Microvenator celer
Adult size uncertain

FOSSIL REMAINS Partial skeleton, juvenile.
ANATOMICAL CHARACTERISTICS Insufficient information.
AGE Early Cretaceous, middle Albian.
DISTRIBUTION AND FORMATION/S Montana; upper Cloverly.
HABITAT Short wet season, otherwise semiarid with floodplain prairies and open woodlands and riverine forests.
NOTES Main enemy *Deinonychus antirrhopus.*

Hagryphus giganteus
2+ m (8 ft) TL, 50 kg (100 lb)

FOSSIL REMAINS Small portion of skeleton.
ANATOMICAL CHARACTERISTICS Insufficient information.
AGE Late Cretaceous, late Campanian.
DISTRIBUTION AND FORMATION/S Utah; middle Kaiparowits.

Caenagnathus collinsi
2.5 m (8 ft) TL, 100 kg (200 lb)

FOSSIL REMAINS Lower jaw.
ANATOMICAL CHARACTERISTICS Lower jaw shallow.
AGE Late Cretaceous, late Campanian.
DISTRIBUTION AND FORMATION/S Alberta; lower Dinosaur Park.
HABITAT Well-watered, forested floodplain with coastal swamps and marshes, cool winters.
NOTES This genus may include *Macrophalangia canadensis* based on even less adequate remains, or the latter may belong to *Chirostenotes pergracilis.*

Chirostenotes pergracilis
2.5 m (8 ft) TL, 100 kg (200 lb)

FOSSIL REMAINS Minority of skull and skeletons.
ANATOMICAL CHARACTERISTICS Lower jaws fairly deep.
AGE Late Cretaceous, late Campanian.
DISTRIBUTION AND FORMATION/S Alberta; middle Dinosaur Park.
HABITAT Well-watered, forested floodplain with coastal swamps and marshes, cool winters.
HABITS Fast runner.
NOTES May include *Chirostenotes elegans*, also *Macrophalangia canadensis.*

Anzu wyliei
3.75 (12 ft) TL, 250 kg (500 lb)

FOSSIL REMAINS Majority of a few skulls and skeletons.
ANATOMICAL CHARACTERISTICS Tall, broad head crest. Leg long.
AGE Late Cretaceous, early Maastrichtian.
DISTRIBUTION AND FORMATION/S South Dakota; Hell Creek, level uncertain.
HABITAT Well-watered coastal woodlands.
NOTES Was usually included in *Chirostenotes pergracilis.*

Elmisaurus (or *Chirostenotes*) *rarus*
1 m (3.3 ft) TL, 4.5 kg (10 lb)

FOSSIL REMAINS Small portion of skeleton.
ANATOMICAL CHARACTERISTICS Insufficient information.
AGE Late Cretaceous, late Campanian and/or early Maastrichtian.
DISTRIBUTION AND FORMATION/S Mongolia; Nemegt.
HABITAT Well-watered woodland with seasonal rain, winters cold.
NOTES Known remains very similar to and possibly same genus as *Chirostenotes.*

Elmisaurus (or *Chirostenotes?*) unnamed species
2.5 m (8 ft) TL, 100 kg (200 lb)

FOSSIL REMAINS Minority of skull and skeletons.
ANATOMICAL CHARACTERISTICS Standard for group.
AGE Late Cretaceous, latest Campanian and/or early Maastrichtian.
DISTRIBUTION AND FORMATION/S Alberta; lower Horseshoe Canyon.

Anzu wyliei

HABITAT Well-watered, forested floodplain with coastal swamps and marshes, cool winters.
NOTES Usually included in *Chirostenotes pergracilis*, more probably the descendant of the earlier species.

Caenagnathasia martinsoni
0.6 m (2 ft) TL, 1.4 kg (3 lb)

FOSSIL REMAINS Small minority of skull and skeleton.
ANATOMICAL CHARACTERISTICS Insufficient information.
AGE Early Late Cretaceous.
DISTRIBUTION AND FORMATION/S Uzbekistan; Nemegt.
HABITAT Well-watered woodland with seasonal rain, winters cold.

Oviraptorids Small oviraptorosaurs limited to the late Late Cretaceous of Asia.

ANATOMICAL CHARACTERISTICS Uniform. Highly pneumatic head subrectangular, snout short, somewhat parrot-like beak deep, nostrils above preorbital opening, blunt pair of pseudoteeth on strongly downward-projecting mouth roof, eyes not especially large, lower jaw deep. Outer two fingers subequal in length and robustness, finger claws well developed. Leg not slender.
HABITS The downward-jutting pseudoteeth indicate a crushing action.
NOTES The number of genera named in this group appears excessive in part because specimens without crests may be juveniles or females of crested species. Head crests were probably enlarged by keratin coverings as they are in emus.

Nomingia gobiensis
1.7 m (5.5 ft) TL, 20 kg (40 lb)

FOSSIL REMAINS Partial skeleton.
ANATOMICAL CHARACTERISTICS Vertebrae at tip of tail fused into a birdlike pygostyle that probably supported a feather fan.
AGE Late Cretaceous, late Campanian and/or early Maastrichtian.
DISTRIBUTION AND FORMATION/S Mongolia; Nemegt.
HABITAT Well-watered woodland with seasonal rain, winters cold.
NOTES It is uncertain whether this is a caenagnathid or oviraptorid; if the latter it may belong to one of the other named species from the Nemegt.

Gigantoraptor erlianensis
8 m (25 ft) TL, 2 tonnes

FOSSIL REMAINS Minority of skull and majority of skeleton.
ANATOMICAL CHARACTERISTICS Hand slender.
AGE Late Cretaceous, Santonian.
DISTRIBUTION AND FORMATION/S Northern China; Iren Dabasu.

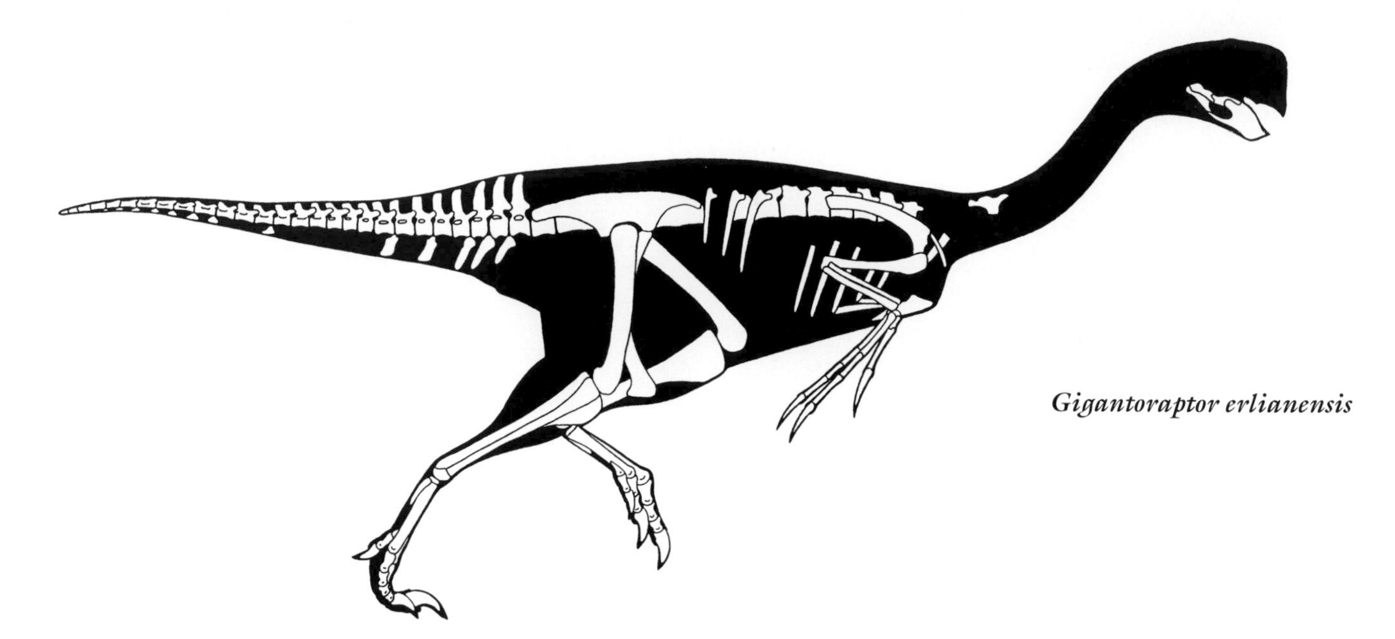

Gigantoraptor erlianensis

HABITAT Seasonally wet-dry woodlands.
HABITS Another example of a large, high-browsing theropod similar to *Deinocheirus* and *Therizinosaurus*. Better able to defend itself against predators than smaller oviraptors, also able to run away from predators.
NOTES Giant eggs up to 0.5 m (1.6 ft) long laid in enormous rings up to 3 m (10 ft) across, found in Asia, probably laid by big oviraptors such as *Gigantoraptor*.

Nankangia jiangxiensis

2.5 m (8 ft) TL, 75 kg (160 kg)

FOSSIL REMAINS Minority of skull and majority of skeleton.
ANATOMICAL CHARACTERISTICS Insufficient information.
AGE Late Cretaceous, late Campanian and/or early Maastrichtian.
DISTRIBUTION AND FORMATION/S Southeast China; Nanxiong.
NOTES Shared its habitat with *Huanansaurus*.

Yulong mini

Adult size uncertain

FOSSIL REMAINS A few complete and partial juvenile skulls and skeletons.
ANATOMICAL CHARACTERISTICS Standard for group.
AGE Late Cretaceous.
DISTRIBUTION AND FORMATION/S Eastern China; Qiupa.

Oviraptor philoceratops

1.6 m (5 ft) TL, 22 kg (50 lb)

FOSSIL REMAINS Majority of poorly preserved skull and minority of skeleton.
ANATOMICAL CHARACTERISTICS Head not as deep as in other oviraptorids, full extent of head crest uncertain. Hand large.
AGE Late Cretaceous, Campanian.
DISTRIBUTION AND FORMATION/S Mongolia; Djadokhta.
HABITAT Desert with dunes and oases.
HABITS Presence of lizard skeleton in gut cavity of the skeleton indicates that the oviraptorid diet included at least some small animals.
NOTES Other oviraptorids were placed in *Oviraptor* until it was realized this is a very distinct genus. Shared its habitat with the more common *Citipati osmolskae*. Main enemies of these oviraptorids were *Saurornithoides mongoliensis* and especially *Tsaagan* and *Velociraptor*.

Oviraptor philoceratops

Huanansaurus ganzhouensis

2.5 m (8 ft) TL, 75 kg (160 kg)

FOSSIL REMAINS Nearly complete skull and minority of skeleton.
ANATOMICAL CHARACTERISTICS Low crest from front to back of head.
AGE Late Cretaceous, late Campanian and/or early Maastrichtian.
DISTRIBUTION AND FORMATION/S Southeast China; Nanxiong.

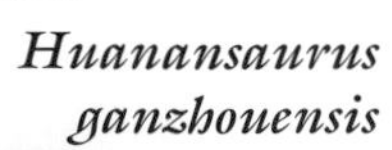

Huanansaurus ganzhouensis

Rinchenia (or *Citipati*) *mongoliensis*

Rinchenia (or *Citipati*) *mongoliensis*
1.7 m (5.5 ft) TL, 25 kg (55 lb)

FOSSIL REMAINS Complete skull and minority of skeleton.
ANATOMICAL CHARACTERISTICS Subtriangular head crest very large.
AGE Late Cretaceous, late Campanian and/or early Maastrichtian.
DISTRIBUTION AND FORMATION/S Mongolia; Nemegt.
HABITAT Well-watered woodland with seasonal rain, winters cold.

NOTES Shared its habitat with *Ajancingenia barsboldi*. Prey of *Saurornithoides junior* and juvenile *Alioramus remotus*, *T. bataar*.

Citipati osmolskae
2.5 m (8 ft) TL, 75 kg (160 kg)

FOSSIL REMAINS Several complete and partial skulls and skeletons from embryo to adult, completely known, nests, some with adults in brooding posture on complete nests.
ANATOMICAL CHARACTERISTICS Well-developed crest projects forward above upper beak. Elongated eggs 18 cm (7 in) long.
AGE Late Cretaceous, Campanian.
DISTRIBUTION AND FORMATION/S Mongolia; Djadokhta.
HABITAT Desert with dunes and oases.

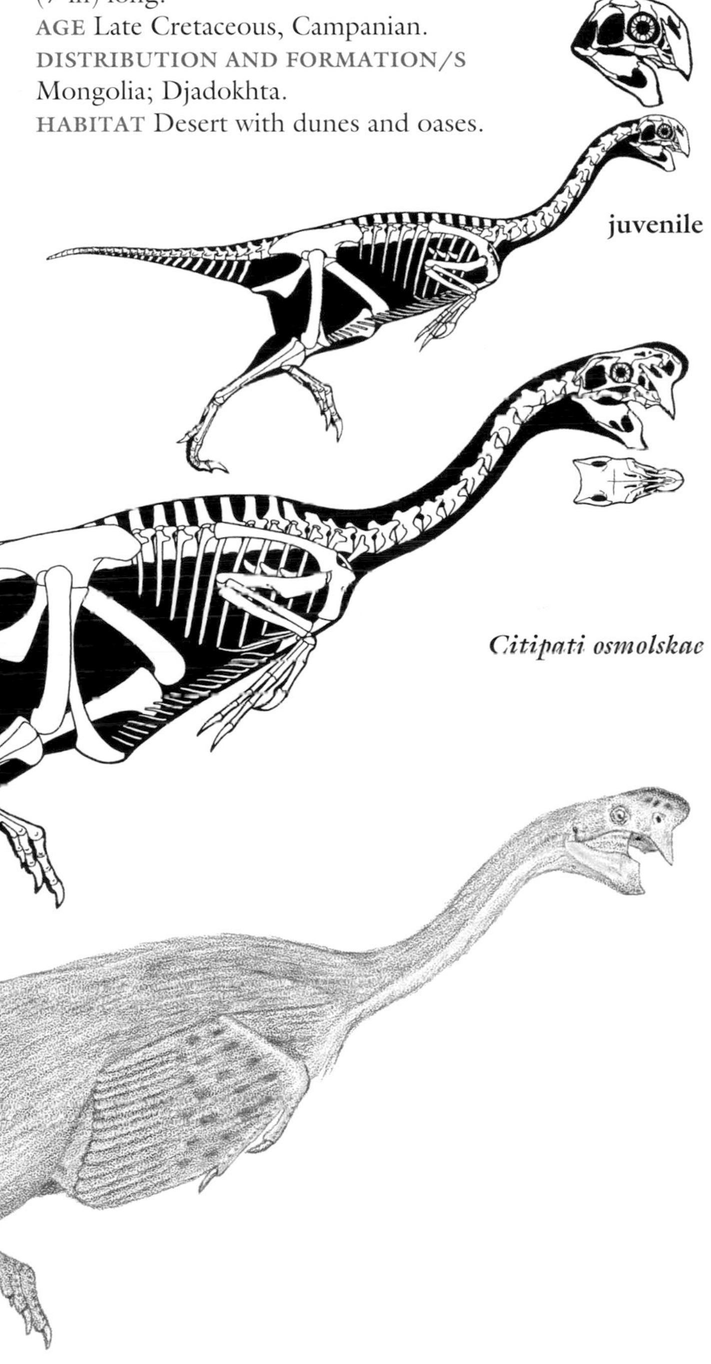

HABITS Presence of remains of juvenile dinosaurs in some nests indicates that the oviraptorid diet included at least some small animals.
NOTES Showing that small, crestless *Khaan mckennai* is or is not the juvenile of this species requires examination of bone microstructure. Shared its habitat with *Protoceratops*.

Wulatelong gobiensis
1.7 m (5.5 ft) TL, 25 kg (55 lb)

FOSSIL REMAINS Majority of poorly preserved skull and minority of skeleton.
ANATOMICAL CHARACTERISTICS Insufficient information.
AGE Late Cretaceous, early Campanian.
DISTRIBUTION AND FORMATION/S Northwest China; Wulansuhai.
NOTES Shared its habitat with *Linhenykus*.

Banji long
Adult size uncertain

FOSSIL REMAINS Nearly complete juvenile skull.
ANATOMICAL CHARACTERISTICS Head crest low arced, but this is probably the juvenile condition.
AGE Late Cretaceous, late Campanian and/or early Maastrichtian.
DISTRIBUTION AND FORMATION/S Southeast China; probably Nanxiong.
NOTES May be juvenile of another named species from the Nanxiong.

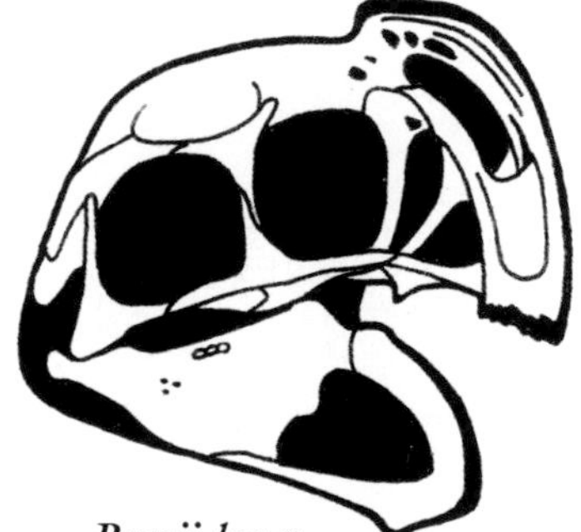

Banji long

Shixinggia oblita
2 m (7 ft) TL, 40 kg (85 lb)

FOSSIL REMAINS Minority of skeleton.
ANATOMICAL CHARACTERISTICS Insufficient information.
AGE Late Cretaceous, Maastrichtian.
DISTRIBUTION AND FORMATION/S Southern China; Pingling.

Conchoraptor (= *Ajancingenia*) *yanshini*
1.5 m (5 ft) TL, 17 kg (45 lb)

FOSSIL REMAINS Complete and partial skulls and skeletons, adult and juvenile.
ANATOMICAL CHARACTERISTICS Very large forward-pointed head crest. Tail deep in at least one morph. Thumb about as long as other fingers, hand robust at least in one morph.
AGE Late Cretaceous, probably middle Campanian.
DISTRIBUTION AND FORMATION/S Mongolia; lower Barun Goyot.
HABITAT Well-watered woodland with seasonal rain.
HABITS Large thumb was probably a weapon and may have been used for feeding in some manner.
NOTES Showing that large, crested *Conchoraptor gracilis* is or is not the adult of this species requires examination of bone microstructure. Original genus title *Ingenia* preoccupied by an invertebrate. *Conchoraptor* has priority over more recent *Ajancingenia* if they are one genus. May be the ancestor of *Conchoraptor barsboldi*.

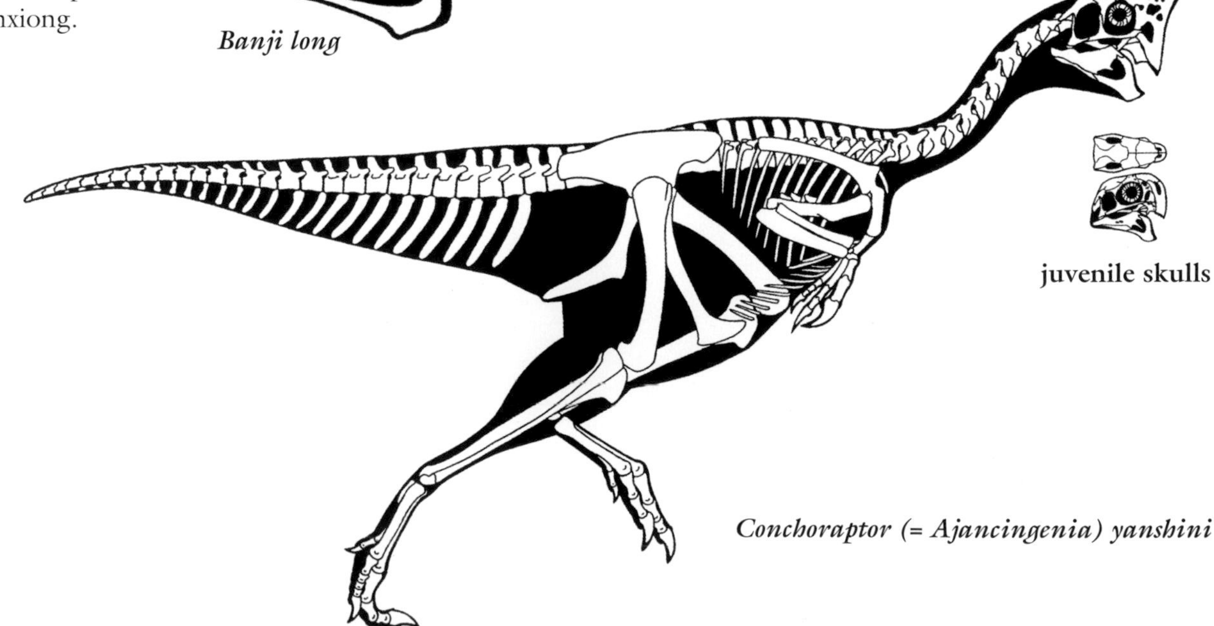

Conchoraptor (= Ajancingenia) yanshini

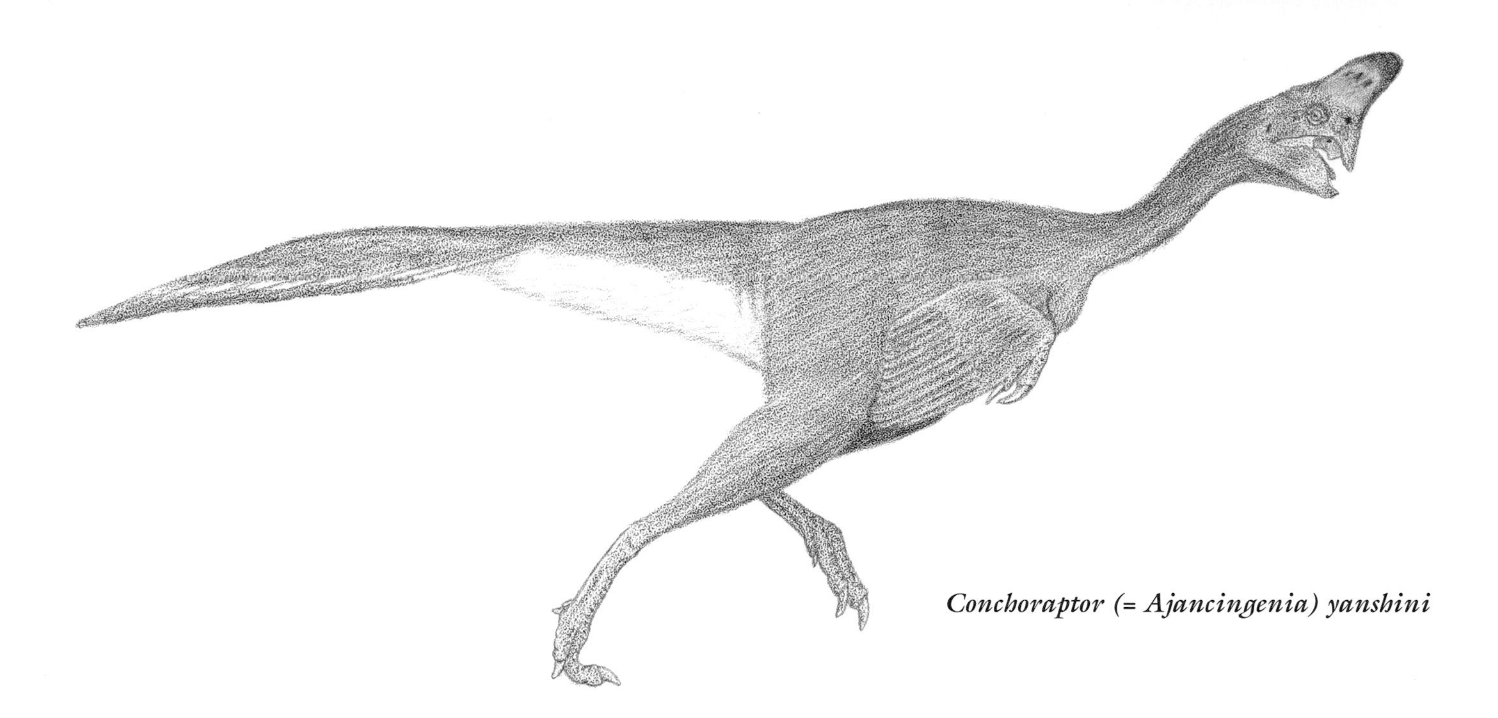

Conchoraptor (= Ajancingenia) yanshini

Conchoraptor (= *Nemegtomaia*) *barsboldi*

2 m (7 ft) TL, 40 kg (85 lb)

FOSSIL REMAINS Two poorly preserved complete skulls and minority of four skeletons.

ANATOMICAL CHARACTERISTICS Large crest above upper beak.

AGE Late Cretaceous, late Campanian and/or early Maastrichtian.

DISTRIBUTION AND FORMATION/S Mongolia; Nemegt.

HABITAT Well-watered woodland with seasonal rain, winters cold.

NOTES Separation from the very similar *Conchoraptor* is not warranted. Shared its habitat with *Citipati mongoliensis*. Prey of *Saurornithoides junior*, juvenile *Alioramus remotus* and *T. bataar*.

Conchoraptor (or *Heyuannia*) *huangi*

1.5 m (5 ft) TL, 20 kg (45 lb)

FOSSIL REMAINS Partial skull and skeleton.

ANATOMICAL CHARACTERISTICS Insufficient information.

AGE Late Late Cretaceous.

DISTRIBUTION AND FORMATION/S Southern China; Dalangshan.

NOTES Available evidence insufficient to make separate genus from *Conchoraptor*.

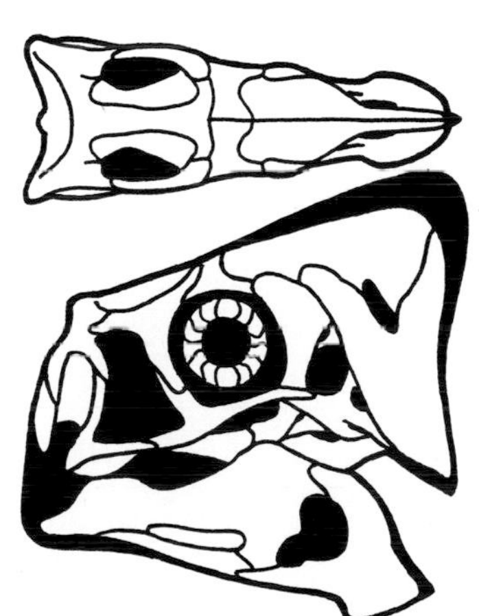

Conchoraptor (= Nemegtomaia) barsboldi

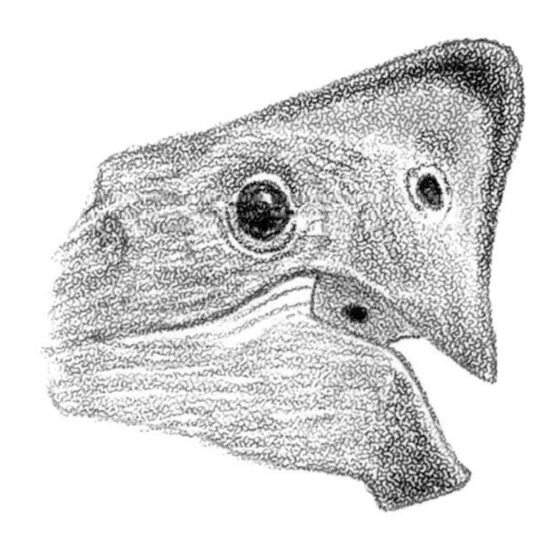

SAUROPODOMORPHS

SMALL TO COLOSSAL HERBIVOROUS AND OMNIVOROUS SAURISCHIAN DINOSAURS FROM THE LATE TRIASSIC TO THE END OF THE DINOSAUR ERA, ALL CONTINENTS.

ANATOMICAL CHARACTERISTICS **Moderately variable. Head small, nostrils enlarged, teeth blunt, nonserrated. Neck long and slender. Tail long. Arm and leg neither elongated nor slender. Five fingers. Pelvis small to large, five to four toes. Largely bipedal to always quadrupedal when moving, all able to rear up on hind legs. Brains reptilian. Gizzard stones sometimes present, used to either help grind or stir up ingested food.**

HABITAT **Very variable, deserts to well-watered forests, tropics to polar regions.**

HABITS **Predominantly herbivorous browsers and grazers, although they may have been prone to pick up and consume small animals; did not extensively chew food before swallowing. Main defense clawed feet and tails.**

PROSAUROPODS

SMALL TO LARGE HERBIVOROUS AND OMNIVOROUS SAUROPODOMORPHS LIMITED TO THE LATE TRIASSIC AND EARLY JURASSIC, ALL CONTINENTS.

ANATOMICAL CHARACTERISTICS Fairly uniform. Skulls lightly built, partial elastic cheeks may have been present in at least some species, same for incipient beaks. Neck moderately long, very slender. Trunk long. Tail long. Skeletons not pneumatic, respiratory system poorly understood except that birdlike system not present. Shoulder girdles not large, hand short and broad, grasping fingers fairly long, large claws on most fingers, especially thumb. Pelvis short, pubis strongly procumbent, lower leg about as long as upper, foot fairly long, toes long and flexible, outermost toe very reduced, large claw on innermost toe. All able to slowly walk quadrupedally, those with long arm mainly quadrupedal, with short arm mainly bipedal, or intermediate. In trackways hands always farther from midline than feet, arm and leg flexed but not elongated or slender, so able to run at modest speeds.

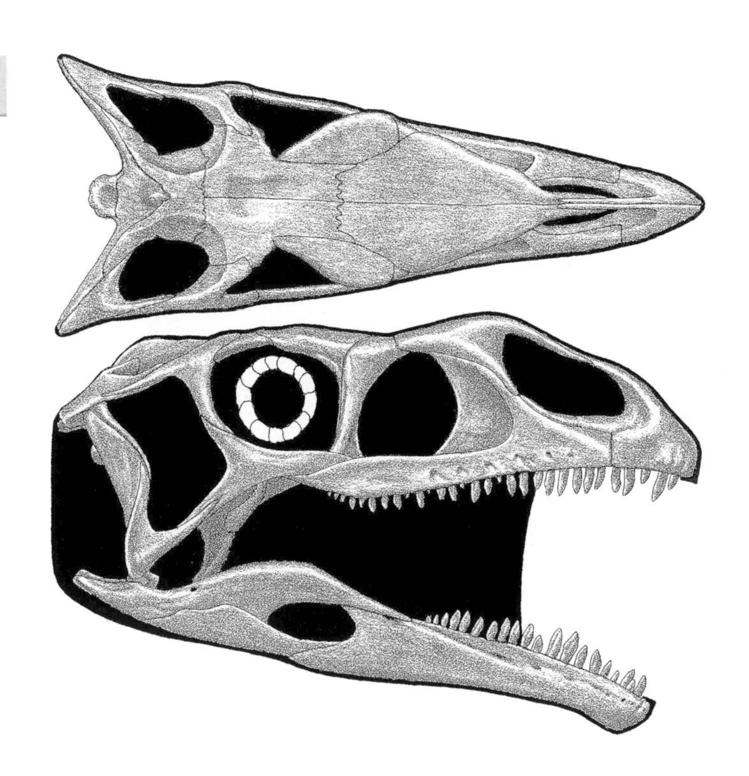

Plateosaurus shaded skull and muscle study

ONTOGENY Growth rates moderate.
HABITAT Very variable, deserts to well-watered forests, tropics to poles.
HABITS The first herbivores able to high browse, especially when rearing; some or all may have been omnivores. Main defense standing and lashing out with clawed hands and feet. Small prosauropods may have used clawed hands to dig burrows.
ENERGETICS Thermophysiology probably intermediate, energy levels and food consumption probably low compared to more-derived dinosaurs.
NOTES Fragmentary Early Jurassic *Glacialisaurus hammeri* demonstrates prosauropods dwelled in Antarctica. Whether all the many genera are justified is doubtful. This group is splittable into a number of subdivisions, but relationships within the group and with sauropods are uncertain. Many researchers consider known prosauropods to be a sister group to sauropods, but others consider some or all of the first five genera to be below the prosauropod-sauropod split, or the latter may have evolved from more-derived prosauropods. *Eoraptor* may be a very basal sauropodomorph. Based on supposedly Early Jurassic inadequate remains, Asian *Eshanosaurus deguchiianus* may be a prosauropod rather than an avepod.

Panphagia protos
1.7 m (5.5 ft) TL, 6 kg (12 lb)

FOSSIL REMAINS Partial skull and skeleton.
ANATOMICAL CHARACTERISTICS Front lower teeth bladed. Arm moderately short. Predominantly bipedal.
AGE Late Triassic, Carnian.
DISTRIBUTION AND FORMATION/S Northern Argentina; Ischigualasto.
HABITAT Seasonally well-watered forests, including dense stands of giant conifers.
HABITS More predaceous than more-derived prosauropods.
NOTES Shared its habitat with *Pisanosaurus*. Prey of *Eoraptor* and *Herrerasaurus*.

Saturnalia tupiniquim
1.5 m (5 ft) TL, 4 kg (8 lb)

FOSSIL REMAINS Partial skull and majority of skeleton.
ANATOMICAL CHARACTERISTICS Arm moderately short. Predominantly bipedal.
AGE Late Triassic, early Carnian.
DISTRIBUTION AND FORMATION/S Southern Brazil; lower Santa Maria.
NOTES Shared its habitat with *Pampadromaeus*. Prey of *Staurikosaurus*.

Pampadromaeus barberenai
1.5 m (5 ft) TL, 4 kg (8 lb)

FOSSIL REMAINS Majority of skull and skeleton.
ANATOMICAL CHARACTERISTICS Most teeth bladed. Arm moderately short. Predominantly bipedal.
AGE Late Triassic, early Carnian.
DISTRIBUTION AND FORMATION/S Southern Brazil; lower Santa Maria.
HABITS More predaceous than more-derived prosauropods.
NOTES Prey of *Staurikosaurus*.

Guaibasaurus candelariensis
2 m (6.5 ft) TL, 10 kg (20 lb)

FOSSIL REMAINS Partial skeletons.
ANATOMICAL CHARACTERISTICS Insufficient information.
AGE Late Triassic, late Carnian and/or early Norian.
DISTRIBUTION AND FORMATION/S Southern Brazil; Caturrita.
NOTES Originally thought to be a baso-theropod. Shared its habitat with *Unaysaurus*.

Asylosaurus yalensis
2 m (6.5 ft) TL, 10 kg (20 lb)

FOSSIL REMAINS Minority of skeleton.
ANATOMICAL CHARACTERISTICS Insufficient information.
AGE Probably Late Triassic, probably Rhaetian.

Pampadromaeus barberenai

DISTRIBUTION AND FORMATION/S Southwest England; unnamed.
NOTES Found in an ancient fissure fill.

Pantydraco caducus
Adult size uncertain

FOSSIL REMAINS Nearly complete skull, majority of a few skeletons.
ANATOMICAL CHARACTERISTICS Head short and subtriangular. Arm probably moderately short. Predominantly bipedal.
AGE Late Triassic or Early Jurassic.
DISTRIBUTION AND FORMATION/S Wales; unnamed.
NOTES Found in an ancient fissure fill; specimens were long assigned to *Thecodontosaurus antiquus.* Skeletal proportions uncertain.

Thecodontosaurus antiquus
2.5 m (8 ft) TL, 20 kg (40 lb)

FOSSIL REMAINS Minority of skull and partial skeletons.
ANATOMICAL CHARACTERISTICS Arm probably moderately short.
AGE Probably Late Triassic, probably Rhaetian.
DISTRIBUTION AND FORMATION/S Wales; unnamed.
NOTES Found in an ancient fissure fill. Some of the remains destroyed in World War II by Axis bombing.

Plateosaurus (= Unaysaurus) tolentinoi
Adult size uncertain

FOSSIL REMAINS Nearly complete skull and minority of skeleton.
ANATOMICAL CHARACTERISTICS Head shallow, subrectangular, snout a little downturned. Arm moderately long.
AGE Late Triassic, late Carnian and/or early Norian.
DISTRIBUTION AND FORMATION/S Southern Brazil; Caturrita.
NOTES Shared its habitat with *Guaibasaurus.*

Plateosaurus (= Sellosaurus) gracilis
5 m (15 ft) TL, 300 kg (600 lb)

FOSSIL REMAINS Majority of two dozen partial skulls and skeletons.
ANATOMICAL CHARACTERISTICS Head shallow, subrectangular, snout a little downturned. Arm moderately long. Bi/quadrupedal.
AGE Late Triassic, middle Norian.
DISTRIBUTION AND FORMATION/S Southern Germany; lower and middle Lowenstein.
NOTES *Efraasia diagnosticus* is probably an immature form of this species. Very similar to and the same genus as the slightly later *P. longiceps.* Prey of pseudosuchian archosaurs.

Plateosaurus longiceps
8 m (26 ft) TL, 1300 kg (2,900 lb)

FOSSIL REMAINS Dozens of complete to partial skulls and skeletons, juvenile to adult, completely known.
ANATOMICAL CHARACTERISTICS Head shallow, subrectangular, narrow, snout a little downturned. Arm moderately long. Bi/quadrupedal.
AGE Late Triassic, middle Norian.
DISTRIBUTION AND FORMATION/S Germany, Switzerland, eastern France; Trossingen, upper Lowenstein, Knollenmergel, Obere Bunte Mergel, Marnes Irisees Superieures.
NOTES The classic prosauropod known from abundant remains. May be the direct descendant of *P. gracilis. Ruehleia bedheimensis* is probably the mature form of this species. Prey of *Liliensternus.*

Plateosaurus (= Unaysaurus) tolentinoi

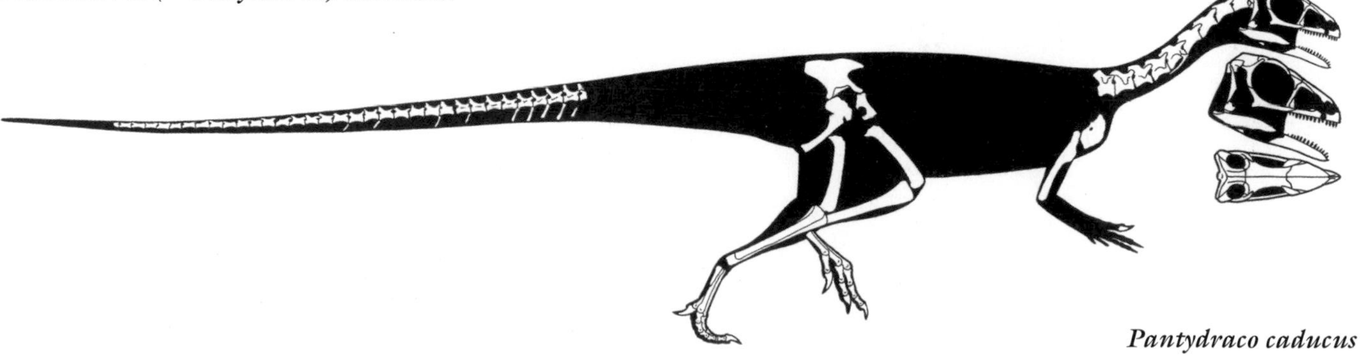

Pantydraco caducus

Plateosaurus (= Sellosaurus) gracilis

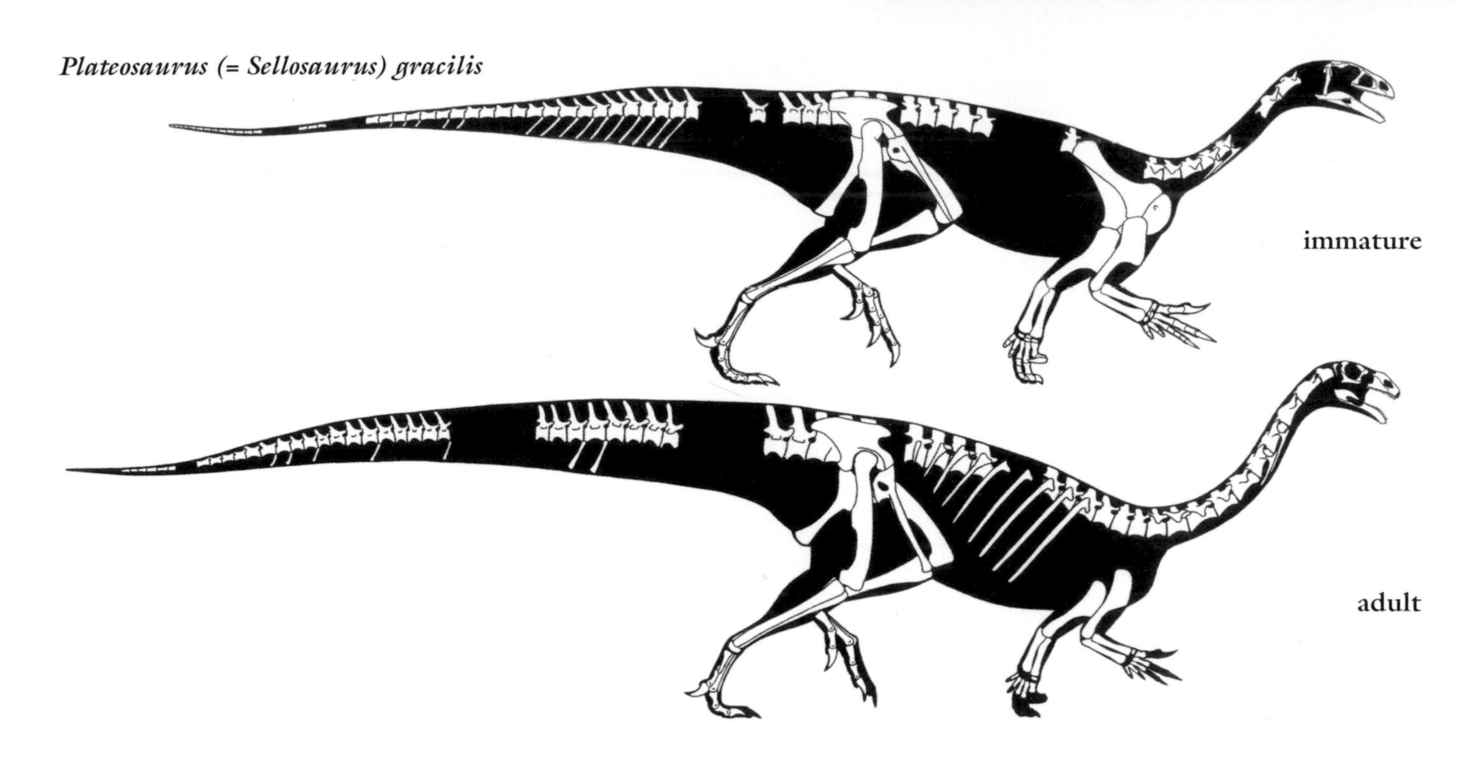

Plateosaurus longiceps
(see also next page)

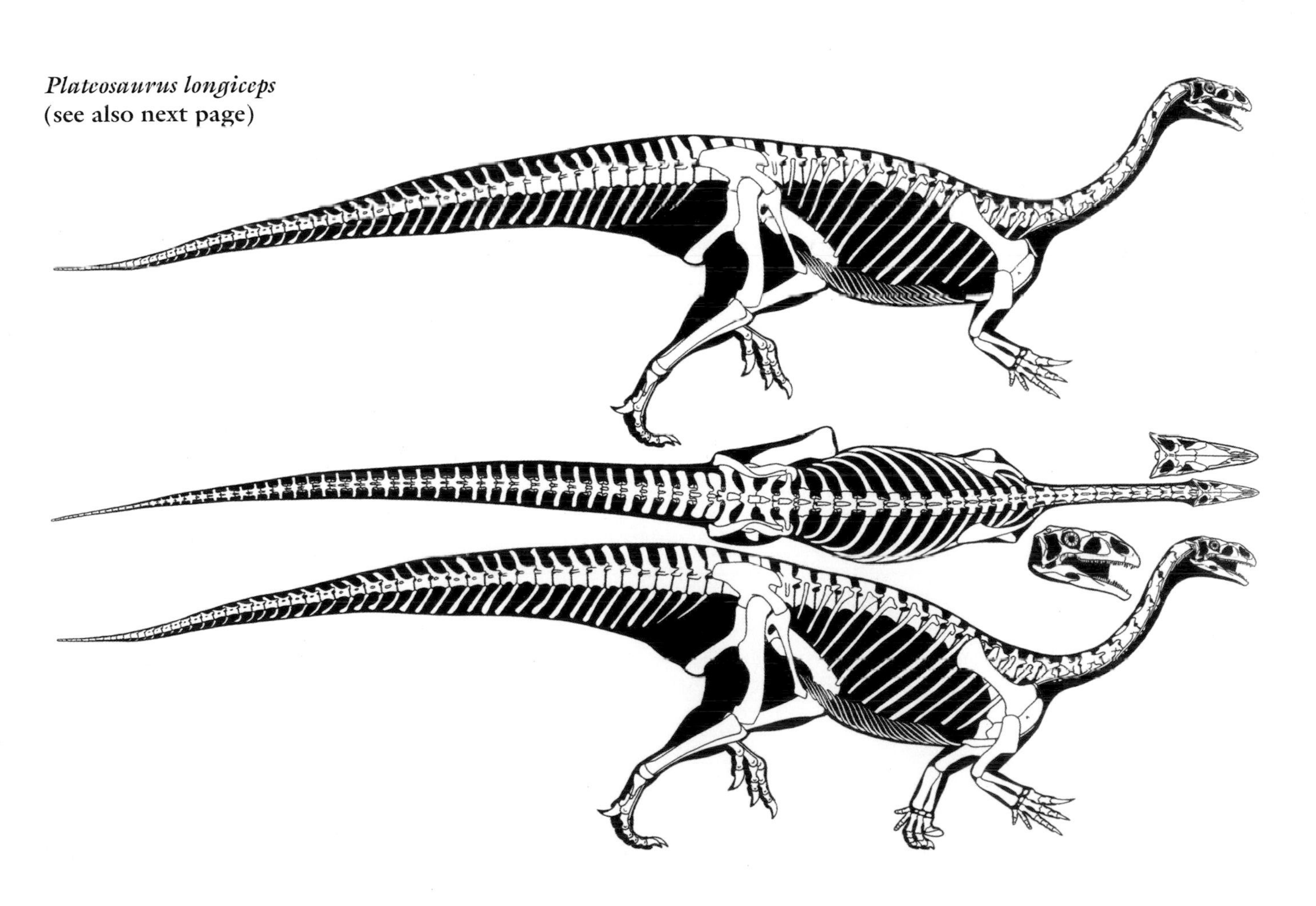

Plateosaurus longiceps

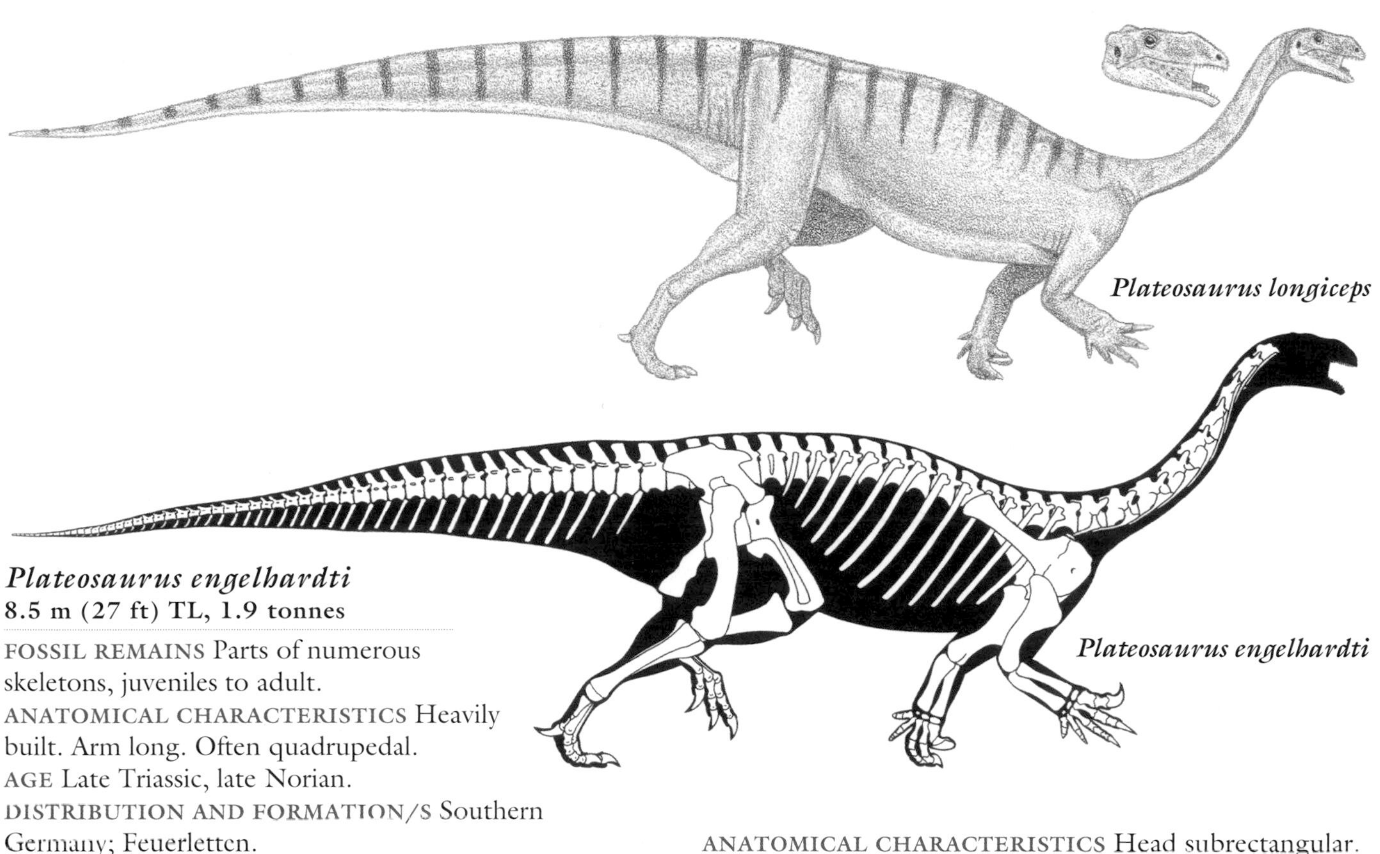

Plateosaurus engelhardti
8.5 m (27 ft) TL, 1.9 tonnes

FOSSIL REMAINS Parts of numerous skeletons, juveniles to adult.
ANATOMICAL CHARACTERISTICS Heavily built. Arm long. Often quadrupedal.
AGE Late Triassic, late Norian.
DISTRIBUTION AND FORMATION/S Southern Germany; Feuerletten.
NOTES May be the direct descendant of *P. longiceps*. Larger size may be a response to being attacked by theropods.

Plateosaurus (or *Massospondylus*) *carinatus*
4.3 m (14 ft) TL, 200 kg (450 lb)

FOSSIL REMAINS Many dozens of skulls and skeletons, many complete, juveniles to adult, completely known, nests with up to three dozen eggs, embryos.
ANATOMICAL CHARACTERISTICS Head subrectangular. Thumb and foot claws large. Arm long in juveniles, moderately long in adults, indicating increasing bipedalism with growth. Eggs spherical, 60 mm in diameter.
AGE Early Jurassic, late Hettangian to perhaps early Pliensbachian.

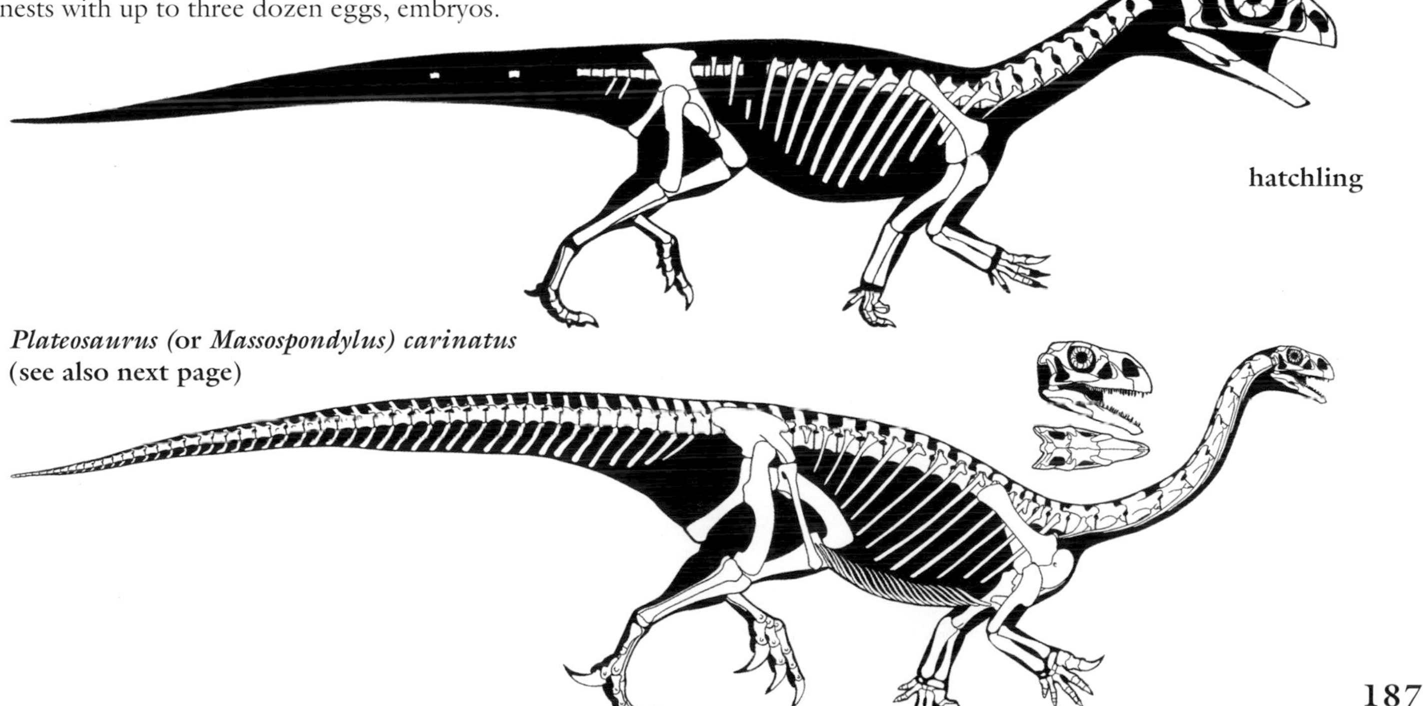

Plateosaurus (or *Massospondylus*) *carinatus*
(see also next page)

Plateosaurus (or *Massospondylus*) *carinatus*

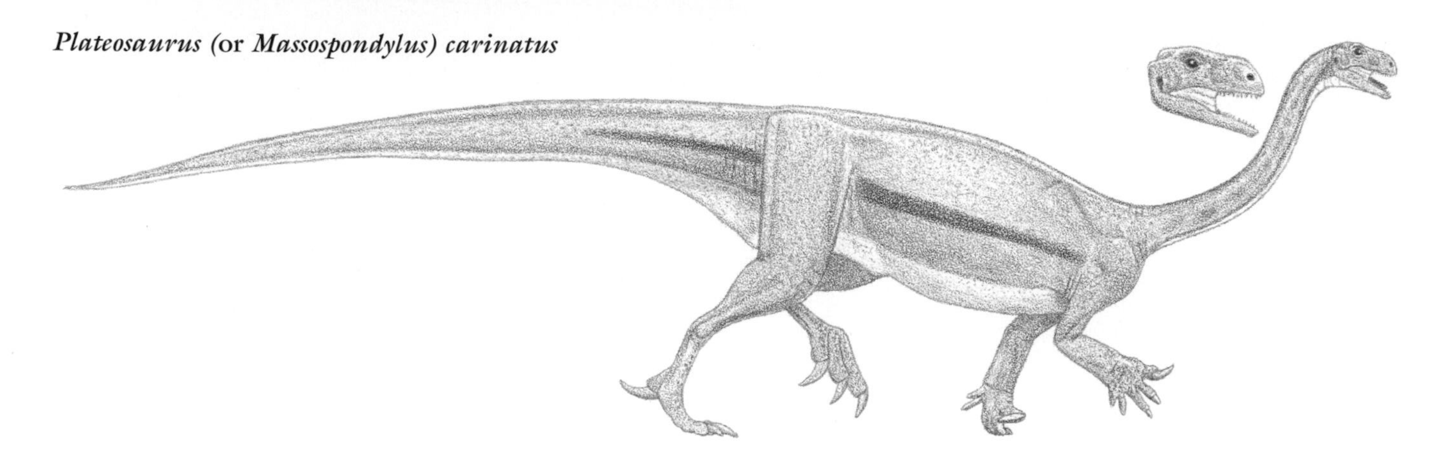

DISTRIBUTION AND FORMATION/S South Africa, Lesotho, Zimbabwe; Upper Elliot, Bushveld Sandstone, Upper Karoo Sandstone, Forest Sandstone.
HABITAT In at least some locations desert.
HABITS Probably fed on vegetation along watercourses and at oases. Eggs probably buried and abandoned.
NOTES The original specimen is inadequate, and the long time span is suspiciously long for a single species. *Massospondylus kaalae* from the Upper Elliot Formation may be a distinct species.

Plateosaurus (or *Adeopapposaurus*) *mognai*

Adult size uncertain

FOSSIL REMAINS Majority of a few skulls and skeletons.
ANATOMICAL CHARACTERISTICS Head shallow, subrectangular, broad. Arm short. Strongly bipedal.
AGE Early Jurassic.
DISTRIBUTION AND FORMATION/S Southern Argentina; Canon del Colorado.
NOTES Skeletal proportions not entirely certain.

Plateosaurus (or *Adeopapposaurus*) *mognai*

Plateosaurus (or *Lufengosaurus*) *huenei*

9 m (30 ft) TL, 1.7 tonnes

FOSSIL REMAINS Over two dozen skulls and skeletons, some complete, juvenile to adult, completely known.
ANATOMICAL CHARACTERISTICS Neck longer than that of most prosauropods. Arm short. Strongly bipedal.
AGE Early Jurassic, Hettangian or Sinemurian.
DISTRIBUTION AND FORMATION/S Southwest China; lower and upper Lower Lufeng.

Plateosaurus (or *Leyesaurus*) *marayensis*

3 m (10 ft) TL, 70 kg (150 lb)

FOSSIL REMAINS Majority of skull and minority of skeleton.
ANATOMICAL CHARACTERISTICS Head shallow, subrectangular, broad.
AGE Early Jurassic.

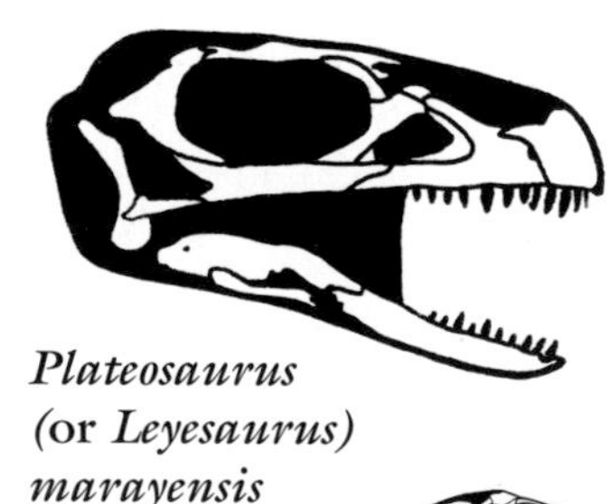

Plateosaurus (or *Leyesaurus*) *marayensis*

Plateosaurus (or *Adeopapposaurus*) *mognai*

Plateosaurus (or *Lufengosaurus*) *huenei*

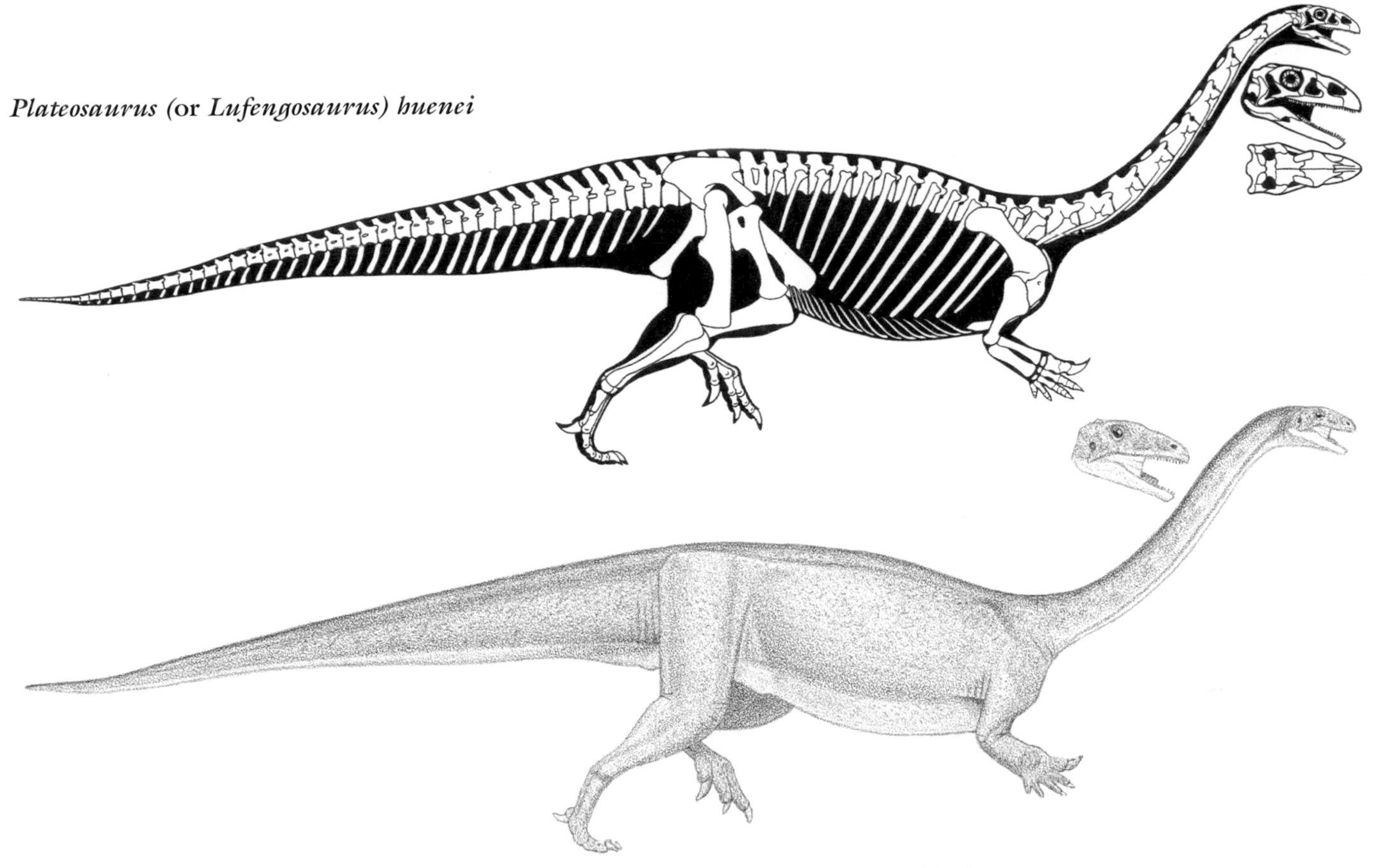

DISTRIBUTION AND FORMATION/S Northwestern Argentina; upper Quebrada del Barro.

Plateosaurus (or *Sarahsaurus*) *aurifontanalis*

4 m (9 ft) TL, 200 kg (400 lb)

FOSSIL REMAINS Majority of skull and two skeletons.
ANATOMICAL CHARACTERISTICS Head shallow, snout probably a little downturned. Arm short. Strongly bipedal.
AGE Late Triassic, Sinemurian or Pliensbachian.
DISTRIBUTION AND FORMATION/S Arizona; Kayenta.
HABITAT Semiarid.

Coloradisaurus brevis

3 m (10 ft) TL, 70 kg (150 lb)

FOSSIL REMAINS Complete skull.
ANATOMICAL CHARACTERISTICS Head short, subtriangular, broad.
AGE Late Triassic, Norian.
DISTRIBUTION AND FORMATION/S Northern Argentina; Los Colorados.
HABITAT Seasonally wet woodlands.

Coloradisaurus brevis

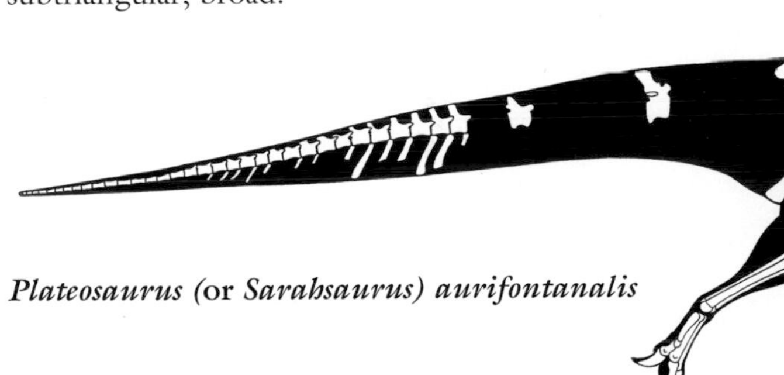

Plateosaurus (or *Sarahsaurus*) *aurifontanalis*

Jingshanosaurus xinwaensis
9 m (30 ft) TL, 1.6 tonnes

FOSSIL REMAINS Complete skull and skeleton.
ANATOMICAL CHARACTERISTICS Head subtriangular, broad, cheeks may have been absent. Arm short. Strongly bipedal.
AGE Early Jurassic, Sinemurian.
DISTRIBUTION AND FORMATION/S Southwest China; upper Lower Lufeng.

Yunnanosaurus huangi
5 m (16 ft) TL, 230 kg (500 lb)

FOSSIL REMAINS Almost two dozen skulls and skeletons, some complete, juvenile to adult.
ANATOMICAL CHARACTERISTICS Head small, subtriangular, cheeks may have been absent. Arm short. Strongly bipedal.
AGE Early Jurassic, Hettangian to Sinemurian.
DISTRIBUTION AND FORMATION/S Southwest China; lower to upper Lower Lufeng.
NOTES May be two species.

Chuxiongosaurus lufengensis
4 m (9 ft) TL, 200 kg (400 lb)

FOSSIL REMAINS Nearly complete skull.
ANATOMICAL CHARACTERISTICS Skull deep, subtriangular, broad.
AGE Early Jurassic, Hettangian.
DISTRIBUTION AND FORMATION/S Southwest China; lower Lufeng.
NOTES Prey of *Sinosaurus.*

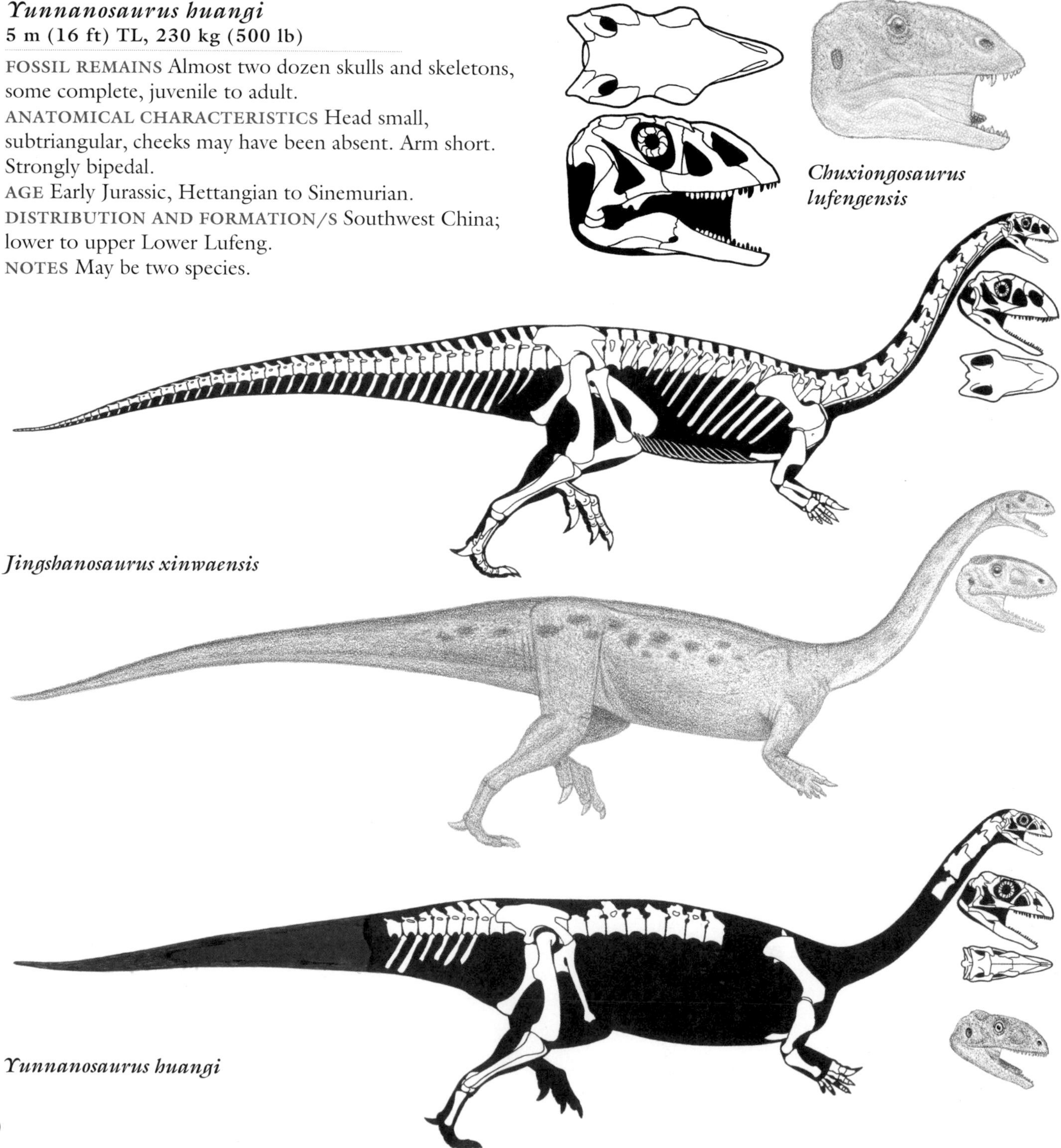

Chuxiongosaurus lufengensis

Jingshanosaurus xinwaensis

Yunnanosaurus huangi

Yimenosaurus youngi
9 m (30 ft) TL, 2 tonnes

FOSSIL REMAINS Majority of skull, numerous partial skeletons.
ANATOMICAL CHARACTERISTICS Skull deep, subrectangular.
AGE Early Jurassic, Pliensbachian and/or Toarcian.
DISTRIBUTION AND FORMATION/S Southwest China; Fengjiahe.

Yimenosaurus youngi

Seitaad ruessi
3 m (10 ft) TL, 70 kg (150 lb)

FOSSIL REMAINS Minority of skeleton.
ANATOMICAL CHARACTERISTICS Arm moderately short. Predominantly bipedal.
AGE Late Triassic, Sinemurian or Pliensbachian.
DISTRIBUTION AND FORMATION/S Utah; lower Navajo Sandstone.
HABITAT Dune desert.

Plateosauravus cullingworthi
9 m (30 ft) TL, 2 tonnes

FOSSIL REMAINS A few partial skeletons.
ANATOMICAL CHARACTERISTICS Insufficient information.
AGE Late Triassic, early Norian.
DISTRIBUTION AND FORMATION/S Southeast Africa, Lower Elliot.
HABITAT Arid.
NOTES Was *Euskelosaurus browni*, which is based on inadequate remains.

Blikanasaurus cromptoni
4 m (13 ft) TL, 250 kg (500 lb)

FOSSIL REMAINS Minority of skeleton.
ANATOMICAL CHARACTERISTICS Leg massively built.
AGE Late Triassic, early Norian.
DISTRIBUTION AND FORMATION/S Southeast Africa; Lower Elliot.
HABITAT Arid.

Camelotia borealis
10 m (33 ft) TL, 2.5 tonnes

FOSSIL REMAINS Minority of skeleton.
ANATOMICAL CHARACTERISTICS Insufficient information.
AGE Late Triassic, Rhaetian.
DISTRIBUTION AND FORMATION/S Southwest England; Westbury.

Lessemsaurus sauropoides
9 m (30 ft) TL, 2 tonnes

FOSSIL REMAINS Minority of skeleton.
ANATOMICAL CHARACTERISTICS Insufficient information.
AGE Late Triassic, Norian.
DISTRIBUTION AND FORMATION/S Northern Argentina; Los Colorados.
HABITAT Seasonally wet woodlands.
NOTES Shared its habitat with *Riojasaurus*.

Anchisaurus polyzelus
3 m (10 ft) TL, 70 kg (150 lb)

FOSSIL REMAINS Nearly complete skull and majority of skeleton.
ANATOMICAL CHARACTERISTICS Skull shallow, subtriangular, broad. Arm moderately long. Bi/quadrupedal.
AGE Early Jurassic, Pliensbachian and/or Toarcian.
DISTRIBUTION AND FORMATION/S Connecticut, Massachusetts; Portland.
HABITAT Semiarid rift valley with lakes.
NOTES *Ammosaurus major* probably the adult of this species.

Unnamed genus and species
4.5 m (18 ft) TL, 250 kg (500 lb)

FOSSIL REMAINS Minority of skeleton.
ANATOMICAL CHARACTERISTICS Insufficient information.
AGE Early Jurassic, Pliensbachian or Toarcian.
DISTRIBUTION AND FORMATION/S Arizona; Navajo Sandstone.

Anchisaurus polyzelus

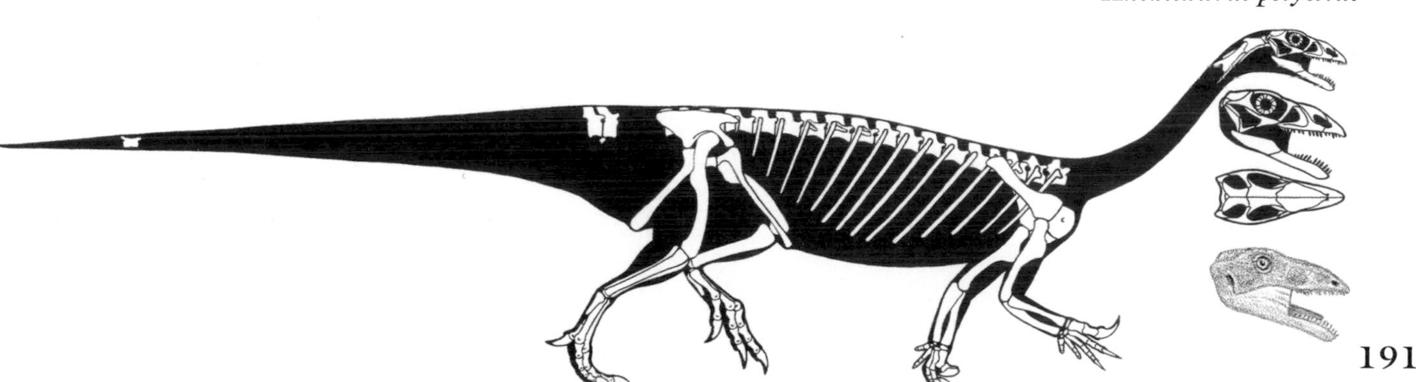

HABITAT Desert with dunes.
HABITS Probably fed on vegetation along watercourses and at oases.
NOTES Has been placed in *Ammosaurus major* and *Massospondylus*, both options highly problematic. Shared its habitat with *Segisaurus*.

Riojasaurus incertus
6.6 m (22 ft) TL, 800 kg (1,800 lb)

FOSSIL REMAINS Complete skull, numerous skeletons of varying completeness, juvenile to adult.
ANATOMICAL CHARACTERISTICS Head subtriangular. Arm long, robust. Strongly quadrupedal.
AGE Late Triassic, Norian.
DISTRIBUTION AND FORMATION/S Northern Argentina; Los Colorados.
HABITAT Seasonally wet woodlands.
NOTES Shared its habitat with *Lessemsaurus*.

Eucnemesaurus fortis
8 m (26 ft) TL, 1300 kg (2,900 lb)

FOSSIL REMAINS Small portion of a few skeletons.
ANATOMICAL CHARACTERISTICS Insufficient information.
AGE Late Triassic, late Carnian or early Norian.
DISTRIBUTION AND FORMATION/S Southeast Africa; Lower Elliot.
HABITAT Arid.
NOTES The scanty remains were once labeled *Aliwalia rex*, which was thought to be a giant baso-theropod. A large partial skeleton labeled *Eucnemesaurus entaxonis* is probably the adult of this species. Shared its habitat with *Melanorosaurus*, *Plateosauravus*, *Blikanasaurus*, and *Antetonitrus*.

Mussaurus patagonicus
8 m (26 ft) TL, 1300 kg (2,900 lb)

FOSSIL REMAINS Almost a dozen complete to partial juvenile skulls and skeletons, partial adult skeletons.
ANATOMICAL CHARACTERISTICS Quadrupedal at least when juvenile.
AGE Late Triassic, probably Norian.
DISTRIBUTION AND FORMATION/S Southern Argentina; Laguna Colorada.
HABITS Small juveniles may have supplemented diet with insects.

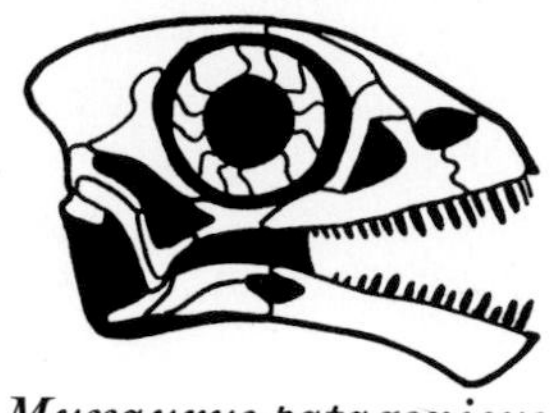

***Mussaurus patagonicus* hatchling**

Aardonyx celestae
Adult size uncertain

FOSSIL REMAINS Two partial skulls and skeletons.
ANATOMICAL CHARACTERISTICS Insufficient information.
AGE Early Jurassic, Hettangian or Sinemurian.
DISTRIBUTION AND FORMATION/S Southeast Africa; Upper Elliot.
HABITAT Arid.
NOTES Prey of *Dracovenator*.

Leonerasaurus taquetrensis
3 m (10 ft) TL, 70 kg (150 lb)

FOSSIL REMAINS Minority of skull and skeleton.
ANATOMICAL CHARACTERISTICS Insufficient information.
AGE Early Jurassic, Pliensbachian or Toarcian.
DISTRIBUTION AND FORMATION/S Southern Argentina; Leoneras.

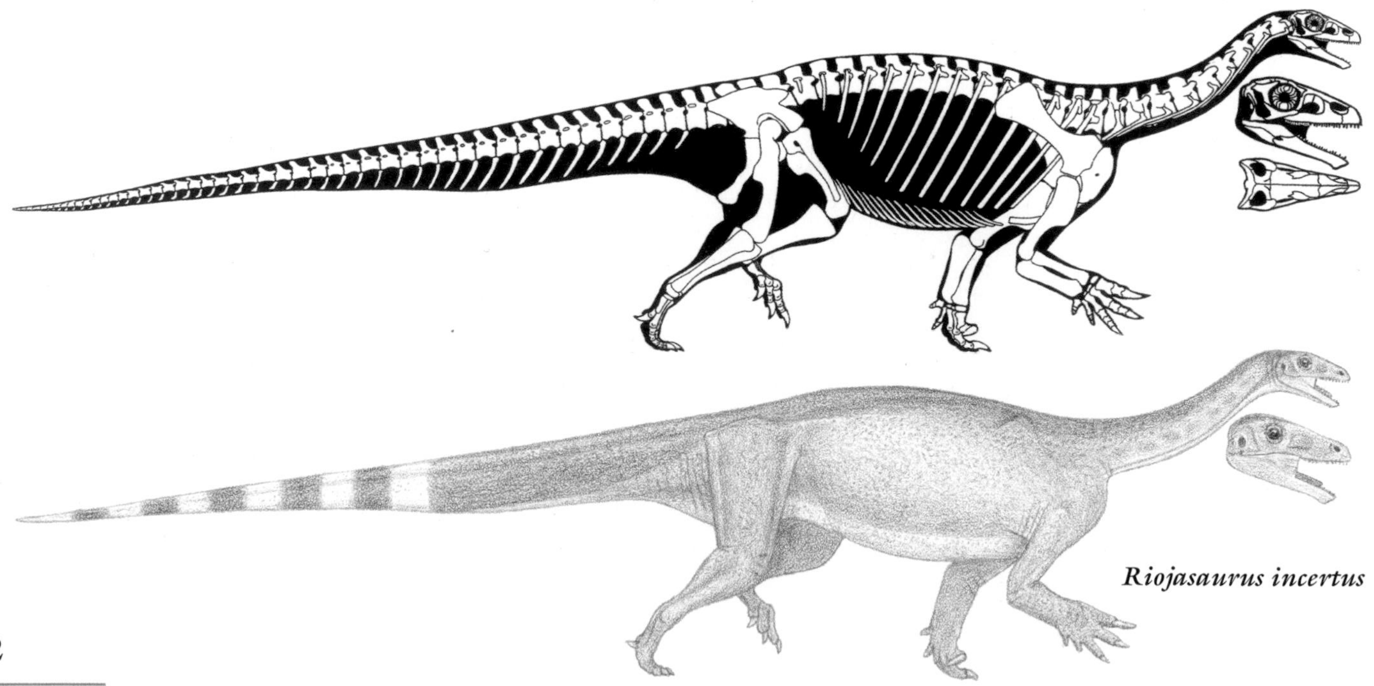

Riojasaurus incertus

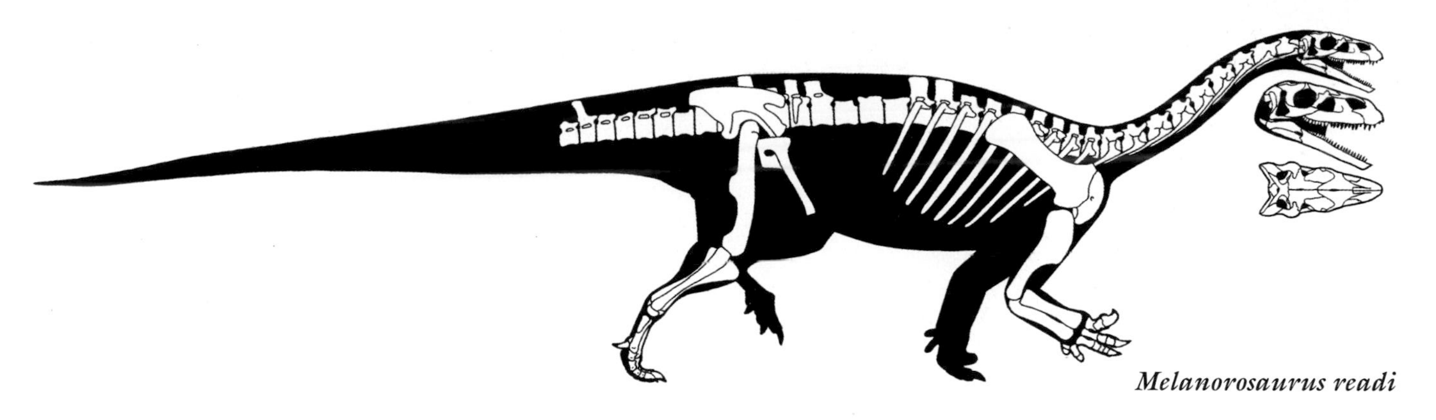

Melanorosaurus readi

Sefapanosaurus zastronensis
4 m (9 ft) TL, 200 kg (400 lb)

FOSSIL REMAINS Four partial skeletons.
ANATOMICAL CHARACTERISTICS Insufficient information.
AGE Late Triassic, Norian.
DISTRIBUTION AND FORMATION/S Southeast Africa; Elliot, level uncertain.
HABITAT Arid.

Melanorosaurus readi
8 m (26 ft) TL, 1300 kg (2,900 lb)

FOSSIL REMAINS Complete distorted skull and majority of skeleton, partial remains.
ANATOMICAL CHARACTERISTICS Arm long, robust. Strongly quadrupedal.
AGE Late Triassic, early Norian.
DISTRIBUTION AND FORMATION/S Southeast Africa, Lower Elliot.
HABITAT Arid.
NOTES Skeletal proportions not entirely certain. Shared its habitat with *Antetonitrus*, *Eucnemesaurus*, *Plateosauravus*, and *Blikanasaurus*.

SAUROPODS

LARGE TO ENORMOUS HERBIVOROUS SAUROPODOMORPHS FROM THE LATE TRIASSIC TO THE END OF THE DINOSAUR ERA, MOST CONTINENTS.

ANATOMICAL CHARACTERISTICS Variable. Skulls not heavily built, nostrils at least somewhat retracted. Skeletons heavily built. Neck moderately to extremely long. Tail moderately to extremely long. Quadrupedal when moving normally, arm and leg less flexed than in prosauropods. Lower leg shorter than upper, foot short and broad. Skeleton at least incipiently pneumatic, birdlike, some degree of air-sac-ventilated respiratory system present.
NOTES Lasting 150 million years and regularly rivaling whales in size, the most successful group of large herbivorous animals that has yet evolved. Absence from Antarctica probably reflects lack of sufficient sampling.

VULCANODONTS

LARGE TO GIGANTIC SAUROPODS LIMITED TO THE LATE TRIASSIC AND EARLY JURASSIC OF THE EASTERN HEMISPHERE.

ANATOMICAL CHARACTERISTICS Fairly uniform. Head short, snout narrow and rounded. Neck and tail moderately long. Limbs moderately flexed. Arm moderately long, so shoulders about as high as hips. Hand not forming an arcade, fingers not extremely abbreviated. Ilium shallow, ankle still markedly flexible. Skeletal pneumaticity partly developed, so birdlike respiratory system developing.
HABITS Probably feeding generalists. Probably able to run slowly, main defense standing and fighting with claws.
ENERGETICS Thermophysiology probably intermediate between that of prosauropods and eusauropods.
NOTES Presence of some of these primitive protosauropods in the Late Triassic shows that this herbivore group and great bulk evolved surprisingly early. Absence from the Western Hemisphere may reflect lack of sufficient sampling. Some anatomical features remain poorly understood. This group may be splittable into a larger number of divisions.

Antetonitrus ingenipes
28 m (25 ft) TL, 1.5 tonnes

FOSSIL REMAINS Minority of skeleton.
ANATOMICAL CHARACTERISTICS Standard for group.
AGE Late Triassic, early Norian.
DISTRIBUTION AND FORMATION/S Southeast Africa; Lower Elliot.
HABITAT Arid.

NOTES *Antetonitrus* indicates that when sauropods first evolved they were the same size as the largest prosauropods. Shared its habitat with *Eucnemesaurus*, *Melanorosaurus*, *Plateosauravus*, and *Blikanasaurus*.

Chinshakiangosaurus chunghoensis
10 m (30 ft) TL, 3 tonnes

FOSSIL REMAINS Minority of skull and skeleton.
ANATOMICAL CHARACTERISTICS Mouth fairly broad, extensive cheeks present.
AGE Early Jurassic.
DISTRIBUTION AND FORMATION/S Southern China; Fengjiahe.

Isanosaurus attavipachi
13 m (43 ft) TL, 7 tonnes

FOSSIL REMAINS Minority of several skeletons, juvenile and adult.
ANATOMICAL CHARACTERISTICS Standard for group.
AGE Late Triassic, late Norian or Rhaetian.
DISTRIBUTION AND FORMATION/S Thailand; Nam Phong.
NOTES The large remains are probably the adult of the juvenile remains named *Isanosaurus. Isanosaurus* shows that giant sauropods evolved just 20 million years after the appearance of dinosaurs.

Gongxianosaurus shibeiensis
11 m (35 ft) TL, 3 tonnes

FOSSIL REMAINS Majority of skeleton.
ANATOMICAL CHARACTERISTICS Base of tail deep.
AGE Early Jurassic.
DISTRIBUTION AND FORMATION/S Central China; Ziliujing.

Vulcanodon karibaensis
11 m (35 ft) TL, 3.5 tonnes

FOSSIL REMAINS Minority of skeleton.
ANATOMICAL CHARACTERISTICS Standard for group.
AGE Early Jurassic, Toarcian.
DISTRIBUTION AND FORMATION/S Zimbabwe; Bakota.
HABITAT Arid.

Pulanesaura eocollum
11 m (35 ft) TL, 3.5 tonnes

FOSSIL REMAINS Minority of skeleton.
ANATOMICAL CHARACTERISTICS Standard for group.
AGE Early Jurassic, late Hettangian or Sinemurian.
DISTRIBUTION AND FORMATION/S South Africa; Upper Elliot.
HABITAT Arid.
NOTES Prey of *Dracovenator*.

Tazoudasaurus naimi
10 m (33 ft) TL, 3.7 tonnes

FOSSIL REMAINS Minority of skull and two skeletons, juvenile and adult.
ANATOMICAL CHARACTERISTICS Neck not elongated. Shoulder somewhat higher than hip.
AGE Early Jurassic, Toarcian.
DISTRIBUTION AND FORMATION/S Morocco; Douar of Tazouda.

Kotasaurus yamanpalliensis
9 m (30 ft) TL, 2.5 tonnes

FOSSIL REMAINS Majority of skeleton.
ANATOMICAL CHARACTERISTICS Standard for group.
AGE Early Jurassic.
DISTRIBUTION AND FORMATION/S Southeast India; Kota.
NOTES Shared its habitat with *Barapasaurus*.

Gongxianosaurus shibeiensis

Tazoudasaurus naimi

EUSAUROPODS

LARGE TO ENORMOUS SAUROPODS FROM THE EARLY JURASSIC TO THE END OF THE DINOSAUR ERA, MOST CONTINENTS.

ANATOMICAL CHARACTERISTICS Fairly variable. Snout broader, rounded or squared off, nostrils further retracted, cheeks absent. Skeletons massively built. Neck moderately to extremely long. Trunk compact, deep, vertebral series usually stiffened. Tail moderately to extremely long. Quadrupedal when moving normally, arm and leg columnar and massively built, so not able to achieve a full run faster than elephants, amble fastest gait. Shoulder girdles large, hand forming a vertical arcade, fingers very short and rigid or lost, no padding, large claw limited to thumb or lost. Pelvis large, ilium deep and strongly arced at top, indicating enlarged upper leg muscles, pubis subvertical, lower leg shorter than upper, mobility of ankle limited, foot very short and broad, five toes short and underlain by large pad, inner toes bearing large claws increasing in size progressing inward. Skeletal pneumaticity and birdlike respiratory system better developed. Skin consisting of small scales in rosette pattern.
ONTOGENY Growth rates moderate in at least some smaller species to moderately rapid, especially in giants; life spans did not exceed 100 years.
HABITATS Seasonally dry, open woodlands and prairies, and coastal wetlands, from tropics to polar regions.
HABITS High-level browsers and low-level grazers. Too slow to flee attackers, main defense standing and lashing out with clawed hands and feet or swinging tails, which often weighed tonnes and matched giant attacking theropods in mass. Long, tall necks may have been used for competitive display within species; delicate construction indicates they were not used as impact weapons for combat within species like giraffe necks. Trackways indicate that small juveniles formed pods of similar-sized individuals separate from the herds of large juveniles and adults over 1 tonne. Numerous trackways laid down along watercourses show that many sauropods of all sizes used shorelines to travel, but ability to move into water was limited because the narrow, padless hands were in danger of getting bogged down in soft sediments, as appears to have happened in some fossils. Probably used clawed hind feet to dig for water in streambeds during droughts.
ENERGETICS Power production probably unusually high in longer-necked examples so that oversized heart could pump blood at very high pressures up to high-held brains.
NOTES The dinosaurs most similar to elephants and giraffes. Fragmentary remains and trackways indicate that some eusauropods approached and perhaps exceeded 100 tonnes.

Cetiosaurs

LARGE TO GIGANTIC SAUROPODS LIMITED TO THE JURASSIC OF THE NORTHERN AND SOUTHERN HEMISPHERES.

ANATOMICAL CHARACTERISTICS Fairly uniform. Head short, snout rounded. Neck rather short to moderately long, able to elevate subvertically. Tail moderately long, sometimes armed with small spikes or club. Arm moderately long, so shoulders about as high as hips.
HABITS Probably feeding generalists.
NOTES The relationships of many of these generalized sauropods are uncertain.

Barapasaurus tagorei
12 m (40 ft) TL, 7 tonnes

FOSSIL REMAINS Majority of skeleton from bone beds.
ANATOMICAL CHARACTERISTICS Neck moderately long.
AGE Early Jurassic.
DISTRIBUTION AND FORMATION/S Southeast India; Kota.
NOTES Shared its habitat with *Kotasaurus.*

Dystrophaeus viaemalae
13 m (43 ft) TL, 7 tonnes

FOSSIL REMAINS Minority of skeleton.
ANATOMICAL CHARACTERISTICS Insufficient information.
AGE Middle and/or Late Jurassic, Callovian and/or Oxfordian.
DISTRIBUTION AND FORMATION/S Utah; Summerville.
NOTES The relationships of *Dystrophaeus* are uncertain.

Rhoetosaurus brownei
15 m (50 ft) TL, 9 tonnes

FOSSIL REMAINS Minority of skeleton.
ANATOMICAL CHARACTERISTICS Insufficient information.
AGE Middle Jurassic, Bajocian.
DISTRIBUTION AND FORMATION/S Northeast Australia; Hutton Sandstone.
HABITAT Polar forests with warm, daylight-dominated summers and cold, dark winters.

Volkheimeria chubutensis
Adult size uncertain

FOSSIL REMAINS Minority of skeleton, juvenile.
ANATOMICAL CHARACTERISTICS Insufficient information.
AGE Middle Jurassic.
DISTRIBUTION AND FORMATION/S Southern Argentina; Canadon Asfalto.
NOTES Shared its habitat with *Patagosaurus, Tehuelchesaurus,* and *Brachytrachelopan.*

Spinophorosaurus nigerensis
13 m (45 ft) TL, 7 tonnes

FOSSIL REMAINS Minority of skulls, majority of skeleton.
ANATOMICAL CHARACTERISTICS Neck moderately long. Pair of small, paired spikes probably near tip of tail.
AGE Probably Middle Jurassic, Bajocian or Bathonian.
DISTRIBUTION AND FORMATION/S Niger; Irhazer.
NOTES The only sauropod known to have tail spikes.

Barapasaurus tagorei

Spinophorosaurus nigerensis

Shunosaurus lii
9.5 m (30 ft) TL, 3 tonnes

FOSSIL REMAINS Numerous skulls and skeletons, completely known.
ANATOMICAL CHARACTERISTICS Neck short by sauropod standards. Tail tipped by small, spiked club. Leg long relative to size of body.
AGE Late Jurassic, Bathonian and/or Callovian.
DISTRIBUTION AND FORMATION/S Central China; Shaximiao.
HABITAT Heavily forested.
HABITS Fed at medium heights. Defense included high-velocity impacts from tail club.
NOTES Almost as short necked as *Brachytrachelopan*.

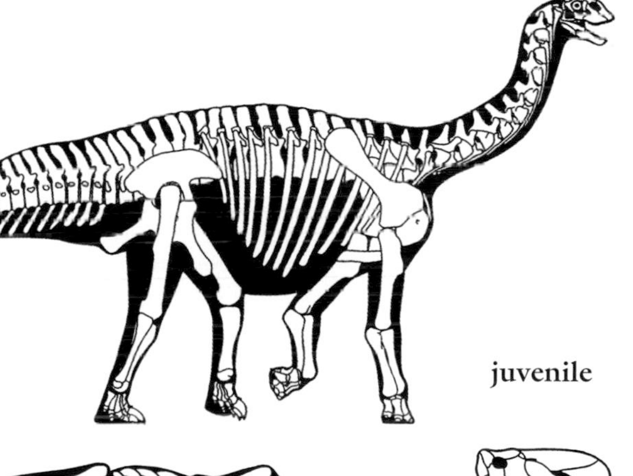

juvenile

adult

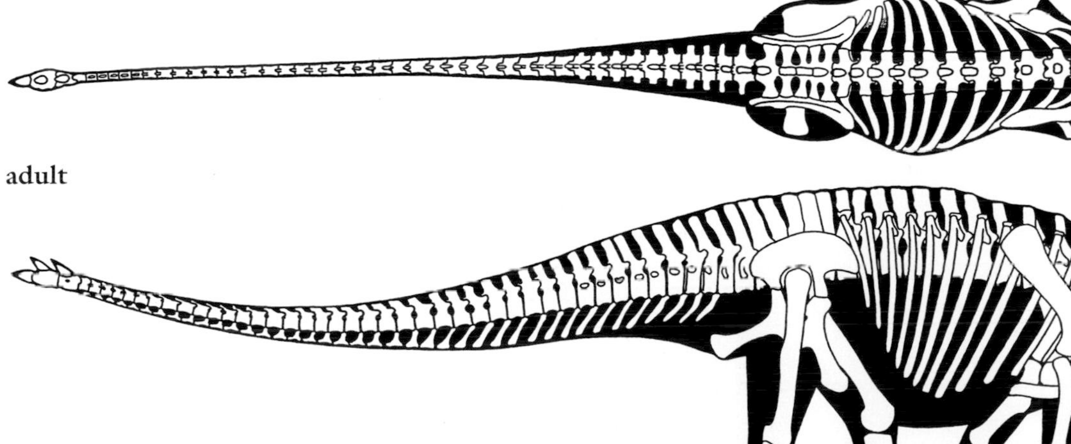

Shunosaurus lii
(see also next page)

Shunosaurus lii and *Gasosaurus constructus*

Shunosaurus lii

Datousaurus bashenesis

Datousaurus bashanensis

10 m (34 ft) TL, 4.5 tonnes

FOSSIL REMAINS Partial skull and skeletons.
ANATOMICAL CHARACTERISTICS Neck moderately long. Shoulder a little higher than hip.
AGE Late Jurassic, Bathonian and/or Callovian.
DISTRIBUTION AND FORMATION/S Central China; Shaximiao.
HABITAT Heavily forested.
HABITS High-level browser.
NOTES Shared its habitat with *Shunosaurus* and *Omeisaurus*.

Amygdalodon patagonicus

12 m (40 ft) TL, 5 tonnes

FOSSIL REMAINS Minority of skeleton.
ANATOMICAL CHARACTERISTICS Insufficient information.
AGE Middle Jurassic, Bajocian.
DISTRIBUTION AND FORMATION/S Southern Argentina; Cerro Carnerero.

Patagosaurus fariasi

16.5 m (53 ft) TL, 8.5 tonnes

FOSSIL REMAINS Minority of skull and numerous skeletons.
ANATOMICAL CHARACTERISTICS Neck moderately long. Tail long.
AGE Middle Jurassic.
DISTRIBUTION AND FORMATION/S Southern Argentina; Canadon Asfalto.
HABITAT Short wet season, otherwise semiarid, riverine forests, open floodplains.
HABITS Long tail facilitated rearing for high browsing.
NOTES Shared its habitat with *Volkheimeria*, *Brachytrachelopan*, and *Tehuelchesaurus*.

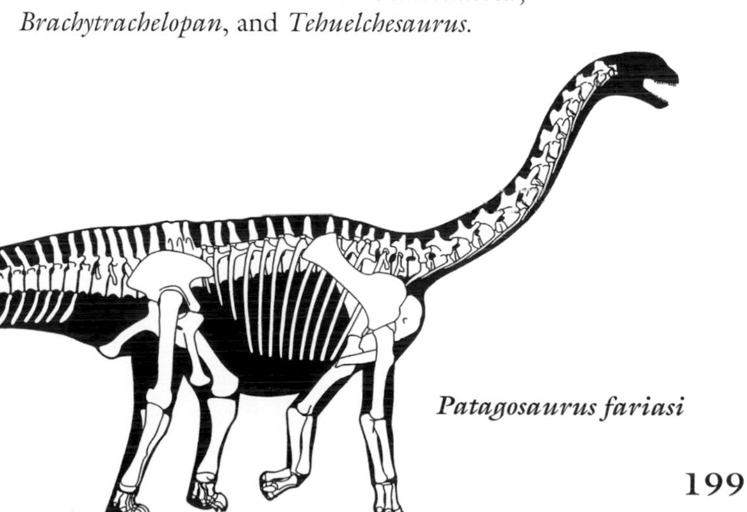

Patagosaurus fariasi

Chebsaurus algeriensis
Adult size uncertain

FOSSIL REMAINS Two partial skeletons, juvenile.
ANATOMICAL CHARACTERISTICS Insufficient information.
AGE Middle Jurassic, probably Callovian.
DISTRIBUTION AND FORMATION/S Algeria; unnamed.

Ferganasaurus verzilini
18 m (60 ft) TL, 15 tonnes

FOSSIL REMAINS Minority of skeleton.
ANATOMICAL CHARACTERISTICS Insufficient information.
AGE Middle Jurassic, Callovian.
DISTRIBUTION AND FORMATION/S Kyrgyzstan; Balabansai.
NOTES The claim that there are two hand claws is problematic.

Cetiosaurus oxoniensis
16 m (50 ft) TL, 11 tonnes

FOSSIL REMAINS Majority of skeleton.
ANATOMICAL CHARACTERISTICS Neck moderately long.
AGE Middle Jurassic, Bathonian.
DISTRIBUTION AND FORMATION/S Central England; Forest Marble.

Cetiosauriscus stewarti
15 m (50 ft) TL, 10 tonnes

FOSSIL REMAINS Partial skeleton.
ANATOMICAL CHARACTERISTICS Insufficient information.
AGE Middle Jurassic, Callovian.
DISTRIBUTION AND FORMATION/S Eastern England; Lower Oxford Clay.
NOTES Not a diplodocoid as was suggested.

Chuanjiesaurus anaensis
17 m (55 ft) TL, 11 tonnes

FOSSIL REMAINS Two partial skeletons.
ANATOMICAL CHARACTERISTICS Neck very long. Shoulders a little higher than hips. Tail not large.
AGE Early Middle Jurassic.
DISTRIBUTION AND FORMATION/S Southwest China; Chuanjie.
NOTES Proportions not entirely certain. May be a mamenchisaur.

Haplocanthosaurus delfsi
16 m (55 ft) TL, 13 tonnes

ANATOMICAL CHARACTERISTICS Insufficient information.
AGE Late Jurassic, late Oxfordian.
DISTRIBUTION AND FORMATION/S Colorado; lower Morrison.
HABITAT Short wet season, otherwise semiarid with open floodplain prairies and riverine forests.
HABITS Probably a feeding generalist.
NOTES Whether this is a different species from the slightly later *H. priscus* is uncertain. *Haplocanthosaurus* may have been a basal diplodocoid.

Haplocanthosaurus priscus
12 m (40 ft) TL, 5 tonnes

FOSSIL REMAINS Majority of two skeletons.
ANATOMICAL CHARACTERISTICS Neck moderately long.
AGE Late Jurassic, late Oxfordian and/or early Kimmeridgian.

Cetiosaurus oxoniensis

Chuanjiesaurus anaensis

Haplocanthosaurus delfsi

DISTRIBUTION AND FORMATION/S Colorado, Wyoming; lower Morrison.
HABITAT Short wet season, otherwise semiarid with open floodplain prairies and riverine forests.
HABITS Probably a feeding generalist.
NOTES May be the direct descendant of *H. delfsi*. The genus does not appear to be present later in the Morrison.

Mamenchisaurs

LARGE TO GIGANTIC SAUROPODS LIMITED TO THE MIDDLE AND LATE JURASSIC OF ASIA.

ANATOMICAL CHARACTERISTICS Variable. Head short, snout rounded. Neck long to extremely long, able to elevate vertically. Tail moderately long. Arm long so shoulders somewhat higher than hips, retroverted pelvis facilitated slow walking when rearing up by keeping hips and tail horizontal when bipedal, sled-shaped chevrons under tail facilitated static rearing with tail as a prop.
HABITS High-level browsers, both when quadrupedal and when bipedal.
NOTES Representing an apparent radiation of Asian sauropods when the continent was isolated; the relationships of these taxa are obscure, with generic designations often problematic; group may be splittable into a number of divisions.

Tonganosaurus hei
Adult size uncertain

FOSSIL REMAINS Minority of skeleton.
ANATOMICAL CHARACTERISTICS Neck long.
AGE Early Jurassic.
DISTRIBUTION AND FORMATION/S Southern China; Yimen.
NOTES Whether this early, 12 m long specimen is a mamenchisaur is uncertain.

Omeisaurus junghsiensis
14 m (45 ft) TL, 4 tonnes

FOSSIL REMAINS Partial skull and skeletons.
ANATOMICAL CHARACTERISTICS Neck very long.
AGE Middle Jurassic, Bathonian and/or Callovian.
DISTRIBUTION AND FORMATION/S Central China; Shaximiao.
HABITAT Heavily forested.

Omeisaurus? maoianus
15 m (50 ft) TL, 5 tonnes

FOSSIL REMAINS Nearly complete skull and partial skeleton.
ANATOMICAL CHARACTERISTICS Neck very long.

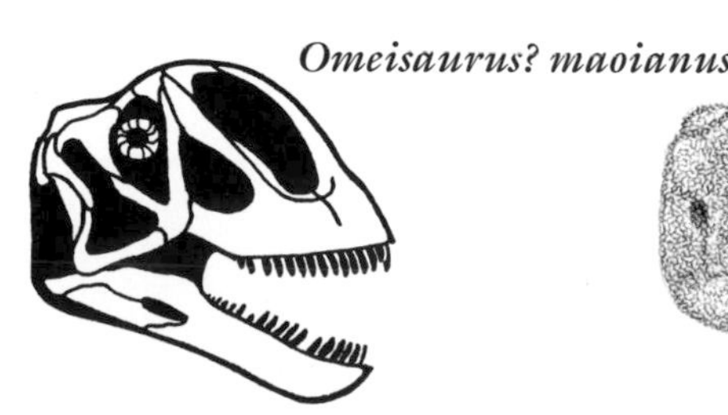

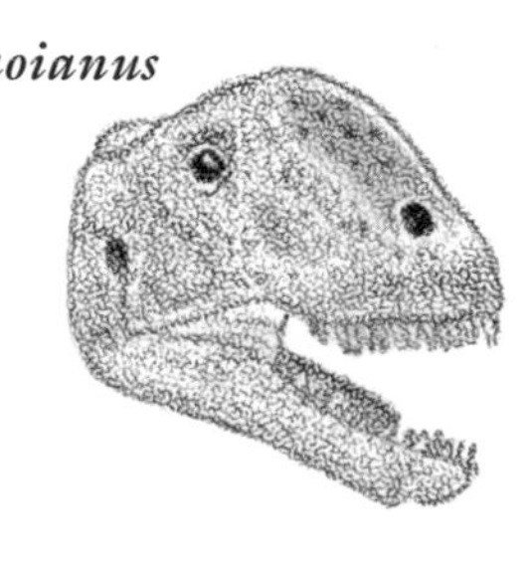

Omeisaurus? maoianus

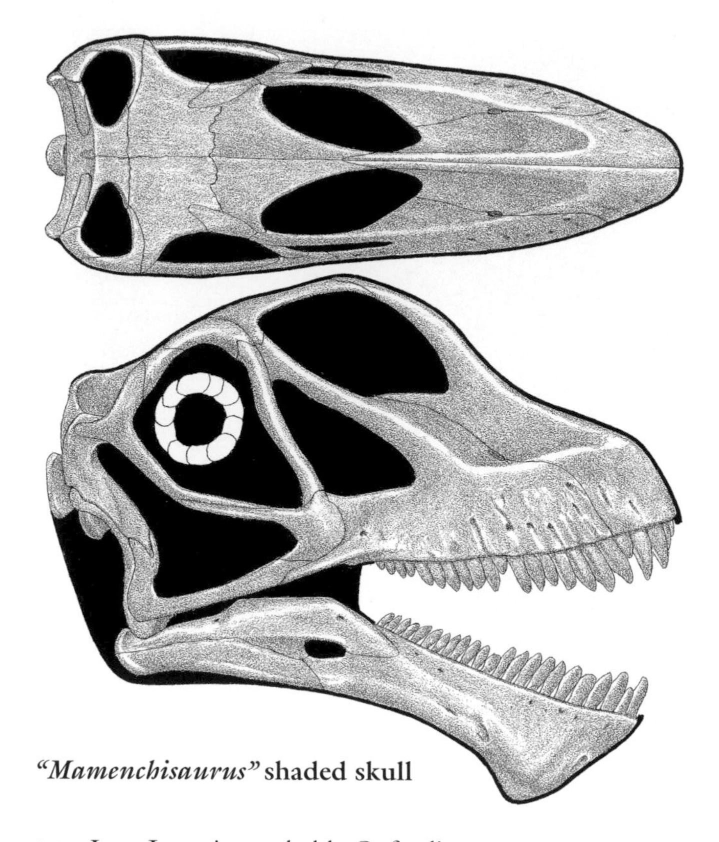

"Mamenchisaurus" **shaded skull**

AGE Late Jurassic, probably Oxfordian.
DISTRIBUTION AND FORMATION/S Central China; lower Shangshaximiao.
HABITAT Heavily forested.

Unnamed genus *tianfuensis*
18 m (60 ft) TL, 8.5 tonnes

FOSSIL REMAINS Majority of skull and skeletons.
ANATOMICAL CHARACTERISTICS Neck extremely long and slender.
AGE Middle Jurassic, Bathonian and/or Callovian.
DISTRIBUTION AND FORMATION/S Central China; Shaximiao.
HABITAT Heavily forested.
NOTES Too different to be placed in *Omeisaurus*. Claims that this sauropod had a tail club problematic. Shared its habitat with *Shunosaurus* and *Omeisaurus*.

Yuanmousaurus jiangyiensis
17 m (60 ft) TL, 12 tonnes

FOSSIL REMAINS Partial skeleton(s).
ANATOMICAL CHARACTERISTICS Neck long.
AGE Middle Jurassic.
DISTRIBUTION AND FORMATION/S Southern China; Zhanghe.
NOTES *Eomamenchisaurus yuanmouensis* may be a juvenile of this species.

Unnamed genus *tianfuensis*
(see also next page)

Unnamed genus *tianfuensis*

"*Mamenchisaurus*" *hochuanensis*

Mamenchisaurus constructus
15 m (50 ft) TL, 5 tonnes

FOSSIL REMAINS Minority of skeleton.
ANATOMICAL CHARACTERISTICS Neck moderately long.
AGE Late Jurassic, probably Oxfordian.
DISTRIBUTION AND FORMATION/S Central China; Shangshaximiao.
HABITAT Heavily forested.
NOTES Based on an inadequate specimen without a very long neck; that so many species have been placed in *Mamenchisaurus*, many from the same formation, indicates that these sauropods are overlumped, being in the wrong genus in some cases, or split into too many species in others.

"*Mamenchisaurus*" *hochuanensis*
21 m (70 ft) TL, 14 tonnes

FOSSIL REMAINS Partial skull and a few skeletons.
ANATOMICAL CHARACTERISTICS Neck extremely long, vertebral spines near base of neck forked. Presence of small tail club problematic. Limbs short.
AGE Late Jurassic, probably Oxfordian.
DISTRIBUTION AND FORMATION/S Central China; Shangshaximiao.
HABITAT Heavily forested.
HABITS Purpose of very small tail club uncertain.
NOTES Main enemy *Yangchuanosaurus shangyuensis*.

"Mamenchisaurus" youngi?

"Mamenchisaurus" youngi?
17 m (55 ft) TL, 7 tonnes

FOSSIL REMAINS Complete skull and majority of skeleton.
ANATOMICAL CHARACTERISTICS Neck extremely long, vertebral spines near base of neck forked. Hip strongly retroverted, and tail directed strongly upward. Limbs short.
AGE Late Jurassic, probably Oxfordian.
DISTRIBUTION AND FORMATION/S Central China; Shangshaximiao.
HABITAT Heavily forested.
NOTES One of the most peculiarly shaped sauropods. Definitely in same genus as *M. hochuanensis*, and may be one of the sexes of the latter species.

"Mamenchisaurus jingyanensis"
20 m (65 ft) TL, 12 tonnes

FOSSIL REMAINS Majority of skull and minority of skeleton.
ANATOMICAL CHARACTERISTICS Neck extremely long.

"Mamenchisaurus jingyanensis"

AGE Late Jurassic, probably Oxfordian.
DISTRIBUTION AND FORMATION/S Central China; Shangshaximiao.
HABITAT Heavily forested.
NOTES Probably belongs to one of the other incomplete species from the Shangshaximiao.

"Mamenchisaurus" anyuensis
25 m (80 ft) TL, 25 tonnes

FOSSIL REMAINS Several partial skeletons.
ANATOMICAL CHARACTERISTICS Neck extremely long.
AGE Late Jurassic.
DISTRIBUTION AND FORMATION/S Central China; Penglaizhen, Suining.

"Mamenchisaurus" sinocanadorum
Adult size uncertain

FOSSIL REMAINS Minority of skull and minority of skeleton(s).
ANATOMICAL CHARACTERISTICS Neck extremely long.
AGE Late Jurassic, probably Oxfordian.
DISTRIBUTION AND FORMATION/S Northwestern China; Shishugou.
NOTES Partial remains that may belong to this species indicate an individual ~35 m (115 ft) long and ~75 tonnes, a contender for the largest known land animal.

Hudiesaurus sinojapanorum
25 m (80 ft) TL, 25 tonnes

FOSSIL REMAINS Minority of skeleton.
ANATOMICAL CHARACTERISTICS Vertebral spines near base of neck forked.
AGE Late Jurassic.
DISTRIBUTION AND FORMATION/S Northwest China; Kalazha.

Xinjiangtitan shanshanesis
30 m (100 ft) TL, 40 tonnes

FOSSIL REMAINS Minority of skeleton.
ANATOMICAL CHARACTERISTICS Neck extremely long.
AGE Middle Jurassic.
DISTRIBUTION AND FORMATION/S Northwest China; Qiju.

TURIASAURS

MEDIUM-SIZED TO ENORMOUS SAUROPODS LIMITED TO THE LATE JURASSIC OF EUROPE.

ANATOMICAL CHARACTERISTICS Neck and tail moderately long. Arm moderately long, so shoulders about as high as hips.

Turiasaurus riodevensis
30 m (100 ft) TL, 50 tonnes

FOSSIL REMAINS Partial skeletons.
ANATOMICAL CHARACTERISTICS Some neck and trunk vertebral spines forked.
AGE Late Jurassic, latest Tithonian.
DISTRIBUTION AND FORMATION/S Eastern Spain; Villar del Arzobispo.
NOTES The largest nonneosauropod. Shared its habitat with *Losillasaurus.*

Losillasaurus giganteus
Adult size uncertain

FOSSIL REMAINS Minority of several skeletons.
ANATOMICAL CHARACTERISTICS Vertebral spines not forked.
AGE Late Jurassic, latest Tithonian.
DISTRIBUTION AND FORMATION/S Eastern Spain; Villar del Arzobispo.
NOTES Subadult remains indicate a very large sauropod.

Galveosaurus herreroi
Adult size uncertain

FOSSIL REMAINS Minority of several skeletons.
ANATOMICAL CHARACTERISTICS Vertebral spines not forked.
AGE Late Jurassic, latest Tithonian.
DISTRIBUTION AND FORMATION/S Eastern Spain; Villar del Arzobispo.
NOTES Subadult remains indicate a very large sauropod.

NEOSAUROPODS

LARGE TO ENORMOUS SAUROPODS OF THE MIDDLE JURASSIC TO THE END OF THE DINOSAUR ERA, MOST CONTINENTS.

ANATOMICAL CHARACTERISTICS Skeletal pneumaticity and birdlike respiratory system well developed.
NOTES Absence from Antarctica probably reflects lack of sufficient sampling.

DIPLODOCOIDS

SMALL (FOR SAUROPODS) TO GIGANTIC NEOSAUROPODS LIMITED TO THE MIDDLE JURASSIC TO THE EARLY LATE CRETACEOUS OF THE AMERICAS, EUROPE, AND AFRICA.

ANATOMICAL CHARACTERISTICS Variable. Head long, shallow, bony nostrils strongly retracted to above the orbits but fleshy nostrils probably still near front of

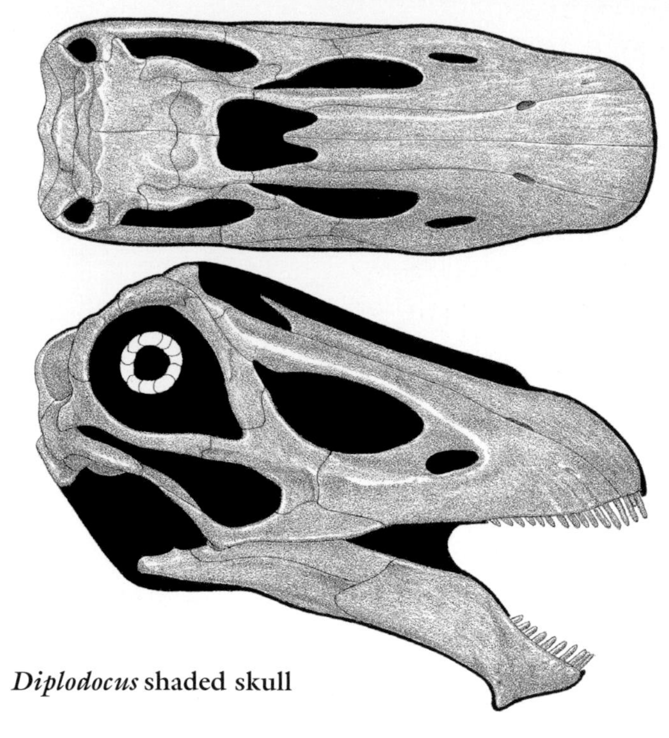

Diplodocus shaded skull

snout, which is broad and squared off, lower jaw short, pencil-shaped teeth limited to front of jaws, head flexed downward relative to neck. Neck short to extremely long, not carried strongly erect. Long tail ending in whip that may have been able to achieve supersonic speeds. Arm and hand short, so shoulders lower than hips, which are heightened by tall vertebral spines. Short arms, large hips, and heavy tails with sled-shaped chevrons facilitated static rearing posture.
HABITS Flexible feeders able to easily browse and graze at all levels from the ground to very high.
NOTES Absence from Australia and Antarctica probably reflects lack of sufficient sampling.

REBBACHISAURIDS Small and medium-sized diplodocoids limited to the Early and early Late Cretaceous of South America and Africa.

ANATOMICAL CHARACTERISTICS Fairly uniform. Neck short by sauropod standards, neck ribs overlapping a little. Vertebral spines not forked. Upper scapula blades very broad.
NOTES The last radiation of diplodocids and nonmacronarian sauropods.

Amazonsaurus maranhensis
12 m (40 ft) TL, 4 tonnes

FOSSIL REMAINS Minority of skeleton.
ANATOMICAL CHARACTERISTICS Insufficient information.
AGE Early Cretaceous, Aptian or Albian.
DISTRIBUTION AND FORMATION/S Northern Brazil; Itapecuru.
NOTES The relationships of *Amazonsaurus* are uncertain.

Zapalasaurus bonapartei
9 m (29 ft) TL, 2 tonnes

FOSSIL REMAINS Partial skeleton.
ANATOMICAL CHARACTERISTICS Insufficient information.
AGE Early Cretaceous, Barremian and/or early Aptian.
DISTRIBUTION AND FORMATION/S Western Argentina; La Amarga.
HABITAT Well-watered coastal woodlands with short dry season.
NOTES Shared its habitat with *Amargasaurus.*

Comahuesaurus windhauseni
12 m (40 ft) TL, 4 tonnes

FOSSIL REMAINS Partial skeleton.
ANATOMICAL CHARACTERISTICS Insufficient information.
AGE Early Cretaceous; late Aptian and/or early Albian.
DISTRIBUTION AND FORMATION/S Western Argentina; Lohan Cura.
HABITAT Well-watered coastal woodlands with short dry season.
NOTES Shared its habitat with *Limaysaurus*, *Ligabuesaurus*, and *Agustinia*.

Rebbachisaurus (or *Rayosaurus*) *agrioensis*
10 m (33 ft) TL, 2.5 tonnes

FOSSIL REMAINS Minority of skeleton.
ANATOMICAL CHARACTERISTICS Insufficient information.
AGE Early Cretaceous, Aptian.
DISTRIBUTION AND FORMATION/S Western Argentina; Rayoso.
NOTES Whether *Rayosaurus*, *Rebbachisaurus*, and *Limaysaurus* are separate genera is uncertain.

Limaysaurus (or *Rebbachisaurus*) *tessonei*
15 m (50 ft) TL, 7 tonnes

FOSSIL REMAINS Minority of skull, majority of skeleton.
ANATOMICAL CHARACTERISTICS Neck fairly deep. Tall vertebral spines over hips form a low sail. Chevrons may be absent from most of underside of tail.
AGE Early Cretaceous, early Cenomanian.
DISTRIBUTION AND FORMATION/S Western Argentina; Candeleros.
HABITAT Short wet season, otherwise semiarid with open floodplains and riverine forests.
NOTES Shared its habitat with *Agustinia* and *Ligabuesaurus.*

Limaysaurus (or *Rebbachisaurus) tessonei*

Rebbachisaurus garasbae
14 m (45 ft) TL, 7 tonnes

FOSSIL REMAINS Partial skeleton.
ANATOMICAL CHARACTERISTICS Hip sail tall.
AGE Early Cretaceous, Albian.
DISTRIBUTION AND FORMATION/S Morocco; Tegana.

Cathartesaura anaerobica
12 m (40 ft) TL, 3 tonnes

FOSSIL REMAINS Partial skeleton.
ANATOMICAL CHARACTERISTICS Insufficient information.
AGE Late Cretaceous, middle Cenomanian.
DISTRIBUTION AND FORMATION/S Western Argentina; lower Huincul.
HABITAT Well-watered woodlands with short dry season.
NOTES The last known diplodocid and nonmacronarian sauropod. Prey of *Skorpiovenator*. This and various *Rebbachisaurus* species may form subfamily Limaysaurine.

Demandasaurus darwini
9 m (30 ft) TL, 2 tonnes

FOSSIL REMAINS Minority of skull and skeletons, many isolated bones.
ANATOMICAL CHARACTERISTICS Neck short. Hip sail present.
AGE Early Cretaceous, late Barremian or early Aptian.
DISTRIBUTION AND FORMATION/S Spain; Castrillo la Reina.
NOTES This, *Tataouinea*, and *Nigersaurus* may form subfamily Nigersaurine.

Tataouinea hannibalis
14 m (45 ft) TL, 8 tonnes

FOSSIL REMAINS Minority of skeleton.
ANATOMICAL CHARACTERISTICS Hip sail present.
AGE Early Cretaceous, early Albian.
DISTRIBUTION AND FORMATION/S Tunisia; Ain el Guetter.
HABITAT Coastal.

Nigersaurus taqueti
9 m (30 ft) TL, 2 tonnes

FOSSIL REMAINS Majority of skull, several partial skeletons, many isolated bones.
ANATOMICAL CHARACTERISTICS Head very lightly built, snout very broad and squared off, teeth limited to front rim of jaws, very numerous and rapidly replaced. Neck short, shallow. No hip sail.
AGE Early Cretaceous, late Aptian.
DISTRIBUTION AND FORMATION/S Niger; upper Elrhaz.
HABITAT Coastal river delta.
HABITS Square muzzle at end of long neck was adaptation for mowing ground cover, also able to rear to high browse.
NOTES The most complex tooth battery among saurischians, mimics in some regards those of ornithischians except teeth were only for cropping plants.

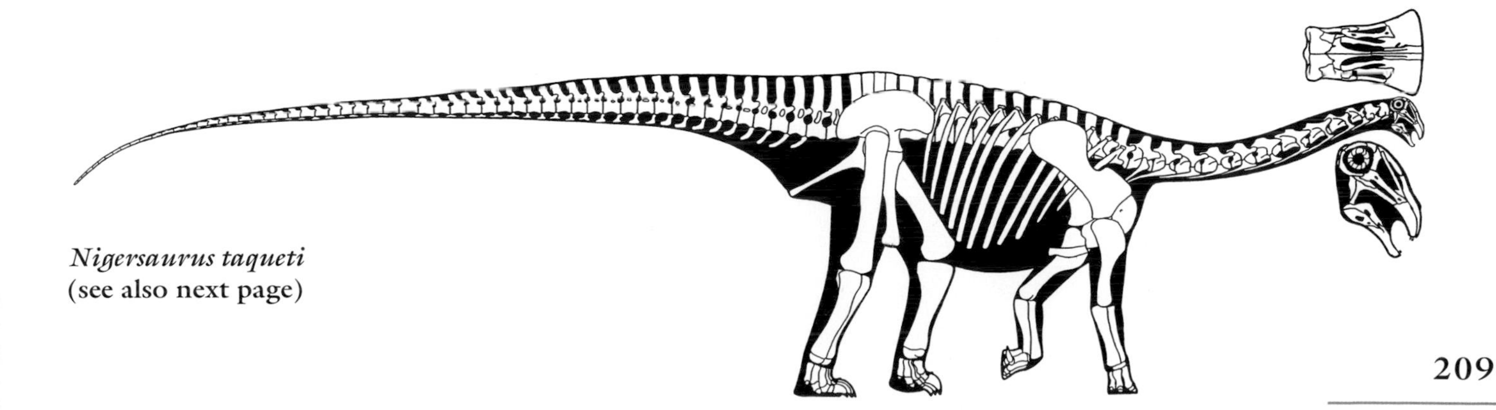

Nigersaurus taqueti
(see also next page)

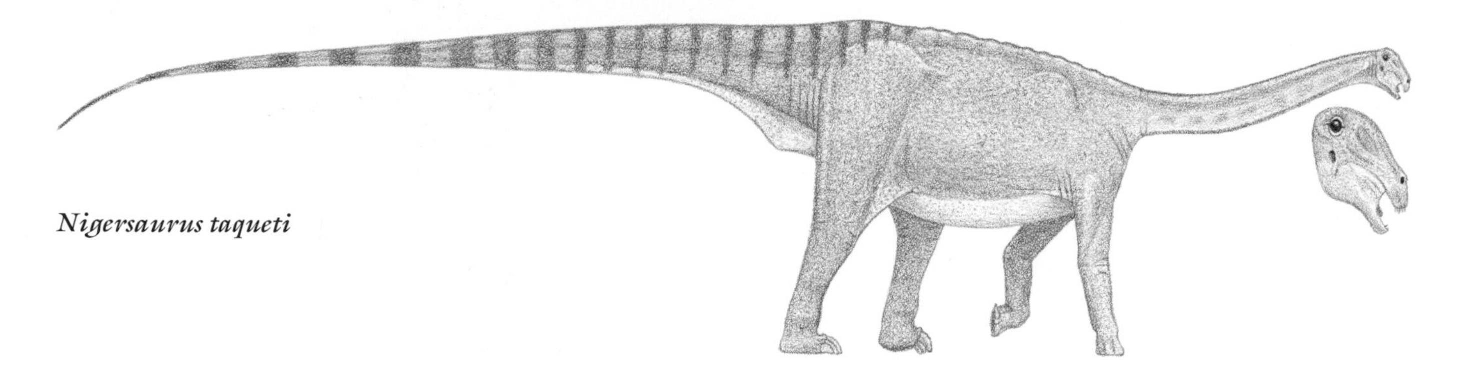

Nigersaurus taqueti

It is not known how many other rebbachisaurids shared these feeding adaptations. The other sauropod known to have a similarly broad and square beak is the titanosaur *Bonitasaura*. Shared its habitat with *Baryonyx tenerensis*, *Lurdusaurus*, and *Ouranosaurus*, of which the last was a competing square-mouthed grazer.

DICRAEOSAURIDS Small (by sauropod standards) diplodocoids limited to the Middle Jurassic to the Early Cretaceous of South America and Africa.

ANATOMICAL CHARACTERISTICS Uniform. Neck short by sauropod standards, spines usually very tall, unable to elevate above shoulder level; ribs so short they do not overlap, increasing flexibility of neck. Tall vertebral spines over hips form a low sail. Most neck and trunk vertebral spines forked.

Suuwassea emilieae

15 m (50 ft) TL, 5 tonnes

FOSSIL REMAINS Minority of skull and skeleton.
ANATOMICAL CHARACTERISTICS Neck spines not tall.
AGE Late Jurassic.
DISTRIBUTION AND FORMATION/S Montana; probably middle Morrison.
HABITAT More coastal and wetter than rest of Morrison.
NOTES Relationships of this diplodocoid are uncertain.

Brachytrachelopan mesai

11 m (35 ft) TL, 5 tonnes

FOSSIL REMAINS Partial skeleton.
ANATOMICAL CHARACTERISTICS Neck spines not tall.
AGE Middle Jurassic.
DISTRIBUTION AND FORMATION/S Southern Argentina; Canadon Asfalto.
NOTES The shortest-necked known sauropod. Shared its habitat with *Volkheimeria*, *Patagosaurus*, and *Tehuelchesaurus*.

Dicraeosaurus hansemanni

14 m (45 ft) TL, 5 tonnes

FOSSIL REMAINS Minority of skull, several skeletons from nearly complete to partial.
ANATOMICAL CHARACTERISTICS Lower jaw did not have contorted lower edge, as usually restored.
AGE Late Jurassic, early Tithonian.

Dicraeosaurus hansemanni

DISTRIBUTION AND FORMATION/S Tanzania; middle Tendaguru.
HABITAT Coastal, seasonally dry with heavier vegetation farther inland.
NOTES May be the direct ancestor of *D. sattleri*. Shared its habitat with *Giraffatitan*.

Dicraeosaurus sattleri
15 m (50 ft) TL, 6 tonnes

FOSSIL REMAINS Minority of skull, several skeletons from nearly complete to partial.
AGE Late Jurassic, middle and/or late Tithonian.
DISTRIBUTION AND FORMATION/S Tanzania; upper Tendaguru.
HABITAT Coastal, seasonally dry with heavier vegetation farther inland.
NOTES Shared its habitat with *Tornieria*.

Amargasaurus cazaui
13 m (43 ft) TL, 4 tonnes

FOSSIL REMAINS Minority of skull and majority of skeleton.
ANATOMICAL CHARACTERISTICS Neck vertebral spines elongated into very long spikes that may have been lengthened by horn sheaths. Hip sail tall.
AGE Late Early Cretaceous, Barremian and/or early Aptian.
DISTRIBUTION AND FORMATION/S Western Argentina; La Amarga.
HABITAT Well-watered coastal woodlands with short dry season.
HABITS Defense included arc of neck spines. Latter may have been used to generate clattering noise display.
NOTES Has been suggested that the neck spikes supported sail fins, but this is not likely. Shared its habitat with *Zapalasaurus*.

Diplodocids Large to gigantic diplodocoids limited to the Middle Jurassic to Early Cretaceous of North America, Europe, and Africa.

ANATOMICAL CHARACTERISTICS Variable. Neck long to extremely long, not able to elevate vertically, ribs so short they do not overlap, increasing flexibility of neck. Most neck and trunk vertebral spines forked. Tall vertebral spines over hips form a low sail. Tail whips long.
NOTES If a diplodocid, fragmentary Argentine *Leikupai laticauda* extends group into early Early Cretaceous. A supposed colossal trunk vertebra (up to 2.6 m [8.5 ft] tall) from Colorado, since lost, has been labeled "*Amphicoelias fragillimus*," but its status and size implications are highly uncertain—may compete with *Bruhathkayosaurus* as the largest known land animal.

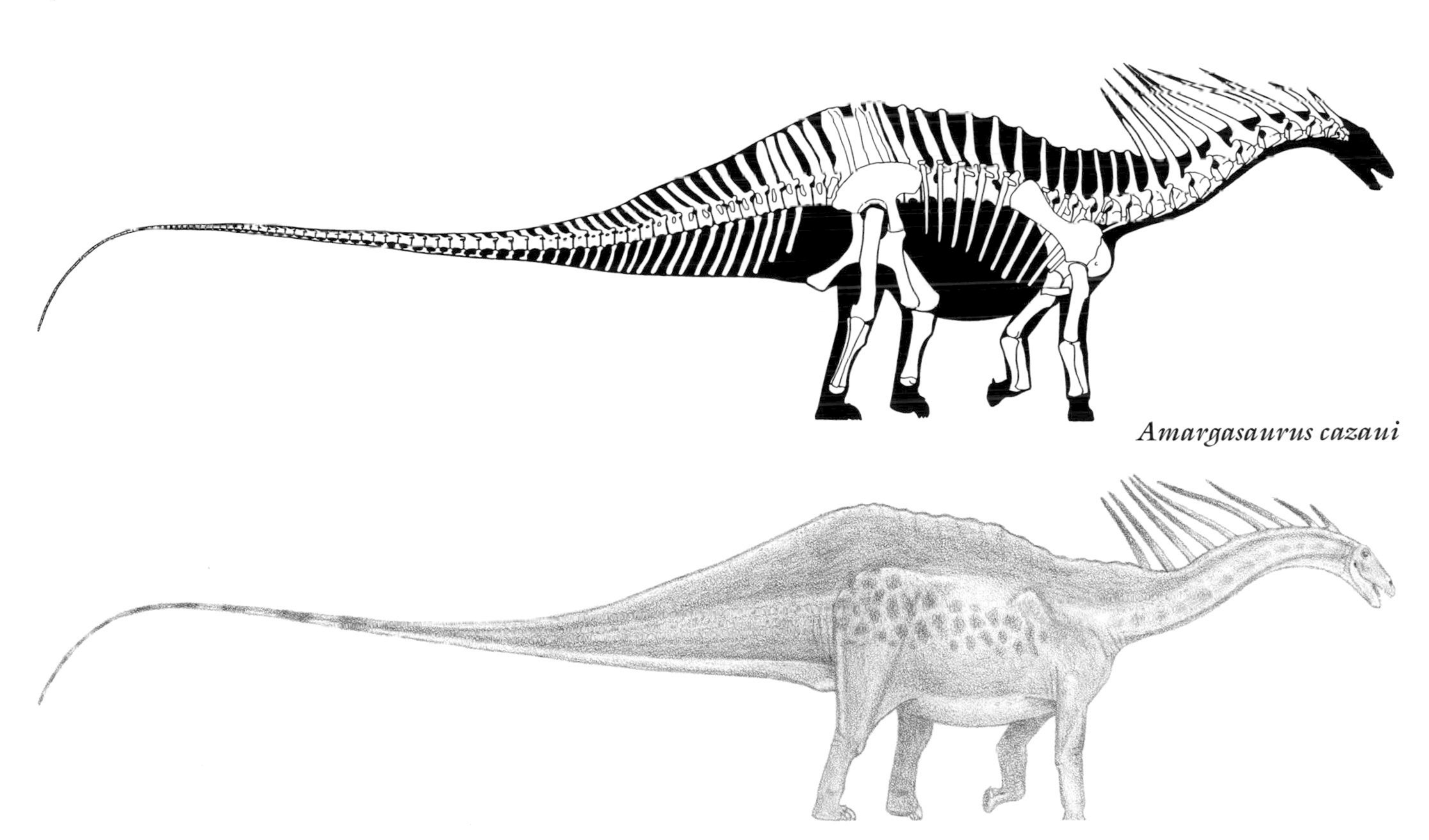

Amargasaurus cazaui

Amphicoelias altus

18 m (60 ft) TL, 15 tonnes

FOSSIL REMAINS Minority of skull and skeleton.
ANATOMICAL CHARACTERISTICS Neck may not be elongated. Leg very slender by sauropod standards.
AGE Late Jurassic, middle Tithonian.
DISTRIBUTION AND FORMATION/S Colorado, Wyoming; upper Morrison.
HABITAT Wetter than earlier Morrison, otherwise semiarid with open floodplain prairies and riverine forests.

Diplodocines Large to gigantic diplodocids limited to the Middle and Late Jurassic of North America, Europe, and Africa.

ANATOMICAL CHARACTERISTICS Fairly uniform. Lightly built. Neck very to extremely long, fairly slender. Tail very long. Femur usually slender. Short vertical spikes appear to run atop vertebral series in at least some diplodocines.

Unnamed genus and species

22 m (71 ft) TL, 8 tonnes

FOSSIL REMAINS Virtually complete skull and skeleton.
ANATOMICAL CHARACTERISTICS Femur robust.
AGE Late Jurassic, late Oxfordian.
DISTRIBUTION AND FORMATION/S Wyoming; lowest Morrison.
HABITAT Short wet season, otherwise semiarid with open floodplain prairies and riverine forests.

Dyslocosaurus polyonychius

18 m (60 ft) TL, 5 tonnes

FOSSIL REMAINS Minority of skeleton.
ANATOMICAL CHARACTERISTICS Insufficient information.
AGE Probably Late Jurassic.
DISTRIBUTION AND FORMATION/S Wyoming; probably Morrison.
HABITAT Short wet season, otherwise semiarid with open floodplain prairies and riverine forests.
NOTES Neither the formation this was found in nor its relationships are entirely certain.

Australodocus bohetii

17 m (55 ft) TL, 4 tonnes

FOSSIL REMAINS Neck vertebrae.
ANATOMICAL CHARACTERISTICS Neck very long.
AGE Late Jurassic, middle and/or late Tithonian.
DISTRIBUTION AND FORMATION/S Tanzania; upper Tendaguru.
HABITAT Coastal, seasonally dry with heavier vegetation farther inland.
NOTES Shared its habitat with *Dicraeosaurus sattleri* and *Janenschia robusta*.

Diplodocus (= Seismosaurus) hallorum

29 m (95 ft) TL, 23 tonnes

FOSSIL REMAINS Possible skull(s), fairly complete and partial skeletons.
ANATOMICAL CHARACTERISTICS Neck very long. Trunk compact. Tail extremely long. Femur slender until mature.
AGE Late Jurassic, late Oxfordian to early Kimmeridgian.
DISTRIBUTION AND FORMATION/S Colorado, Utah, New Mexico; middle Morrison.
HABITAT Short wet season, otherwise semiarid with open floodplain prairies and riverine forests.
NOTES Many of the remains are usually placed in lower Morrison *D. longus*, but that is based on very fragmentary

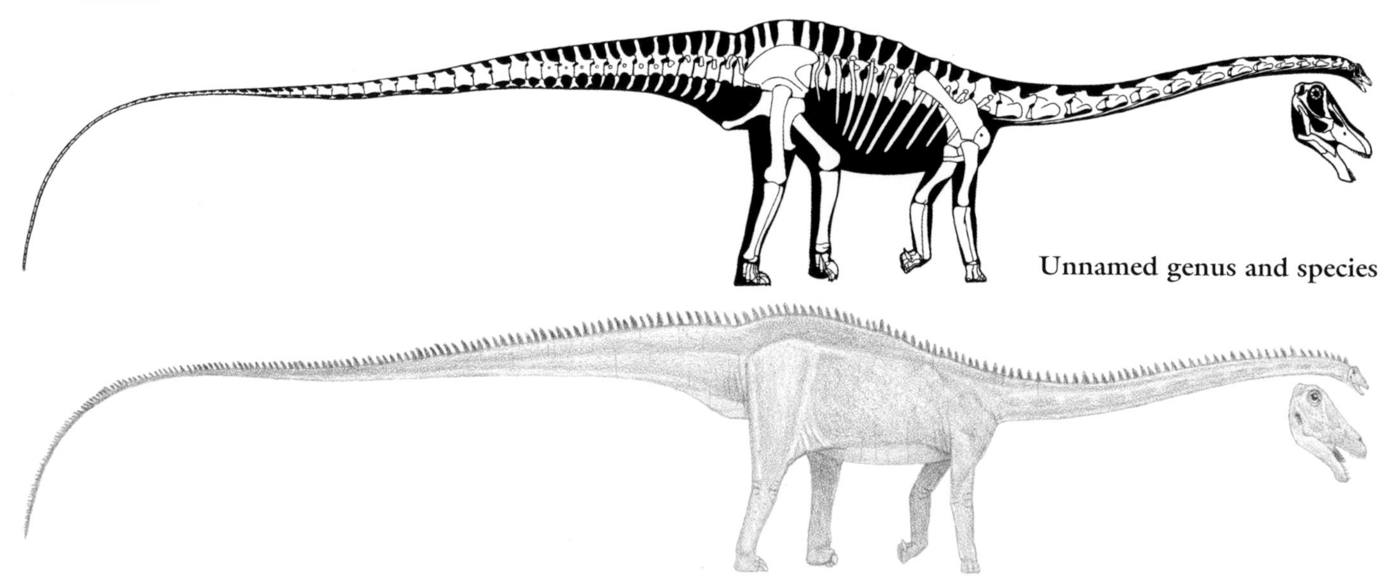

Unnamed genus and species

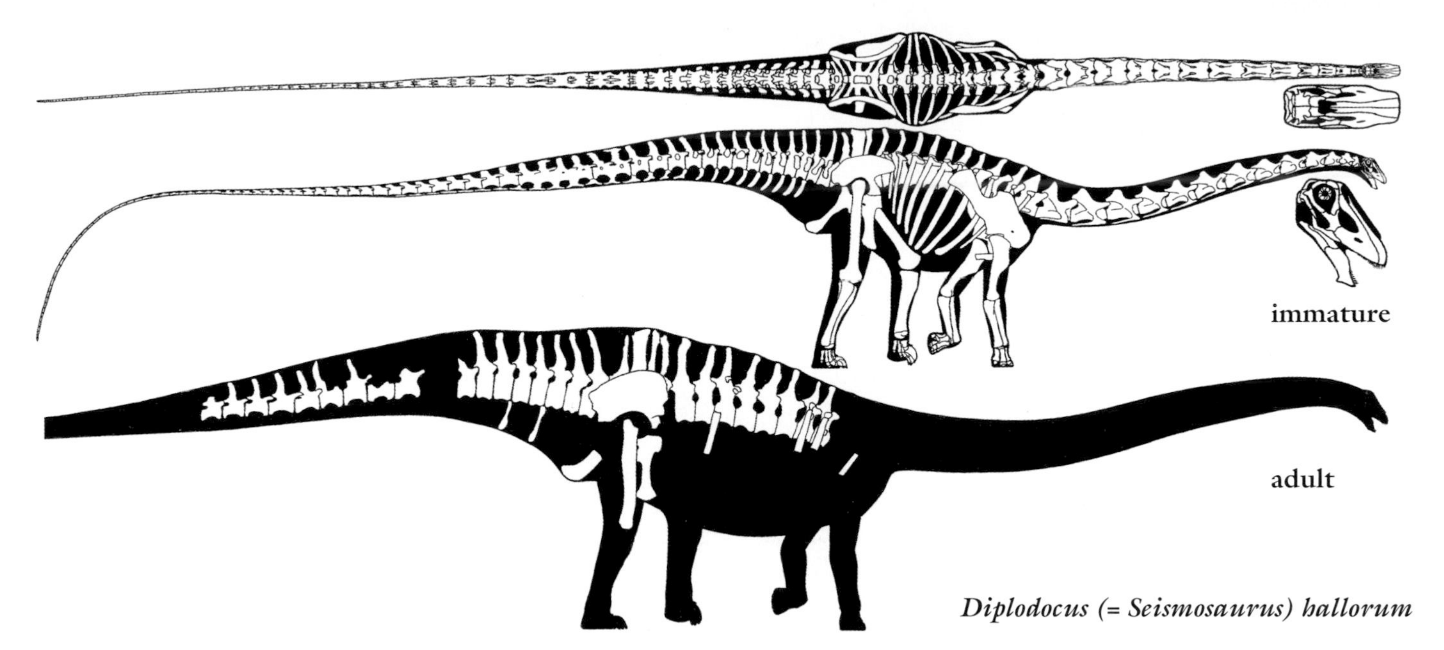

Diplodocus (= Seismosaurus) hallorum

remains that may not be of the same type as skeletons commonly placed in *Diplodocus*. The size of the largest "*Seismosaurus*" skeleton was greatly exaggerated. It is possible that the 22 m long, gracile-legged skeletons from the northern Morrison are a species distinct from the 29 m, apparently shorter-limbed *S. hallorum* from the southern Morrison.

Diplodocus carnegii
24 m (80 ft) TL, 12 tonnes

FOSSIL REMAINS Majority of several skeletons.
ANATOMICAL CHARACTERISTICS Neck very long. Trunk fairly long. Tail extremely long. Femur slender, at least at known sizes.
AGE Late Jurassic, early Tithonian.
DISTRIBUTION AND FORMATION/S Wyoming; middle Morrison.
HABITAT Short wet season, otherwise semiarid with open floodplain prairies and riverine forests.
NOTES It is possible this species grew as large as *D. hallorum*. Main enemy *Allosaurus*.

Galeamopus (or *Diplodocus*) *hayi*
Adult size uncertain

FOSSIL REMAINS Majority of an immature skeleton.
ANATOMICAL CHARACTERISTICS Neck very long. Tail extremely long.
AGE Late Jurassic.
DISTRIBUTION AND FORMATION/S Wyoming; Morrison, level uncertain.
HABITAT Short wet season, otherwise semiarid with open floodplain prairies and riverine forests.

Tornieria (or *Barosaurus*) *africana*
25 m (80 ft) TL, 10 tonnes

FOSSIL REMAINS Minority of skull and several skeletons.
ANATOMICAL CHARACTERISTICS Neck extremely long.
AGE Late Jurassic, middle and/or late Tithonian.
DISTRIBUTION AND FORMATION/S Tanzania; upper Tendaguru.
HABITAT Coastal, seasonally dry with heavier vegetation farther inland.
NOTES Shared its habitat with *Dicraeosaurus sattleri* and *Australodocus bohetii*.

Barosaurus lentus
27 m (88 ft) TL, 12 tonnes

FOSSIL REMAINS Possible partial skull, a few partial skeletons.
ANATOMICAL CHARACTERISTICS Neck extremely long. Tail moderately long.
AGE Late Jurassic, probably early Tithonian.
DISTRIBUTION AND FORMATION/S South Dakota, possibly Wyoming and Utah; probably middle Morrison.

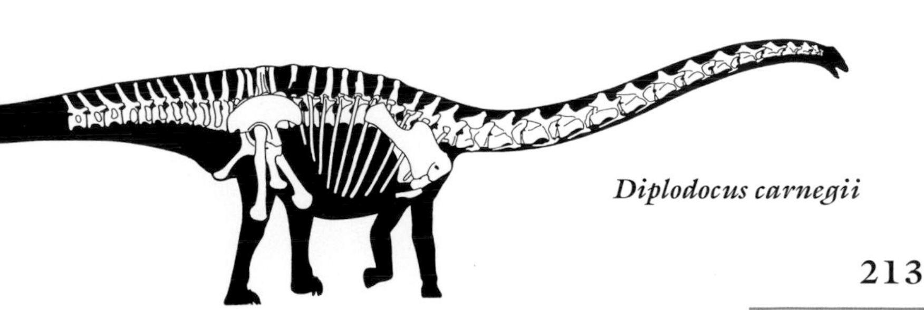
Diplodocus carnegii

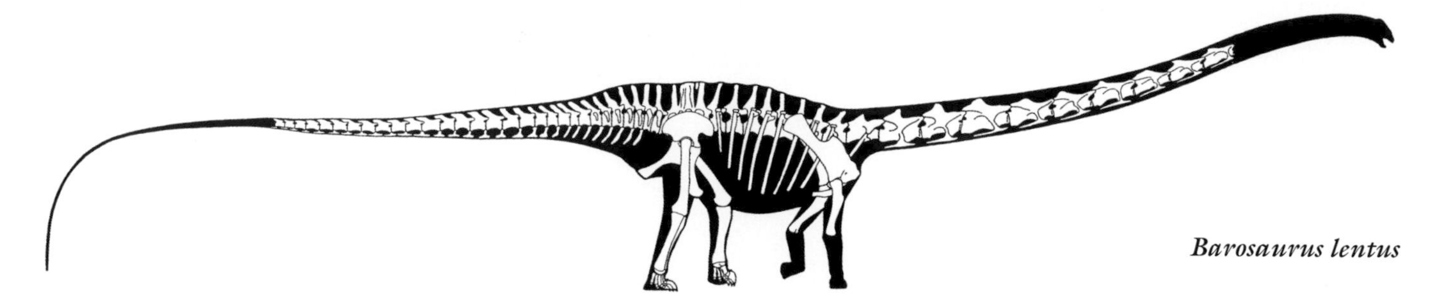
Barosaurus lentus

HABITAT Northern near coastal portion of range, wetter than rest of Morrison.
HABITS High-level browser, although easily able to graze.
NOTES *Kaatedocus siberi* may belong to this taxon. Presence in more coastal portion of Morrison may be because of presence of taller trees.

Supersaurus vivianae
35 m (110 ft) TL, 35 tonnes

FOSSIL REMAINS Minority of several skeletons.
ANATOMICAL CHARACTERISTICS More robustly built than other diplodocines. Neck very long.
AGE Late Jurassic, early Tithonian.
DISTRIBUTION AND FORMATION/S Colorado; middle Morrison.
HABITAT Short wet season, otherwise semiarid with open floodplain prairies and riverine forests.
NOTES Relationships to other diplodocids not entirely certain. Originally incorrectly thought to be the brachiosaur *Ultrasauros* (= *Ultrasaurus*).

Supersaurus (= Lourinhasaurus) alenquerensis
18 m (60 ft) TL, 5 tonnes

FOSSIL REMAINS Minority of several skeletons.
ANATOMICAL CHARACTERISTICS
Insufficient information.
AGE Late Jurassic, late Kimmeridgian or early Tithonian.
DISTRIBUTION AND FORMATION/S
Portugal; Camadas de Alcobaca.
HABITAT Large, seasonally dry island with open woodlands.
NOTES May include *Dinheirosaurus lourinhanensis.*

Apatosaurines Gigantic diplodocids limited to the Late Jurassic of North America.

ANATOMICAL CHARACTERISTICS
Uniform. Skeleton massively constructed. Neck moderately long. Trunk very short. Tail whips very long. Pelvis large to very large.
HABITS Flexible feeder from ground to highest levels. Built for pushing down trees. Powerful build indicates strong defense against predators.

Apatosaurus **muscle study**

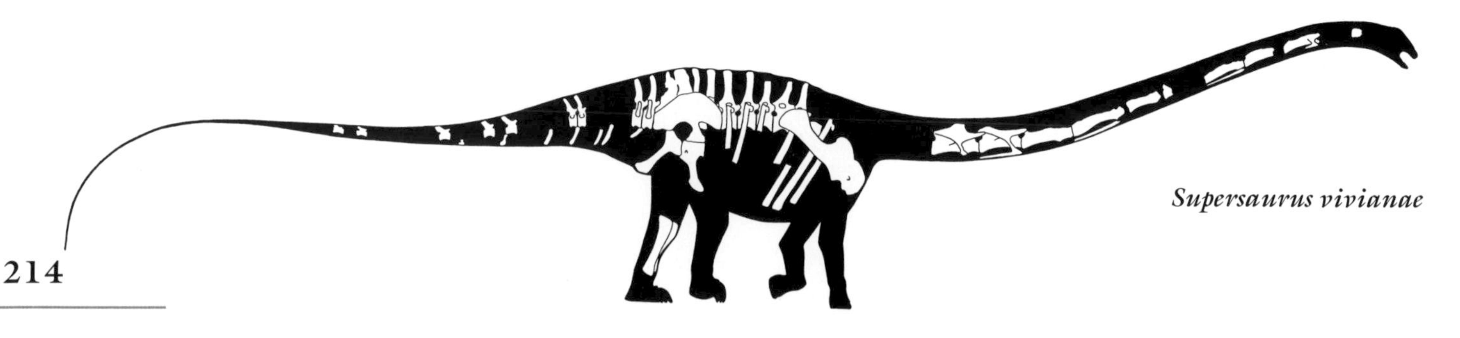
Supersaurus vivianae

Unnamed genus and species

Unnamed genus and species

23 m (75 ft) TL, 13 tonnes

FOSSIL REMAINS Nearly complete skeleton.
ANATOMICAL CHARACTERISTICS Neck fairly long, not very broad. Hip sail not especially tall.
AGE Late Jurassic, late Oxfordian.
DISTRIBUTION AND FORMATION/S Wyoming; lowest Morrison.
HABITAT Short wet season, otherwise semiarid with open floodplain prairies and riverine forests.

Apatosaurus ajax

23 m (75 ft) TL, 20 tonnes

FOSSIL REMAINS One or two partial skeletons.
ANATOMICAL CHARACTERISTICS Neck broad.
AGE Late Jurassic, middle Tithonian.
DISTRIBUTION AND FORMATION/S Colorado; upper Morrison.
HABITAT Wetter than earlier Morrison, otherwise semiarid with open floodplain prairies and riverine forests.
NOTES Validity of fragmentary original specimen problematic. Shared its habitat with *Camarasaurus supremus* and *Amphicoelias.* Main enemy *Allosaurus maximus.*

Apatosaurus unnamed species

Adult size uncertain

FOSSIL REMAINS Majority of a large juvenile skeleton.
ANATOMICAL CHARACTERISTICS Neck fairly long, shallow, very broad. Hip sail not especially tall. Arm and leg long. Pelvis not especially large.
AGE Late Jurassic, middle Tithonian.
DISTRIBUTION AND FORMATION/S Colorado; upper Morrison.
HABITAT Wetter than earlier Morrison, otherwise semiarid with open floodplain prairies and riverine forests.
HABITS Broad neck best adapted for horizontal movements.
NOTES Probably somewhat later than *A. ajax,* which it has been problematically placed within; may be direct descendant of latter.

Brontosaurus or *Apatosaurus* or unnamed genus *parvus* (Illustrated overleaf)

22 m (72 ft) TL, 14 tonnes

FOSSIL REMAINS Majority of skeleton.
ANATOMICAL CHARACTERISTICS Neck moderately broad. Hip sail tall. Pelvis very large.
AGE Late Jurassic, late Oxfordian and/or early Kimmeridgian.
DISTRIBUTION AND FORMATION/S Wyoming; lower Morrison.
HABITAT Short wet season, otherwise semiarid with open floodplain prairies and riverine forests.
NOTES Intermediate anatomy of this early apatosaurine makes generic placement problematic. The same applies to *A. yahnahpin,* which may be the same species.

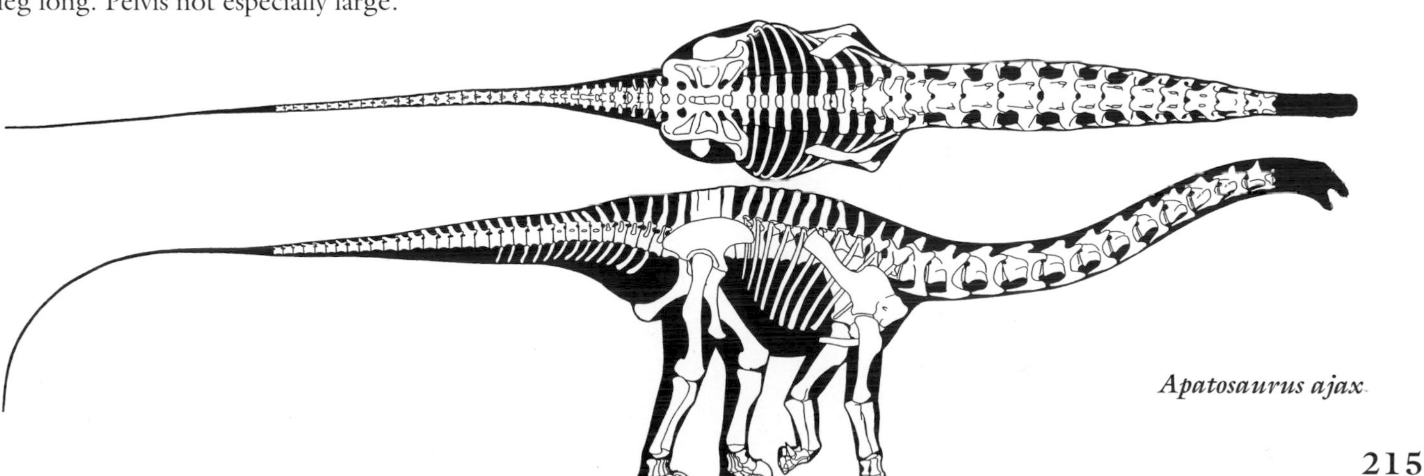
Apatosaurus ajax

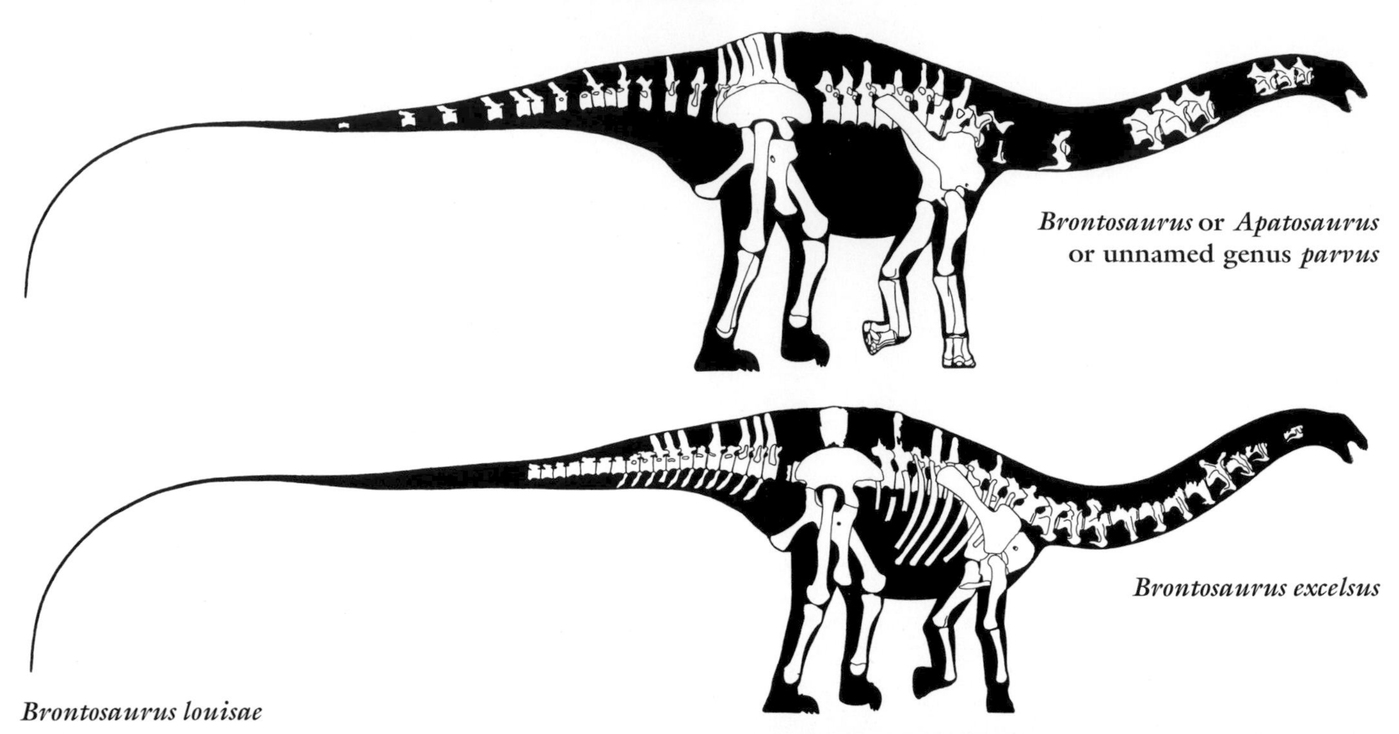

Brontosaurus or *Apatosaurus* or unnamed genus *parvus*

Brontosaurus excelsus

Brontosaurus louisae

Brontosaurus excelsus
22 m (72 ft) TL, 15 tonnes

FOSSIL REMAINS Majority of skeleton.
ANATOMICAL CHARACTERISTICS Neck very deep, moderately broad. Hip sail tall. Pelvis very large.
AGE Late Jurassic, late Kimmeridgian and/or early Tithonian.
DISTRIBUTION AND FORMATION/S Wyoming, Colorado; middle Morrison.
HABITAT Short wet season, otherwise semiarid with open floodplain prairies and riverine forests.
HABITS Deep neck best adapted for vertical movements.
NOTES The classic sauropod. Main enemy *Allosaurus*.

Brontosaurus louisae
23 m (75 ft) TL, 18 tonnes

FOSSIL REMAINS Possible nearly complete skull and a few skeletons, one almost complete.
ANATOMICAL CHARACTERISTICS Neck very deep, moderately broad. Hip sail very tall. Pelvis exceptionally large.
AGE Late Jurassic, early Tithonian.
DISTRIBUTION AND FORMATION/S Utah; middle Morrison.
HABITAT Short wet season; otherwise semiarid with open floodplain prairies and riverine forests.
HABITS Deep neck best adapted for vertical movements.
NOTES Skull usually assigned to this species may belong to a different apatosaurine. Very deep neck with highly unusual ventral projections of cervical ribs indicates this is *Brontosaurus* rather than *Apatosaurus*.

MACRONARIANS

LARGE TO ENORMOUS NEOSAUROPODS OF THE MIDDLE JURASSIC TO THE END OF THE DINOSAUR ERA, MOST CONTINENTS.

ANATOMICAL CHARACTERISTICS Variable. Nostrils enlarged. Neck able to elevate subvertically. Hand elongated. Pubis broad.
NOTES Absence from Antarctica probably reflects lack of sufficient sampling.

MACRONARIAN MISCELLANEA

NOTES The relationships of these macronarians are uncertain.

Brontosaurus louisae

Abrosaurus dongpoi
11 m (35 ft) TL, 5 tonnes

FOSSIL REMAINS Skull.
ANATOMICAL CHARACTERISTICS Insufficient information.
AGE Middle Jurassic, Bathonian or Callovian.
DISTRIBUTION AND FORMATION/S Central China; Shaximiao.
HABITAT Heavily forested.

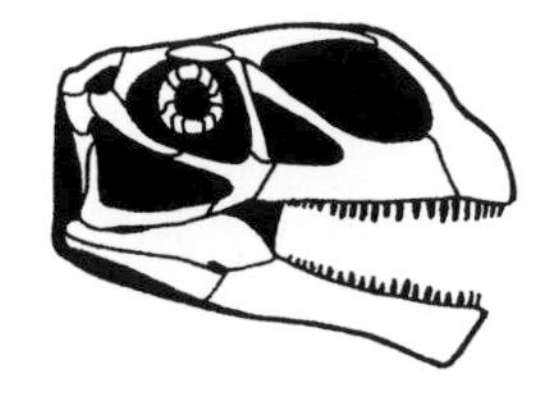

Abrosaurus dongpoi

Tehuelchesaurus benitezii
15 m (50 ft) TL, 9 tonnes

FOSSIL REMAINS Majority of skeleton, skin patches.
ANATOMICAL CHARACTERISTICS Insufficient information.
AGE Middle Jurassic.
DISTRIBUTION AND FORMATION/S Southern Argentina; Canadon Asfalto.
HABITAT Short wet season, otherwise semiarid, riverine forests, open floodplains.

Atlasaurus imelakei
15 m (50 ft) TL, 14 tonnes

FOSSIL REMAINS Partial skull and majority of skeleton.
ANATOMICAL CHARACTERISTICS Head broad and fairly shallow. Neck rather short. Tail not large. Arm and hand very long, and humerus almost as long as femur, so shoulder much higher than hips. Limbs long relative to size of body.
AGE Middle Jurassic, late Bathonian.
DISTRIBUTION AND FORMATION/S Morocco; Douar of Tazouda.

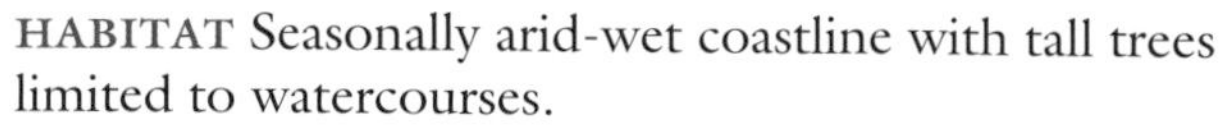

HABITAT Seasonally arid-wet coastline with tall trees limited to watercourses.
HABITS Medium- and high-level browser, not able to easily feed at ground level.
NOTES Its limbs proportionally longer than those of any other known sauropod, *Atlasaurus* emphasized leg over neck length to increase vertical reach to a greater extent than any other known member of the group.

Jobaria tiguidensis
16 m (52 ft) TL, 16 tonnes

FOSSIL REMAINS Complete skull and several skeletons, nearly completely known.
ANATOMICAL CHARACTERISTICS Head not broad. Neck rather short. Tail moderately long. Arm and hand long, so shoulder higher than hips.
AGE Late Middle or early Late Jurassic.
DISTRIBUTION AND FORMATION/S Niger; Tiouraren.
HABITAT Well-watered woodlands.
HABITS Medium- and high-level browser, not able to feed easily at ground level.
NOTES Originally thought to be from the Early Cretaceous, the Tiouraren is from the later Jurassic. Relationships of *Jobaria* are uncertain, may not be a neosauropod.

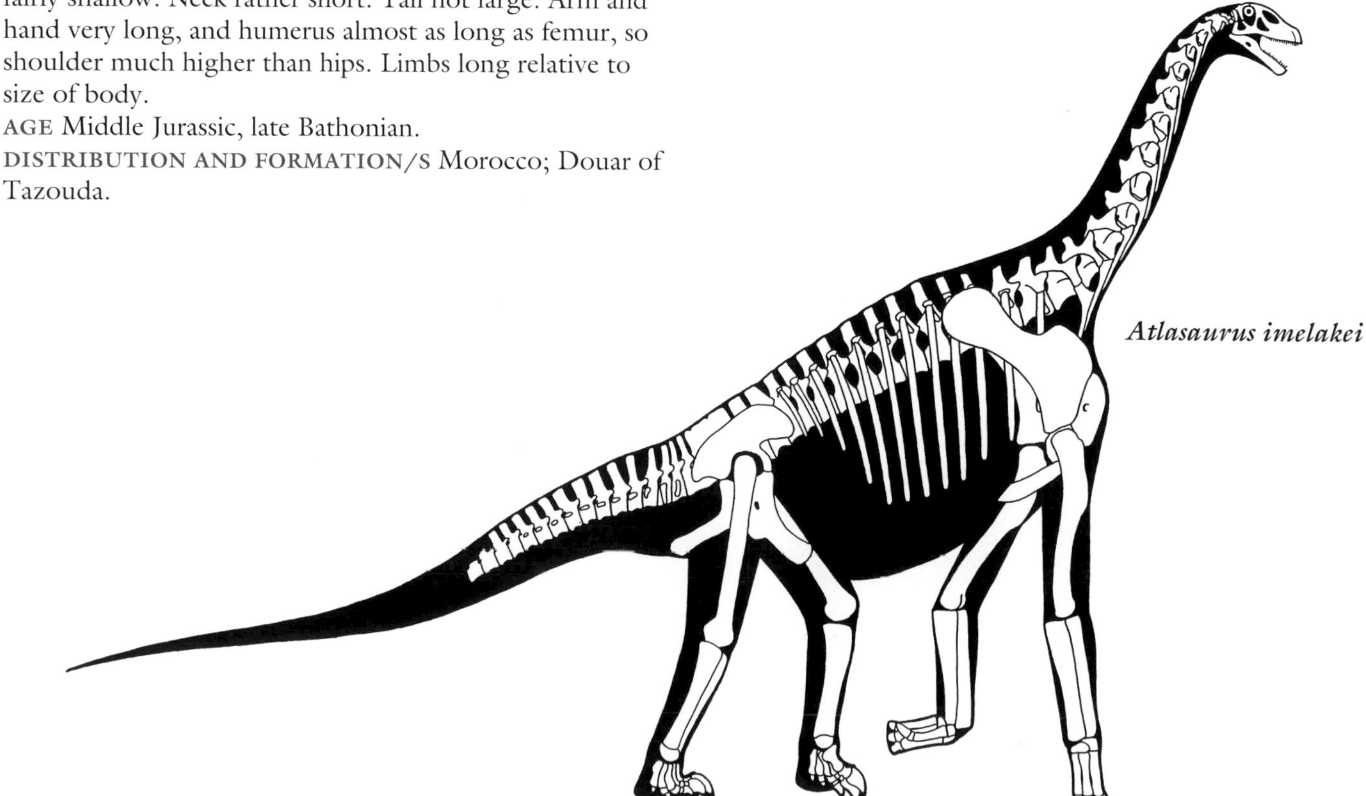

Atlasaurus imelakei

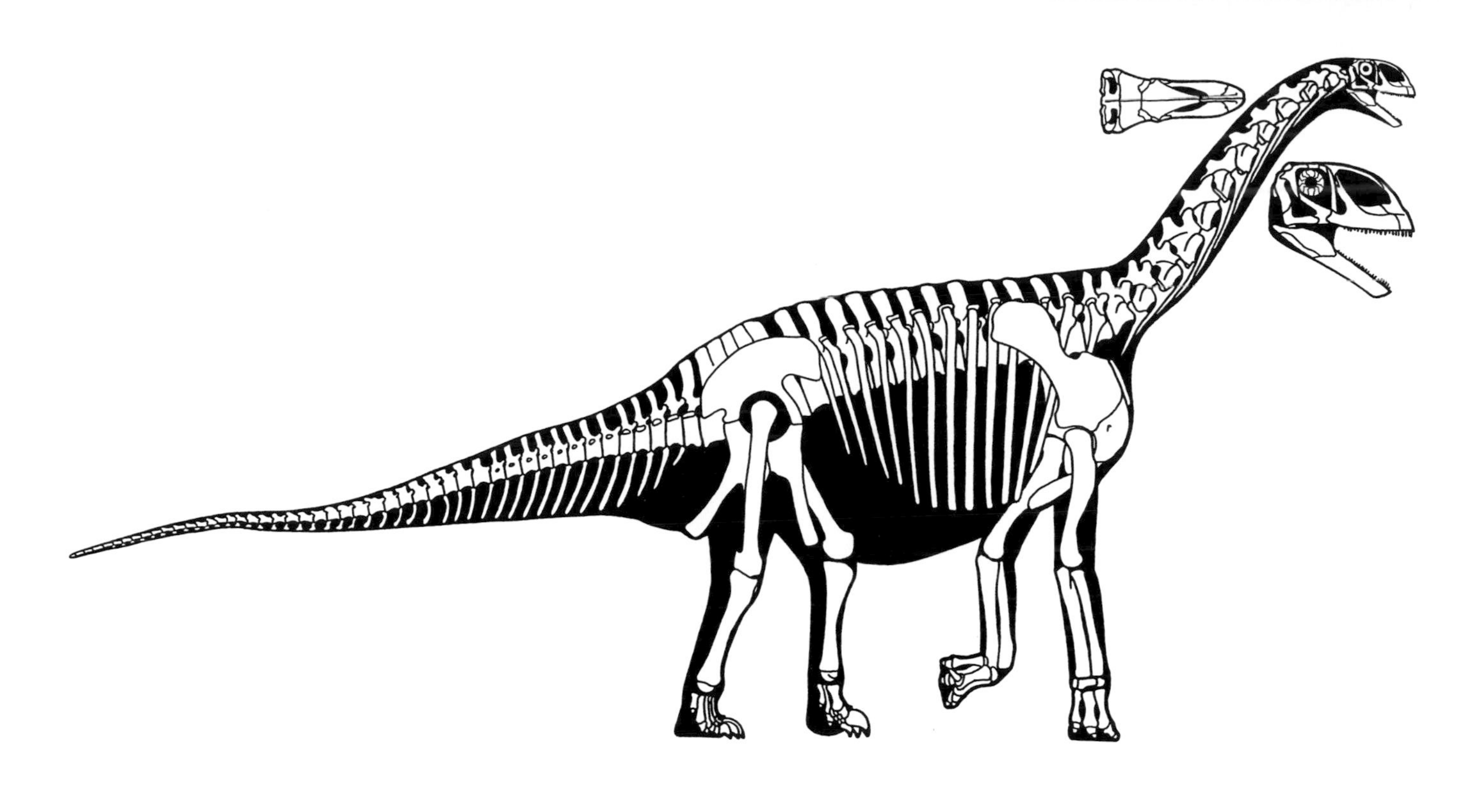

Jobaria tiguidensis

CAMARASAURIDS Large to gigantic macronarian sauropods limited to the Late Jurassic to perhaps the Early Cretaceous of North America and Europe.

ANATOMICAL CHARACTERISTICS Uniform. Head large for sauropods, deep, teeth fairly large. Neck rather short, shallow, broad. Most neck and trunk vertebral spines forked. Tail moderately long. Arm and hand long, so shoulders a little higher than hips. Front of pelvis and belly ribs flare very strongly sideways so belly is very broad and large. Retroverted pelvis facilitated slow walking when rearing up by keeping hips and tail horizontal when bipedal.
HABITS Medium- and high-level browsers, unable to feed easily at ground level. Able to consume coarse vegetation.
NOTES Whether camarasaurs survived into the Early Cretaceous is uncertain.

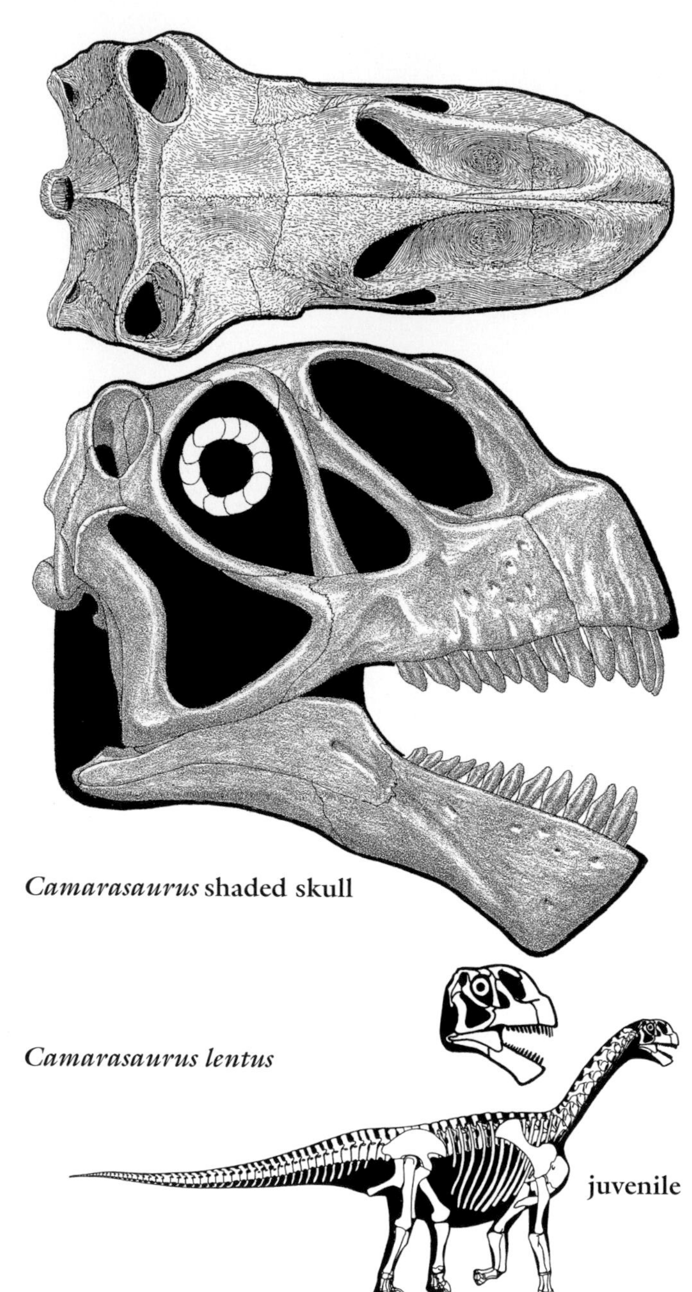

Camarasaurus shaded skull

Camarasaurus grandis
14 m (45 ft) TL, 13 tonnes

FOSSIL REMAINS A few skulls and majority of skeletons.
ANATOMICAL CHARACTERISTICS Standard for group.
AGE Late Jurassic, late Kimmeridgian and/or early Tithonian.
DISTRIBUTION AND FORMATION/S Wyoming, Colorado, Montana; middle Morrison.
HABITAT Short wet season, otherwise semiarid with open floodplain prairies and riverine forests.
NOTES Shared its habitat with *Camarasaurus lewisi*, *Brontosaurus*, *Diplodocus*, *Barosaurus*, and *Stegosaurus*. Main enemy *Allosaurus*.

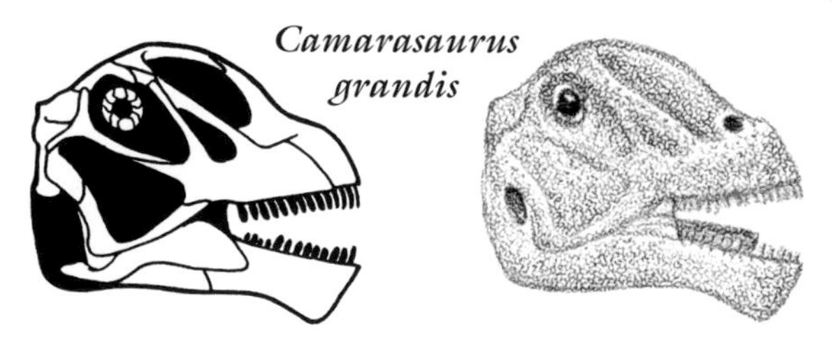

Camarasaurus lentus
15 m (50 ft) TL, 15 tonnes

FOSSIL REMAINS A number of skulls and skeletons including juveniles, completely known.
ANATOMICAL CHARACTERISTICS Standard for group.
AGE Late Jurassic, late Kimmeridgian and/or early Tithonian.
DISTRIBUTION AND FORMATION/S Wyoming, Colorado, Utah; middle Morrison.
HABITAT Short wet season, otherwise semiarid with open floodplain prairies and riverine forests.
NOTES Apparently present later in the Middle Morrison than *C. grandis*, may be direct descendant of the latter.

Camarasaurus supremus
18 m (60 ft) TL, 23 tonnes

FOSSIL REMAINS Some skulls and skeletons.
ANATOMICAL CHARACTERISTICS Standard for group.
AGE Late Jurassic, middle Tithonian.
DISTRIBUTION AND FORMATION/S Wyoming, Colorado, New Mexico; upper Morrison.
HABITAT Wetter than earlier Morrison, otherwise semiarid with open floodplain prairies and riverine forests.
NOTES Anatomically very similar to *C. lentus*, may be the same species or probably the direct descendant of the earlier camarasaurid. Shared its habitat with *Apatosaurus* and *Amphicoelias*. Main enemy *Allosaurus maximus*.

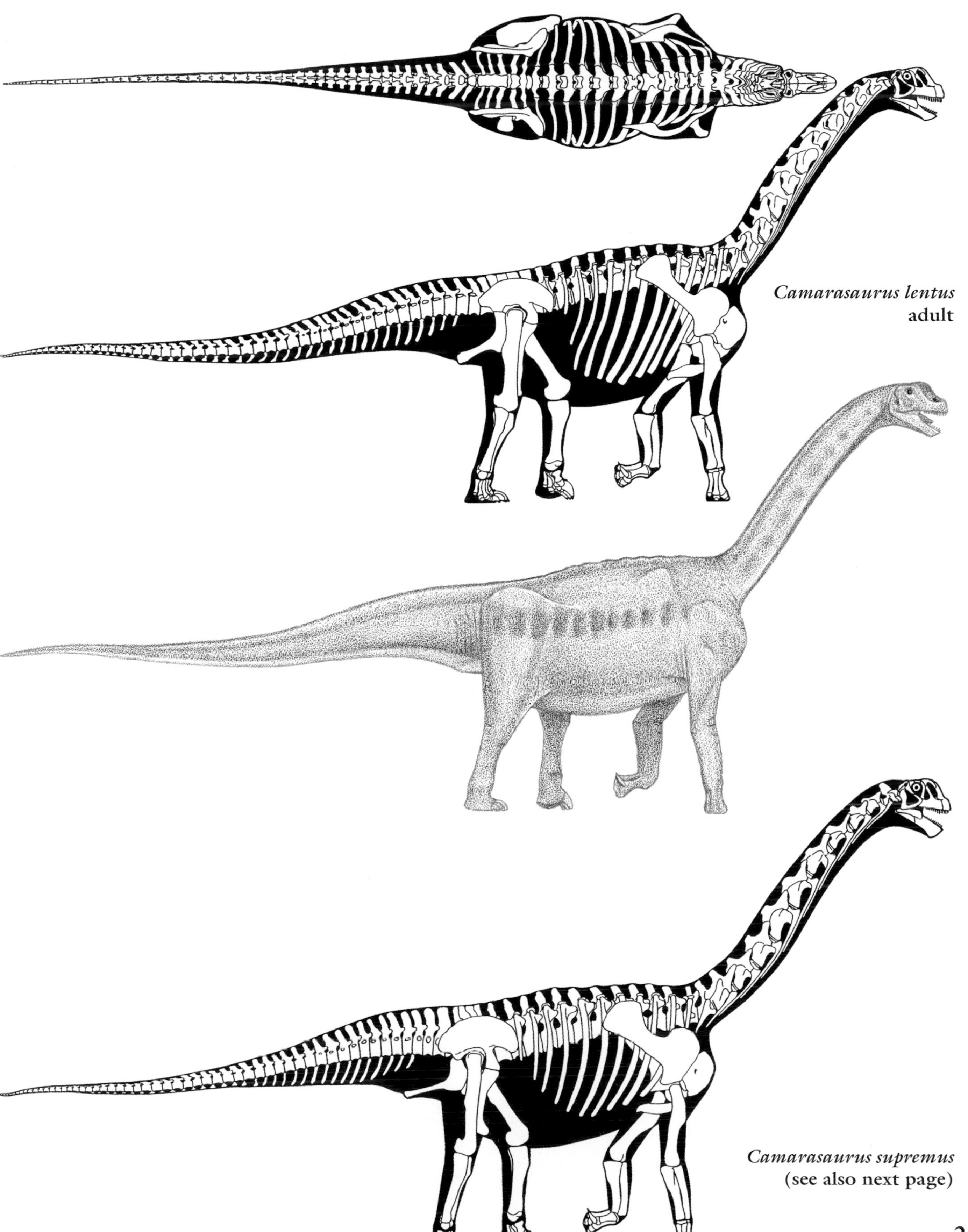

Camarasaurus lentus
adult

Camarasaurus supremus
(see also next page)

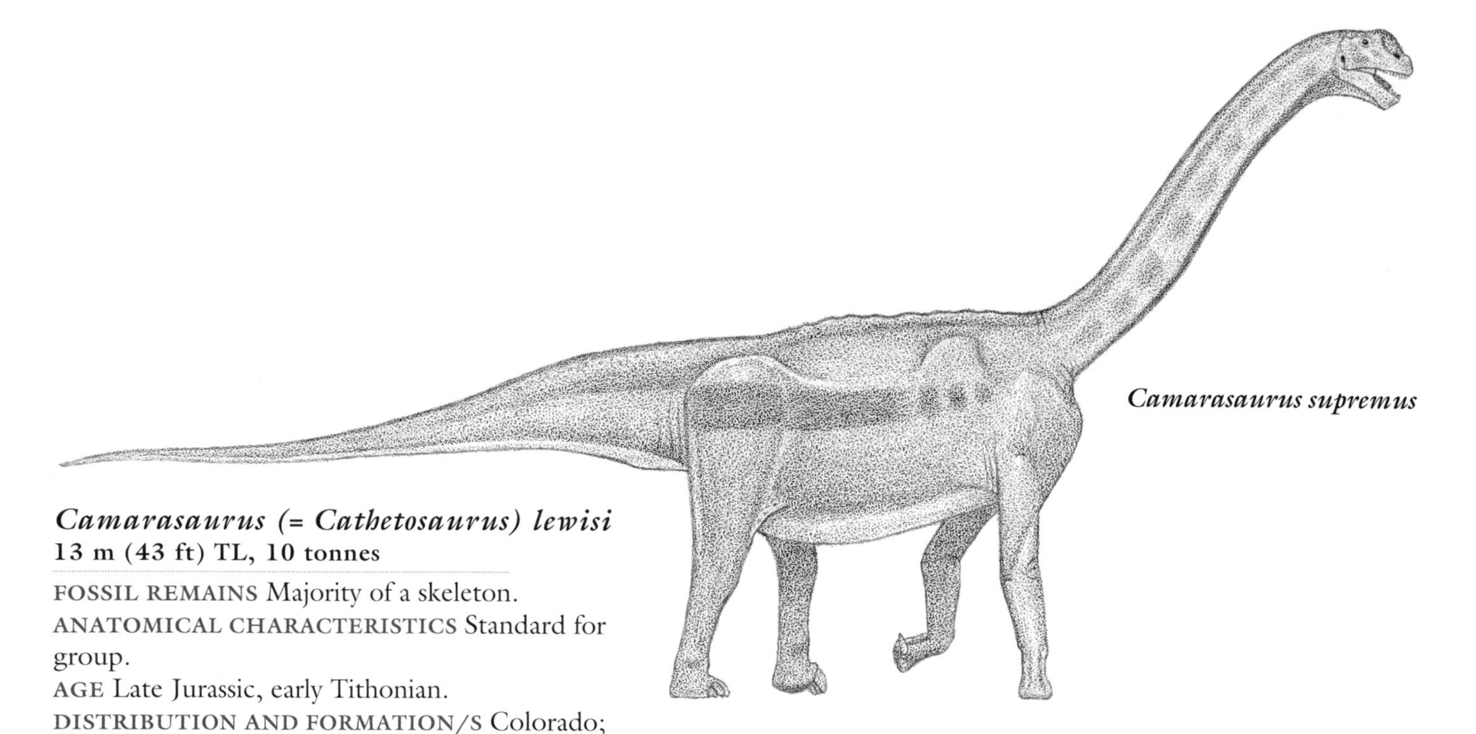
Camarasaurus supremus

Camarasaurus (= Cathetosaurus) lewisi
13 m (43 ft) TL, 10 tonnes

FOSSIL REMAINS Majority of a skeleton.
ANATOMICAL CHARACTERISTICS Standard for group.
AGE Late Jurassic, early Tithonian.
DISTRIBUTION AND FORMATION/S Colorado; middle Morrison.
HABITAT Short wet season, otherwise semiarid with open floodplain prairies and riverine forests.

Aragosaurus ischiaticus
18 m (60 ft) TL, 25 tonnes

FOSSIL REMAINS Minority of skeleton.
ANATOMICAL CHARACTERISTICS Arm longer than that of *Camarasaurus*, so shoulder higher.
AGE Early Cretaceous, late Hauterivian and/or early Barremian.
DISTRIBUTION AND FORMATION/S Northern Spain; Castellar.
HABITS High-level browser.
NOTES Relationships uncertain.

Titanosauriforms

LARGE TO ENORMOUS MACRONARIAN SAUROPODS OF THE MIDDLE OR LATE JURASSIC TO THE END OF THE DINOSAUR ERA, MOST CONTINENTS.

ANATOMICAL CHARACTERISTICS Variable. Teeth elongated. Gauge of trackways broader than those of other sauropods. Front of pelvis and belly ribs flare very strongly sideways, so belly is very broad and large. Fingers further reduced or absent, thumb claw reduced or absent.
NOTES Absence from Antarctica probably reflects lack of sufficient sampling.

TITANOSAURIFORM MISCELLANEA
NOTES The relationships of these titanosauriforms are uncertain.

Fusuisaurus zhaoi
22 m (70 ft) TL, 35 tonnes

FOSSIL REMAINS Minority of skeleton.
ANATOMICAL CHARACTERISTICS Insufficient information.
AGE Early Cretaceous.
DISTRIBUTION AND FORMATION/S Southern China; Napai.

Huanghetitan liujiaxiaensis
12 m (40 ft) TL, 3 tonnes

FOSSIL REMAINS Minority of skeleton.
ANATOMICAL CHARACTERISTICS Insufficient information.
AGE Late Early Cretaceous.
DISTRIBUTION AND FORMATION/S Northern China; Hekou Group.
NOTES Shared its habitat with *Daxiatitan*.

Dongbeititan dongi
15 m (59 ft) TL, 7 tonnes

FOSSIL REMAINS Minority of skeleton.
ANATOMICAL CHARACTERISTICS Heavily constructed. Neck broad, moderately long.
AGE Early Cretaceous, early Aptian.
DISTRIBUTION AND FORMATION/S Northeast China; Yixian.

HABITAT Well-watered forests and lakes, winters chilly with some snow.
NOTES Prey of *Yutyrannus.*

Tastavinsaurus sanzi
16 m (50 ft) TL, 8 tonnes

FOSSIL REMAINS Minority of skeleton.
ANATOMICAL CHARACTERISTICS Insufficient information.
AGE Early Cretaceous, early Aptian.
DISTRIBUTION AND FORMATION/S Eastern Spain; Xert.

Fukuititan nipponensis
Adult size uncertain

FOSSIL REMAINS Minority of skeleton.
ANATOMICAL CHARACTERISTICS Insufficient information.
AGE Early Cretaceous, Barremian.
DISTRIBUTION AND FORMATION/S Japan; Kitadani.

Wintonotitan wattsi
15 m (50 ft) TL, 10 tonnes

FOSSIL REMAINS Minority of two skeletons.
ANATOMICAL CHARACTERISTICS Insufficient information.
AGE Early Cretaceous, latest Albian.
DISTRIBUTION AND FORMATION/S Northeast Australia; Winton.
HABITAT Well-watered areas, cold winter.
NOTES Shared its habitat with *Diamantinasaurus.*

EUHELOPIDS Small (for sauropods) to gigantic titanosauriform sauropods of the Middle or Late Jurassic to the Early or Late Cretaceous of Asia.

ANATOMICAL CHARACTERISTICS Fairly uniform. Head fairly broad, snout forms shelf below nostrils, which are very large and arced. Skeleton rather lightly built. Neck moderately to very long. Tail not large. Arm and hand very to exceptionally long, so shoulder much higher than hips. Thumb claw reduced or absent. Pelvis rather small, retroverted.
HABITS High-level browsers, not able to feed easily near ground level.

Bellusaurus sui (= Klamelisaurus gobiensis?)
13 m (45 ft) TL, 6 tonnes

FOSSIL REMAINS Dozen and a half partial skeletons, juvenile and possibly adult.
ANATOMICAL CHARACTERISTICS Neck long in adults. Shoulders a little higher than hips. Tail not large.
AGE Middle Jurassic, late Callovian.
DISTRIBUTION AND FORMATION/S Northwest China; lower Shishugou.
NOTES *Klamelisaurus gobiensis* may be the adult of this species. Relationships uncertain, if a titanosauriform indicates presence of group in Middle Jurassic. Prey of *Monolophosaurus.*

Qiaowanlong kangxii
12 m (40 ft) TL, 6 tonnes

FOSSIL REMAINS Minority of skeleton.
ANATOMICAL CHARACTERISTICS Neck long, vertebral spines forked.
AGE Early Cretaceous, Aptian or Albian.
DISTRIBUTION AND FORMATION/S Central China; middle Xinminpu.

Erketu ellisoni
15 m (50 ft) TL, 5 tonnes

FOSSIL REMAINS Minority of skeleton.
ANATOMICAL CHARACTERISTICS Neck extremely long, with vertebrae more elongated than in any other sauropod.

Bellusaurus sui
(= Klamelisaurus gobiensis?)

AGE Early Cretaceous.
DISTRIBUTION AND FORMATION/S Mongolia; Bayanshiree.
HABITS Probably a high browser.

Euhelopus zdanskyi
11 m (35 ft) TL, 3.5 tonnes

FOSSIL REMAINS Majority of skull and two skeletons.
ANATOMICAL CHARACTERISTICS Neck long. Vertebral spines near base of neck forked. Arm probably very long, so shoulder higher than hips.
AGE Early Cretaceous, Barremian or Aptian.
DISTRIBUTION AND FORMATION/S Eastern China; Meng-Yin.
NOTES Proportions of arm and leg are uncertain. Long thought to be a Late Jurassic mamenchisaur relative, now recognized as an Early Cretaceous titanosauriform.

Daxiatitan binglingi
Adult size not available

FOSSIL REMAINS Minority of skeleton.
ANATOMICAL CHARACTERISTICS Insufficient information.
AGE Late Early Cretaceous.
DISTRIBUTION AND FORMATION/S Northern China; Hekou Group.
NOTES Shared its habitat with *Huanghetitan*.

Euhelopus zdanskyi

Tangvayosaurus hoffeti
19 m (62 ft) TL, 17 tonnes

FOSSIL REMAINS Two partial skeletons.
ANATOMICAL CHARACTERISTICS Skeleton robustly built.
AGE Early Cretaceous, late Barremian or early Cenomanian.
DISTRIBUTION AND FORMATION/S Laos; Gres Superieurs.
NOTES Shared its habitat with *Ichthyovenator*.

Phuwiangosaurus sirindhornae
19 m (62 ft) TL, 17 tonnes

FOSSIL REMAINS Partial skeletons, juvenile and adult.
ANATOMICAL CHARACTERISTICS Neck moderately long, some vertebral spines forked.
AGE Early Cretaceous, Valanginian or Hauterivian.
DISTRIBUTION AND FORMATION/S Thailand; Sao Khua.
HABITAT Short wet season, otherwise semiarid with open floodplain prairies and riverine forests.

Tambatitanis amicitiae
14 m (45 ft) TL, 4 tonnes

FOSSIL REMAINS Minority of skeleton.
ANATOMICAL CHARACTERISTICS Insufficient information.
AGE Early Cretaceous, probably early Albian.
DISTRIBUTION AND FORMATION/S Japan; lower Sasayama Group.

Yunmenglong ruyangensis
20 m (65 ft) TL, 30 tonnes

FOSSIL REMAINS Minority of skeleton.
ANATOMICAL CHARACTERISTICS Neck very long.
AGE Early Cretaceous, Aptian or Albian.
DISTRIBUTION AND FORMATION/S Eastern China; Haoling.

Huabeisaurus allocotus
17 m (55 ft) TL, 15 tonnes

FOSSIL REMAINS Majority of skeleton.
ANATOMICAL CHARACTERISTICS Neck long. Shoulders level with hips.
AGE Late Cretaceous.
DISTRIBUTION AND FORMATION/S Northern China; upper Huiquanpu.
NOTES Relationships uncertain; if a euhelopid, indicates presence of group in Late Cretaceous. Shared its habitat with *Tianzhenosaurus*.

Huabeisaurus allocotus

BRACHIOSAURIDS Small (for sauropods) to enormous titanosauriform sauropods of the Late Jurassic to the Early Cretaceous of the Americas, Europe, and Africa.

ANATOMICAL CHARACTERISTICS Fairly uniform. Head fairly broad, snout forms shelf below nostrils, which are very large and arced. Skeleton rather lightly built. Neck moderately to very long. Tail not large. Arm and hand very to exceptionally long, so shoulder much higher than hips. Thumb claw reduced or absent. Pelvis rather small, retroverted.
HABITS High-level browsers, not able to feed easily near ground level. Reared up less often than other sauropods. Fragmentary Early Cretaceous Colombian *Padillasaurus leivaensis* indicates presence of group in South America.

Lapparentosaurus madagascariensis
Adult size uncertain

FOSSIL REMAINS A few partial skeletons, subadult to juvenile.
ANATOMICAL CHARACTERISTICS Insufficient information.
AGE Middle Jurassic, Bathonian.
DISTRIBUTION AND FORMATION/S Madagascar; Isalo III.
NOTES Relationships of *Lapparentosaurus* are uncertain.

Giraffatitan shaded skull

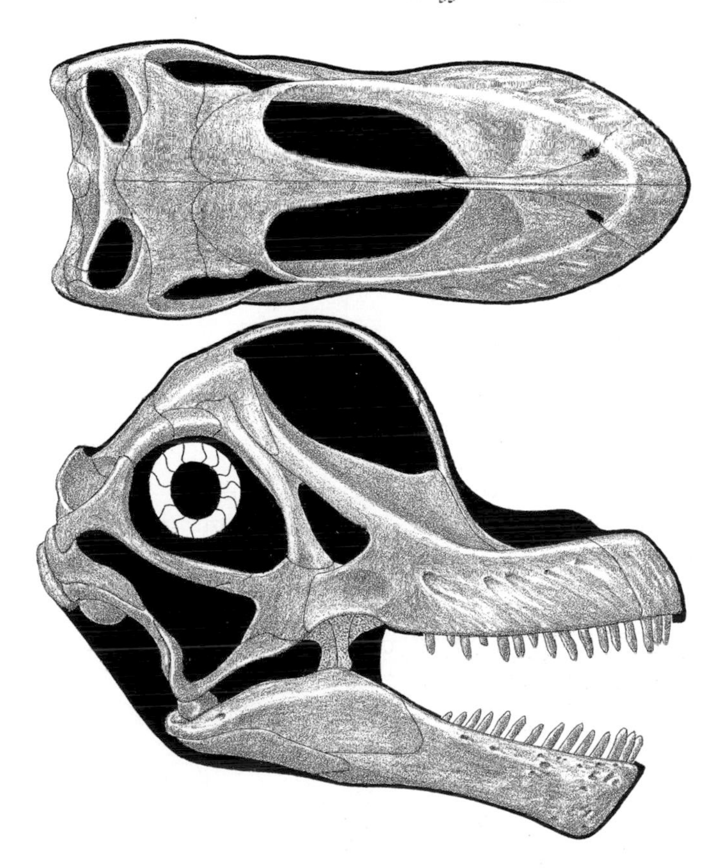

Giraffatitan **muscle study**

Daanosaurus zhangi
Adult size uncertain

FOSSIL REMAINS Partial skeleton, juvenile.
ANATOMICAL CHARACTERISTICS Insufficient information.
AGE Late Jurassic.
DISTRIBUTION AND FORMATION/S Southern China; Shangshaximiao.
NOTES Relationships of *Daanosaurus* are uncertain.

Europasaurus holgeri
5.7 m (19 ft) TL, 750 kg (1,700 lb)

FOSSIL REMAINS Majority of skull and a number of skeletons.
ANATOMICAL CHARACTERISTICS Snout shelf short. Neck moderately long. Thumb claw small.
AGE Late Jurassic, middle Kimmeridgian.
DISTRIBUTION AND FORMATION/S Northern Germany; Mittlere Kimmeridge-Stufe.
HABITS Small size limited browsing height.
NOTES Found as drift in nearshore marine deposits set amid islands then immediately off the northeast coast of North America; small size is probably dwarfism forced by limited food resources.

Brachiosaurus altithorax
22 m (72 ft) TL, 35 tonnes

FOSSIL REMAINS Minority of skeleton and other bones.
ANATOMICAL CHARACTERISTICS Tail short (for sauropods). Arm and hand exceptionally long, and humerus longer than femur, so shoulders very high.
AGE Late Jurassic, early Tithonian.

DISTRIBUTION AND FORMATION/S Colorado; middle Morrison.
HABITAT Short wet season, otherwise semiarid with open floodplain prairies and riverine forests.
NOTES Probably includes *Dystylosaurus edwini*. A partial skull from the lower Morrison may belong to this or another genus but probably to another species, and some other Morrison remains are more similar to *Giraffatitan*.

Giraffatitan brancai (Illustrated overleaf)
23 m (75 ft) TL, 40 tonnes

FOSSIL REMAINS At least one partial skull and skeleton, possibly other skulls and skeletons.
ANATOMICAL CHARACTERISTICS Snout shelf long. Neck very long. Tall withers at shoulder anchored unusually deep neck tendons. Back trunk vertebrae relatively small. Tail short (for sauropods). Arm and hand exceptionally long, and humerus longer than femur, so shoulders very high, limbs long relative to body. Thumb claw small.
AGE Late Jurassic, late Kimmeridgian and/or early Tithonian.
DISTRIBUTION AND FORMATION/S Tanzania; middle Tendaguru.
HABITAT Coastal, seasonally dry with heavier vegetation farther inland.
NOTES The most giraffe-like dinosaur known, both neck and limb length used to increase vertical reach. Not placeable in the *Brachiosaurus* it was long assigned to; a considerable portion of remains placed in *G. brancai* from middle and upper Tendaguru may be different taxa. Shared its habitat with *Dicraeosaurus*.

Lusotitan atalaiensis
21 m (70 ft) TL, 30 tonnes

FOSSIL REMAINS Minority of skeletons.
ANATOMICAL CHARACTERISTICS Humerus longer than femur, so shoulders very high.
AGE Late Jurassic, late Kimmeridgian or early Tithonian.
DISTRIBUTION AND FORMATION/S Portugal; Lourinha.
HABITAT Large, seasonally dry island with open woodlands.
NOTES Relationships of *Lusotitan* uncertain. The presence of this and other gigantic sauropods on a

Giraffatitan brancai

Portuguese island shows that dwarfism was not occurring, perhaps because of intermittent immigration from nearby continents.

Abydosaurus mcintoshi
Adult size uncertain

FOSSIL REMAINS Complete skull and partial skull and skeletal remains.
ANATOMICAL CHARACTERISTICS Snout shelf long, nasal opening and projection moderately developed.
AGE Early Cretaceous, Aptian.
DISTRIBUTION AND FORMATION/S Utah; Middle Cedar Mountain.
HABITAT Short wet season, otherwise semiarid with floodplain prairies, open woodlands, and riverine forests.

Abydosaurus mcintoshi

Cedarosaurus weiskopfae
15 m (50 ft) TL, 10 tonnes

FOSSIL REMAINS Majority of skeleton.
ANATOMICAL CHARACTERISTICS Neck length uncertain.
AGE Early Cretaceous, probably early Barremian.
DISTRIBUTION AND FORMATION/S Utah, Lower Cedar Mountain.
HABITAT Short wet season, otherwise semiarid with floodplain prairies and open woodlands, and riverine forests.
NOTES Shared its habitat with *Iguanacolossus.*

Sonorasaurus thompsoni
15 m (50 ft) TL, 10 tonnes

FOSSIL REMAINS Small minority of skeleton(s).
ANATOMICAL CHARACTERISTICS Toe claws reduced.
AGE Early Cretaceous, late Albian.
DISTRIBUTION AND FORMATION/S Arizona; Turney Ranch.

Venenosaurus dicrocei
12 m (40 ft) TL, 6 tonnes

FOSSIL REMAINS Minority of skeleton.
ANATOMICAL CHARACTERISTICS Insufficient information.
AGE Early Cretaceous, early Aptian.
DISTRIBUTION AND FORMATION/S Utah; Middle Cedar Mountain.
HABITAT Short wet season, otherwise semiarid with floodplain prairies and open woodlands, and riverine forests.

Pleurocoelus nanus
Adult size uncertain

FOSSIL REMAINS Minority of a few juvenile skulls and skeletons.
ANATOMICAL CHARACTERISTICS Neck moderately long in juveniles.
AGE Early Cretaceous, middle or late Aptian or early Albian.
DISTRIBUTION AND FORMATION/S Maryland; Arundel.
NOTES Originally *Astrodon johnstoni* based on inadequate remains.

Pleurocoelus nanus

Sauroposeidon proteles
27 m (90 ft) TL, 40 tonnes

FOSSIL REMAINS Several partial skeletons.
ANATOMICAL CHARACTERISTICS Neck very long.
AGE Early Cretaceous, Aptian.
DISTRIBUTION AND FORMATION/S Texas, Oklahoma; Antlers, Paluxy, Glen Rose.
HABITAT Floodplain with coastal swamps and marshes.
NOTES Includes *Paluxysaurus jonesi*. May not be a brachiosaurid. Main enemy *Acrocanthosaurus*.

TITANOSAURIDS Large to enormous titanosauriforms of the Late Jurassic to the end of the dinosaur era, most continents.

ANATOMICAL CHARACTERISTICS Variable. Trunk vertebrae more flexible, possibly aiding rearing. Tail moderately long, very flexible especially upward, ending in a short whip. Arm at least fairly long, so shoulders as high as or higher than hips. Often armored, usually lightly in adults.
HABITS Often used armor as the passive side of their defense strategy, may have been most important in the more vulnerable juveniles. Flexible tail may have been used as display organ by arcing it over the back. Fossil dung indicates titanosaurs consumed flowering plants, including early grasses, as well as nonflowering plants.
NOTES Absence from Antarctica probably reflects lack of sufficient sampling. The last of the sauropod groups, titanosaurs are the only sauropods known to have survived into the late Late Cretaceous. Armor may have assisted them in surviving in a world of increasingly sophisticated and gigantic predators. The relationships of the numerous but often incompletely preserved titanosaurs are not well understood; the group is potentially splittable into a number of subdivisions. Poorly documented Indian fossils labeled *Bruhathkayosaurus matleyi* may or may not be a titanosaur well over 100 tonnes and the largest known land animal.

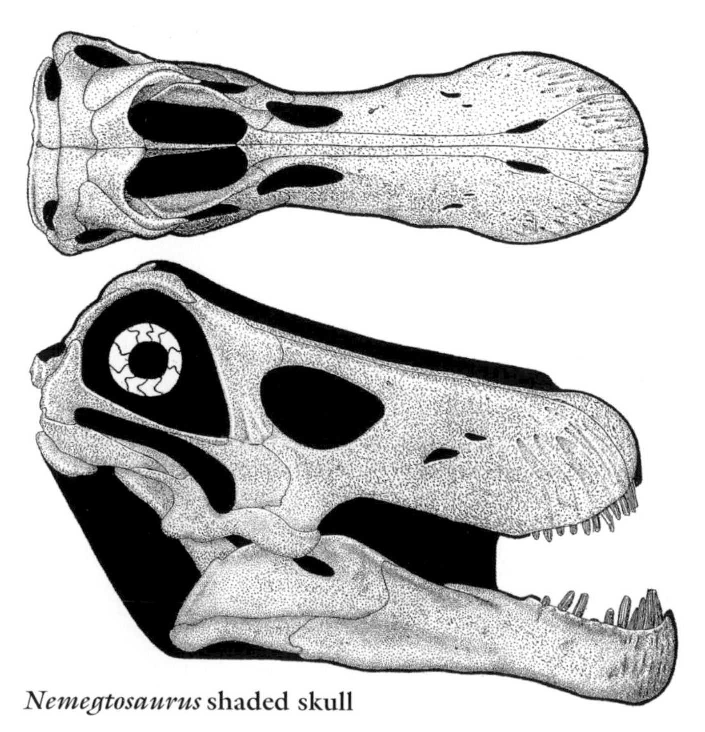

Nemegtosaurus shaded skull

Baso-titanosaurids Large to enormous titanosaurids of the Late Jurassic to the end of the dinosaur era, most continents.

ANATOMICAL CHARACTERISTICS Variable.

Janenschia robusta
17 m (53 ft) TL, 10 tonnes

FOSSIL REMAINS Minority of a few skeletons.
ANATOMICAL CHARACTERISTICS Fingers and thumb claw present.
AGE Late Jurassic, middle and/or late Tithonian.
DISTRIBUTION AND FORMATION/S Tanzania; upper Tendaguru.
HABITAT Coastal, seasonally dry with heavier vegetation farther inland.
NOTES The earliest known titanosaur, and the only one yet named from the Jurassic. Shared its habitat with *Dicraeosaurus sattleri* and *Tornieria africana*.

Ligabuesaurus leanzi
18 m (60 ft) TL, 20 tonnes

FOSSIL REMAINS Minority of skull and partial skeleton.
ANATOMICAL CHARACTERISTICS Neck moderately long. Spines of neck and trunk vertebrae very broad. Arm long, so shoulders high.
AGE Early Cretaceous, late Aptian or early Albian.
DISTRIBUTION AND FORMATION/S Western Argentina; Lohan Cura.
HABITAT Well-watered coastal woodlands with short dry season.
NOTES Shared its habitat with *Comahuesaurus*, *Agustinia*, and *Limaysaurus*.

Malarguesaurus florenciae
Adult size uncertain

FOSSIL REMAINS Minority of large juvenile skeleton.
ANATOMICAL CHARACTERISTICS Insufficient information.
AGE Late Cretaceous, late Turonian.
DISTRIBUTION AND FORMATION/S Western Argentina; Portezuelo.
HABITAT Well-watered woodlands with short dry season.

NOTES Shared its habitat with *Futalognkosaurus* and *Muyelensaurus.*

Gobititan shenzhouensis

20 m (65 ft) TL, 20 tonnes

FOSSIL REMAINS Minority of skeleton.
ANATOMICAL CHARACTERISTICS Insufficient information.
AGE Early Cretaceous, Albian.
DISTRIBUTION AND FORMATION/S Central China; Xinminbo.

Andesaurus delgadoi

15 m (50 ft) TL, 7 tonnes

FOSSIL REMAINS Minority of skeleton.
ANATOMICAL CHARACTERISTICS Insufficient information.
AGE Late Cretaceous, early Cenomanian.
DISTRIBUTION AND FORMATION/S Western Argentina; Candeleros.
HABITAT Short wet season, otherwise semiarid with open floodplains and riverine forests.
NOTES Shared its habitat with *Limaysaurus.* Main enemy *Giganotosaurus.*

Muyelensaurus pecheni

11 m (36 ft) TL, 3 tonnes

FOSSIL REMAINS Minority of skeleton.
ANATOMICAL CHARACTERISTICS Lightly built.
AGE Late Cretaceous, late Turonian.
DISTRIBUTION AND FORMATION/S Western Argentina; Portezuelo.
HABITAT Well-watered woodlands with short dry season.
NOTES Shared its habitat with *Futalognkosaurus* and *Malarguesaurus.*

Rinconsaurus caudamirus

11 m (36 ft) TL, 3 tonnes

FOSSIL REMAINS Parts of several skeletons.
ANATOMICAL CHARACTERISTICS Neck moderately long.
AGE Late Cretaceous, Turonian or Coniacian.
DISTRIBUTION AND FORMATION/S Western Argentina; Rio Neuquen.
HABITAT Well-watered woodlands with short dry season.
NOTES Shared its habitat with *Mendozasaurus.*

Agustinia ligabuei

15 m (50 ft) TL, 8 tonnes

FOSSIL REMAINS Minority of skeleton.
ANATOMICAL CHARACTERISTICS Long rows of spiked armor along top of body.
AGE Early Cretaceous, late Aptian or early Albian.
DISTRIBUTION AND FORMATION/S Western Argentina; Lohan Cura.
HABITAT Well-watered coastal woodlands with short dry season.
NOTES The most heavily armored sauropod. Shared its habitat with *Comahuesaurus, Limaysaurus,* and *Ligabuesaurus.*

Epachthosaurus sciuttoi

13 m (45 ft) TL, 5 tonnes

FOSSIL REMAINS Majority of skeleton.
ANATOMICAL CHARACTERISTICS Insufficient information.
AGE Late Cretaceous, late Cenomanian or Turonian.
DISTRIBUTION AND FORMATION/S Southern Argentina; lower Bajo Barreal.

Aegyptosaurus baharijensis

15 m (50 ft) TL, 7 tonnes

FOSSIL REMAINS Minority of skeleton.
ANATOMICAL CHARACTERISTICS Insufficient information.
AGE Late Cretaceous, Cenomanian.
DISTRIBUTION AND FORMATION/S Egypt; Bahariya.
HABITAT Coastal mangroves.
NOTES Shared its habitat with *Paralititan.* Main enemy *Carcharodontosaurus.*

Ruyangosaurus giganteus

30 m (100 ft) TL, 50+ tonnes

FOSSIL REMAINS Minority of skeleton.
ANATOMICAL CHARACTERISTICS Insufficient information.
AGE Early Late Cretaceous.
DISTRIBUTION AND FORMATION/S Eastern China; Shangdonggou.
NOTES Shows that Asian titanosaurs reached the same dimensions as those of South America.

Atacamatitan chilensis

Adult size uncertain

FOSSIL REMAINS Minority of skeleton.
ANATOMICAL CHARACTERISTICS Insufficient information.
AGE Late Cretaceous.
DISTRIBUTION AND FORMATION/S Northern Chile; Tolar.

Argentinosaurus huinculensis

30 m (100 ft) TL, 50+ tonnes

FOSSIL REMAINS Partial skeleton.
ANATOMICAL CHARACTERISTICS Insufficient information.

AGE Late Cretaceous, Cenomanian.
DISTRIBUTION AND FORMATION/S Western Argentina; Huincul, level uncertain.
HABITAT Short wet season, otherwise semiarid with open floodplains and riverine forests.

Puertosaurus roulli
30 m (100 ft) TL, 50+ tonnes

FOSSIL REMAINS Small portion of skeleton.
ANATOMICAL CHARACTERISTICS Neck moderately long.
AGE Late Cretaceous, early Maastrichtian.
DISTRIBUTION AND FORMATION/S Southern Argentina; Pari Aike.
HABITAT Short wet season, otherwise semiarid with open floodplains and riverine forests.
NOTES In the same size class as *Argentinosaurus*, *Futalognkosaurus*, *Pellegrinisaurus*, and *Ruyangosaurus*, this titanosaur shows that supersized sauropods survived until the close of the dinosaur era.

Chubutisaurus insignis
18 m (60 ft) TL, 12 tonnes

FOSSIL REMAINS Two partial skeletons.
ANATOMICAL CHARACTERISTICS Insufficient information.
AGE Early Cretaceous, Albian.
DISTRIBUTION AND FORMATION/S Southern Argentina; Cerro Barcino.
NOTES Prey of *Tyrannotitan*.

Austrosaurus mckillopi
20 m (65 ft) TL, 16 tonnes

FOSSIL REMAINS Minority of a few skeletons.
ANATOMICAL CHARACTERISTICS Insufficient information.
AGE Early Cretaceous, Albian.
DISTRIBUTION AND FORMATION/S Northeast Australia; Allaru.

Baotianmansaurus henanensis
20 m (65 ft) TL, 16 tonnes

FOSSIL REMAINS Minority of skeleton.
ANATOMICAL CHARACTERISTICS Insufficient information.
AGE Late Cretaceous.
DISTRIBUTION AND FORMATION/S Eastern China; Gaogou.
NOTES Position within titanosaurs uncertain.

Dreadnoughtus schrani
25 m (80 ft) TL, 25 tonnes

FOSSIL REMAINS Majority of skeleton.
ANATOMICAL CHARACTERISTICS Standard for group.
AGE Late Cretaceous, Campanian and/or Maastrichtian.
DISTRIBUTION AND FORMATION/S Southern Argentina; Cerro Fortaleza.
NOTES Initial claim of exceptional size was greatly exaggerated.

Isisaurus colberti
18 m (60 ft) TL, 15 tonnes

FOSSIL REMAINS Partial skeleton.
ANATOMICAL CHARACTERISTICS Neck moderately long. Arm and hand very long, so shoulder much higher than hips.
AGE Late Cretaceous, Maastrichtian.
DISTRIBUTION AND FORMATION/S Central India; Lameta.
NOTES Approaches brachiosaurs in its giraffe-like form. Shared its habitat with *Jainosaurus*.
Main enemy *Rajasaurus*.

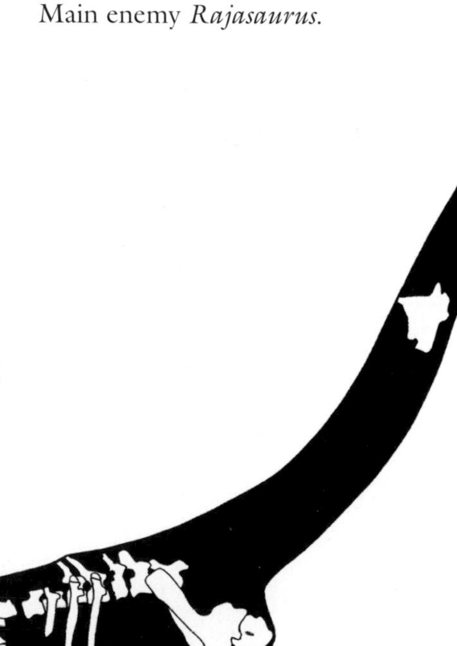

Dreadnoughtus schrani

Uberabatitan ribeiroi
Adult size uncertain

FOSSIL REMAINS Minority of juvenile skeleton.
ANATOMICAL CHARACTERISTICS Neck moderately long.
AGE Late Cretaceous, late Maastrichtian.
DISTRIBUTION AND FORMATION/S Southeast Brazil; Marilia.
NOTES Shared its habitat with *Trigonosaurus*.

Lithostrotians Large to enormous titanosaurids of the late Early Cretaceous to the end of the dinosaur era, most continents.

ANATOMICAL CHARACTERISTICS Variable. Neck short to long. Tail more flexible. Fingers and thumb claw absent. "Egg tooth" at tip of snout of hatchlings.
HABITS Dozens of spherical 0.15 m (6 in) eggs deposited in irregular, shallow nests 1 to 1.5 m (3–5 ft) across. Nests were probably covered with vegetation that generated heat through fermentation, or they were placed near geothermal heat sources; nests formed large nesting areas. Parents probably abandoned nests.
NOTES The last of the sauropods.

Malawisaurus dixeyi
16 m (50 ft) TL, 10 tonnes

FOSSIL REMAINS Minority of skull and skeleton.
ANATOMICAL CHARACTERISTICS Skull short and deep. Neck long, deep, and broad.
AGE Early Cretaceous, Aptian.
DISTRIBUTION AND FORMATION/S Malawi; unnamed formation.
NOTES Indicates that some titanosaurs retained short heads, although may be a partly juvenile feature.

Malawisaurus dixeyi

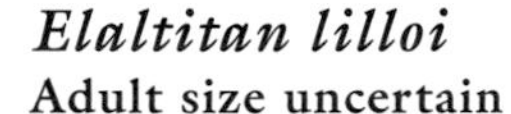

Elaltitan lilloi
Adult size uncertain

FOSSIL REMAINS Minority of skeleton.
ANATOMICAL CHARACTERISTICS Insufficient information.
AGE Late Cretaceous, late Cenomanian or Turonian.
DISTRIBUTION AND FORMATION/S Southern Argentina; lower Bajo Barreal.

Diamantinasaurus matildae
Adult size uncertain

FOSSIL REMAINS Minority of skeleton.
ANATOMICAL CHARACTERISTICS Limbs may have been unusually short.
AGE Early Cretaceous, latest Albian.
DISTRIBUTION AND FORMATION/S Northeast Australia; Winton.
HABITAT Well-watered, cold winter.
NOTES Shared its habitat with *Wintonotitan*.

Futalognkosaurus dukei
30 m (100 ft) TL, 50+ tonnes

FOSSIL REMAINS Majority of skeleton.
ANATOMICAL CHARACTERISTICS Neck long, deep.
AGE Late Cretaceous, late Turonian.
DISTRIBUTION AND FORMATION/S Western Argentina; Portezuelo.
HABITAT Well-watered woodlands with short dry season.
HABITS Probable high-level browser.
NOTES Some initial uncertainty about extreme size has been resolved, the largest dinosaur known from a majority of the skeleton, in the same size class as *Argentinosaurus*, *Puertosaurus*, *Pellegrinisaurus*, and *Ruyangosaurus*, and shows that some past titanosaur mass estimates have been excessive. Shared its habitat with *Malarguesaurus* and *Muyelensaurus*.

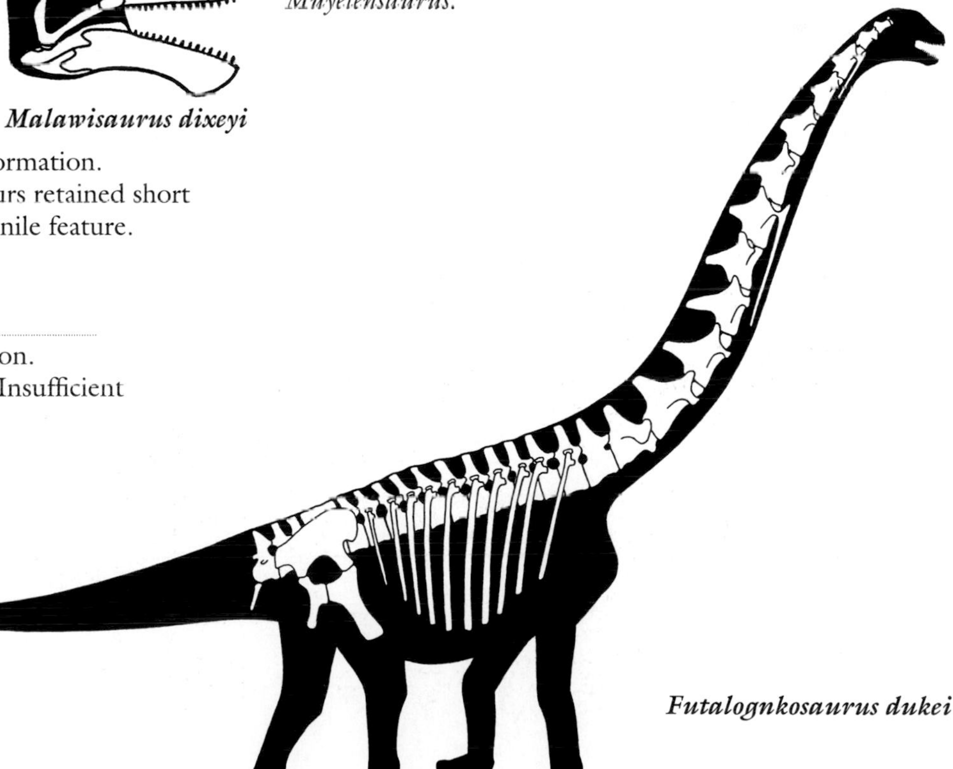

Futalognkosaurus dukei

Mendozasaurus neguyelap
20 m (65 ft) TL, 16 tonnes

FOSSIL REMAINS Minority of a few skeletons.
ANATOMICAL CHARACTERISTICS Neck fairly short. Vertebral spines very broad.
AGE Late Cretaceous, Turonian to Coniacian.
DISTRIBUTION AND FORMATION/S Western Argentina; Rio Neuquen.
HABITAT Well-watered woodlands with short dry season.
NOTES Shared its habitat with *Rinconsaurus.*

Ampelosaurus atacis
16 m (50 ft) TL, 8 tonnes

FOSSIL REMAINS Minority of a few skeletons.
ANATOMICAL CHARACTERISTICS Teeth broad, line most of length of dentary.
AGE Late Cretaceous, early Maastrichtian.
DISTRIBUTION AND FORMATION/S France; Gres de Labarre, Marnes Rouges Inferieures, Gres de Saint-Chinian.
NOTES *Ampelosaurus* shows that broad-toothed sauropods survived until the last stage of the dinosaur era.

Jiangshanosaurus lixianensis
11 m (35 ft) TL, 2.5 tonnes

FOSSIL REMAINS Minority of skeleton.
ANATOMICAL CHARACTERISTICS Insufficient information.
AGE Early Cretaceous, Albian.
DISTRIBUTION AND FORMATION/S Southeast China; Jinhua.

Jainosaurus septentrionalis
18 m (60 ft) TL, 15 tonnes

FOSSIL REMAINS Partial skeleton.
ANATOMICAL CHARACTERISTICS Insufficient information.
AGE Late Cretaceous, middle to late Maastrichtian.
DISTRIBUTION AND FORMATION/S Central India; Lameta.
NOTES It is possible that this is the same genus as *Titanosaurus indicus*, which is based on inadequate material. Shared its habitat with *Isisaurus.* Prey of *Indosuchus* and *Rajasaurus.*

Aeolosaurus rionegrinus
14 m (45 ft) TL, 6 tonnes

FOSSIL REMAINS Minority of skeleton.
ANATOMICAL CHARACTERISTICS Insufficient information.
AGE Late Cretaceous, probably Campanian or Maastrichtian.
DISTRIBUTION AND FORMATION/S Southern Argentina; Angostura Colorada.

Gondwanatitan faustoi
7 m (23 ft) TL, 1 tonne

FOSSIL REMAINS Minority of skeleton.
ANATOMICAL CHARACTERISTICS Insufficient information.
AGE Late Cretaceous, probably Campanian or Maastrichtian.
DISTRIBUTION AND FORMATION/S Southern Brazil; Adamantina.
NOTES Shared its habitat with *Adamantisaurus* and *Maxakalisaurus.*

Adamantisaurus mezzalirai
13 m (43 ft) TL, 5 tonnes

FOSSIL REMAINS Minority of skeleton.
ANATOMICAL CHARACTERISTICS Insufficient information.
AGE Late Cretaceous, probably Campanian or Maastrichtian.
DISTRIBUTION AND FORMATION/S Southern Brazil; Adamantina.

Lirainosaurus astibiae
7 m (23 ft) TL, 1 tonne

FOSSIL REMAINS Minority of several skeletons.
ANATOMICAL CHARACTERISTICS Insufficient information.
AGE Late Cretaceous, late Campanian.
DISTRIBUTION AND FORMATION/S Northern Spain; unnamed formation.

Paralititan stromeri
20+ m (65+ ft) TL, 20 tonnes

FOSSIL REMAINS Minority of skeleton.
ANATOMICAL CHARACTERISTICS Insufficient information.
AGE Late Cretaceous, Cenomanian.
DISTRIBUTION AND FORMATION/S Egypt; Bahariya.
HABITAT Coastal mangroves.
NOTES Early claims that *Paralititan* rivaled the largest titanosaurs in size were incorrect. Shared its habitat with *Aegyptosaurus.* Main enemy *Carcharodontosaurus.*

Laplatasaurus araukanicus
18 m (60 ft) TL, 14 tonnes

FOSSIL REMAINS Minority of skeletons.
ANATOMICAL CHARACTERISTICS Neck moderately long.
AGE Late Cretaceous, late Campanian.
DISTRIBUTION AND FORMATION/S Central Argentina; Allen.
HABITAT Semiarid coastline.
NOTES Shared its habitat with *Saltasaurus robustus*; *Rocasaurus muniozi* may be juvenile of one of these taxa.

Trigonosaurus pricei
Adult size uncertain

FOSSIL REMAINS Two partial skeletons.
ANATOMICAL CHARACTERISTICS Neck long.
AGE Late Cretaceous, late Maastrichtian.
DISTRIBUTION AND FORMATION/S Southeast Brazil; Marilia.
NOTES Shared its habitat with *Uberabatitan*.

Pellegrinisaurus powelli
25 m (80 ft) TL, 50 tonnes

FOSSIL REMAINS Minority of skeleton.
ANATOMICAL CHARACTERISTICS Insufficient information.
AGE Late Cretaceous, late Santonian and/or early Campanian.
DISTRIBUTION AND FORMATION/S Central Argentina; Anacleto.
HABITAT Semiarid coastline.
NOTES Shared its habitat with *Antarctosaurus*. A problematically large number of Anacleto titanosaurs have been named, and smaller examples including *Neuquensaurus*, *Barrosasaurus casamiquelai*, *Narambuenatitan palomoi*, *Pitekunsaurus macayai*, and *Overosaurus paradasorum* may be juveniles of other titanosaurs from this formation. At least one of the known species may have laid the numerous eggs found in this formation. Main enemy *Abelisaurus*.

Antarctosaurus wichmannianus
17 m (55 ft) TL, 12 tonnes

FOSSIL REMAINS Lower jaw and minority of skeleton.
ANATOMICAL CHARACTERISTICS Head probably long, shallow, front of snout broad and squared off, pencil-shaped teeth limited to front of jaws.
AGE Late Cretaceous, late Santonian or early Campanian.
DISTRIBUTION AND FORMATION/S Western Argentina; Anacleto.
HABITS Jaws adapted to browse swaths of plant material, perhaps at ground level.

Alamosaurus sanjuanensis
20 m (65 ft) TL, 16 tonnes

FOSSIL REMAINS Partial skeletons.
ANATOMICAL CHARACTERISTICS Neck long.
AGE Late Cretaceous, Maastrichtian.
DISTRIBUTION AND FORMATION/S New Mexico, Utah, Texas; lower and upper Kirtland, North Horn, Javelina, El Picacho, Black Peaks.
HABITAT Seasonally dry open woodlands.
HABITS High-level browser.
NOTES Largest known specimens may not be fully mature. The last known of North American sauropods, *Alamosaurus* may represent a reinhabitation of the continent by sauropods from South America or Asia after a hiatus.

Alamosaurus sanjuanensis

Rapetosaurus krausei
Adult size uncertain

FOSSIL REMAINS Majority of skulls and a skeleton, large juvenile.
ANATOMICAL CHARACTERISTICS Head long, shallow, bony nostrils strongly retracted to above the orbits but fleshy nostrils probably still near front of snout, which is broad and rounded, lower jaws short, pencil-shaped teeth limited to front of jaws, head flexed downward relative to neck. Neck long.
AGE Late Cretaceous, Campanian.
DISTRIBUTION AND FORMATION/S Madagascar; Maevarano.
HABITAT Seasonally dry floodplain with coastal swamps and marshes.
HABITS High-level browser.
NOTES Main enemy *Majungasaurus.* Herbivorous ornithischians apparently absent from habitat.

Unnamed genus *giganteus*
30+ m (100+ ft) TL, 80+ tonnes

FOSSIL REMAINS Minority of skeleton.
ANATOMICAL CHARACTERISTICS Arm very long, so shoulders high. Limbs elongated.
AGE Late Cretaceous, Turonian or Coniacian.
DISTRIBUTION AND FORMATION/S Western Argentina; Rio Neuquen.
HABITAT Well-watered woodlands with short dry season.
NOTES Originally placed in *Antarctosaurus.* In same size class as *Argentinosaurus, Puertosaurus, Futalognkosaurus, Pellegrinisaurus,* and *Ruyangosaurus.*

Bonitasaura salgadoi
10 m (33 ft) TL, 5 tonnes

FOSSIL REMAINS Minority of skull and skeleton.
ANATOMICAL CHARACTERISTICS Head probably long, shallow, front of snout broad and squared off, pencil-shaped teeth limited to front of jaws; behind lower teeth a short, cutting beak appears to be present. Neck moderately long.
AGE Late Cretaceous, Santonian.
DISTRIBUTION AND FORMATION/S Central Argentina; Bajo de la Carpa.
HABITS Predominantly grazed ground cover, also able to rear to high browse. Appears to have complemented the cropping ability of its front teeth with a supplementary beak immediately behind.
NOTES Apparently the only beaked sauropod yet known. Other sauropods known to have had similarly broad, square, ground-grazing beaks were rebbachisaurid diplodocoids like *Nigersaurus.*

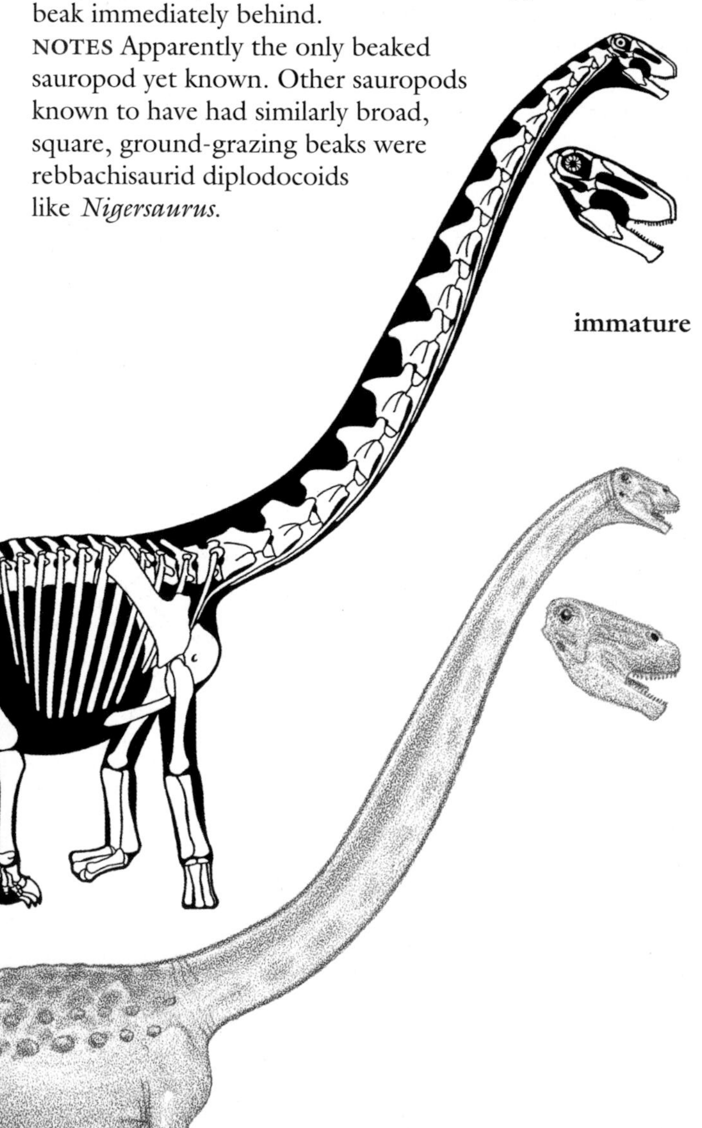

Rapetosaurus krausei

Dongyangosaurus sinensis
15 m (50 ft) TL, 7 tonnes

FOSSIL REMAINS Minority of skeleton.
ANATOMICAL CHARACTERISTICS Insufficient information.
AGE Early Late Cretaceous.
DISTRIBUTION AND FORMATION/S Eastern China; Fangyan.

Tapuiasaurus macedoi
Adult size uncertain

FOSSIL REMAINS Nearly complete skull and minority of skeleton.
ANATOMICAL CHARACTERISTICS Head long, shallow, bony nostrils strongly retracted to above the orbits but fleshy nostrils probably still near front of snout, which is broad and rounded, lower jaws short, pencil-shaped teeth limited to front of jaws.
AGE Early Cretaceous, Aptian.
DISTRIBUTION AND FORMATION/S Southeast Brazil; Quirico.

Tapuiasaurus madedoi

Nemegtosaurus (= Quaesitosaurus) orientalis
Size uncertain

FOSSIL REMAINS Partial skull.
ANATOMICAL CHARACTERISTICS Head long, shallow, bony nostrils strongly retracted to above the orbits but fleshy nostrils probably still near front of snout, which is broad and rounded, lower jaws short, pencil-shaped teeth limited to front of jaws, head flexed downward relative to neck.
AGE Late Cretaceous, late Campanian or early Maastrichtian.
DISTRIBUTION AND FORMATION/S Mongolia; Barun Goyot or Nemegt equivalent.

Nemegtosaurus (= Quaesitosaurus) orientalis

Nemegtosaurus mongoliensis (= Opisthocoelocaudia skarzynskii)
13+ m (43+ ft) TL, 8.5 tonnes

FOSSIL REMAINS A nearly complete skull and the majority of a skeleton.
ANATOMICAL CHARACTERISTICS Head long, shallow, bony nostrils strongly retracted to above the orbits but fleshy nostrils probably still near front of snout, which is broad and rounded, lower jaws short; pencil-shaped teeth limited to front of jaws, head flexed downward relative to neck. Skeleton massively constructed.
AGE Late Cretaceous, late Campanian and/or early Maastrichtian.
DISTRIBUTION AND FORMATION/S Mongolia; Nemegt.
HABITAT Well-watered woodland with seasonal rains, winters cold.
NOTES *Nemegtosaurus* and *Opisthocoelocaudia* are often considered entirely different sauropods, but the two are known only from a skull and from a skeleton, respectively, and no other titanosaurs are known from the Nemegt, so they are probably the same dinosaur. Main enemy *T. bataar*.

Maxakalisaurus topai
13 m (45 ft) TL, 5 tonnes

FOSSIL REMAINS Minority of skeleton.
ANATOMICAL CHARACTERISTICS Neck moderately long.
AGE Late Cretaceous, probably Campanian or Maastrichtian.
DISTRIBUTION AND FORMATION/S Southern Brazil; Adamantina.
NOTES Shared its habitat with *Adamantisaurus* and *Gondwanatitan*.

Nemegtosaurus mongoliensis (= Opisthocoelocaudia skarzynskii)

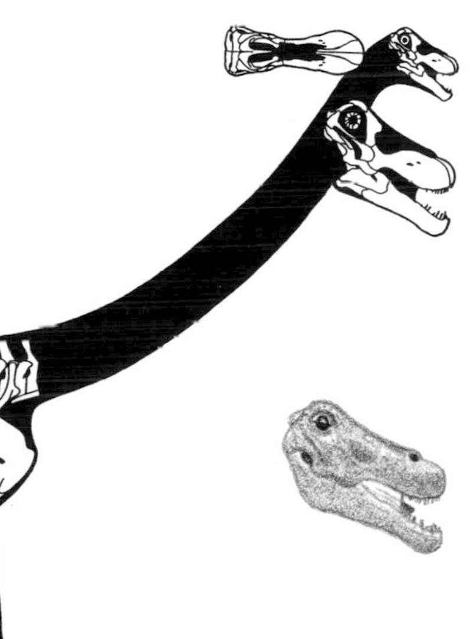

Saltasaurs Medium-sized lithostrotians of the Late Cretaceous of Eurasia and South America.

ANATOMICAL CHARACTERISTICS Uniform. Neck short by sauropod standards.

Magyarosaurus dacus or *Paludititan nalatzensis*

6 m (20 ft), 1 tonne

FOSSIL REMAINS A dozen partial skeletons.
ANATOMICAL CHARACTERISTICS Standard for group.
AGE Late Cretaceous, late Maastrichtian.
DISTRIBUTION AND FORMATION/S Romania; Sanpetru.
HABITAT Forested island.
NOTES Whether the remains represent one or more taxa, and which has priority, await further research. Small size of most individuals implies island dwarfism, but some researchers cite larger sauropod specimens and higher estimate of size of island as contrary evidence. Shared its habitat with *Struthiosaurus*, *Rhabdodon robustus*, and *Telmatosaurus*.

Saltasaurus (= *Neuquensaurus*) *australis*

7.5 m (24 ft) TL, 1.8 tonnes

FOSSIL REMAINS Partial skeletons.
ANATOMICAL CHARACTERISTICS Standard for group.
AGE Late Cretaceous, early Campanian.
DISTRIBUTION AND FORMATION/S Western Argentina; upper Anacleto.
NOTES May include *Microcoelus patagonicus* juveniles. Shared its habitat with *Pelligrinisaurus* and *Antarctosaurus*. Main enemy *Abelisaurus*.

Saltasaurus robustus

8 m (25 ft) TL, 2 tonnes

FOSSIL REMAINS A few partial skeletons.
ANATOMICAL CHARACTERISTICS Standard for group.
AGE Late Cretaceous, late Campanian.
DISTRIBUTION AND FORMATION/S Central Argentina; Allen.
NOTES Shared its habitat with *Rocasaurus* and *Laplatasaurus*.

Saltasaurus loricatus

8.5 m (27 ft) TL, 2.5 tonnes

FOSSIL REMAINS Minority of skull and half a dozen partial skeletons.
ANATOMICAL CHARACTERISTICS Standard for group.
AGE Late Cretaceous, probably early Maastrichtian.
DISTRIBUTION AND FORMATION/S Northern Argentina; Lecho.
NOTES May be direct descendant of *S. robustus*.

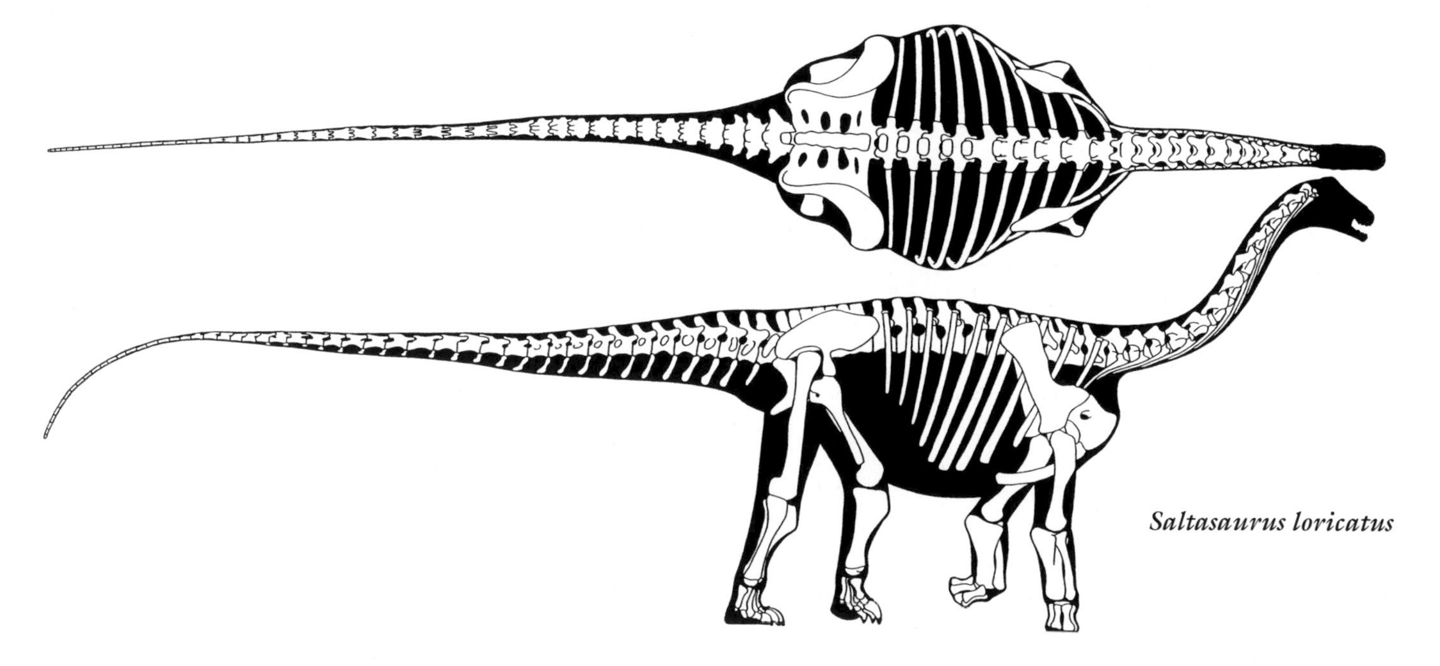

Saltasaurus loricatus

SMALL TO GIGANTIC HERBIVOROUS DINOSAURS FROM THE LATE TRIASSIC TO THE END OF THE DINOSAUR ERA, ALL CONTINENTS.

ANATOMICAL CHARACTERISTICS **Extremely variable. Head size and shape variable, skull heavily built, beaks at front of jaws, anchored on toothless predentary bone on lower jaw, vertical coranoid projection at back end of lower tooth row increased leverage of jaw muscles, rows of leaf-shaped teeth covered by presumably elastic cheeks, jaw gap limited. Neck not long. Trunk stiff. Tail short to moderately long. Bipedal to quadrupedal. Arm short to long, usually five fingers, sometimes four or three. Pubis strongly retroverted to accommodate large belly, pelvis large, ilium shallow. Usually four toes, sometimes three. Skeletons not pneumatic, birdlike respiratory system not present. Brains reptilian in size and form.**
HABITAT **Very variable, from sea level to highlands, from tropics to polar winters, from arid to wet.**
HABITS **Predominantly herbivorous browsers and grazers, although smaller examples may have been prone to pick up and consume small animals, and others may have scavenged; extensively chewed food before swallowing. Defense ranged from passive armor to running to aggressive combat. Smaller species potentially or actually able to burrow.**

BASO-ORNITHISCHIANS

SMALL ORNITHISCHIANS LIMITED TO THE LATE TRIASSIC AND EARLY JURASSIC OF SOUTH AMERICA AND AFRICA.

ANATOMICAL CHARACTERISTICS Head modest sized, subtriangular, beaks narrow and not hooked, teeth at front of upper jaw, main tooth rows not deeply inset, large eyes shaded by overhanging rim. Tail moderately long, stiffened by ossified tendons. Bipedal except could move quadrupedally at slow speeds. Arm fairly short, hand small, five grasping fingers tipped with small blunt claws. Leg long, flexed, and gracile so speed potential high, four long toes tipped with blunt claws.
HABITS Low-level browsers, probably picked up insects and small vertebrates. Predominantly terrestrial, probably some climbing ability. Main defense high speed.
NOTES Closest living analogs small kangaroos, deer, and antelope. The relationships of these generalized ornithischians are uncertain, ultimately splittable into a number of divisions; some may belong to other groups. Absence from other continents may be due to lack of sufficient sampling.

Pisanosaurus mertii

1.3 m (4.5 ft) TL, 2 kg (5 lb)

FOSSIL REMAINS Small minority of skull and skeleton.
ANATOMICAL CHARACTERISTICS Standard for group.
AGE Late Triassic, Carnian.
DISTRIBUTION AND FORMATION/S Northern Argentina; Ischigualasto.
HABITAT Seasonally well-watered forests, including dense stands of giant conifers.
NOTES The earliest known ornithisichian. Shared its habitat with *Panphagia*; prey of *Eoraptor* and *Herrerasaurus*.

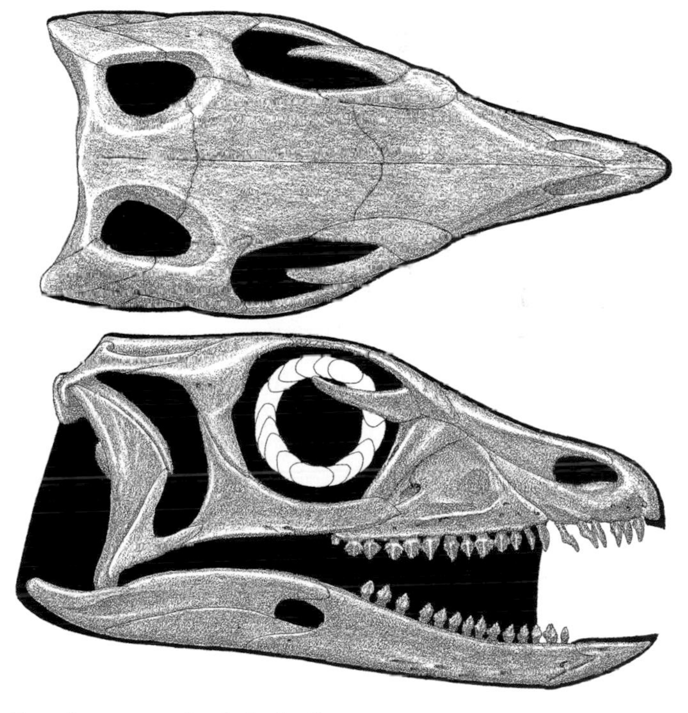

***Lesothosaurus* shaded skull**

Eocursor parvus
1+ m (3.5+ ft) TL, 1 kg (2 lb)

FOSSIL REMAINS Partial skull and skeleton, large juvenile.
ANATOMICAL CHARACTERISTICS Standard for group.
AGE Late Triassic, early Norian.
DISTRIBUTION AND FORMATION/S Southeast Africa; Lower Elliot.
HABITAT Arid.

Laquintasaura venezuelae
1 m (3 ft) TL, 1 kg (2 lb)

FOSSIL REMAINS Bone bed.
ANATOMICAL CHARACTERISTICS Standard for group.
AGE Early Jurassic, early Hettangian.
DISTRIBUTION AND FORMATION/S Venezuela; La Quinta.

Lesothosaurus diagnosticus
2 m (6.5 ft) TL, 6 kg (12 lb)

FOSSIL REMAINS Majority of some skulls and skeletons, juvenile to adult.
ANATOMICAL CHARACTERISTICS Standard for group.
AGE Early Jurassic, late Hettangian or Sinemurian.
DISTRIBUTION AND FORMATION/S Southeast Africa; Upper Elliot.
HABITAT Arid.
NOTES Originally known as *Fabrosaurus australis* based on inadequate remains; *Stormbergia dangershoeki* probably the adult of this species. May be a genasaur, possibly thyreophoran. Shared its habitat with *Heterodontosaurus*. Prey of *Dracovenator*.

GENASAURS

SMALL TO GIGANTIC ORNITHISCHIANS FROM THE EARLY JURASSIC TO THE END OF THE DINOSAUR ERA, ALL CONTINENTS.

ANATOMICAL CHARACTERISTICS Main tooth rows deeply inset, enlarging capacity of cheek spaces.
HABITAT Very variable, from sea level to highlands, from tropics to polar winters, from arid to wet.

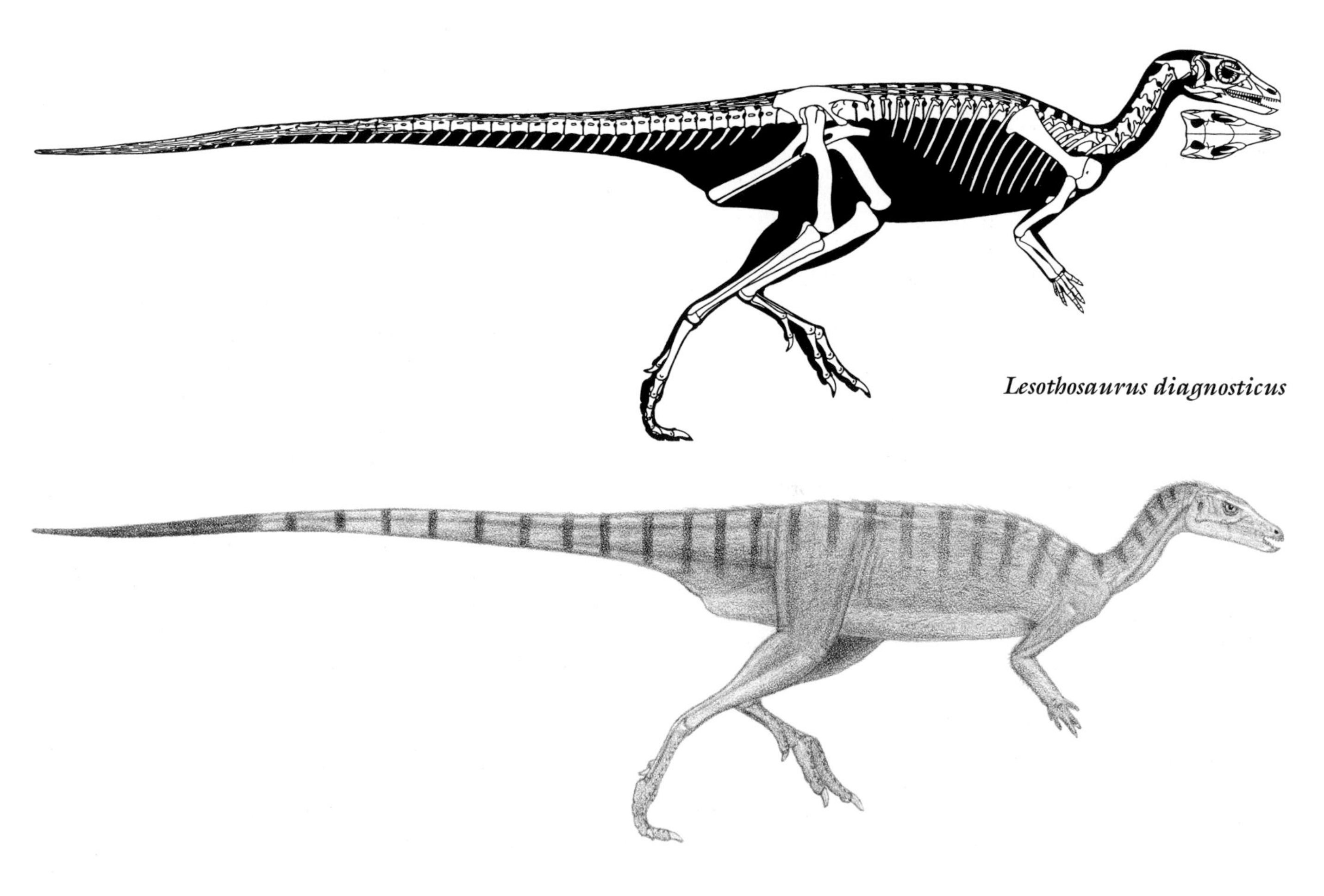

Lesothosaurus diagnosticus

Lesothosaurus diagnosticus

THYREOPHORANS

SMALL TO VERY LARGE ARMORED GENASAUR ORNITHISCHIANS FROM THE EARLY JURASSIC TO THE END OF THE DINOSAUR ERA, ALL CONTINENTS.

ANATOMICAL CHARACTERISTICS Variable. Head not large, solidly constructed, eyes not large. Skeleton heavily built. Tail moderately long. Bipedal to fully quadrupedal, able to rear on hind legs. Arm short to long, five fingers. Four to three toes. Substantial armor always present, dense pavement of ossicles under throat in at least some examples.
ONTOGENY Growth apparently slower than in most dinosaurs.
HABITAT Very variable, from deserts to well-watered forests.
HABITS Low-level browsers and grazers. Generally not fast moving. Main defense passive armor, some may have used armor spines and clubs as weapons.
NOTES *Lesothosaurus* may be at base of group. The only known armored ornithischians, among saurischians armor paralleled only by titanosaurs.

Scelidosaurs

SMALL TO MEDIUM-SIZED THYREOPHORANS LIMITED TO THE EARLY JURASSIC OF EUROPE AND AFRICA.

ANATOMICAL CHARACTERISTICS Head not large, solidly constructed, beaks narrow, eyes not large, teeth at front of upper jaw. Belly and hips moderately broad. Tail long. Bipedal to fully quadrupedal, arm and leg flexed so could run. Arm short to long, five fingers tipped with blunt claws. Four long toes tipped with blunt claws. Armor substantial but simple, generally scutes set in long rows, including top of vertebral series and bottom of tail.
HABITS Low-level browsers and grazers. Generally not fast moving. Main defense passive armor, some may have used armor spines and clubs as weapons.
NOTES Absence from other continents probably reflects lack of sufficient sampling. May be splittable into a larger number of divisions or subdivisions.

Scutellosaurus lawleri
1.3 m (4.2 ft) TL, 3 kg (7 lb)

FOSSIL REMAINS Small portion of skull and majority of two skeletons with loose armor.

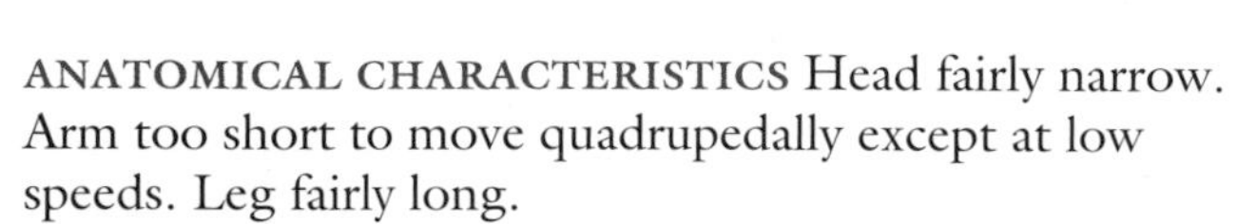

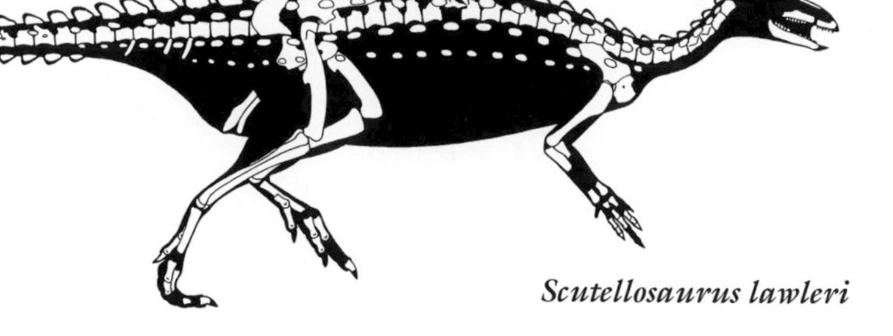

Scutellosaurus lawleri

ANATOMICAL CHARACTERISTICS Head fairly narrow. Arm too short to move quadrupedally except at low speeds. Leg fairly long.
AGE Early Jurassic, Sinemurian or Pliensbachian.
DISTRIBUTION AND FORMATION/S Arizona; middle Kayenta.
HABITAT Semiarid.
HABITS Defense included running.
NOTES Only certain thyreophoran known to be strongly bipedal and have a very long tail; distribution of armor uncertain. Prey of *Coelophysis kayentakatae*.

Emausaurus ernsti
2.5 m (8 ft) TL, 50 kg (100 lb)

FOSSIL REMAINS Majority of skull and minority of skeleton.
ANATOMICAL CHARACTERISTICS Head broad.
AGE Early Jurassic, Toarcian.
DISTRIBUTION AND FORMATION/S Germany; unnamed formation.

Scelidosaurus harrisonii
3.8 m (12 ft) TL, 270 kg (600 lb)

FOSSIL REMAINS Two complete skulls and a few skeletons, juvenile to adult, some with armor in place.
ANATOMICAL CHARACTERISTICS Head fairly narrow. Trunk and hips moderately broad. Arm long, so fully quadrupedal. Armor well developed, triple-pronged piece immediately behind head.
AGE Early Jurassic, late Sinemurian.
DISTRIBUTION AND FORMATION/S England; Lower Lias.
NOTES Skeleton restored here is the first complete dinosaur fossil. Some consider this the earliest basal ankylosaur. Prey of *Sarcosaurus*.

Emausaurus ernsti

Emausaurus ernsti

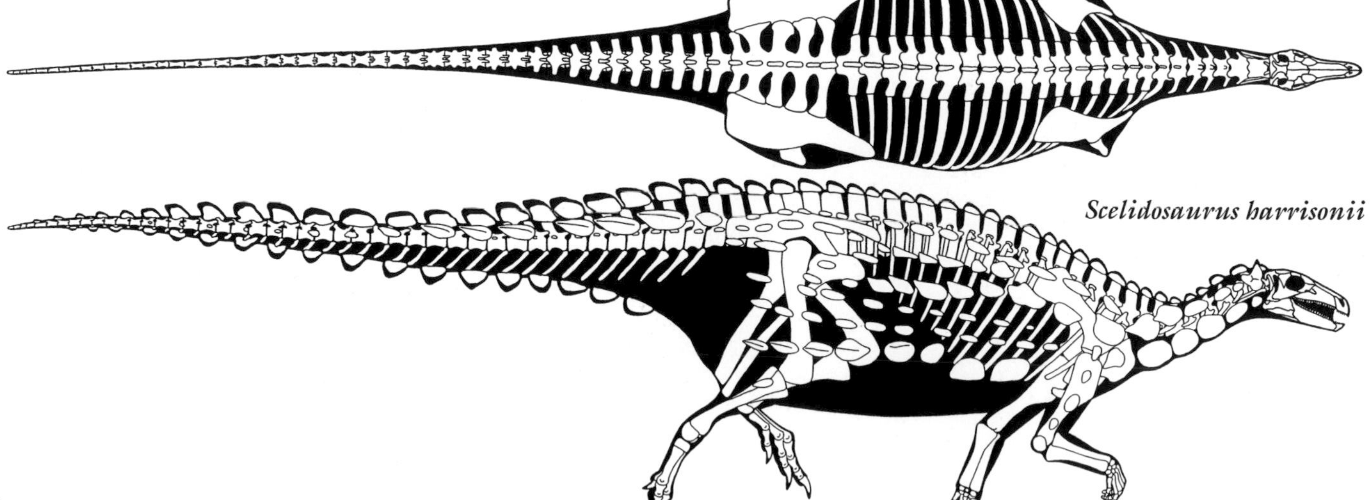

Scelidosaurus harrisonii

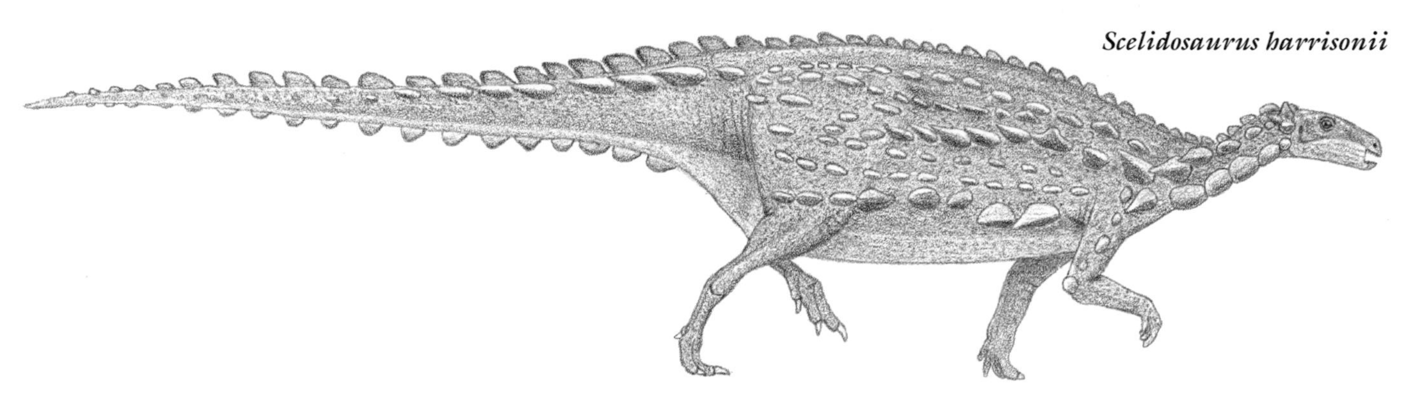
Scelidosaurus harrisonii

EURYPODS

MEDIUM-SIZED TO VERY LARGE THYREOPHORANS FROM THE MIDDLE JURASSIC TO THE LATE CRETACEOUS OF NORTH AMERICA, EURASIA, AND AFRICA.

ANATOMICAL CHARACTERISTICS Teeth small. Hand, fingers, foot, and toes short, limiting speed, fingers and toes tipped with hooves.
ENERGETICS Energy levels and food consumption probably low for dinosaurs.

STEGOSAURS

MEDIUM-SIZED TO VERY LARGE THYREOPHORANS FROM THE MIDDLE JURASSIC TO THE EARLY CRETACEOUS, MOST CONTINENTS.

ANATOMICAL CHARACTERISTICS Beaks narrow. Neck U-curved. Tail moderately long. Largely quadrupedal. Foot and three toes short, limiting speed. Armor predominantly parallel rows of tall plates and spikes running atop vertebral column.
HABITAT Semiarid to well-watered forests.
HABITS Low- to medium-level browsers. Main defense swinging tail to puncture flanks of theropods with spine arrays. In addition to protection, plates and spines also for display, possibly thermoregulation.
NOTES Absence from some other continents may reflect insufficient sampling.

HUAYANGOSAURIDS Medium-sized stegosaurs limited to the Middle Jurassic of Asia.

ANATOMICAL CHARACTERISTICS Head fairly deep, broad, teeth at front of upper jaw. Belly and hips moderately broad. Arm moderately long, so shoulders as high as hips. Arm and leg flexed, so able to run.

HABITS Low-level browsers. Defense included running while swinging tail.

Huayangosaurus taibaii
4 m (13 ft) TL, 500 kg (1,000 lb)

FOSSIL REMAINS Complete skull and skeleton and partial skeletons.
AGE Late Jurassic, Bathonian and/or Callovian.
ANATOMICAL CHARACTERISTICS Long spine on shoulder, small club at tip of tail.

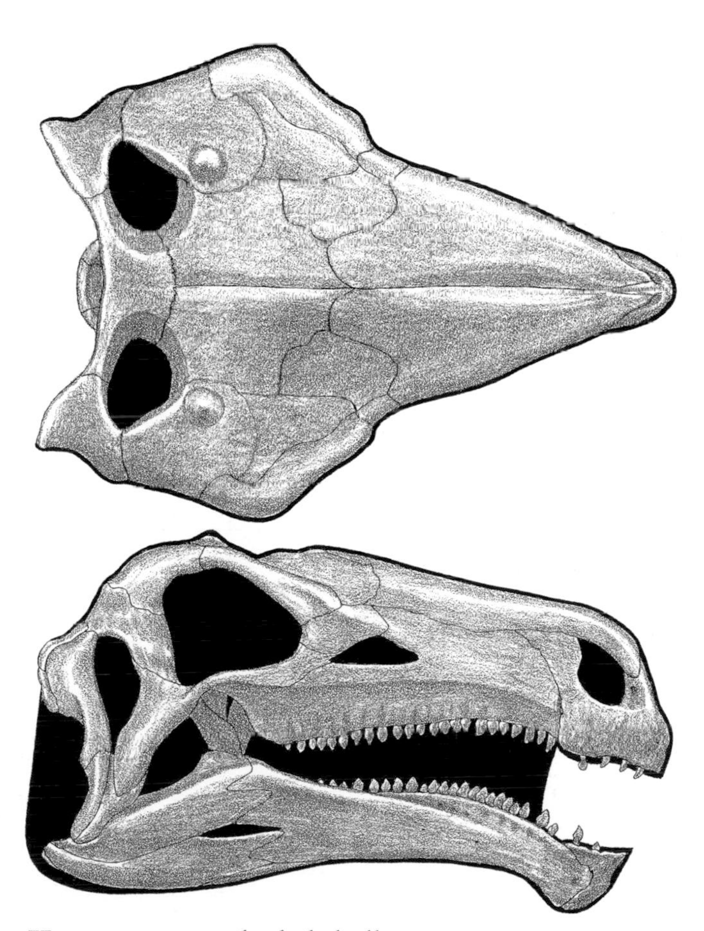
Huayangosaurus shaded skull

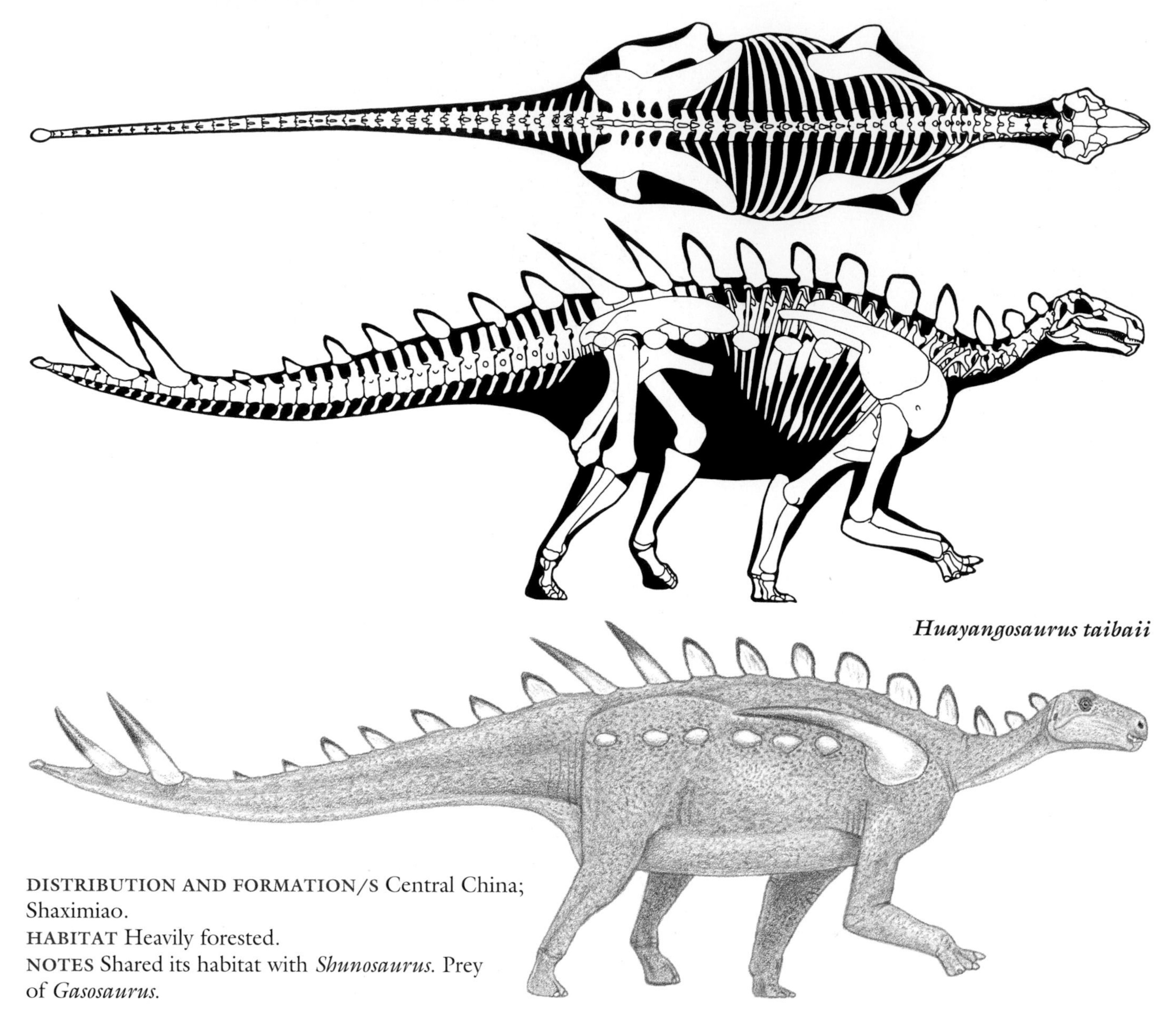

Huayangosaurus taibaii

DISTRIBUTION AND FORMATION/S Central China; Shaximiao.
HABITAT Heavily forested.
NOTES Shared its habitat with *Shunosaurus*. Prey of *Gasosaurus*.

STEGOSAURIDS Large stegosaurs limited to the Late Jurassic and Early Cretaceous, most continents.

ANATOMICAL CHARACTERISTICS Fairly uniform. Head small, slender, no teeth at front of upper jaw, teeth smaller. Neck slender. Trunk vertebral series downcurved, and arm fairly short, so shoulders lower than hips. Tail held high above ground. Arm and leg columnar and massively built, so not able to achieve a full run faster than elephants. Short arms, large hips, and stout tails with sled-shaped chevrons facilitated static rearing posture.
HABITS Medium- to high-level browsers. Too slow to flee attackers, so spun around to keep spiny tail toward enemies.
NOTES Late Cretaceous *Dravidosaurus blanfordi* is probably a plesiosaur rather than a stegosaur.

***Stegosaurus* shaded skull**

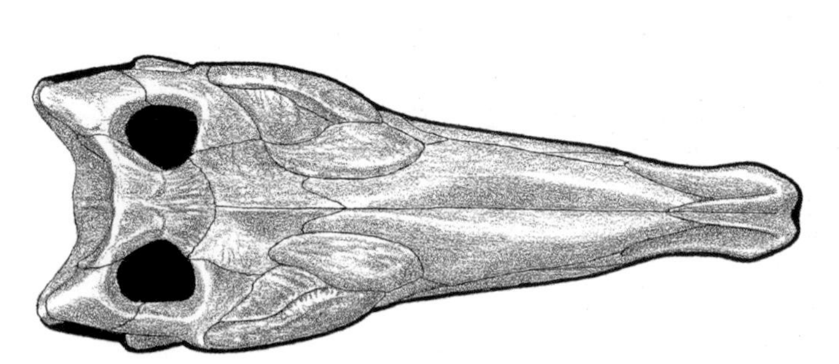

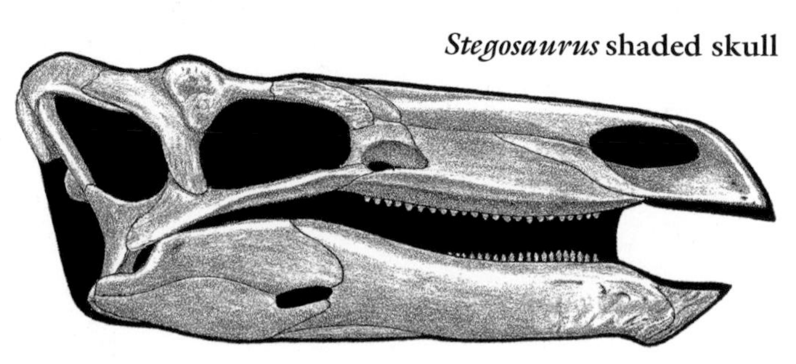

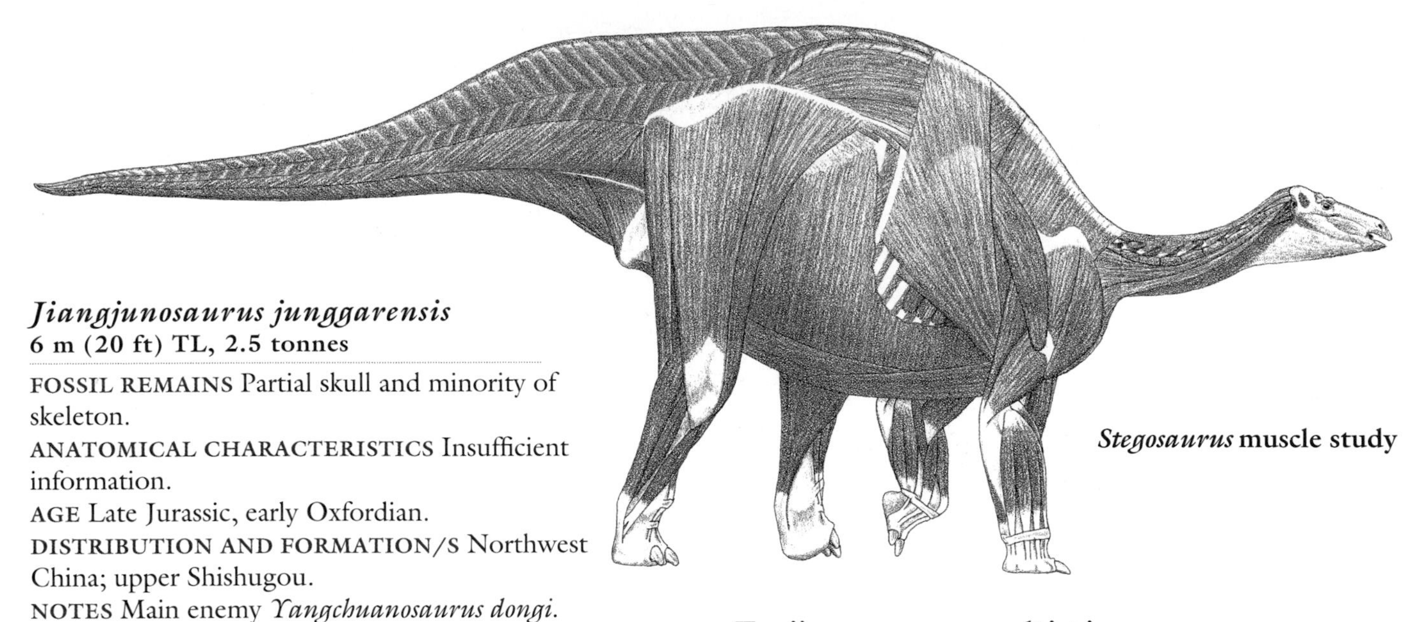

Stegosaurus muscle study

Jiangjunosaurus junggarensis
6 m (20 ft) TL, 2.5 tonnes

FOSSIL REMAINS Partial skull and minority of skeleton.
ANATOMICAL CHARACTERISTICS Insufficient information.
AGE Late Jurassic, early Oxfordian.
DISTRIBUTION AND FORMATION/S Northwest China; upper Shishugou.
NOTES Main enemy *Yangchuanosaurus dongi.*

Lexovisaurus durobrivensis
6 m (20 ft) TL, 2 tonnes

FOSSIL REMAINS A few partial skeletons.
ANATOMICAL CHARACTERISTICS Belly and hip broad. Limbs fairly short. Long spine on shoulder. Main armor intermediate between plates and spines in shape.
AGE Middle Jurassic, Callovian.
DISTRIBUTION AND FORMATION/S Eastern England; Lower Oxford Clay.
NOTES Shared its habitat with *Cetiosauriscus.*

Tuojiangosaurus multispinus
6.5 m (22 ft) TL, 2.8 tonnes

FOSSIL REMAINS Minority of skulls and majority of a few skeletons, juvenile to adult.
ANATOMICAL CHARACTERISTICS Head shallow. Belly and hip broad. Limbs fairly short. Front armor medium sized plates, middle and tail plates taller, sharp tipped. Three pairs of terminal tail spines with first two erect, last pair directed backward forming pincushion array.
AGE Late Jurassic, probably Oxfordian.

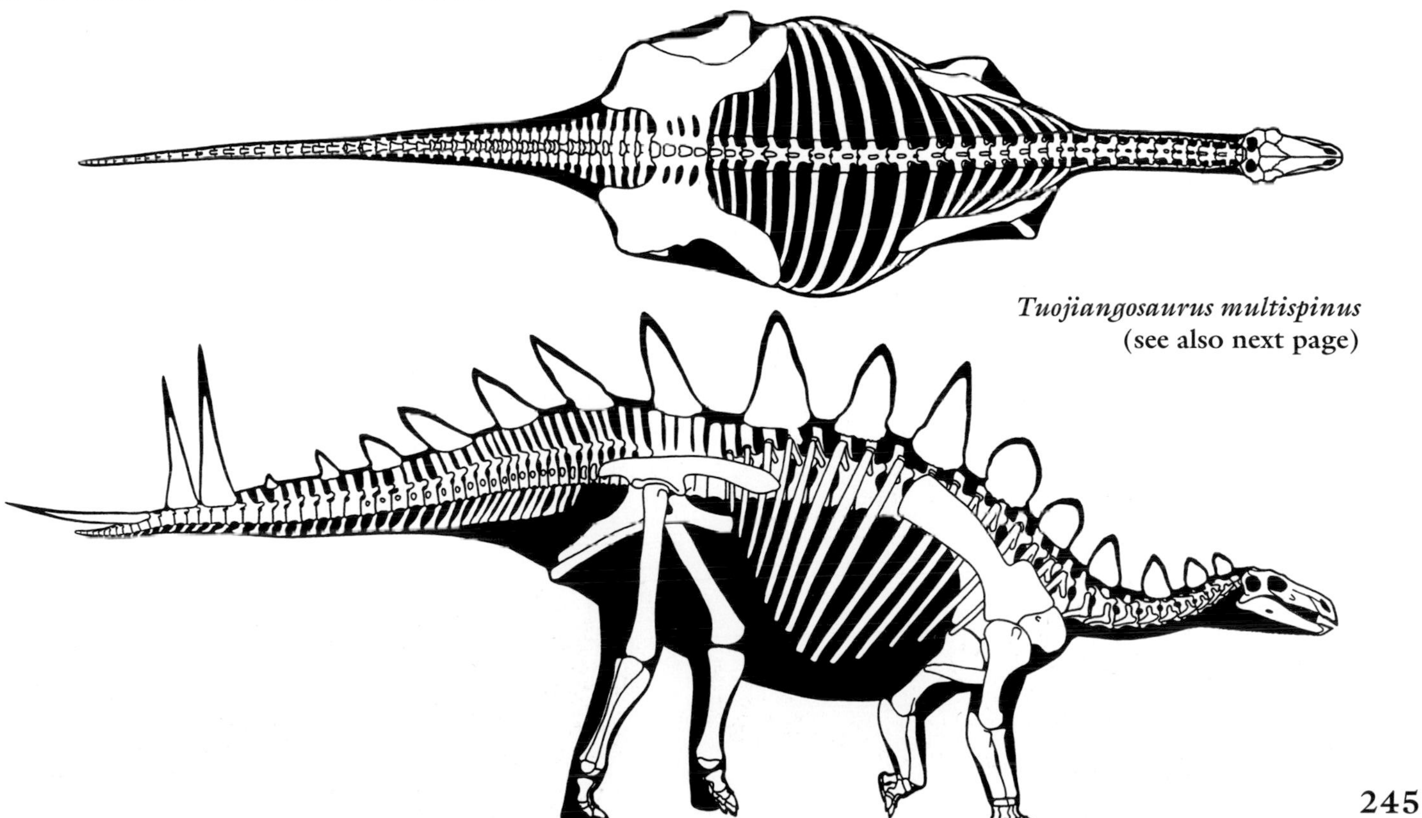

Tuojiangosaurus multispinus
(see also next page)

Tuojiangosaurus multispinus

DISTRIBUTION AND FORMATION/S Central China; Shangshaximiao.
HABITAT Heavily forested.
NOTES *Chungkingosaurus jiangbeiensis* and *Chialingosaurus kuani* are probably juveniles of this species. Shared its habitat with *Gigantspinosaurus*. Main enemy *Yangchuanosaurus shangyuensis*.

Gigantspinosaurus sichuanensis
4.2 m (14 ft) TL, 700 kg (1,500 lb)

FOSSIL REMAINS Minority of skull and majority of skeleton.
ANATOMICAL CHARACTERISTICS Belly and hip very broad. Limbs short. Top armor includes small plates and spikes, arrangement of those of tail uncertain, exact orientation of enormous shoulder spike uncertain.
AGE Late Jurassic, probably Oxfordian.
DISTRIBUTION AND FORMATION/S Central China; Shangshaximiao.
HABITAT Heavily forested.

Kentrosaurus aethiopicus
4 m (13 ft) TL, 700 kg (1,500 lb)

FOSSIL REMAINS A large number of partial skeletons and bones, juvenile to adult.
ANATOMICAL CHARACTERISTICS Belly and hip broad. Limbs fairly short. Front armor medium-sized plates, transitioning into long tail spines, long spine on shoulder.
AGE Late Jurassic, late Kimmeridgian and/or early Tithonian.
DISTRIBUTION AND FORMATION/S Tanzania; middle Tendaguru.

Gigantspinosaurus sichuanensis

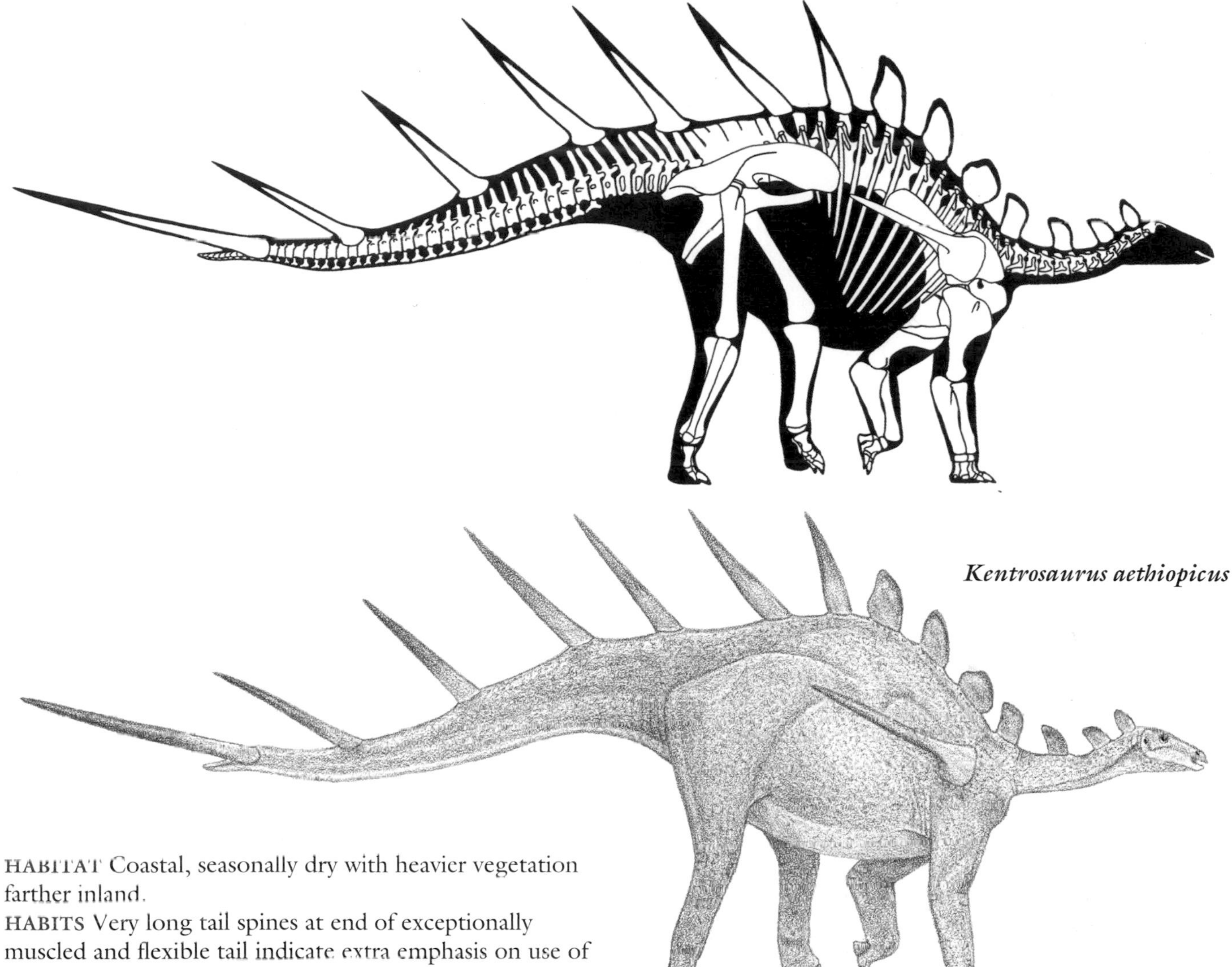

Kentrosaurus aethiopicus

HABITAT Coastal, seasonally dry with heavier vegetation farther inland.
HABITS Very long tail spines at end of exceptionally muscled and flexible tail indicate extra emphasis on use of spikes for defense and display.
NOTES Remains from the upper Tendaguru placed in this species probably belong to a different taxon. Shoulder spine was long thought to be on hip. Shared its habitat with *Giraffatitan* and *Dryosaurus lettowvorbecki*.

Paranthodon africanus

Size uncertain

FOSSIL REMAINS Minority of skull.
ANATOMICAL CHARACTERISTICS Head shallow.
AGE Late Jurassic or Early Cretaceous.
DISTRIBUTION AND FORMATION/S Southern South Africa; Upper Kirkwood.

Dacentrurus armatus

8 m (25 ft) TL, 5 tonnes

FOSSIL REMAINS Partial skeletons.
ANATOMICAL CHARACTERISTICS Belly and hip very broad. Limbs fairly short.
AGE Late Jurassic, Kimmeridgian, possibly Tithonian.
DISTRIBUTION AND FORMATION/S England, possibly other parts of western Europe; Kimmeridge Clay, possibly a number of other formations.
NOTES Whether all the specimens assigned to this species actually belong is uncertain. Shared its habitat with *Camptosaurus prestwichii*.

Miragaia longicollum

6.5 m (21 ft) TL, 2 tonnes

FOSSIL REMAINS Partial skull and skeleton.
ANATOMICAL CHARACTERISTICS Head long and low. Neck longer than body. Main armor includes small plates.
AGE Late Jurassic, late Kimmeridgian or early Tithonian.
DISTRIBUTION AND FORMATION/S Portugal; Lourinha.
HABITAT Large, seasonally dry island with open woodlands.
HABITS Well adapted for high browsing; long neck may also have been used for display.

Wuerhosaurus homheni
7 m (23 ft) TL, 4 tonnes

FOSSIL REMAINS Partial skeleton.
ANATOMICAL CHARACTERISTICS Belly and hip very broad. Limbs fairly short. Plates long and low.
AGE Early Cretaceous.
DISTRIBUTION AND FORMATION/S Northwest China; Lianmuging.
NOTES Not a species of *Stegosaurus*, as has been suggested.

Wuerhosaurus? ordosensis
5 m (16 ft) TL, 1.2 tonnes

FOSSIL REMAINS Partial skeleton.
ANATOMICAL CHARACTERISTICS Neck probably elongated, trunk short. Belly and hip very broad.
AGE Early Cretaceous.
DISTRIBUTION AND FORMATION/S Northern China; Ejinhoro.
NOTES Whether this is the same taxon as *W. homheni* or a different genus is uncertain.

Hesperosaurus mjosi
6.5 m (21 ft) TL, 3.5 tonnes

FOSSIL REMAINS Complete skull and majority of skeleton.
ANATOMICAL CHARACTERISTICS Head shallow. Trunk short and deep, belly and hips narrow and slab sided. Leg long. Alternating armor plates large over trunk and tail, two pairs of spines form subhorizontal pincushion array.
AGE Late Jurassic, late Oxfordian.
DISTRIBUTION AND FORMATION/S Wyoming; lowest Morrison.
HABITAT Short wet season, otherwise semiarid with open floodplain prairies and riverine forests.
NOTES Although very similar in overall form to *Stegosaurus*, many differing details of skull anatomy require generic distinction. Sexual differences may be reflected in differing plate size and sharpness of tips.

Stegosaurus stenops
6.5 m (21 ft) TL, 3.5 tonnes

FOSSIL REMAINS Two complete skulls and several skeletons, completely known.
ANATOMICAL CHARACTERISTICS Head shallow. Trunk short and deep, belly and hips narrow and slab sided. Leg long. Alternating armor plates very large over trunk and tail, two pairs of spines form subhorizontal pincushion array at end of S-curved tail tip.
AGE Late Jurassic, late Oxfordian to middle or late Kimmeridgian.
DISTRIBUTION AND FORMATION/S Colorado; lower and perhaps middle Morrison.
HABITAT Short wet season, otherwise semiarid with open floodplain prairies and riverine forests.
HABITS Very well adapted for rearing, may have been able to walk slowly bipedally. Broken and healed spines and wound in *Allosaurus* tail show it used tail spine array as a weapon.
NOTES Shared its habitat with diplodocids and camarasaurs. Main enemy *Allosaurus*.

Stegosaurus ungulatus
7 m (23 ft) TL, 3.8 tonnes

FOSSIL REMAINS Two partial skeletons.
ANATOMICAL CHARACTERISTICS Trunk short and

Hesperosaurus mjosi

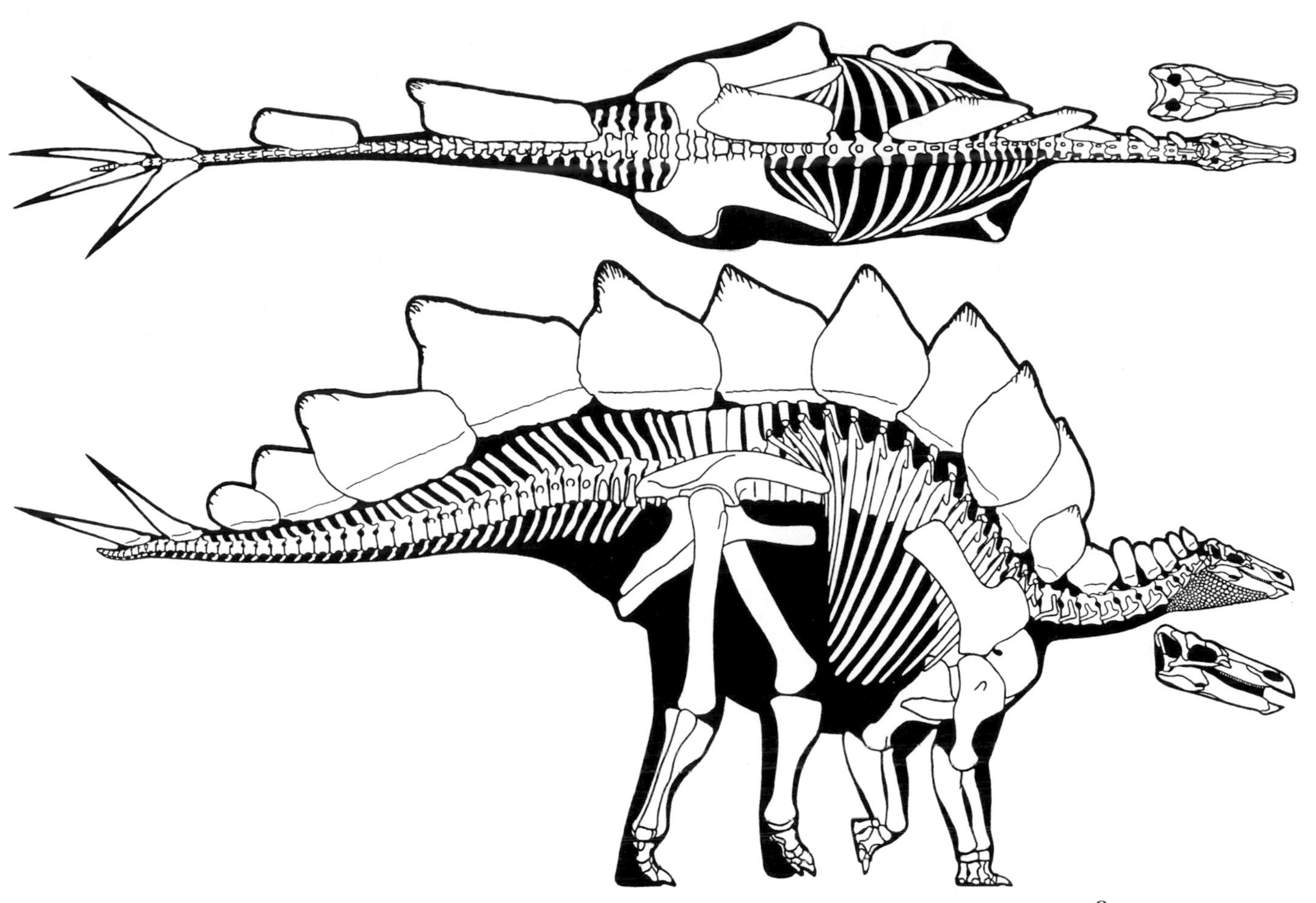

Stegosaurus stenops
(see also next page)

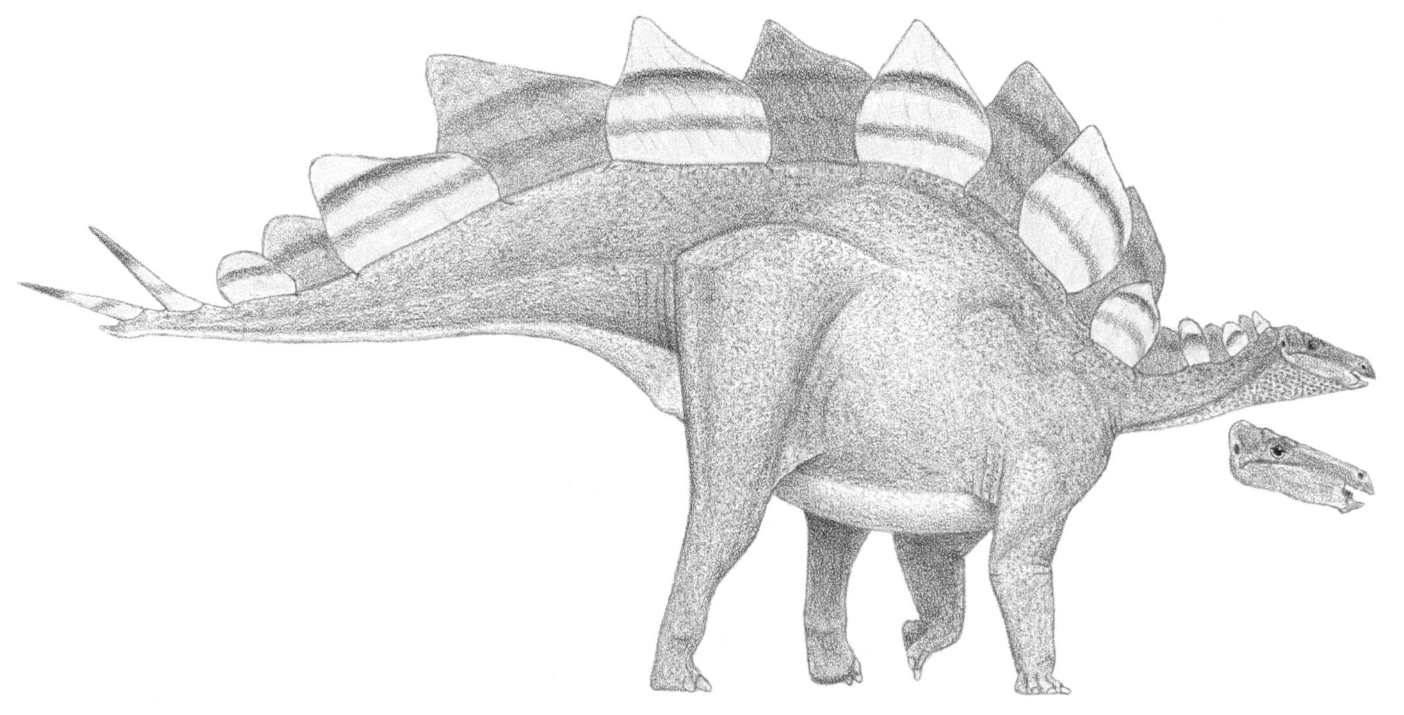

Allosaurus fragilis and *Stegosaurus stenops*

deep, belly and hips narrow and slab sided. Leg very long. Alternating armor plates fairly large over trunk and tail, two pairs of tail spines.
AGE Late Jurassic, middle Tithonian.
DISTRIBUTION AND FORMATION/S Wyoming; upper Morrison.
HABITAT Wetter than earlier Morrison, otherwise semiarid with open floodplain prairies and riverine forests.
HABITS Very well adapted for rearing, may have been able to walk slowly bipedally.
NOTES Once thought to have four pairs of tail spines.

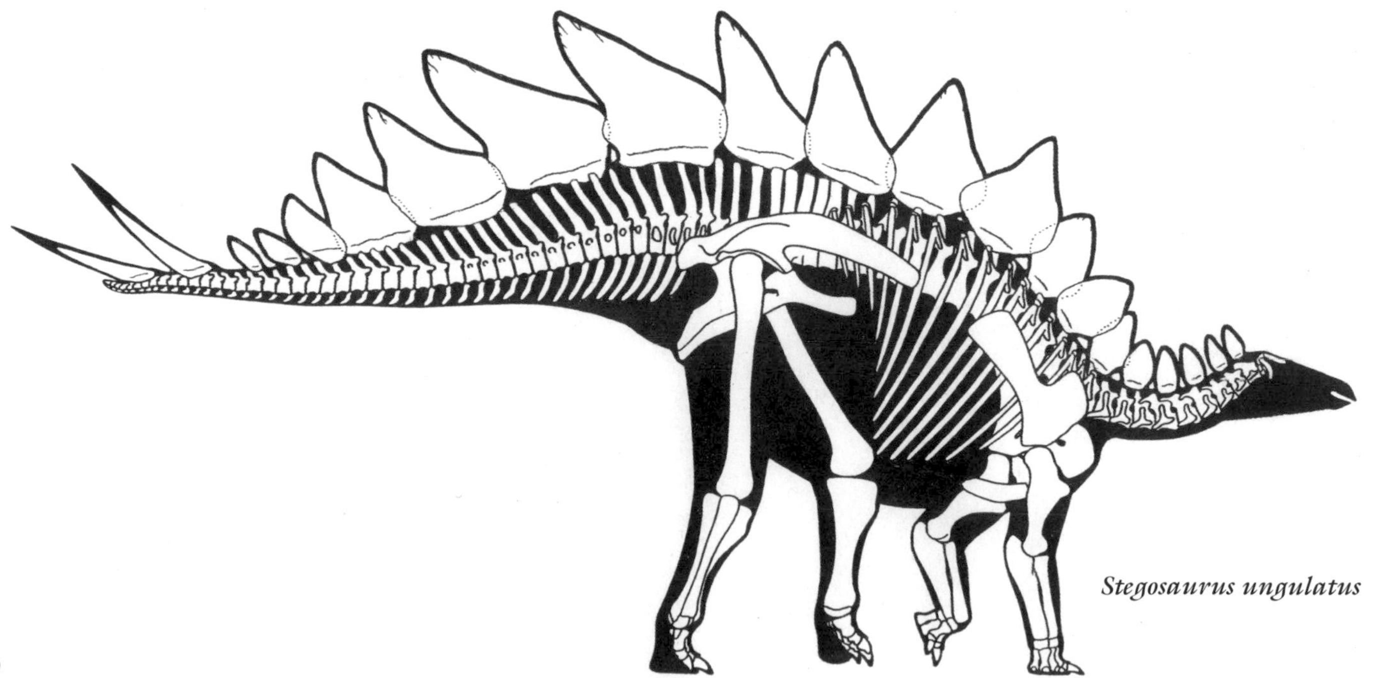

Stegosaurus ungulatus

Alcovasurus (or *Stegosaurus*) *longispinus*
6.5 m (21 ft) TL, 3.5 tonnes

FOSSIL REMAINS Minority of skeleton.
ANATOMICAL CHARACTERISTICS Terminal tail spines extremely long and slender.
AGE Late Jurassic, middle Tithonian.
DISTRIBUTION AND FORMATION/S Wyoming; upper Morrison.
HABITAT Wetter than earlier Morrison, otherwise semiarid with open floodplain prairies and riverine forests.
HABITS Very long tail spines at end of exceptionally muscled and flexible tail indicate extra emphasis on use of spikes for defense and display.

Ankylosaurs

MEDIUM-SIZED TO VERY LARGE EURYPOD THYREOPHORANS FROM THE LATE JURASSIC TO THE END OF THE DINOSAUR ERA, ALL CONTINENTS.

ANATOMICAL CHARACTERISTICS Fairly uniform. Head massively constructed, broad, nasal passages large, main tooth rows short, teeth smaller. Neck short and straight. Aft ribs fused to vertebrae, trunk long and shallow, belly and hips very to extremely broad. Shoulder and hip usually about same height, or shoulder a little lower. Tail base sweeps down from hips and tail held low above ground, moderately long. Quadrupedal, shoulders as high as hips, arm and leg short but flexed, so able to run slowly. Four or three toes. Body armor variable but always extensive, always includes longitudinal and transverse rows of large scutes covering most of top of neck, trunk, and tail. Unarmored skin standard for dinosaurs.
HABITAT Very variable, from deserts to well-watered forests.
HABITS Usually low-level browsers and grazers.
NOTES The most heavily armored dinosaurs, and the dinosaurs most similar to turtles and glyptodonts, although not as extensively armored. Skull armor sometimes coalesced, obscuring details of skull.

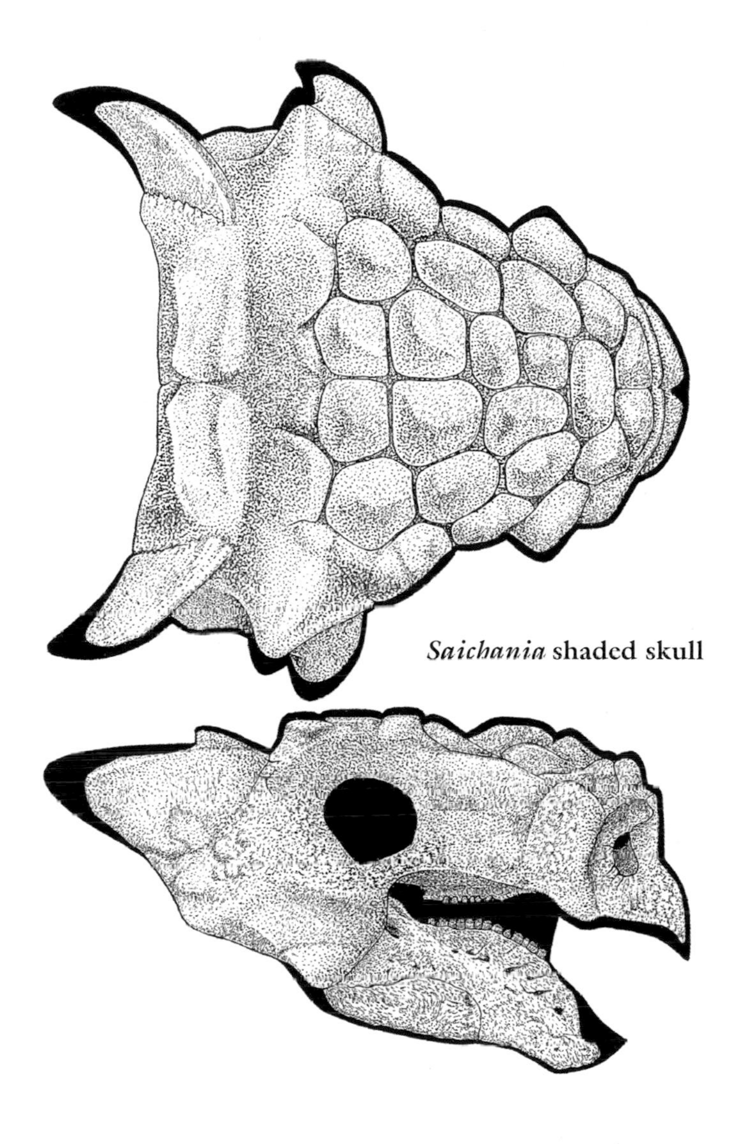

Saichania shaded skull

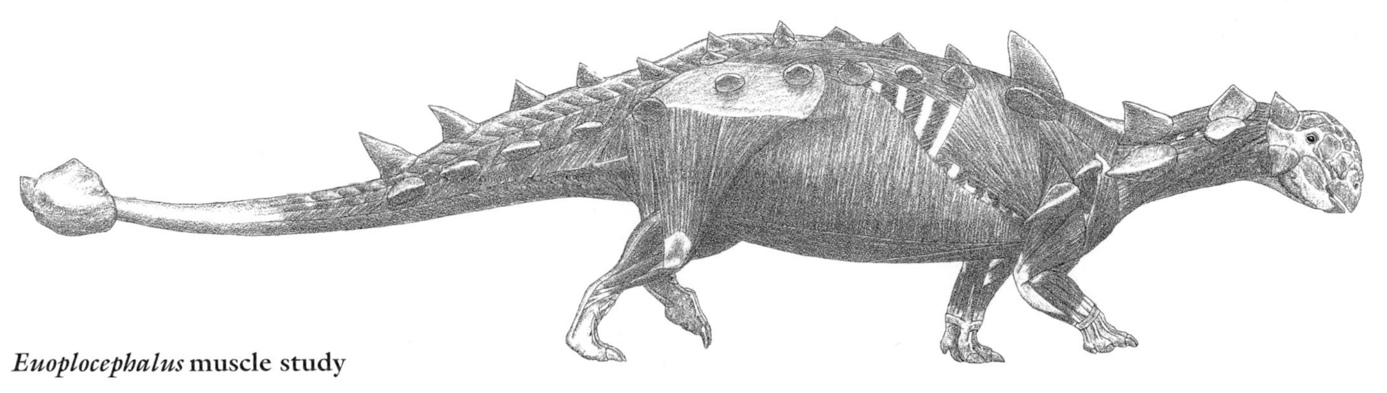

Euoplocephalus muscle study

Nodosaurids Medium-sized to large ankylosaurs of the Late Jurassic to the end of the dinosaur era, both hemispheres.

ANATOMICAL CHARACTERISTICS Some have an armor pelvic shield. Tail club absent.
HABITS Defense included hunkering down on the belly and using armor plates and spikes to avoid being wounded while using the great breadth of the body to prevent being overturned, and dashing into heavy brush when possible.

Antarctopelta oliveroi
6 m (20 ft) TL, 350 kg (800 lb)

FOSSIL REMAINS Minority of skull and skeleton.
ANATOMICAL CHARACTERISTICS Insufficient information.
AGE Late Cretaceous, late Campanian.
DISTRIBUTION AND FORMATION/S Western Antarctica; Santa Marta.
HABITAT Polar forests with warm, daylight-dominated summers and cold, dark winters.
NOTES The only ankylosaur yet named from Antarctica; this is an artifact of difficult conditions and the lack of more extensive exposed deposits.

Mymoorapelta maysi
3 m (9 ft) TL, 300 kg (600 lb)

FOSSIL REMAINS Minority of skull and several skeletons.
ANATOMICAL CHARACTERISTICS Insufficient information.
AGE Late Jurassic.
DISTRIBUTION AND FORMATION/S Colorado; Morrison, level uncertain.
HABITAT Short wet season, otherwise semiarid with open floodplain prairies and riverine forests.

Dracopelta zbyszewskii
3 m (9 ft) TL, 300 kg (600 lb)

FOSSIL REMAINS Minority of skeleton.
ANATOMICAL CHARACTERISTICS No shoulder spines.
AGE Late Jurassic, late Kimmeridgian or Tithonian.
DISTRIBUTION AND FORMATION/S Portugal; Lourinha.
HABITAT Large, seasonally dry island with open woodlands.

Hylaeosaurus armatus
5 m (17 ft) TL, 2 tonnes

FOSSIL REMAINS Two partial skeletons.
ANATOMICAL CHARACTERISTICS Long, backward-directed shoulder spines.
AGE Early Cretaceous, Valanginian.
DISTRIBUTION AND FORMATION/S Southeast England; Hastings Beds.

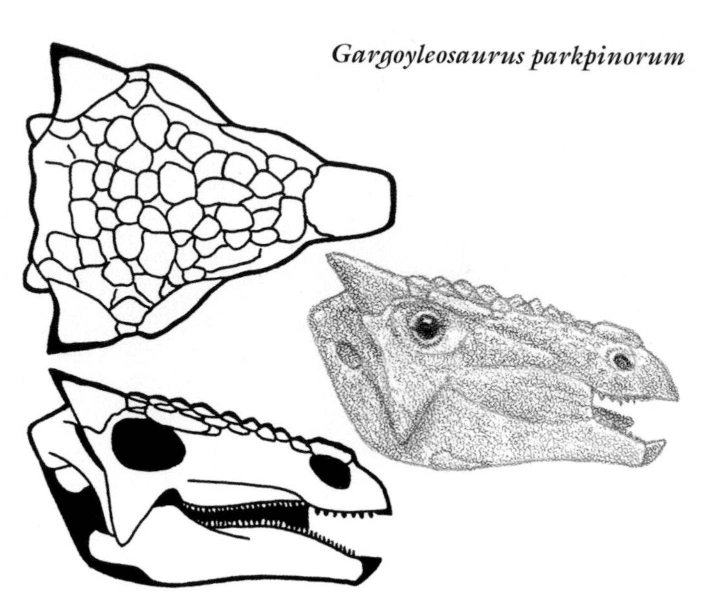

Gargoyleosaurus parkpinorum

Gargoyleosaurus parkpinorum
3 m (9 ft) TL, 300 kg (600 lb)

FOSSIL REMAINS Complete skull, minority of skeleton.
ANATOMICAL CHARACTERISTICS Head heavily armored, teeth near front of upper jaw.
AGE Late Jurassic, probably middle Tithonian.
DISTRIBUTION AND FORMATION/S Wyoming; probably upper Morrison.
HABITAT Short wet season, otherwise semiarid with open floodplain prairies and riverine forests.
NOTES Shared its habitat with *Stegosaurus*. Main enemy *Allosaurus*.

Hoplitosaurus marshi
4.5 m (15 ft) TL, 1.5 tonnes

FOSSIL REMAINS Skeletons.
ANATOMICAL CHARACTERISTICS Armor includes spines, arrangement uncertain.
AGE Early Cretaceous, probably Barremian.
DISTRIBUTION AND FORMATION/S South Dakota; Lakota.
NOTES Shared its habitat with *Dakotadon* and *Planicoxa*.

Gastonia burgei
5 m (17 ft) TL, 1.9 tonnes

FOSSIL REMAINS A few skulls and skeletons from nearly complete to partial.
ANATOMICAL CHARACTERISTICS Head very small, heavily armored, no teeth on front of upper jaw. Arm and leg very short. Belly extremely broad. Rows of large vertical and horizontal shoulder spines, no lateral spines at hip, modest spines on side of tail.
AGE Early Cretaceous, probably early Barremian.
DISTRIBUTION AND FORMATION/S Utah; Lower Cedar Mountain.

Gastonia burgei

HABITAT Short wet season, otherwise semiarid with floodplain prairies, open woodlands, and riverine forests.

Peloroplites cedrimontanus
6 m (20 ft) TL, 2 tonnes

FOSSIL REMAINS Minority of skull and almost complete skeleton known from partial specimens.
ANATOMICAL CHARACTERISTICS Head shallow.
AGE Early Cretaceous, early Albian.
DISTRIBUTION AND FORMATION/S Utah; Lower Cedar Mountain.
HABITAT Short wet season, otherwise semiarid with floodplain prairies, open woodlands, and riverine forests.
NOTES Shared its habitat with *Cedarpelta*.

Polacanthus foxii
5 m (17 ft) TL, 2 tonnes

FOSSIL REMAINS Minority of skull and two skeletons.
ANATOMICAL CHARACTERISTICS Belly very broad. Armor includes spines, arrangement uncertain.
AGE Early Cretaceous, Aptian.
DISTRIBUTION AND FORMATION/S Southeast England; Lower Greensand, Wessex, Vectis.
NOTES Shared its habitat with *Mantellisaurus*.

Zhejiangosaurus lishuiensis
4.5 m (15 ft) TL, 1.4 tonnes

FOSSIL REMAINS Partial skeleton.
ANATOMICAL CHARACTERISTICS Head shallow.
AGE Late Cretaceous, Cenomanian.

DISTRIBUTION AND FORMATION/S Eastern China; Chaochuan.

Hungarosaurus tormai
4.8 m (16 ft) TL, 1 tonne

FOSSIL REMAINS Partial skulls and skeletons.
ANATOMICAL CHARACTERISTICS Arm long, so shoulder higher than hip. Two rows of four large spines over and on side of shoulder, pair of modest pelvic spines set on conjoined plate.
AGE Late Cretaceous, Santonian.
DISTRIBUTION AND FORMATION/S Hungary; Csehbanya.
HABITS High browser rather than ground grazer.
NOTES Only ankylosaur known to have high shoulder.

Struthiosaurus austriacus
3 m (9 ft) TL, 300 kg (600 lb)

FOSSIL REMAINS Minority of skull and skeletons.
ANATOMICAL CHARACTERISTICS Belly and hip very broad.
AGE Late Cretaceous, Campanian.
DISTRIBUTION AND FORMATION/S Austria, Southern France; Gosau, unnamed formation.
HABITAT Forested island.
NOTES *S. languedocensis* may be the adult of this species. Small size may be dwarfism due to island habitat.

Struthiosaurus transylvanicus
3 m (9 ft) TL, 300 kg (600 lb)

FOSSIL REMAINS Partial skull and skeletons.
ANATOMICAL CHARACTERISTICS Insufficient information.
AGE Late Cretaceous, late Maastrichtian.
DISTRIBUTION AND FORMATION/S Romania; Sanpetru.
HABITAT Forested island.
NOTES May be the descendant of *S. austriacus*. Shared its habitat with *Magyarosaurus*.

Animantarx ramaljonesi
3 m (9 ft) TL, 300 kg (600 lb)

FOSSIL REMAINS Partial skull and skeleton.
ANATOMICAL CHARACTERISTICS Insufficient information.
AGE Late Cretaceous, early Cenomanian.
DISTRIBUTION AND FORMATION/S Utah; Upper Cedar Mountain.
HABITAT Short wet season, otherwise semiarid with floodplain prairies, open woodlands, and riverine forests.
NOTES Shared its habitat with *Eolambia*.

Niobrarasaurus coleii
6.5 m (21 ft) TL, 4 tonnes

FOSSIL REMAINS Minority of skull and partial skeleton.
ANATOMICAL CHARACTERISTICS Insufficient information.
AGE Late Late Cretaceous.
DISTRIBUTION AND FORMATION/S Kansas; Niobrara.
NOTES Found as drift in marine deposits.

Pawpawsaurus campbelli
Adult size uncertain

FOSSIL REMAINS Skull and minority of a skeleton, juveniles.
ANATOMICAL CHARACTERISTICS Standard for group.
AGE Early Cretaceous, late Albian.

Pawpawsaurus campbelli

Hungarosaurus tormai

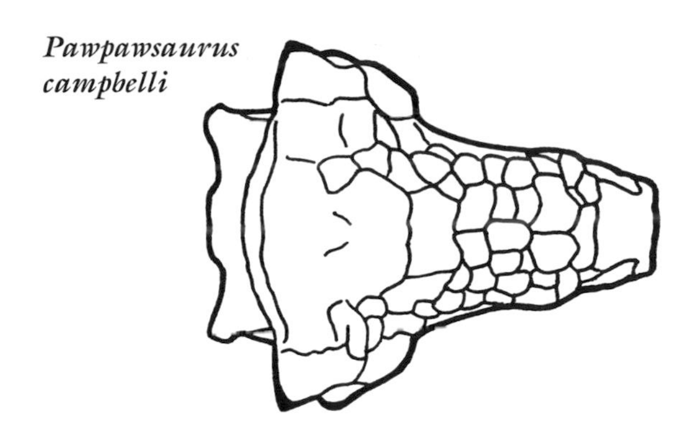
Pawpawsaurus campbelli

DISTRIBUTION AND FORMATION/S Texas; Paw Paw.
NOTES Probably includes *Texasetes pleurohalio*. Found as drift in marine deposits.

Nodosaurus textilis
6 m (20 ft) TL, 3.5 tonnes

FOSSIL REMAINS Partial skeleton.
ANATOMICAL CHARACTERISTICS Insufficient information.
AGE Early Cretaceous, late Albian.
DISTRIBUTION AND FORMATION/S Wyoming; lower Frontier.
NOTES Found as drift in marine deposits.

Europelta carbonensis
5 m (16 ft) TL, 1.3 tonnes

FOSSIL REMAINS Partial skulls and skeletons.
ANATOMICAL CHARACTERISTICS Belly and hip moderately broad. Limbs fairly long. Two prominent horizontal spikes on shoulder.
AGE Early Cretaceous, early Albian.
DISTRIBUTION AND FORMATION/S Spain; Escucha.
HABITAT Coastal.

Sauropelta edwardsi (Illustrated overleaf)
6 m (20 ft) TL, 2 tonnes

FOSSIL REMAINS Several partial skulls and skeletons.
ANATOMICAL CHARACTERISTICS Belly and hip very broad. Tail long. Two rows of three large spines on side of neck, very large and slightly split spine on shoulder.
AGE Early Cretaceous, late Aptian to lower Albian.
DISTRIBUTION AND FORMATION/S Wyoming; Cloverly.
HABITAT Short wet season, otherwise semiarid with floodplain prairies, open woodlands, and riverine forests.
NOTES *Tatankacephalus cooneyorum* is probably the juvenile of this taxon. Shared its habitat with *Tenontosaurus tilletti*.

Stegopelta landerensis
4 m (12 ft) TL, 1 tonne

FOSSIL REMAINS Minority of skull and partial skeleton.
ANATOMICAL CHARACTERISTICS Insufficient information.
AGE Late Cretaceous, early Cenomanian.
DISTRIBUTION AND FORMATION/S Wyoming; Frontier.
NOTES Found as drift in marine sediments.

Europelta carbonensis

Sauropelta edwardsi

Silvisaurus condrayi
4 m (12 ft) TL, 1 tonne

FOSSIL REMAINS Skull and minority of skeleton.
ANATOMICAL CHARACTERISTICS No teeth on front of upper jaw.
AGE Late Cretaceous, Cenomanian.
DISTRIBUTION AND FORMATION/S Kansas; Dakota.
HABITAT Well-watered, forested floodplain with coastal swamps and marshes.
NOTES One of the few dinosaurs found on the eastern shore of the interior seaway.

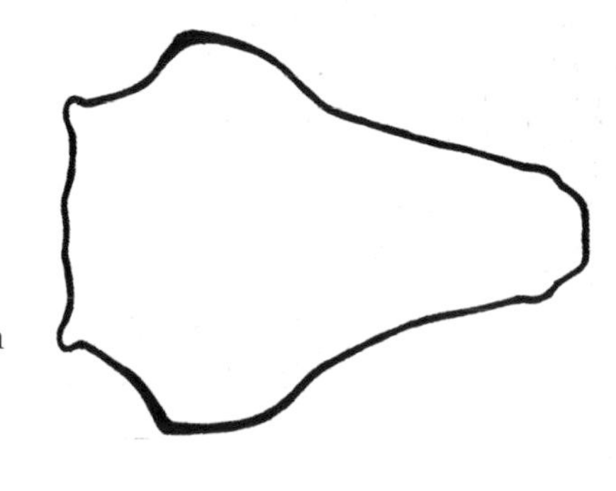

Silvisaurus condrayi

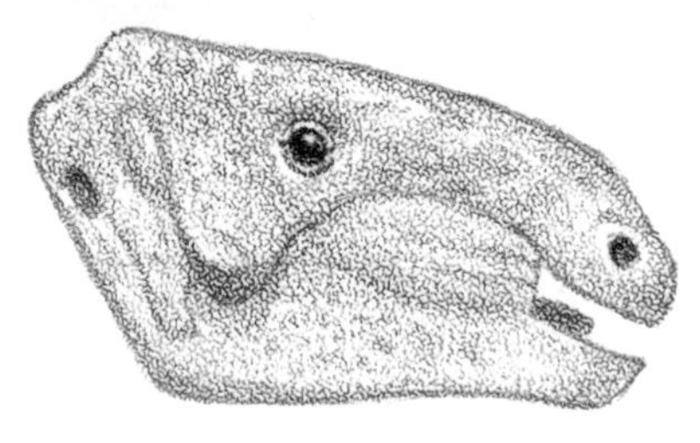

Propanoplosaurus marylandicus
Adult size uncertain

FOSSIL REMAINS Complete skull and half of skeleton, poorly preserved.
ANATOMICAL CHARACTERISTICS Insufficient information.
AGE Early Cretaceous, early Aptian.
DISTRIBUTION AND FORMATION/S Maryland; Patuxent.
HABITAT Well-watered, forested floodplain with coastal swamps and marshes.

Panoplosaurus mirus
5 m (15 ft) TL, 1.5 tonnes

FOSSIL REMAINS Complete skull and majority of skeletons.
ANATOMICAL CHARACTERISTICS Armor plate covered cheeks, large spines absent.
AGE Late Cretaceous, late Campanian.
DISTRIBUTION AND FORMATION/S Alberta; at least middle Dinosaur Park.
HABITAT Well-watered, forested floodplain with coastal swamps and marshes, cool winters.
NOTES Main enemies *Daspletosaurus* and *Albertosaurus libratus*.

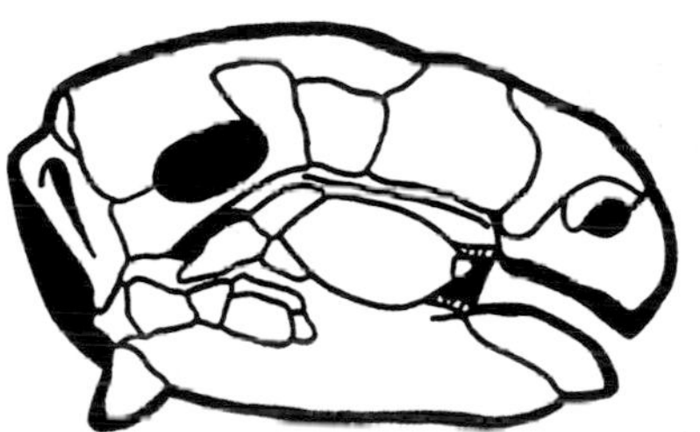

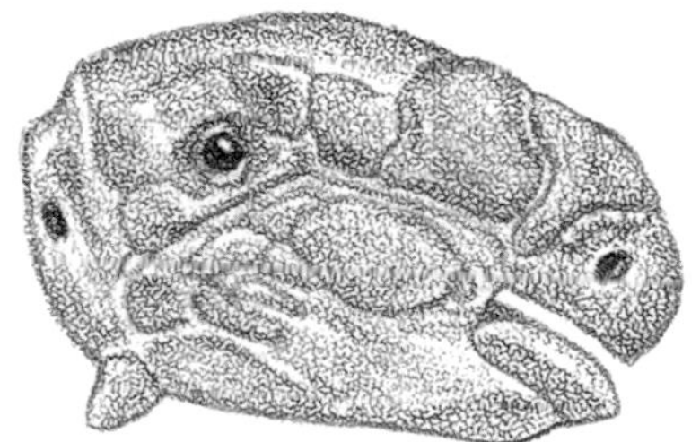

Panoplosaurus mirus

Edmontonia rugosidens
6 m (19 ft) TL, 3 tonnes

FOSSIL REMAINS Several complete skulls and majority of skeletons.
ANATOMICAL CHARACTERISTICS Belly and hip extremely broad. Armor plate covered cheeks, large, forward-directed spikes on flank of neck and shoulder, one spine partly split, no spines on main trunk or hips.
AGE Late Cretaceous, middle and/or late Campanian.
DISTRIBUTION AND FORMATION/S Montana, Alberta?; Upper Two Medicine, middle Dinosaur Park? Judith River?
HABITAT Well-watered, forested floodplain with coastal swamps and marshes and drier upland woodlands.
HABITS May have charged at opponents within species and tyrannosaurids with shoulder spikes, also hunkered down on belly and used armor to avoid being wounded while using the great breadth of the body to prevent being overturned.

Edmontonia longiceps
6 m (19 ft) TL, 3 tonnes

FOSSIL REMAINS Several complete skulls and partial skeletons.
ANATOMICAL CHARACTERISTICS Shoulder spikes modest sized, directed sideways.

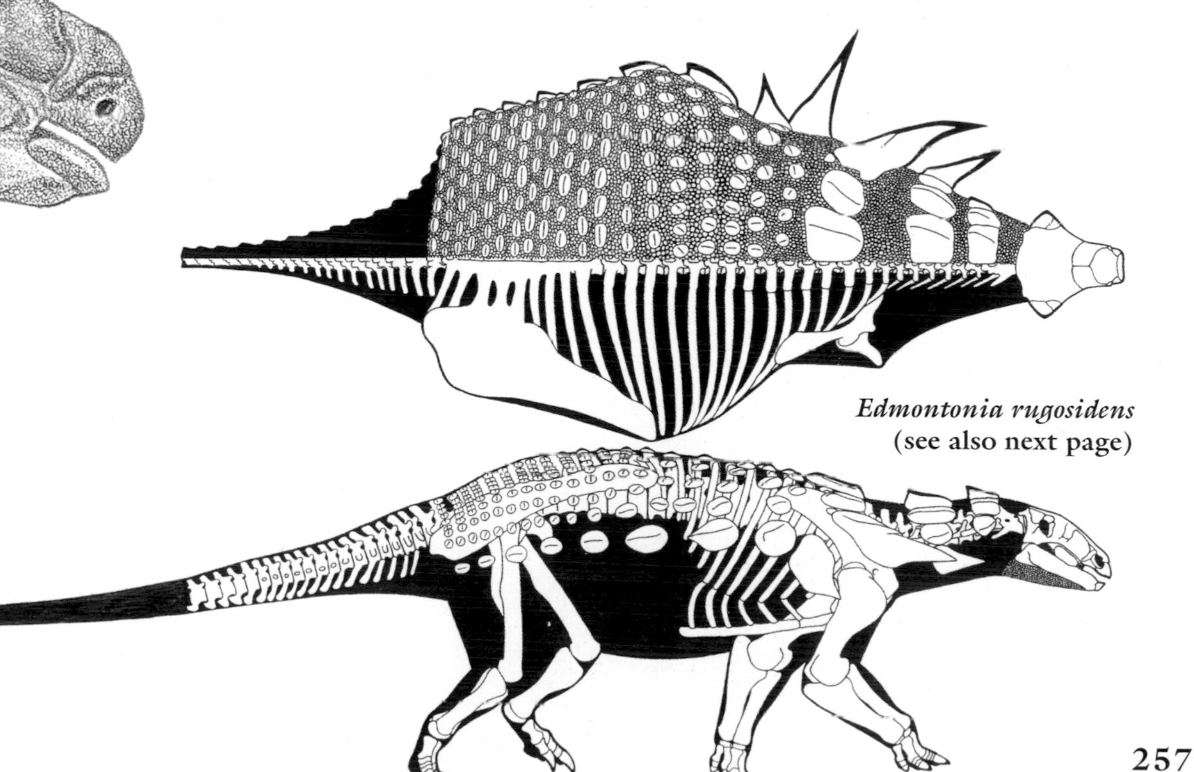

Edmontonia rugosidens
(see also next page)

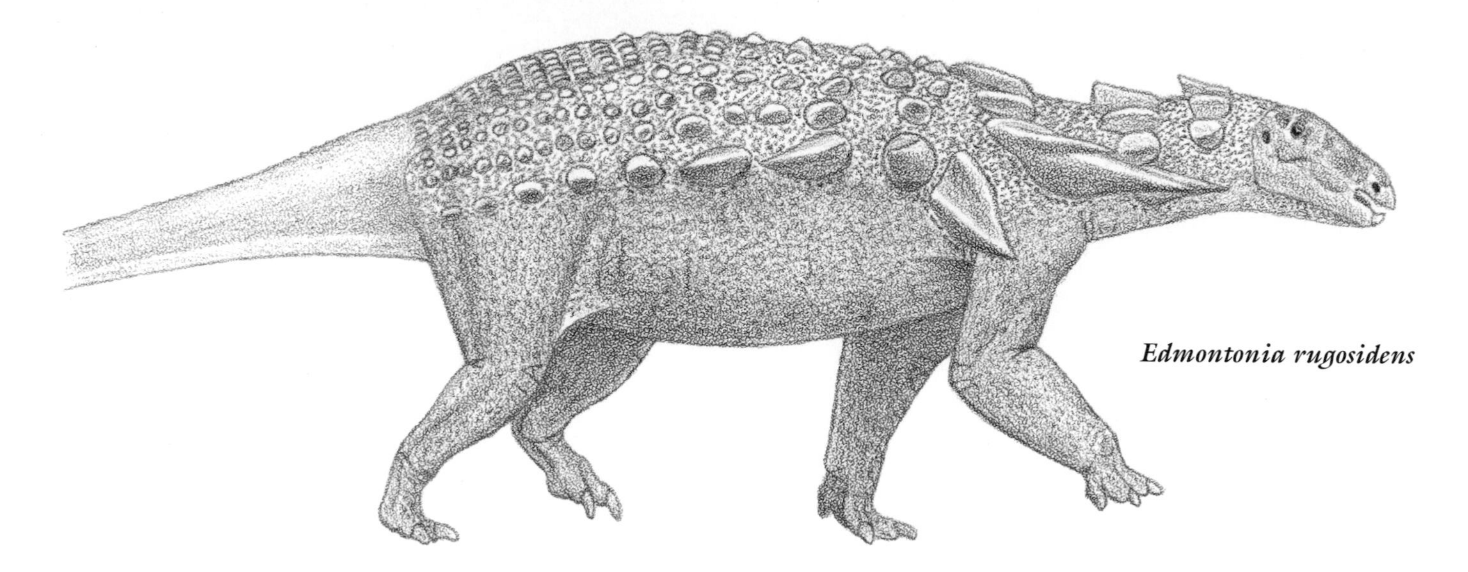

Edmontonia rugosidens

AGE Late Cretaceous, early Maastrichtian.
DISTRIBUTION AND FORMATION/S Alberta, Montana; lower Horseshoe Canyon.
HABITAT Well-watered, forested floodplain with coastal swamps and marshes, cool winters.
NOTES May be direct descendant of *E. rugosidens*.

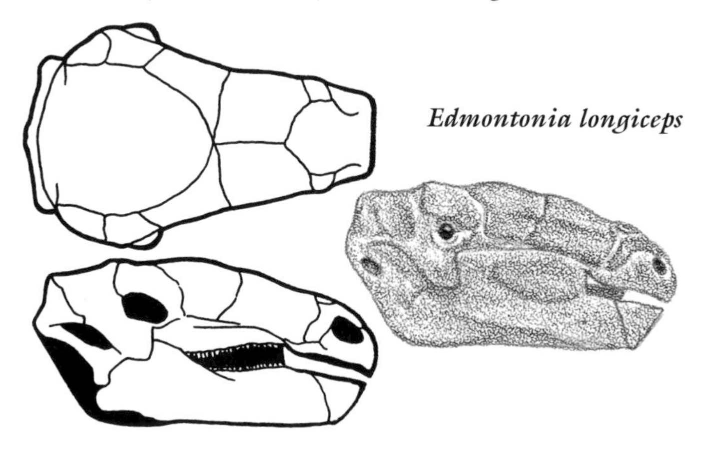

Edmontonia longiceps

Edmontonia (= *Denversaurus*) *schlessmani*
6 m (19 ft) TL, 3 tonnes

FOSSIL REMAINS Complete skull, minority of skeletons.
ANATOMICAL CHARACTERISTICS Insufficient information.
AGE Late Cretaceous, late Maastrichtian.
DISTRIBUTION AND FORMATION/S Montana, South Dakota, Wyoming; Hell Creek, Lance, levels uncertain.
HABITAT Well-watered coastal woodlands.
NOTES May be a direct descendant of *E. longiceps*. Shared its habitat with *Ankylosaurus*; main enemy *Tyrannosaurus*.

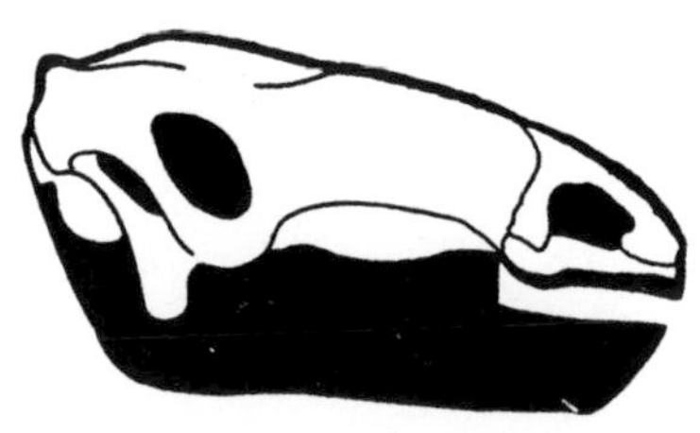

Edmontonia (=Denversaurus) schlessmani

Zhongyuansaurus luoyangensis
5 m (15 ft) TL, 2 tonnes

FOSSIL REMAINS Majority of crushed skull, minority of skeleton.
ANATOMICAL CHARACTERISTICS Head probably shallow, heavily armored, no teeth on front of upper jaw.
AGE Late Cretaceous.
DISTRIBUTION AND FORMATION/S Eastern China; Liu Dianxiang Sichuan.

ANKYLOSAURIDS Medium-sized to large ankylosaurs of the Early Cretaceous to the end of the dinosaur era of the Northern Hemisphere.

Minmi paravertebra
3 m (9 ft) TL, 300 kg (600 lb)

FOSSIL REMAINS Minority of skeleton.
ANATOMICAL CHARACTERISTICS Standard for group.
AGE Early Cretaceous, Aptian.
DISTRIBUTION AND FORMATION/S Northeast Australia; Bungil.
HABITAT Polar forests with warm, daylight-dominated summers and cold, dark winters.

Minmi? unnamed species
3 m (9 ft) TL, 300 kg (600 lb)

FOSSIL REMAINS A few skulls and skeletons from nearly complete to partial.
ANATOMICAL CHARACTERISTICS Standard for group.
AGE Early Cretaceous, Albian.
DISTRIBUTION AND FORMATION/S Northeast Australia; Bungil.

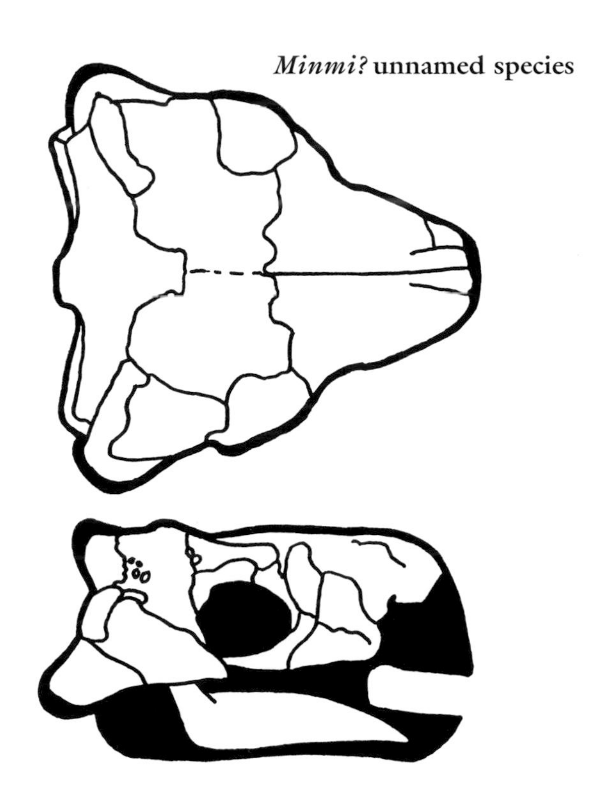

Minmi? unnamed species

NOTES Whether this is the same genus as *M. paravertebra* is uncertain.
HABITAT Polar forests with warm, daylight-dominated summers and cold, dark winters.

Liaoningosaurus paradoxus
Adult size uncertain

FOSSIL REMAINS Almost complete distorted skull and skeleton, juvenile.
ANATOMICAL CHARACTERISTICS Head heavily armored. Belly and pelvis very broad. Belly covered by solid armor plate.
AGE Early Cretaceous, early Aptian.
DISTRIBUTION AND FORMATION/S Northeast China; Yixian.
HABITAT Well-watered forests and lakes, winters chilly with some snow.

Gobisaurus domoculus
6 m (20 ft) TL, 3.5 tonnes

FOSSIL REMAINS Nearly complete skull and skeleton.
ANATOMICAL CHARACTERISTICS Head heavily armored.
AGE Late Cretaceous, Turonian.
DISTRIBUTION AND FORMATION/S Northern China; Ulansuhai.
NOTES Probable prey of *Chilantaisaurus.*

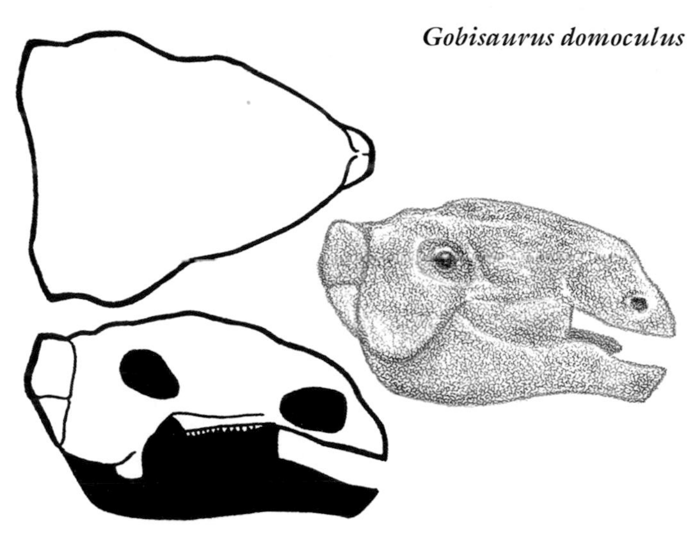

Gobisaurus domoculus

Shamosaurus scutatus
5 m (15 ft) TL, 2 tonnes

FOSSIL REMAINS Two skulls, partial skeleton.
ANATOMICAL CHARACTERISTICS Head shallow, heavily armored, no teeth on front of upper jaw. Neck armor forming cervical half rings.
AGE Early Cretaceous, Aptian or Albian.
DISTRIBUTION AND FORMATION/S Mongolia; Huhteeg Svita.
NOTES Shared its habitat with *Altirhinus.*

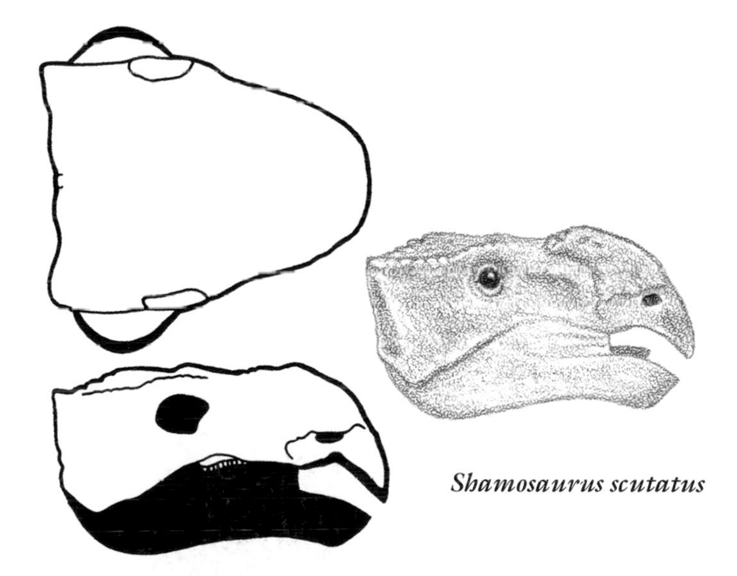

Shamosaurus scutatus

Cedarpelta bilbeyhallorum
7 m (23 ft) TL, 5 tonnes

FOSSIL REMAINS Minority of skull and almost complete skeleton known from very many partial specimens.
ANATOMICAL CHARACTERISTICS Head shallow, not heavily armored, teeth near front end of upper jaw.
AGE Early Cretaceous, Barremian.

DISTRIBUTION AND FORMATION/S Utah; Lower Cedar Mountain.
HABITAT Short wet season, otherwise semiarid with floodplain prairies, open woodlands, and riverine forests.
NOTES Shared its habitat with *Gastonia* and *Peloroplites.*

Crichtonpelta benxiensis
Adult size uncertain

FOSSIL REMAINS A few partial skulls and skeletons, possibly juvenile.
ANATOMICAL CHARACTERISTICS Insufficient information.
AGE Early Cretaceous, Albian.
DISTRIBUTION AND FORMATION/S Northeast China; Sunjiawan.
HABITAT Well-watered forests and lakes, winters chilly with some snow.

Ankylosaurines Medium-sized to very large ankylosaurids limited to the Cretaceous of North America and Asia.

ANATOMICAL CHARACTERISTICS Beak and overall head broad, head heavily armored with large triangular hornlets at the back corners, no teeth on front of upper jaw. Belly and hips very broad. Limbs short. Three toes. Large spines absent, two short spikes flanking neck, last half of tail stiffened by ossified rods and tipped with a club.
HABITS Low-level browsers and grazers. Defense included running while swinging clubbed tail to keep theropod from getting close to its victim, spinning around to keep tail toward attacker while using club to damage legs or flanks or to topple theropod, and dashing into heavy brush when possible. Tail clubs may also have been used for display and combat within a species.

Tsagantegia longicranialis
3.5 m (12 ft) TL, 500 kg (1,000 lb)

FOSSIL REMAINS Skull.
ANATOMICAL CHARACTERISTICS Head shallow.
AGE Late Cretaceous, Santonian.
DISTRIBUTION AND FORMATION/S Mongolia; Bayenshiree Svita.
NOTES Shared its habitat with *Talarurus*.

Talarurus plicatospineus
5 m (16 ft) TL, 2 tonnes

FOSSIL REMAINS Partial skull, majority of skeleton.
ANATOMICAL CHARACTERISTICS Standard for group.
AGE Early Late Cretaceous.
DISTRIBUTION AND FORMATION/S Mongolia; Bayenshiree Svita.

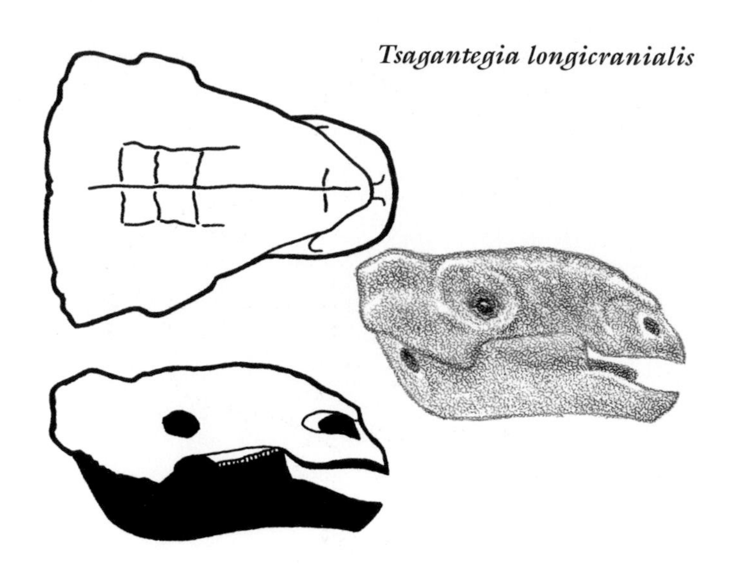

Tsagantegia longicranialis

Saichania chulsanensis
5 m (16 ft) TL, 2 tonnes

FOSSIL REMAINS Complete and partial skulls and skeletons.
ANATOMICAL CHARACTERISTICS Head spines well developed.

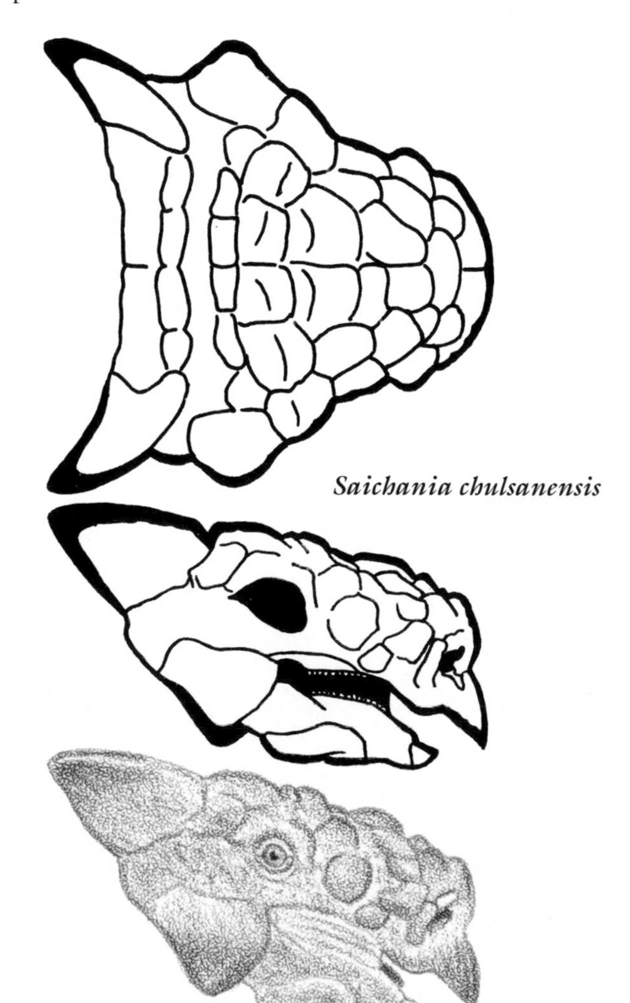

Saichania chulsanensis

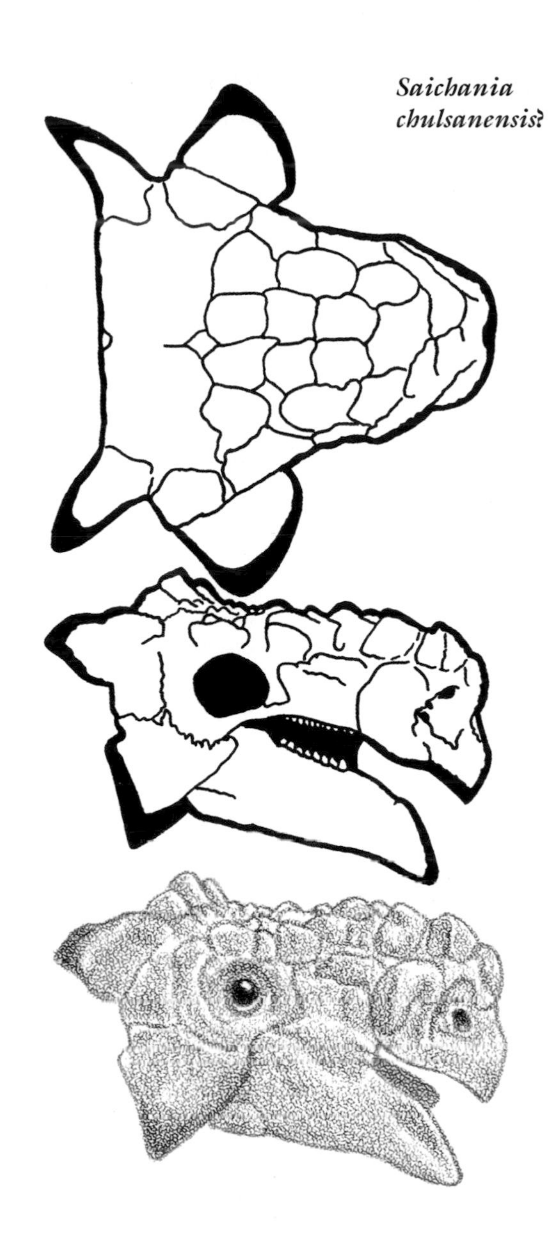

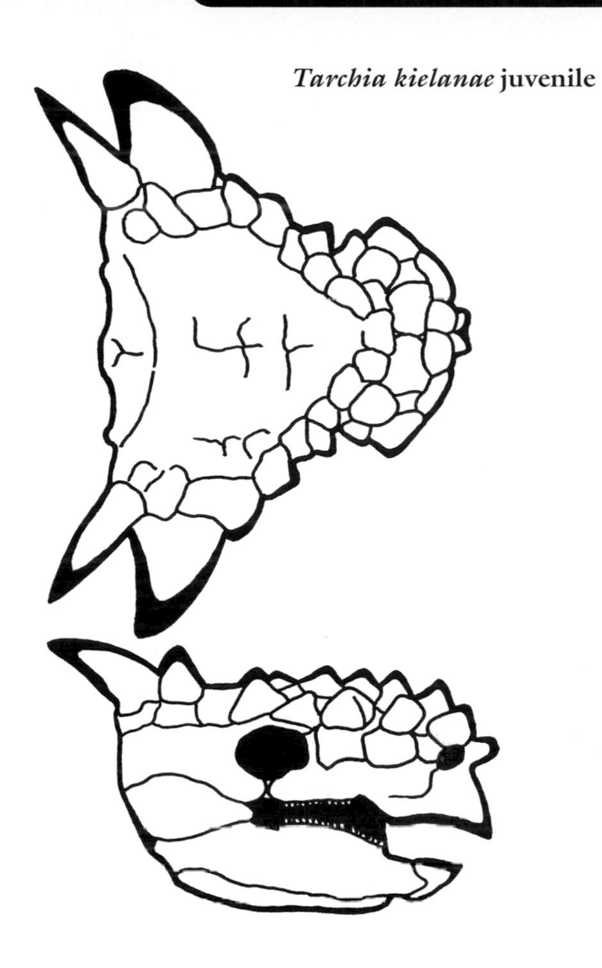

AGE Late Cretaceous, probably late Campanian, possibly early Maastrichtian.
DISTRIBUTION AND FORMATION/S Mongolia; Barun Goyot, Nemegt?
HABITAT Semidesert with some dunes and oases.
NOTES Probably includes *Tarchia gigantea*, *Tianzhenosaurus youngi* and *Shanxia tianzhenensis*, latter two as juveniles. Complete skeleton not yet available for restoration.

Tarchia kielanae

5.5 m (18 ft) TL, 2.5 tonnes

FOSSIL REMAINS A number of complete skulls and skeletons, completely known.
ANATOMICAL CHARACTERISTICS Insufficient information.
AGE Late Cretaceous, late Campanian and/or early Maastrichtian.
DISTRIBUTION AND FORMATION/S Mongolia; Nemegt, Baruungoyot Svita, upper Barun Goyot.
HABITAT Well-watered woodland with seasonal rain.
NOTES Not yet well described. *Minotaurasaurus ramachandrani* is a juvenile of this species. *Zaraapelta nomadis* may belong within this species, or genus. Main enemy *Tyrannosaurus bataar*.

Pinacosaurus grangeri

5 m (17 ft) TL, 1.9 tonnes

FOSSIL REMAINS Numerous complete and partial skulls and skeletons from juvenile to adult, completely known.
ANATOMICAL CHARACTERISTICS Head very small, nostrils with multiple exits on side of snout. Short spines along flanks of body and hips, arm heavily armored, tail club small.
AGE Late Cretaceous, Campanian.
DISTRIBUTION AND FORMATION/S Mongolia, northern China; Djadokhta.
HABITAT Desert with dunes and oases.
HABITS Probably fed on vegetation along watercourses and at oases. Small club was high-velocity weapon for use

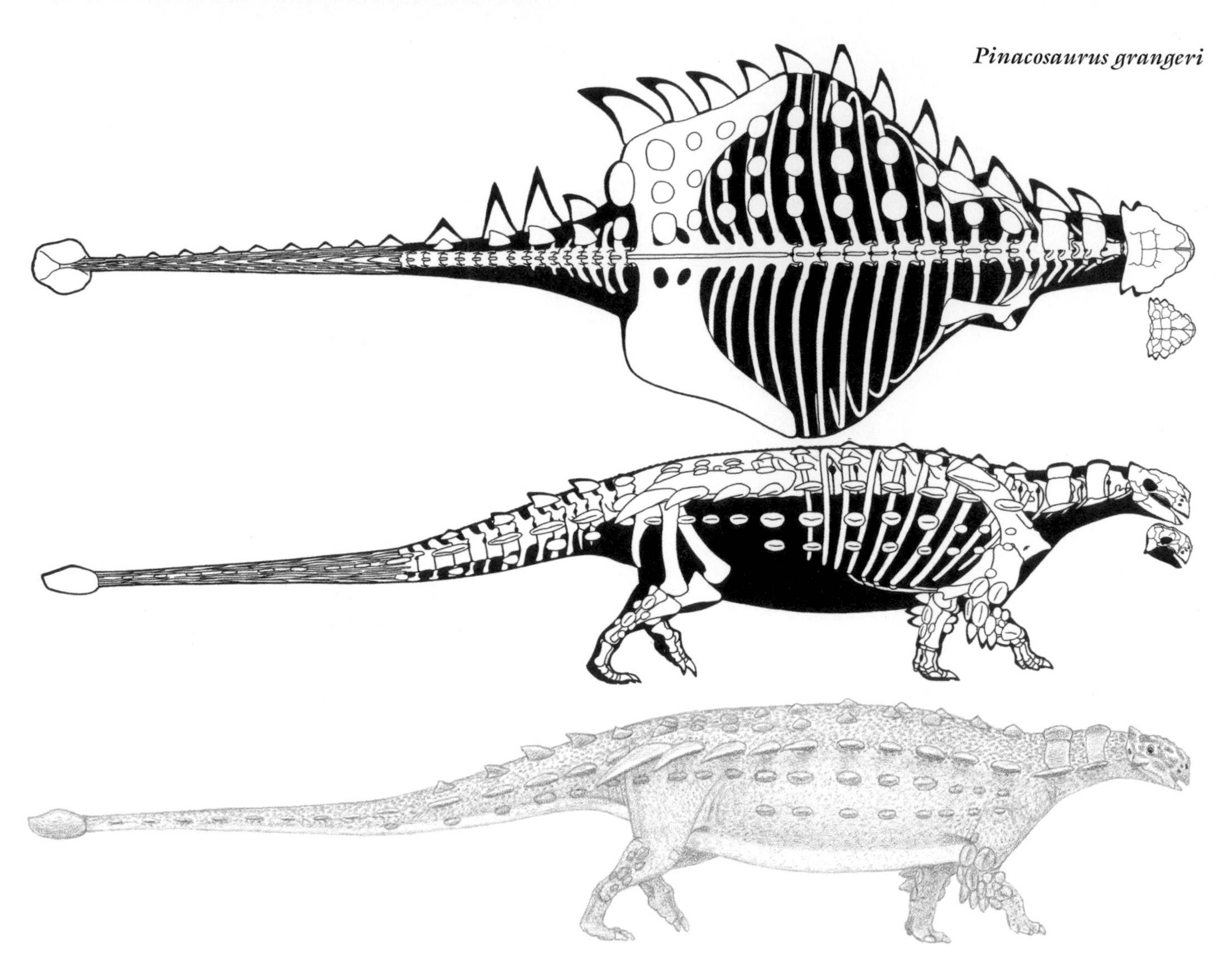

Pinacosaurus grangeri

on small theropods. A group of six juveniles buried at the same time by sand slide indicates they formed pods.
NOTES Probably includes *P. mephistocephalus.* Habitat probably lacked theropods large enough to attack adult *Pinacosaurus.*

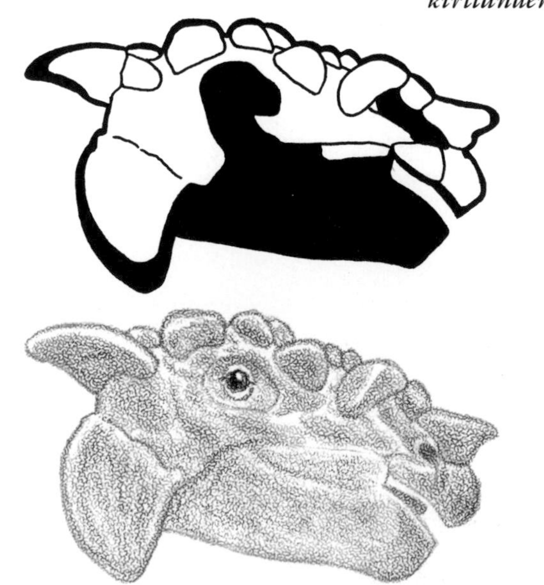

Nodocephalosaurus kirtlandensis

Nodocephalosaurus kirtlandensis

4.5 m (15 ft) TL, 1.5 tonnes

FOSSIL REMAINS Skull.
ANATOMICAL CHARACTERISTICS Bulbous osteoderms on snout, hornlets at back of skull very prominent.
AGE Late Cretaceous, late Campanian and/or early Maastrichtian.
DISTRIBUTION AND FORMATION/S New Mexico; lower Kirtland.
HABITAT Moderately watered floodplain woodlands, coastal swamps and marshes.
NOTES More similar to Asian than to other American ankylosaurids.

Aletopelta coombsi

5 m (16 ft) TL, 2 tonnes

FOSSIL REMAINS Partial skeleton.
ANATOMICAL CHARACTERISTICS Insufficient information.
AGE Late Cretaceous, late Campanian.
DISTRIBUTION AND FORMATION/S California; Point Loma.
HABITAT Found as drift in marine deposits near rugged terrain.

Ziapelta sanjuanensis

5 m (16 ft) TL, 2 tonnes

FOSSIL REMAINS Majority of skull and small minority of skeleton.
ANATOMICAL CHARACTERISTICS Hornlets at side and back of head very large.
AGE Late Cretaceous, late Campanian.
DISTRIBUTION AND FORMATION/S New Mexico; upper Kirtland.
HABITAT Well-watered, forested floodplain with coastal swamps and marshes and possibly drier upland woodlands.

Dyoplosaurus acutosquameus?

4 m (13 ft) TL, 1.2 tonne

FOSSIL REMAINS Minority of skull and skeleton.
ANATOMICAL CHARACTERISTICS Tail club small. Pelvis very broad.
AGE Late Cretaceous, late Campanian.
DISTRIBUTION AND FORMATION/S Alberta; lower Dinosaur Park.
HABITAT Well-watered, forested floodplain with coastal swamps and marshes and possibly drier upland woodlands.
NOTES May be an immature *Euoplocephalus tutus.*

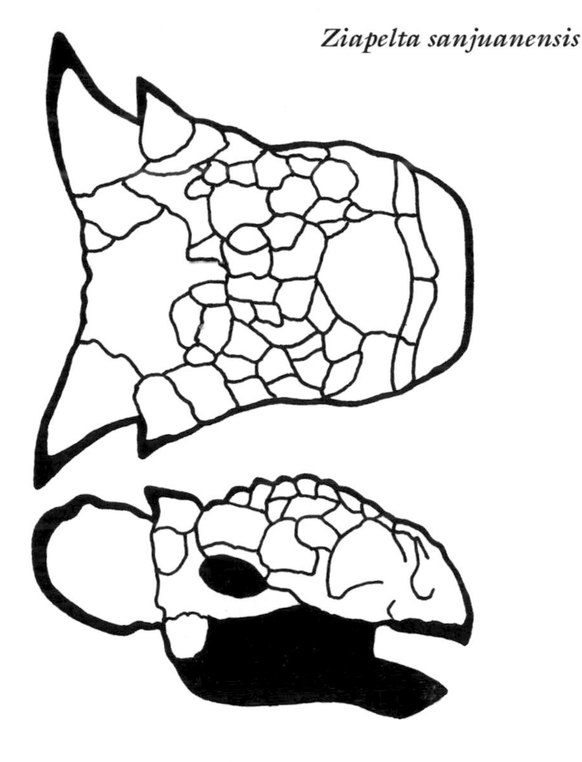

Ziapelta sanjuanensis

Euoplocephalus (or *Scolosaurus*) *cutleri?*

5.6 m (18 ft) TL, 2.2 tonnes

FOSSIL REMAINS Complete and partial skulls and skeletons.
ANATOMICAL CHARACTERISTICS Upper eyelid armored. Pelvis moderately broad. Tail club large. Short spines on forequarters and base of tail.
AGE Late Cretaceous, late Campanian.
DISTRIBUTION AND FORMATION/S Alberta, Montana?; lower Dinosaur Park, Upper Two Medicine?

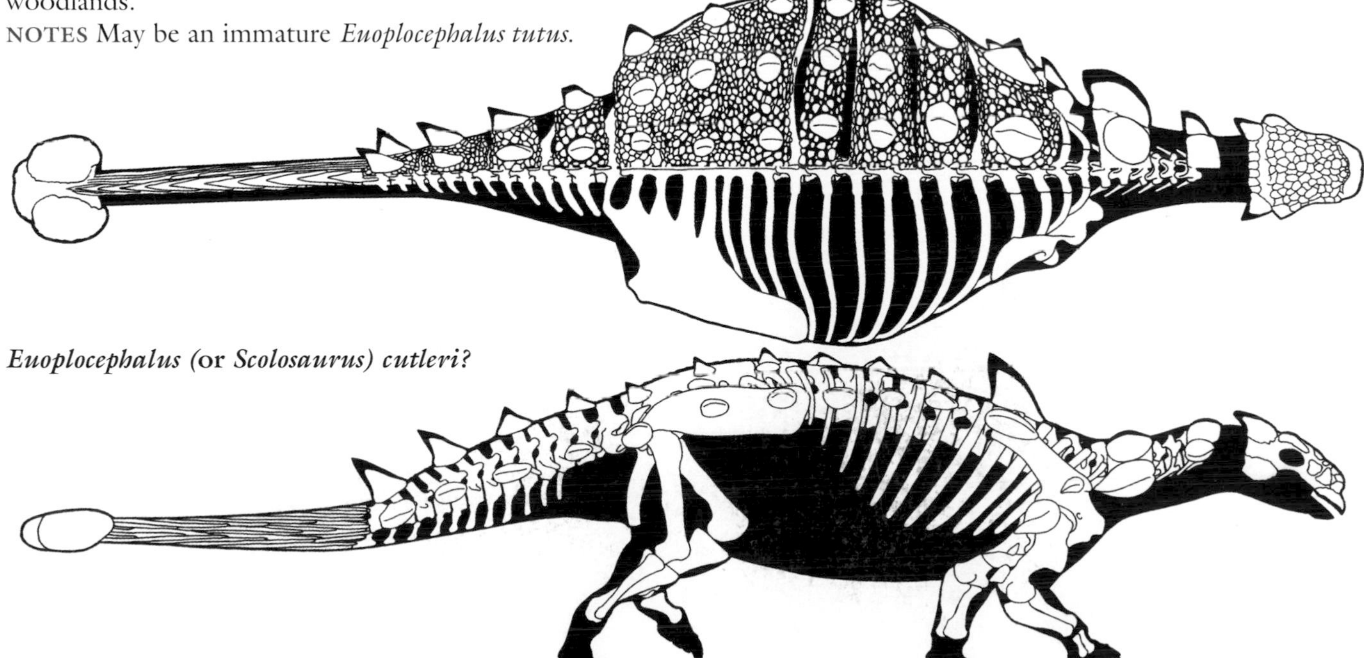

Euoplocephalus (or *Scolosaurus*) *cutleri?*

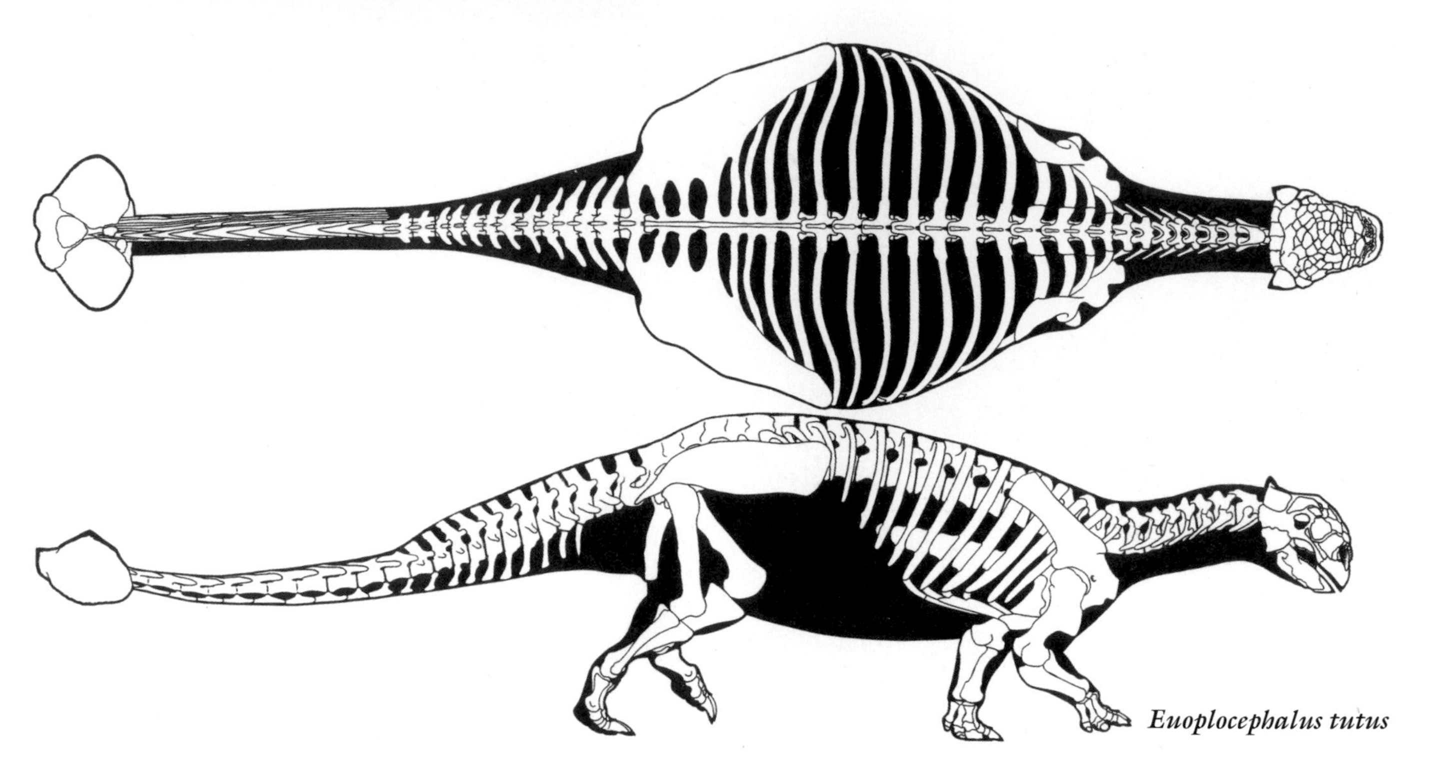

Euoplocephalus tutus

HABITAT Well-watered, forested floodplain with coastal swamps and marshes and possibly drier upland woodlands.
NOTES That three ankylosaurids inhabited the lower Dinosaur Park Formation is problematic; it is possible this is one of the sexes of *E. tutus*.

Euoplocephalus tutus
5.3 m (17) TL, 2 tonnes

FOSSIL REMAINS Complete and partial skulls and skeleton.
ANATOMICAL CHARACTERISTICS Upper eyelid armored. Tail club large. Pelvis very broad.
AGE Late Cretaceous, late Campanian.
DISTRIBUTION AND FORMATION/S Alberta; lower and middle Dinosaur Park.
HABITAT Well-watered, forested floodplain with coastal swamps and marshes and possibly drier upland woodlands.
NOTES Prey of *Daspletosaurus* and *Albertosaurus libratus*.

Euoplocephalus (or *Oohkotokia*) *horneri*
5 m (16 ft) TL, 2 tonnes

FOSSIL REMAINS Partial, crushed skull and skeletons.
ANATOMICAL CHARACTERISTICS Rear skull hornlets very prominent.
AGE Late Cretaceous, late Campanian.
DISTRIBUTION AND FORMATION/S Montana; Upper Two Medicine.
HABITAT Well-watered, forested floodplain with coastal swamps and marshes, cool winters.

Euoplocephalus (or *Anodontosaurus*) *lambei*
5 m (16 ft) TL, 2 tonnes

FOSSIL REMAINS Several complete and partial skulls and skeletons.
ANATOMICAL CHARACTERISTICS Short spines on forequarters and base of tail.
AGE Late Cretaceous, late Campanian, early/middle Maastrichtian.
DISTRIBUTION AND FORMATION/S Alberta; upper Dinosaur Park, middle Horseshoe Canyon.

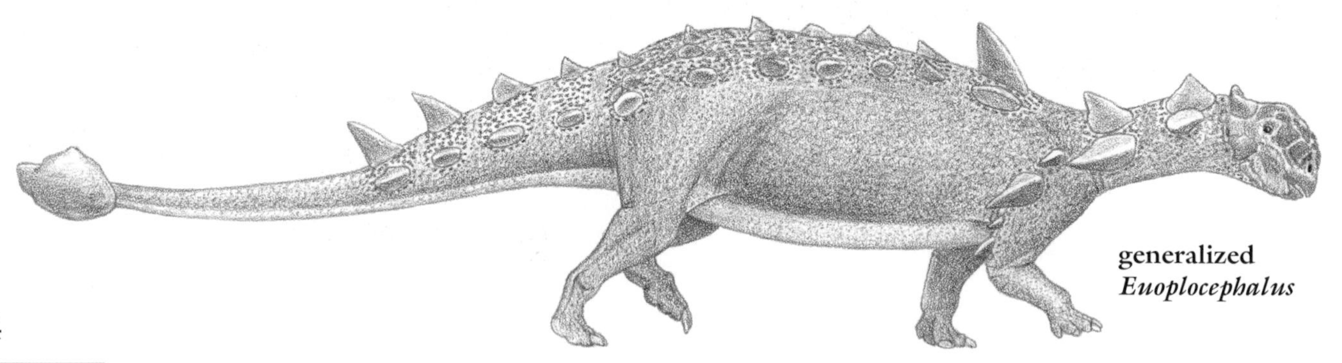

generalized *Euoplocephalus*

Euoplocephalus

HABITAT Well-watered, forested floodplain with coastal swamps and marshes, cool winters.

Ankylosaurus magniventris

7 m (23 ft) TL, 6 tonnes

FOSSIL REMAINS Several skulls and partial skeletons.
ANATOMICAL CHARACTERISTICS Nostrils on side of snout, hornlets at back of skull very prominent. Tail rather short.
AGE Late Cretaceous, late Maastrichtian.
DISTRIBUTION AND FORMATION/S Wyoming, Montana, Alberta; Lance, Hell Creek, Scollard, levels uncertain
HABITAT Well-watered coastal woodlands.
NOTES May be more than one stratigraphic species. Shared its habitat with far more common *Edmontosaurus* and *Triceratops*; main enemy *Tyrannosaurus*.

HETERODONTOSAURIFORMES

SMALL TO GIGANTIC GENASAUR ORNITHISCHIANS OF THE EARLY JURASSIC TO THE END OF THE DINOSAUR ERA, MOST CONTINENTS.

ANATOMICAL CHARACTERISTICS Very variable. Large jugal boss on cheeks, beaks narrow. Neck S-curved. Tail long to fairly short. Fully bipedal to quadrupedal, arm and leg flexed and able to run. Fingers five to three.
HABITS Low- to medium-level browsers. Defense included running and aggressive combat.
NOTES This very large group is valid only if heterodontosaurids are close relatives of marginocephalians.

HETERODONTOSAURS

SMALL HETERODONTOSAURIFORMES OF THE JURASSIC AND EARLY CRETACEOUS OF AFRICA, EURASIA, AND THE AMERICAS.

ANATOMICAL CHARACTERISTICS Highly uniform. Head modest sized, fairly deep, subtriangular, teeth at front of upper jaw, teeth in main rows large and chisel shaped, large eyes shaded by overhanging rim. Trunk and tail stiffened by ossified tendons. Tail long. Bipedal and semiquadrupedal. Arm and hand fairly long, three of the grasping fingers tipped with large claws. Leg long, flexed, and gracile so speed potential high, toes long and tipped with blunt claws. Much or most of body covered by long hollow fibers.

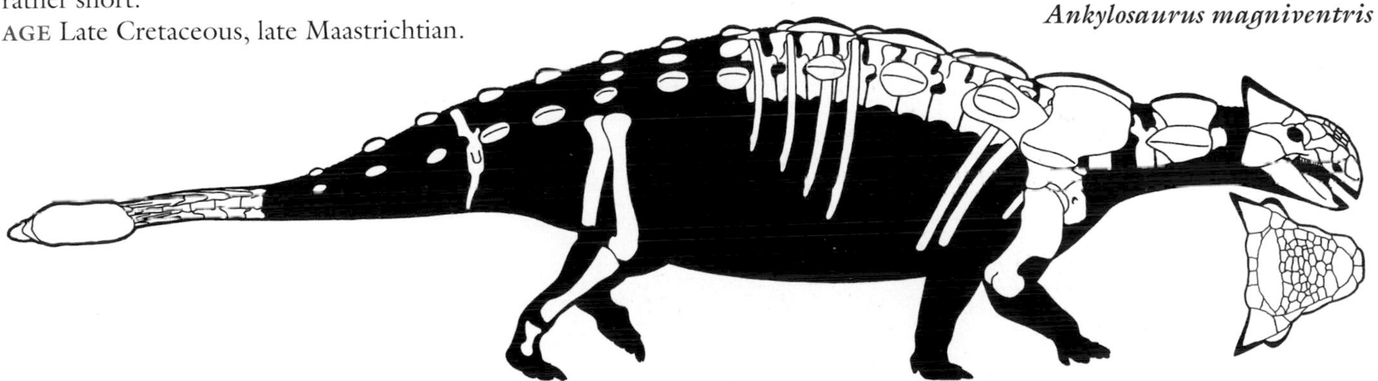

Ankylosaurus magniventris

HABITS Able to consume coarse vegetation, probably omnivores that hunted small vertebrates and scavenged. Significant climbing ability. Main defense high speed, also biting. Jugal bosses on cheeks probably for combat and/or display within species. Longer dorsal fibers probably for display, shorter probably for insulation.
NOTES Relationships of heterodontosaurs are uncertain; they have been considered a distinct group at the base of ornithischians (but they have a more sophisticated form), or ornithopods (but they lack key attributes), or close to marginocephalians; the cheek boss and chisel teeth favor the latter. In any case, they show that sophisticated ornithischians evolved by the early Jurassic. Closest living analogs are kangaroos and small, tusked deer and antelope. Whether Early Cretaceous *Echinodon becklesii*, based on inadequate remains from England, is a heterodontosaur is uncertain. Distribution probably greater than yet known.

HETERODONTOSAURIDS Small heterodontosauriformes of the Jurassic and Early Cretaceous of Africa, Eurasia, and North America.

Heterodontosaurus tucki
1.2 m (4 ft) TL, 3.5 kg (7.5 lb)

FOSSIL REMAINS Several complete and partial skulls and a complete skeleton, juvenile to adult.
ANATOMICAL CHARACTERISTICS Short fangs in front of main tooth rows in at least one sex.
AGE Early Jurassic, late Hettangian or Sinemurian.
DISTRIBUTION AND FORMATION/S Southeast Africa; Upper Elliot.
HABITAT Arid.
HABITS Fangs may have been limited to males, probably used for combat within species and for defense against predators.

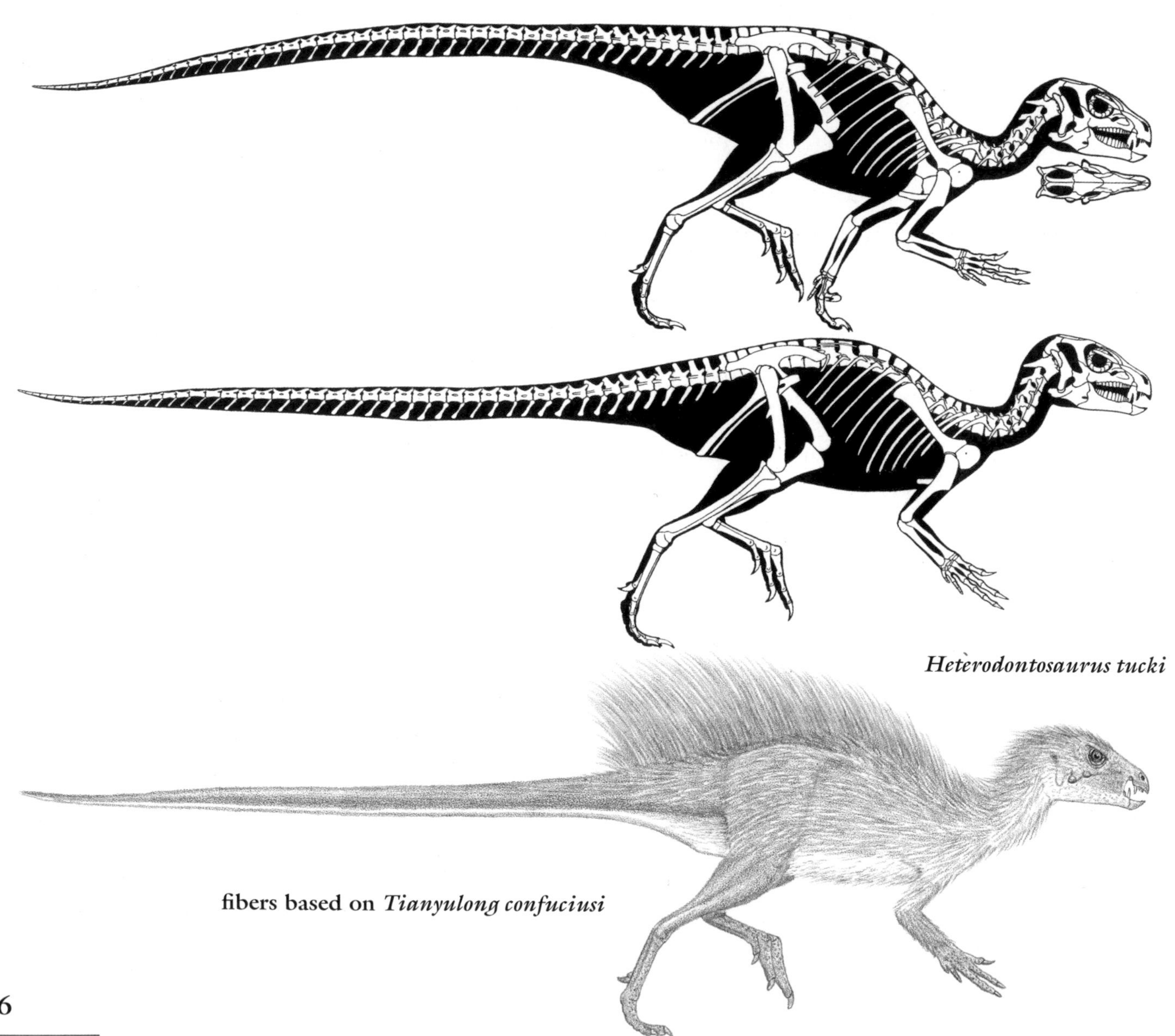

Heterodontosaurus tucki

fibers based on *Tianyulong confuciusi*

NOTES Probably includes the smaller, tuskless *Abrictosaurus consors* as well as *Lycorhinus angustidens*. Shared its habitat with *Lesothosaurus*, *Aardonyx*, and *Pulanesaura*. Prey of *Dracovenator*.

Manidens condorensis

0.6 m (2 ft) TL, 0.5 kg (1.8 lb)

FOSSIL REMAINS Minority of skeleton.
ANATOMICAL CHARACTERISTICS Insufficient information.
AGE Middle Jurassic.
DISTRIBUTION AND FORMATION/S Southern Argentina; Canadon Asfalto.
NOTES If an adult, the smallest known ornithischian.

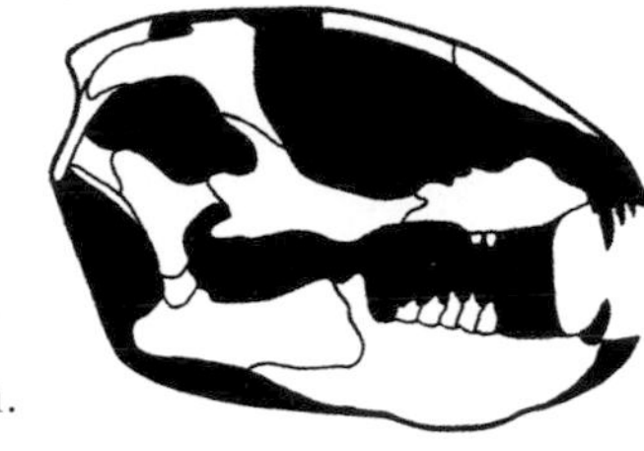

Fruitadens haagarorum

0.75 m (2.5 ft) TL, 0.8 kg (1.8 lb)

FOSSIL REMAINS Partial skulls and skeletons.
ANATOMICAL CHARACTERISTICS Short fang in front of lower tooth row.
AGE Late Jurassic, late Kimmeridgian or early or middle Tithonian.
DISTRIBUTION AND FORMATION/S Colorado; middle Morrison.
HABITAT Short wet season, otherwise semiarid with open floodplain prairies and riverine forests.

Tianyulong confuciusi

Adult size uncertain

FOSSIL REMAINS Partial skull and skeleton with external fibers, subadult.
ANATOMICAL CHARACTERISTICS Short fangs in front of main tooth rows. Fibers dense, thick and very long over trunk and especially tail.
AGE Early Cretaceous, early Aptian.
DISTRIBUTION AND FORMATION/S Northeast China; Yixian.
HABITAT Well-watered forests and lakes, winters chilly with some snow.
NOTES This fossil shows that at least some small ornithischians bore external fibers and that such body coverings evolved early in the group. Shared its habitat with *Psittacosaurus*, prey of *Sinornithosaurus millenii*.

MARGINOCEPHALIANS

SMALL TO GIGANTIC HETERODONTOSAURIFORM ORNITHISCHIANS OF THE LATE JURASSIC TO THE END OF THE DINOSAUR ERA, MOST CONTINENTS.

ANATOMICAL CHARACTERISTICS Very variable. Head large, heavily constructed, back of head broad, forming at least incipient crest, beaks narrow, eyes not very large. Tail long to fairly short. Fully bipedal to quadrupedal, arm and leg flexed and able to run. Fingers five to three, tipped with small blunt claws or hooves. Leg not gracile, so speed potential moderate, four long toes.
HABITS Low- to medium-level browsers. Defense included running and aggressive combat.

PACHYCEPHALOSAURS

SMALL TO LARGE MARGINOCEPHALIANS OF THE CRETACEOUS OF THE NORTHERN HEMISPHERE.

ANATOMICAL CHARACTERISTICS Fairly uniform. Head deep, massively constructed, beak small, skull roof thickened, may be domed in all adults, dome may

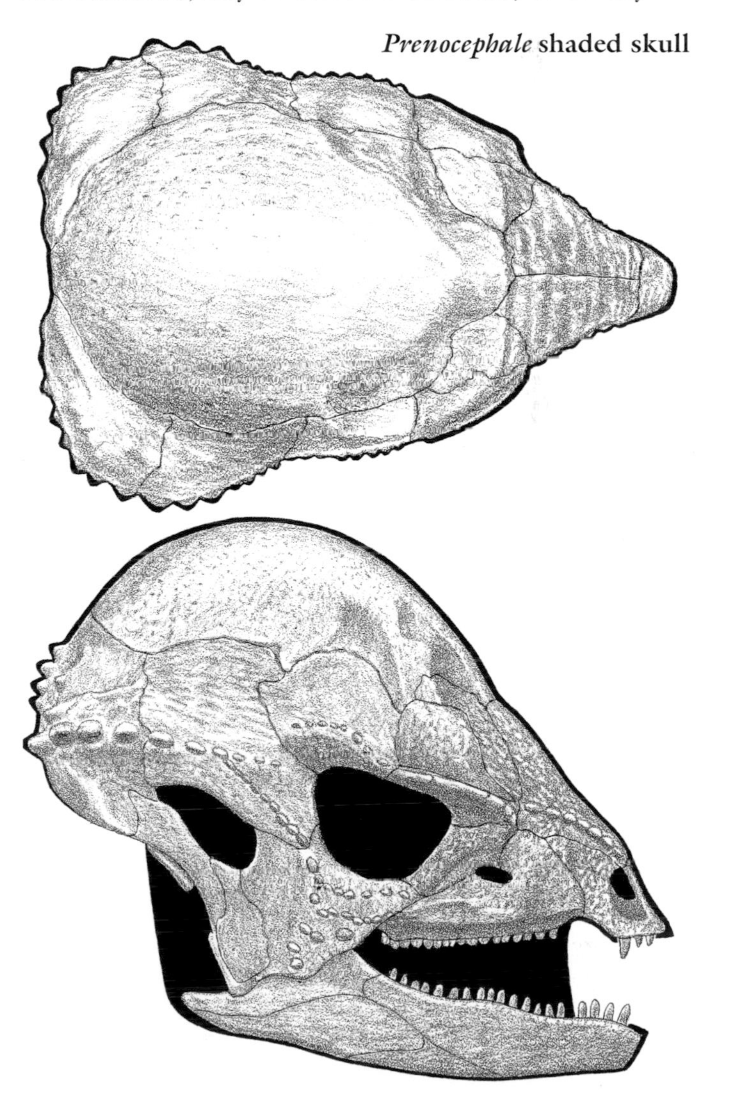

Prenocephale shaded skull

Prenocephale muscle study

be better developed in males in at least some species; adorned with rows of small hornlets, main tooth rows short, teeth small. Vertebrae heavily constructed. Trunk fairly long. Belly, hips, and base of tail very broad to accommodate enlarged belly. Tail long, base sweeps down from hips, last two-thirds stiffened by a dense basket-weave lattice of ossified tendons. Fully bipedal. Arm and hand small, five grasping fingers tipped with small blunt claws. Four long toes tipped with blunt claws.
HABITAT Variable, from semiarid to well-watered forests.
HABITS Defense included running and possible head butting. Males may have used domes for competitive displays. May have butted flanks of competitors and predators, high-speed head-to-head butting unlikely because of lack of a broad impact surface to provide stability.
NOTES Pachycephalosaurs are sometimes segregated into two groups, flat-headed and dome-headed, but it is probable that the former are immature examples of the dome-headed group, or females.

PACHYCEPHALOSAURIDS Small to large marginocephalians of the Cretaceous of the Northern Hemisphere.

Stenopelix valdensis
1.4 m (4.5 ft) TL, 10 kg (20 lb)

FOSSIL REMAINS Majority of skeleton.
ANATOMICAL CHARACTERISTICS Tail base not as broad as in other pachycephalosaurs.
AGE Early Cretaceous, Berriasian.
DISTRIBUTION AND FORMATION/S Central Germany; Obernkirchen Sandstein.

Wannanosaurus yansiensis
Adult size uncertain

FOSSIL REMAINS Majority of skull and minority of skeleton, probably immature.
ANATOMICAL CHARACTERISTICS Insufficient information.
AGE Late Cretaceous, Campanian.
DISTRIBUTION AND FORMATION/S Eastern China; Xiaoyan.
NOTES Probably an immature specimen.

Goyocephale (or *Stegoceras*) *lattimorei*
Adult size uncertain

FOSSIL REMAINS Partial skull and skeleton.
ANATOMICAL CHARACTERISTICS Standard for group.
AGE Late Cretaceous, late Santonian and/or early Campanian.
DISTRIBUTION AND FORMATION/S Mongolia; unnamed.
HABITAT Well-watered woodland with seasonal rain.

Stegoceras? brevis
1.5 m (5 ft) TL, 10 kg (20 lb)

FOSSIL REMAINS Skull domes.
ANATOMICAL CHARACTERISTICS Head not heavily adorned.
AGE Late Cretaceous, early and/or middle Campanian.
DISTRIBUTION AND FORMATION/S Alberta; Oldman, possibly Foremost.
HABITAT Well-watered, forested floodplain with coastal swamps and marshes, cool winters.
NOTES May include *Colepiocephale lambei.*

Stegoceras validum
2.2 m (7 ft) TL, 40 kg (80 lb)

FOSSIL REMAINS Skull domes, possibly complete skull and minority of skeleton.
ANATOMICAL CHARACTERISTICS Head not heavily adorned.
AGE Late Cretaceous, late Campanian.
DISTRIBUTION AND FORMATION/S Alberta; Dinosaur Park.
HABITAT Well-watered, forested floodplain with coastal swamps and marshes, cool winters.

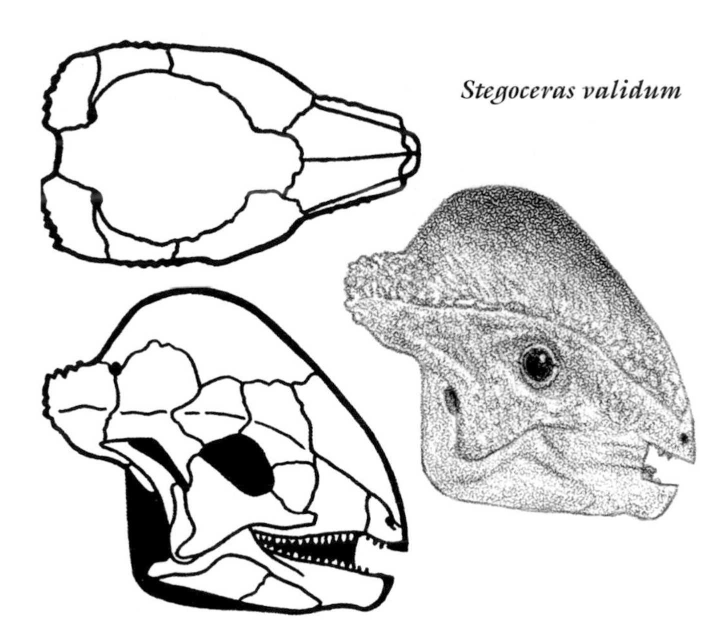

NOTES Because of inadequate remains, and because the level of many specimens is uncertain, it is possible that more than one species existed; may include *Hanssuesia sternbergi*, probably not present earlier or later than Dinosaur Park Formation.

Stegoceras? edmontonensis
2 m (6.5 ft) TL, 40 kg (80 lb)

FOSSIL REMAINS Skull domes.
ANATOMICAL CHARACTERISTICS Head not heavily adorned.
AGE Late Cretaceous, early Maastrichtian.
DISTRIBUTION AND FORMATION/S Alberta; Horseshoe Canyon.
HABITAT Well-watered, forested floodplain with coastal swamps and marshes, cool winters.

Tylocephale (or *Stegoceras*) *gilmorei*
2 m (6.5 ft) TL, 40 kg (80 lb)

FOSSIL REMAINS Partial skull.
ANATOMICAL CHARACTERISTICS Insufficient information.
AGE Late Cretaceous, Santonian or Campanian.
DISTRIBUTION AND FORMATION/S Mongolia; Barun Goyot.
HABITAT Semidesert with some dunes and oases.
NOTES Shared its habitat with *Ceratonykus* and *Bagaceratops*.

Prenocephale (or *Stegoceras*) *prenes*
2.2 m (7 ft) TL, 40 kg (80 lb)

FOSSIL REMAINS Complete skull with minority of skeleton.
ANATOMICAL CHARACTERISTICS Head not heavily adorned.
AGE Late Cretaceous, late Campanian and/or early Maastrichtian.

Prenocephale (or *Stegoceras*) *prenes*

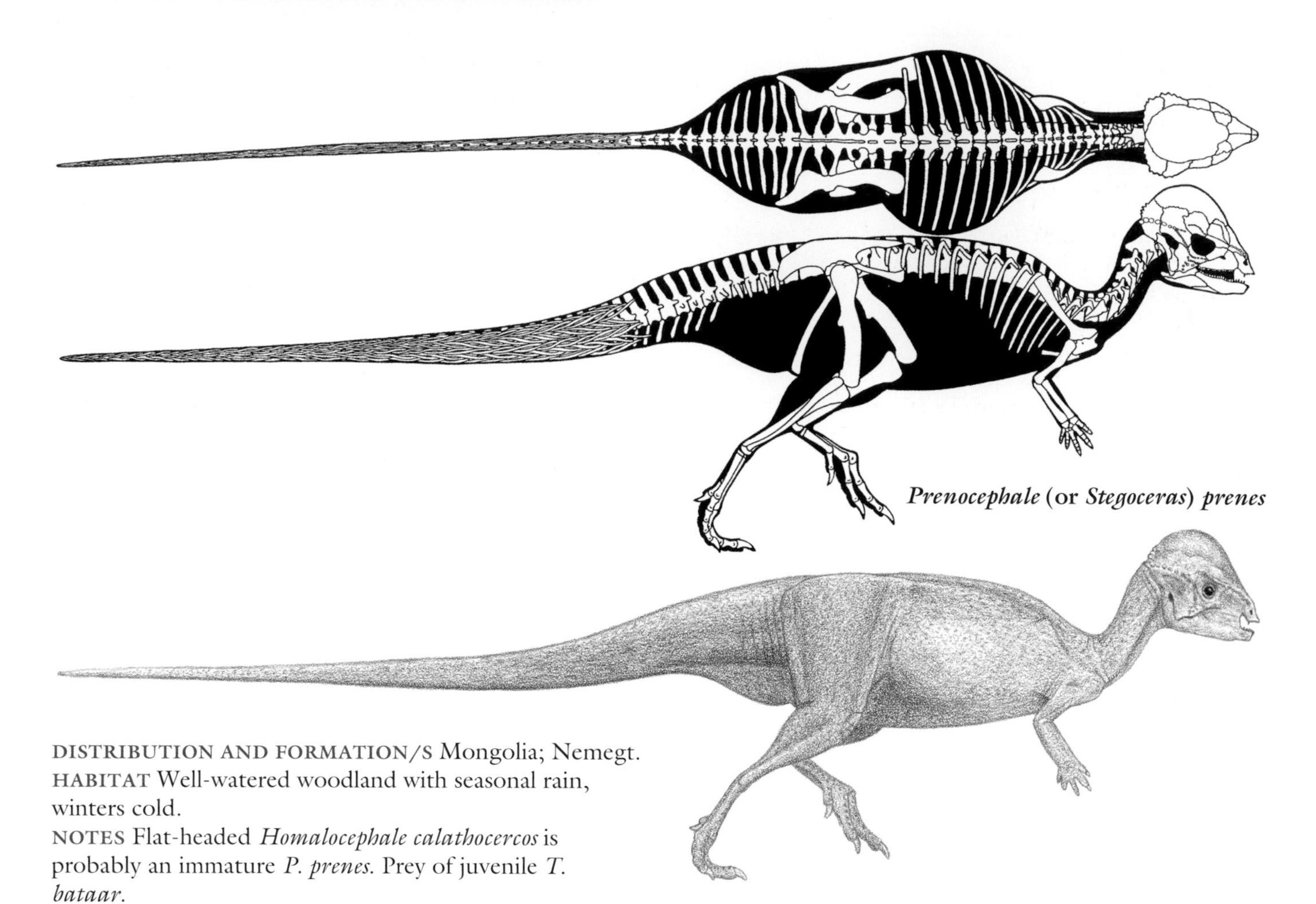
Prenocephale (or *Stegoceras*) *prenes*

DISTRIBUTION AND FORMATION/S Mongolia; Nemegt.
HABITAT Well-watered woodland with seasonal rain, winters cold.
NOTES Flat-headed *Homalocephale calathocercos* is probably an immature *P. prenes*. Prey of juvenile *T. bataar*.

Pachycephalosaurus wyomingensis
4.5 m (15 ft) TL, 450 kg (1,000 lb)

FOSSIL REMAINS A few skulls and majority of skeleton, juvenile to adult.
ANATOMICAL CHARACTERISTICS Large spikes on back of head in at least some adults.
AGE Late Cretaceous, late Maastrichtian.
DISTRIBUTION AND FORMATION/S Montana, South Dakota, Wyoming, Saskatchewan; Hell Creek, Lance, Scollard, levels uncertain.
HABITAT Well-watered coastal woodlands.
HABITS May have used blunt head spikes as additional impact weapons during combat within species.
NOTES *Dracorex hogwartsia* and *Stygimoloch spinifer* possible juveniles of this species and/or genus, in which case the spikes are a sexual characteristic, or there may be two species, the other being *P. spinifer*; species may be partly stratigraphic. Main enemy *Tyrannosaurus*.

Pachycephalosaurus wyomingensis

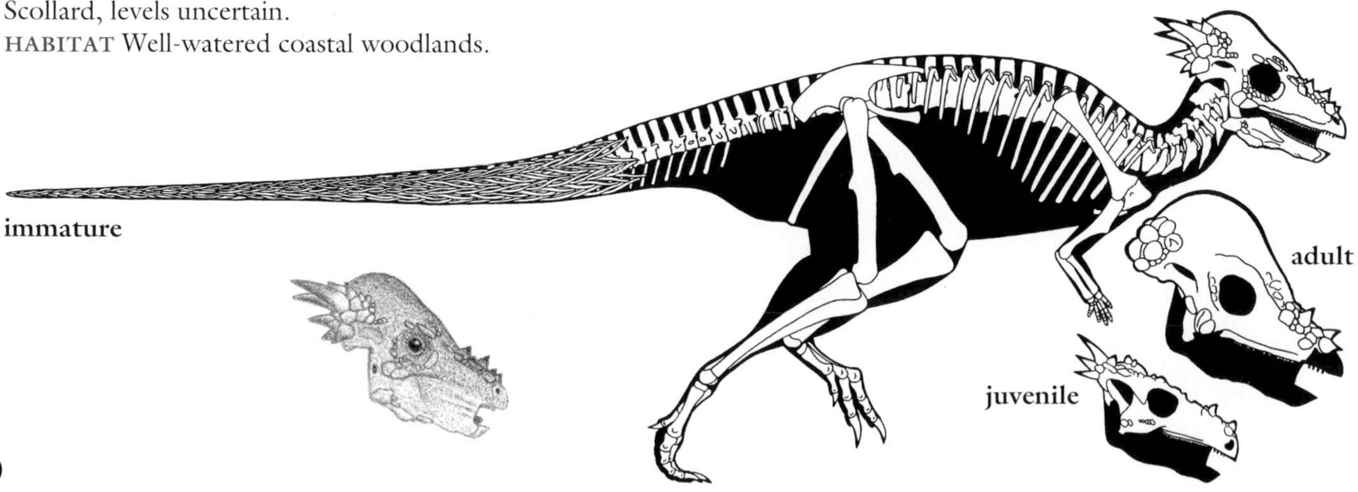

Ceratopsians

SMALL TO GIGANTIC MARGINOCEPHALIANS OF THE LATE JURASSIC TO THE END OF THE DINOSAUR ERA OF NORTH AMERICA AND ASIA.

ANATOMICAL CHARACTERISTICS Variable. Upper beaks set on rostral bone, teeth in main tooth rows form cutting and slicing edges.
HABITAT Very variable, from deserts to well-watered forests.
HABITS Probably omnivores that hunted small vertebrates and/or scavenged. Defense included biting. Very fragmentary, Early Cretaceous *Serendipaceratops arthurcclarkei* from Australia may indicate group's presence in Southern Hemisphere.

CHAOYANGSAURIDS Small ceratopsians limited to the Late Jurassic of Asia.

ANATOMICAL CHARACTERISTICS Head deep, eyes shaded by overhanging rim, beaks small, slightly hooked, teeth at front of upper jaw large, teeth in main rows chisel shaped. Bipedal except could move quadrupedally at slow speeds. Arm short, grasping fingers tipped with small blunt claws. Toes long and tipped with blunt claws.
HABITS Defense included biting with beak and front teeth.

Yinlong downsi
1.2 m (4 ft) TL, 10 kg (20 lb)

FOSSIL REMAINS Complete skull and nearly complete skeleton.

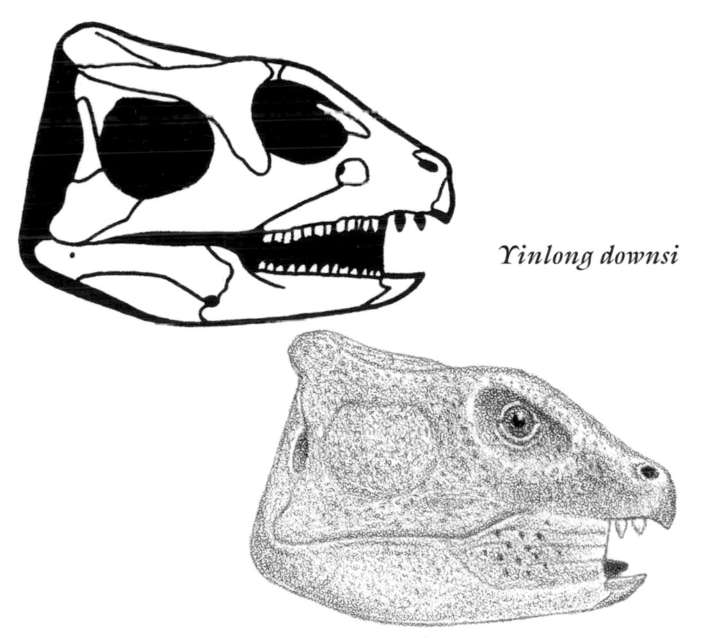
Yinlong downsi

ANATOMICAL CHARACTERISTICS Head moderately broad, back half very large.
AGE Late Jurassic, early Oxfordian.
DISTRIBUTION AND FORMATION/S Northwest China; upper Shishugou.
NOTES The earliest known ceratopsian. Prey of *Zuolong*.

Chaoyangsaurus youngi
1 m (3 ft) TL, 6 kg (13 lb)

FOSSIL REMAINS Partial skull and minority of skeleton.
ANATOMICAL CHARACTERISTICS Head very broad.
AGE Late Jurassic, Tithonian.
DISTRIBUTION AND FORMATION/S Northeast China; Tuchengzi.
HABITAT Well-watered forests and lakes, winters chilly with some snow.

Xuanhuaceratops niei
1 m (3 ft) TL, 6 kg (13 lb)

FOSSIL REMAINS Several partial skulls and skeletons.
ANATOMICAL CHARACTERISTICS Insufficient information.
AGE Probably Late Jurassic.
DISTRIBUTION AND FORMATION/S Northeast China; Houcheng.
HABITAT Well-watered forests and lakes, winters chilly with some snow.

Paxceratopsians

SMALL TO GIGANTIC CERATOPSIANS OF THE CRETACEOUS OF NORTH AMERICA AND ASIA.

ANATOMICAL CHARACTERISTICS Variable. Head deep, massively constructed, beaks deep, parrot shaped, jugal bosses very large, teeth in main tooth rows form cutting edges. Skin consists of large, rosette-pattern scales.
HABITAT Very variable, from deserts to well-watered forests.
HABITS Able to consume coarse vegetation, probably omnivores that hunted small vertebrates and/or scavenged with parrot beaks and slicing teeth. Defense included running, biting with beak.
NOTES Ceratopsians including psittacosaurs and neoceratopsians and their common ancestor.

PSITTACOSAURIDS Small ceratopsians limited to the Early to Mid-Cretaceous of Asia.

ANATOMICAL CHARACTERISTICS Highly uniform. Head extremely broad, beak not hooked, nostrils small, jugal bosses exceptionally large, eyes facing partly upward,

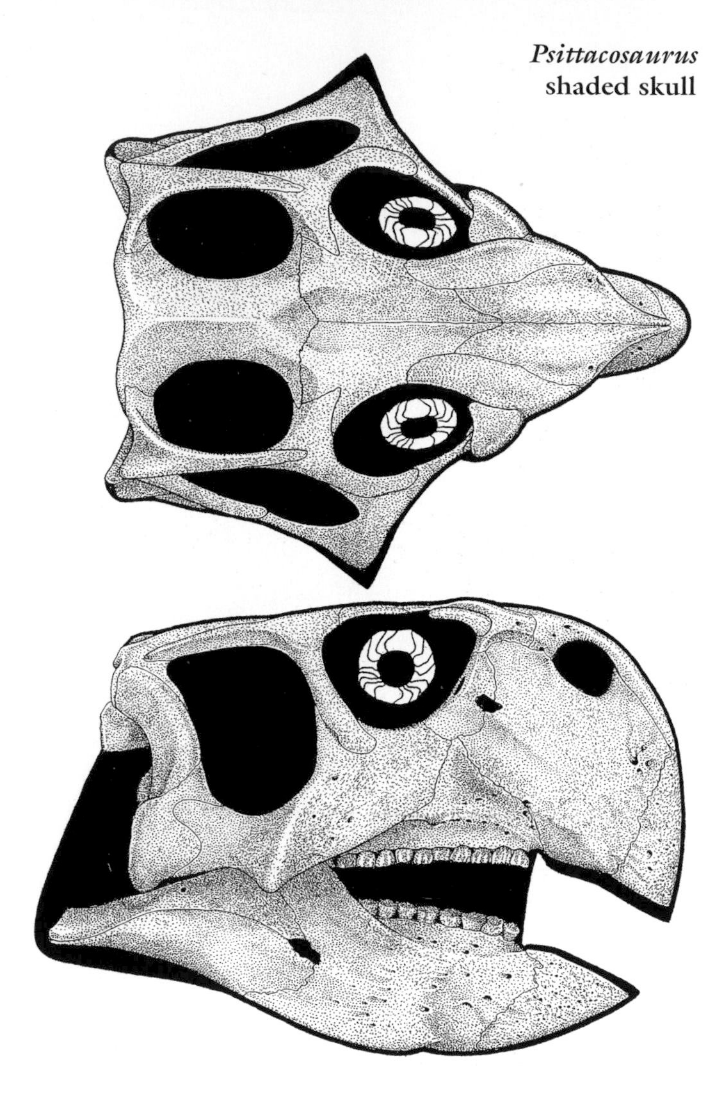

Psittacosaurus shaded skull

lower jaws very deep with flange on lower rim, no teeth on front of upper jaw, main tooth rows short, teeth chisel shaped. Tail fairly long. Bipedal except could move quadrupedally at slow speeds. Arm short, three grasping fingers tipped with small blunt claws. Toes long and tipped with blunt claws. Long bristle fibers atop at least first half of tail in at least some species. Gizzard stone bundles present.
ONTOGENY Growth rates moderate.
HABITAT Very variable, from deserts to well-watered forests.
HABITS May have consumed hard nuts as parrots do.

Hongshanosaurus houi

1.5 m (5 ft) TL, 15 kg (35 lb)

FOSSIL REMAINS Two nearly complete skulls, probably juvenile and adult.
ANATOMICAL CHARACTERISTICS Head subtriangular, snout large.
AGE Early Cretaceous, early Aptian.

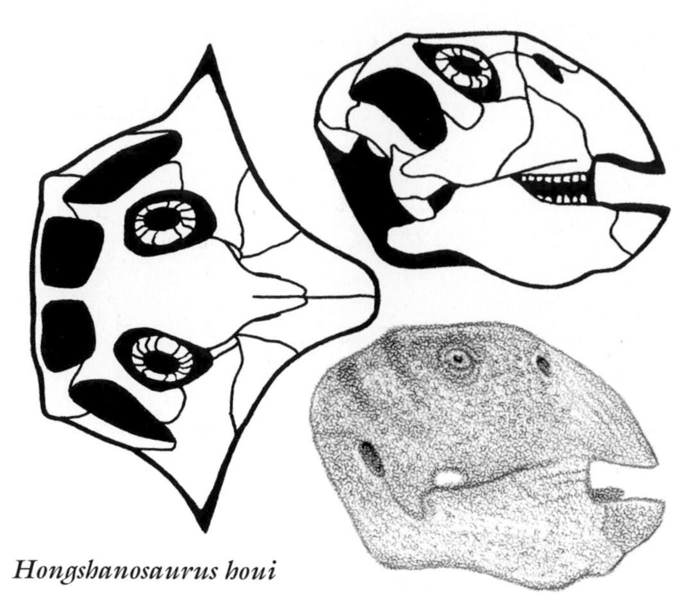

Hongshanosaurus houi

DISTRIBUTION AND FORMATION/S Northeast China; Yixian.
HABITAT Well-watered forests and lakes, winters chilly with some snow.

Psittacosaurus lujiatunensis

0.9 m (3 ft) TL, 5 kg (10 lb)

FOSSIL REMAINS Several skulls and small portion of skeleton.
ANATOMICAL CHARACTERISTICS Head subrectangular, snout short, wider than long.
AGE Early Cretaceous, probably late Hauterivian.
DISTRIBUTION AND FORMATION/S Northeast China; lowest Yixian.
HABITAT Well-watered forests and lakes, winters chilly with some snow.
NOTES The large number of distinctive yet similar species in this one genus appears to be correct.

Psittacosaurus lujiatuensis

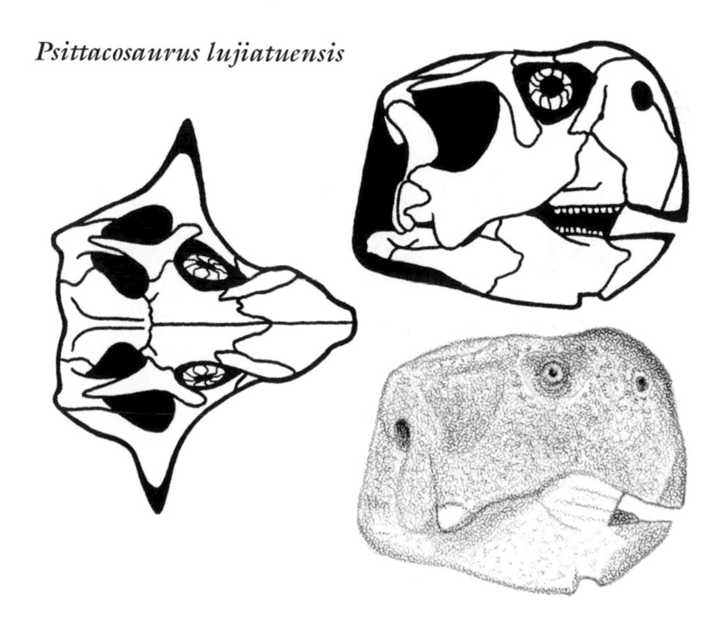

Psittacosaurus unnamed species?

1.2 m (4 ft) TL, 10 kg (20 lb)

FOSSIL REMAINS Nearly complete distorted skull and skeleton, skin and tail bristles.
ANATOMICAL CHARACTERISTICS Larger scales on upper arm, tail adorned with long bristles.
AGE Early Cretaceous, probably early Aptian.
DISTRIBUTION AND FORMATION/S Northeast China; Yixian.
HABITAT Well-watered forests and lakes, winters chilly with some snow.
NOTES Original discovery not documented, probably from higher in the Yixian than *P. lujiatunensis*. Prey of *Sinornithosaurus millenii*. May have been direct ancestor of *P. meileyingensis*.

Psittacosaurus meileyingensis

1.1 m (3.5 ft) TL, 8 kg (18 lb)

FOSSIL REMAINS Two complete skulls and partial skeleton.
ANATOMICAL CHARACTERISTICS Head subrectangular, snout short, eyes shaded by overhanging rim.
AGE Early Cretaceous, early or middle Aptian.
DISTRIBUTION AND FORMATION/S Northeast China; Jiufotang.
HABITAT Well-watered forests and lakes, winters chilly with some snow.
NOTES Prey of *Sinornithosaurus zhaoianus*.

Psittacosaurus sinensis

1 m (3 ft) TL, 6 kg (13 lb)

FOSSIL REMAINS Numerous skulls and skeletons, some complete, completely known.
ANATOMICAL CHARACTERISTICS Head subrectangular, snout short, jugal bosses are large spikes.

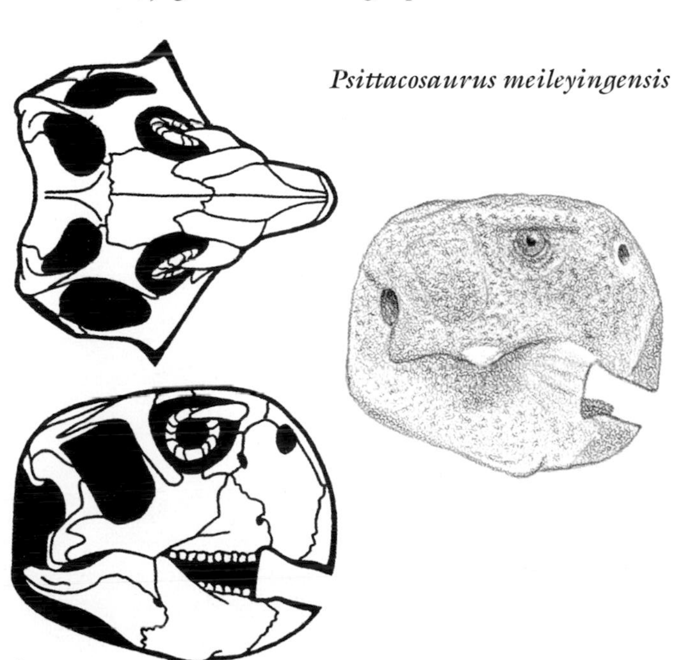
Psittacosaurus meileyingensis

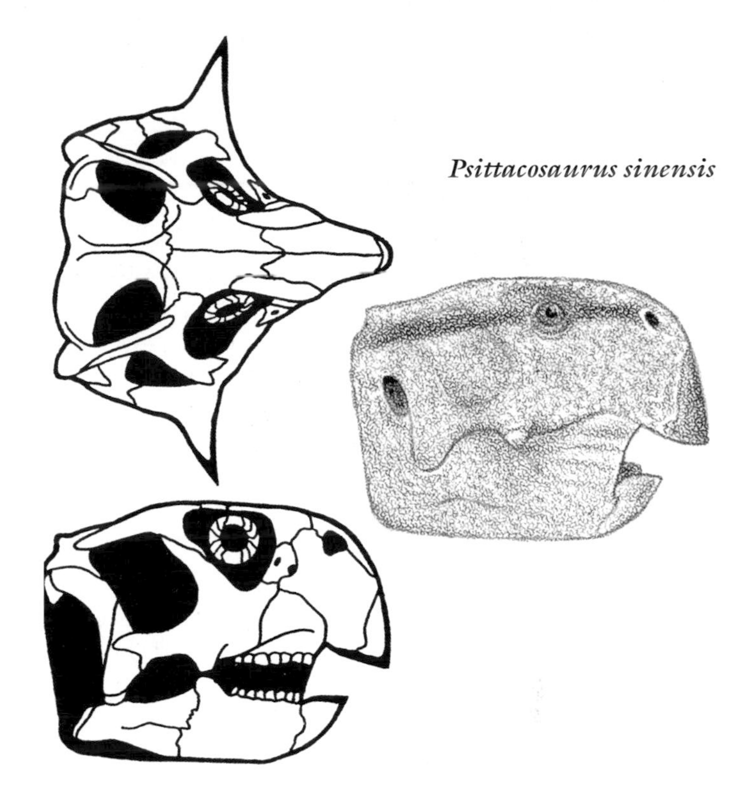
Psittacosaurus sinensis

AGE Early Cretaceous, Aptian or Albian.
DISTRIBUTION AND FORMATION/S Eastern China; Qingshan.
NOTES Probably includes *P. youngi*.

Psittacosaurus gobiensis

1 m (3 ft) TL, 6 kg (13 lb)

FOSSIL REMAINS Complete skull and majority of skeletons.
ANATOMICAL CHARACTERISTICS Head subrectangular, snout short, jugal bosses are large spikes, small triangular hornlet behind orbit.

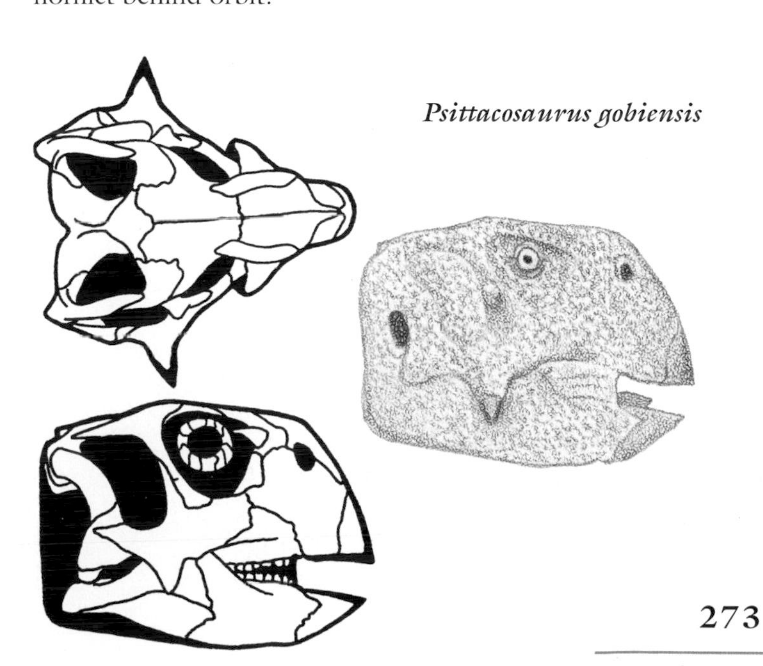
Psittacosaurus gobiensis

AGE Early Cretaceous, Aptian.
DISTRIBUTION AND FORMATION/S Northern China; Bayan Gobi.

Psittacosaurus sibiricus
1.5 m (5 ft) TL, 15 kg (35 lb)

FOSSIL REMAINS Complete skull and majority of skeleton, partial remains.
ANATOMICAL CHARACTERISTICS Head subrectangular, snout short, wider than long, exceptionally deep, large projections above and forward of orbits, jugal bosses are large spikes, flange on lower jaw enlarged.
AGE Early Cretaceous, Aptian or Albian.
DISTRIBUTION AND FORMATION/S Central Siberia; Ilek.

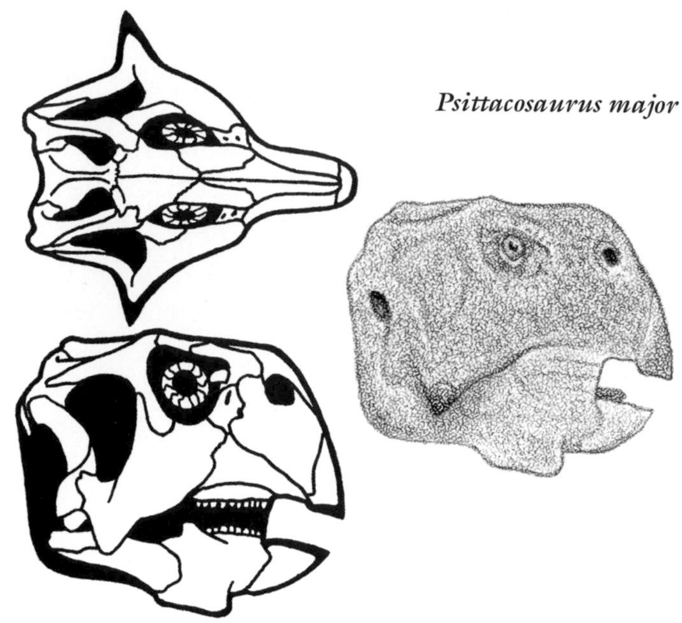

Psittacosaurus major

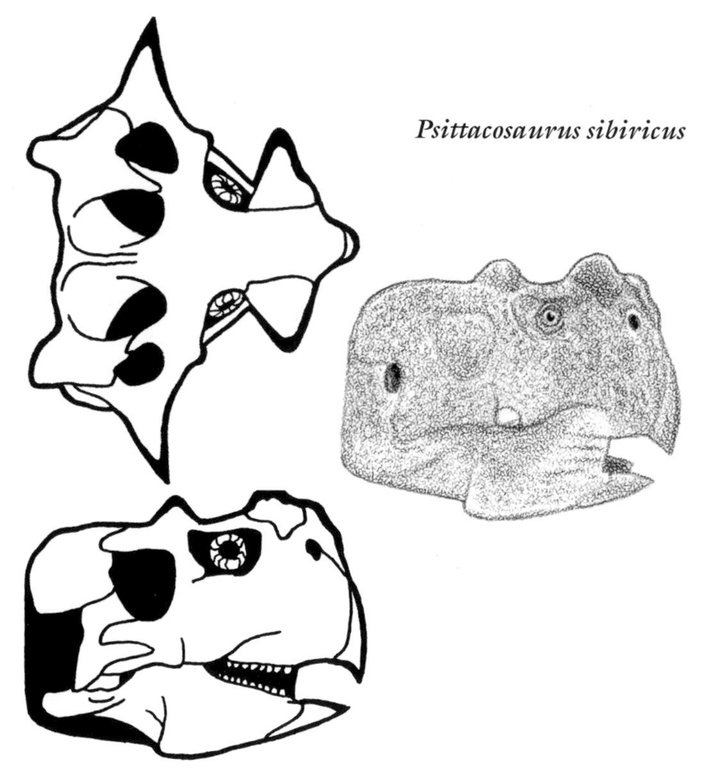

Psittacosaurus sibiricus

Psittacosaurus major
1.6 m (5 ft) TL, 18 kg (40 lb)

FOSSIL REMAINS Complete skull and nearly complete skeleton.
ANATOMICAL CHARACTERISTICS Head exceptionally large but rather narrow, deep, subrectangular, snout short.
AGE Early Cretaceous, early Aptian.
DISTRIBUTION AND FORMATION/S Northeast China; Yixian.
HABITAT Well-watered forests and lakes, winters chilly with some snow.
NOTES Narrowness of the large head may have reduced its weight in this biped.

Psittacosaurus neimongoliensis
1.1 m (3.5 ft) TL, 8 kg (18 lb)

FOSSIL REMAINS Nearly complete skull and skeleton.
ANATOMICAL CHARACTERISTICS Head subrectangular, snout short.

Psittacosaurus neimongoliensis

scales and tail fibers based on Yixian *Psittacosaurus*

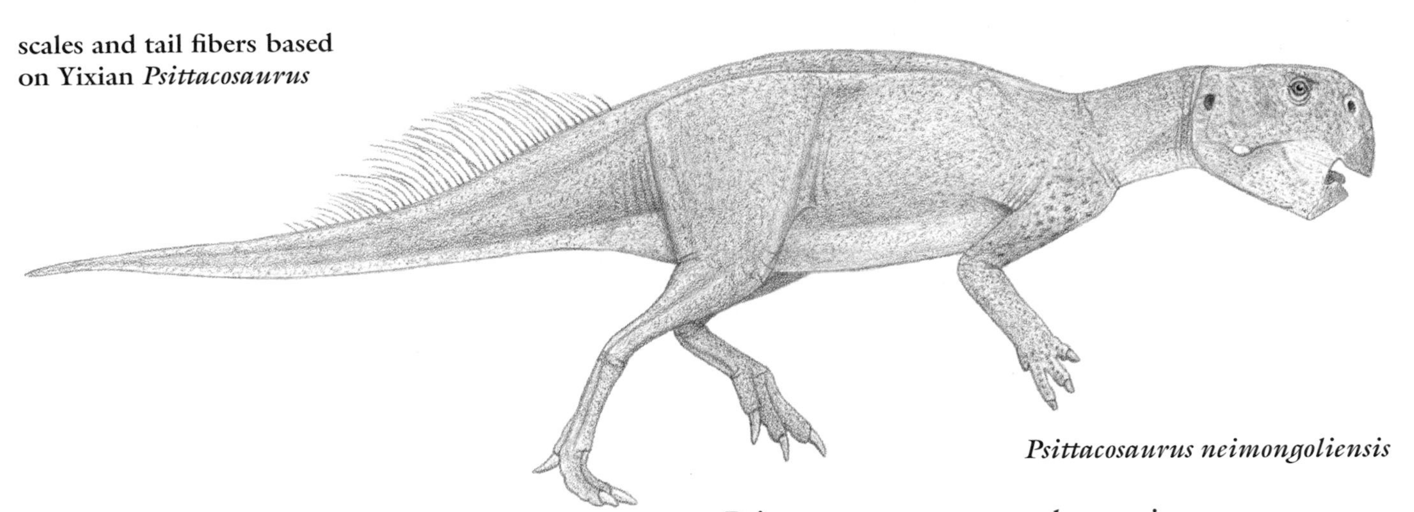

Psittacosaurus neimongoliensis

AGE Late Early Cretaceous.
DISTRIBUTION AND FORMATION/S Northern China; Ejinhoro.
NOTES Probably includes *P. ordosensis*. Prey of *Sinornithoides*.

Psittacosaurus mongoliensis
1.5 m (5 ft) TL, 15 kg (35 lb)

FOSSIL REMAINS Dozens of skulls and skeletons, many complete, juvenile to adult, completely known.
ANATOMICAL CHARACTERISTICS Head subrectangular, snout short.
AGE Early Cretaceous, Aptian and/or Albian.
DISTRIBUTION AND FORMATION/S Southern Siberia, Mongolia, northern China; Khukhtekskaya Svita, Khulsyngolskaya Svita, Shestakovskaya Svita.
NOTES The classic psittacosaur.

Psittacosaurus mazongshanensis
Size uncertain

FOSSIL REMAINS Nearly complete skull and partial skeleton.
ANATOMICAL CHARACTERISTICS Snout rather long.
AGE Late Early Cretaceous.
DISTRIBUTION AND FORMATION/S Central China; Xinminbo Group.

Psittacosaurus xinjiangensis
Adult size uncertain

FOSSIL REMAINS Minority of skull and majority of skeleton, large juvenile.
ANATOMICAL CHARACTERISTICS Head subrectangular, snout short, eyes shaded by overhanging rim.
AGE Late Early Cretaceous.

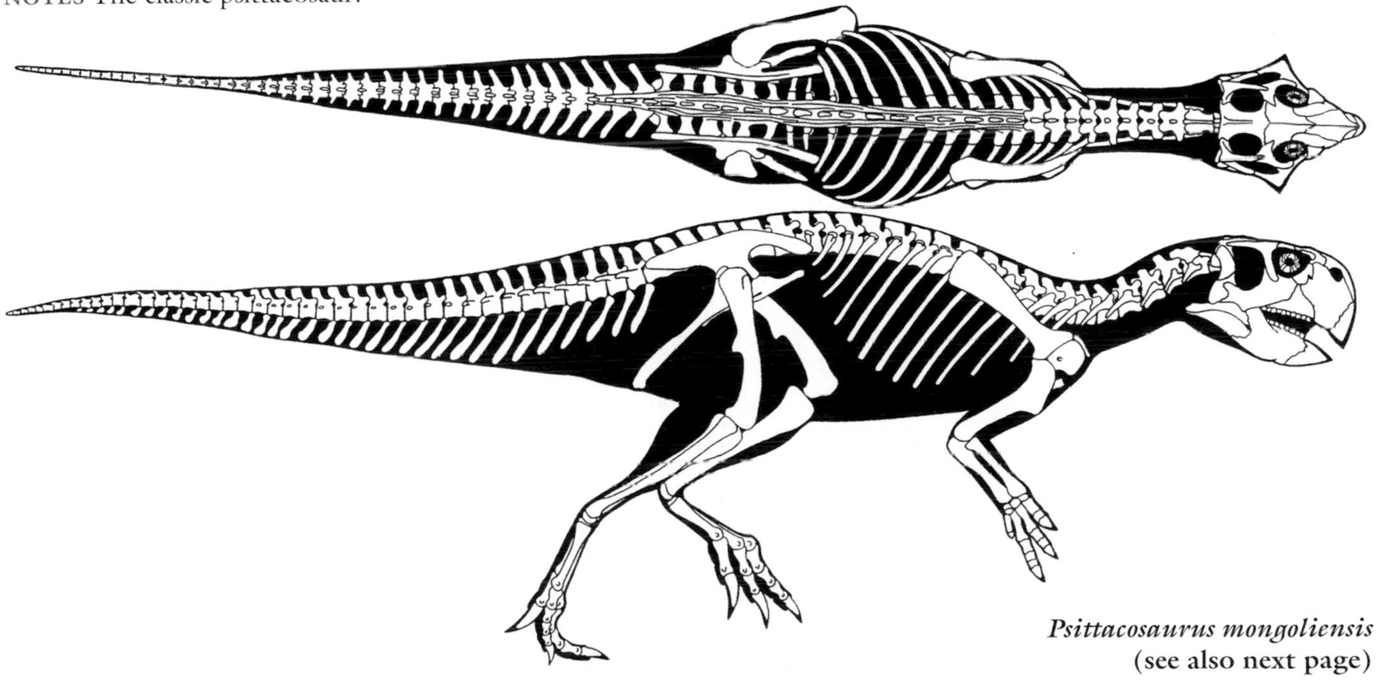

Psittacosaurus mongoliensis
(see also next page)

Psittacosaurus mongoliensis

DISTRIBUTION AND FORMATION/S Northwest China; Tugulu Group.

HABITS Oversized heads may have been at least in part for competitive display within species. Defense probably often aggressive as in suids and rhinos.

Neoceratopsians Small to gigantic ceratopsians of the Cretaceous of North America and Asia.

ANATOMICAL CHARACTERISTICS Moderately variable. Head exceptionally large, jugal bosses moved back toward jaw joint, upper beaks at least slightly hooked. Neck straight. Trunk stiffened by ossified tendons. Tail not long. Largely quadrupedal, may have been able to gallop as well as trot, shoulders somewhat lower than hips. Hand and five fingers short, tipped with hooves. Toes tipped with hooves.
HABITAT Highly variable, from deserts to well-watered forests.

Protoceratopsids Small to fairly large neoceratopsians of the Cretaceous of North America and Asia.

ANATOMICAL CHARACTERISTICS Nasal openings set high and not very large, main tooth rows short, teeth chisel shaped. Pelvic ilium horizontal. May have been able to run bipedally as well as quadrupedally. Toes long.
HABITAT Highly variable, from deserts to well-watered forests.
NOTES Group may be splittable into a number of subdivisions. Closest living analogs suids.

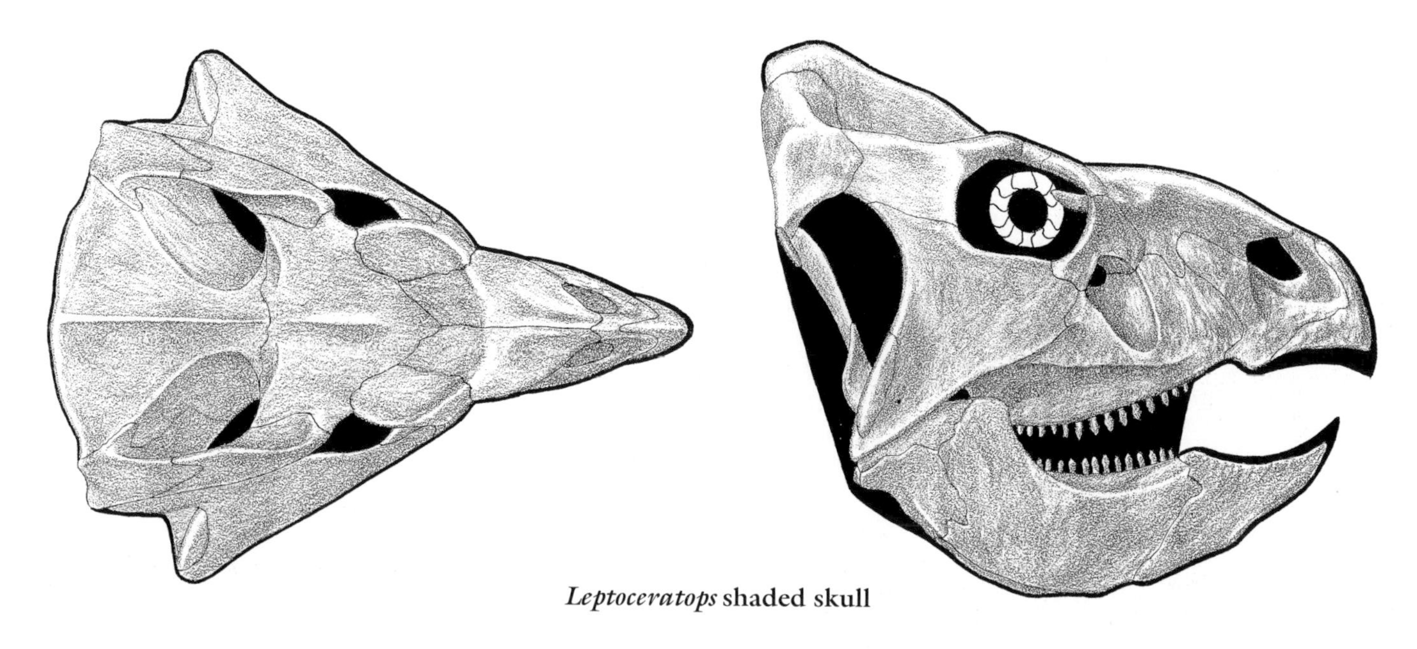

Leptoceratops shaded skull

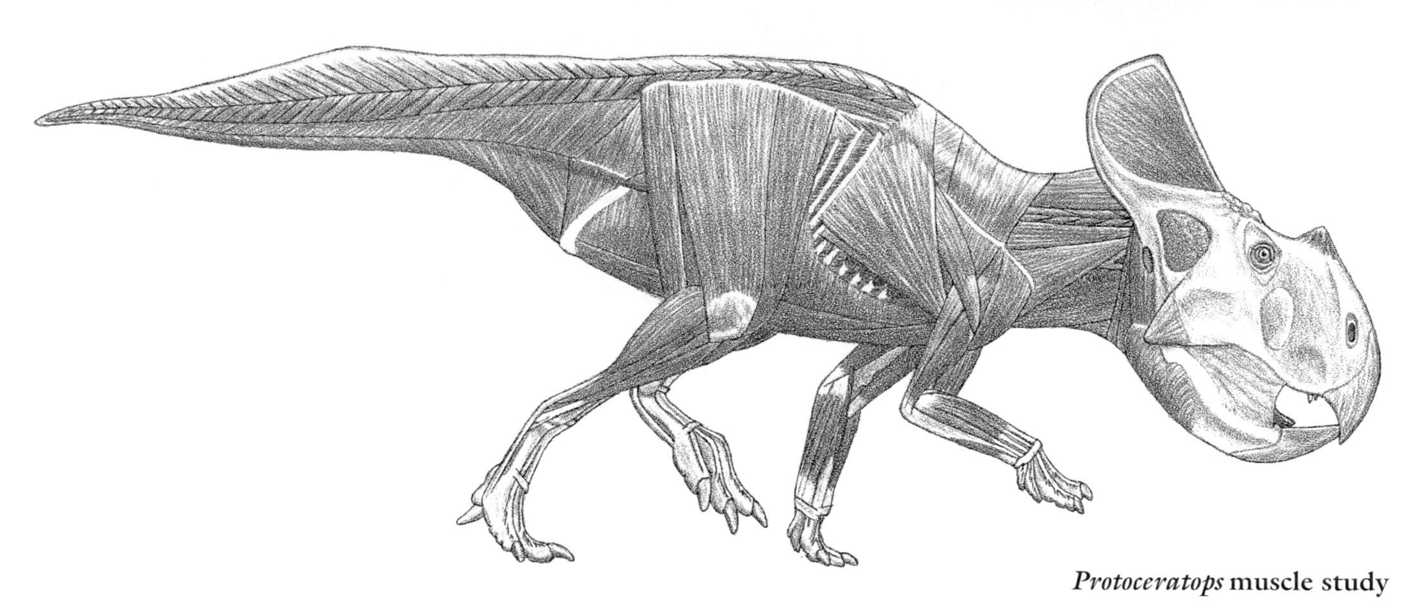

Protoceratops muscle study

Mosaiceratops azumai

1 m (3 ft) TL, 10 kg (20 lb)

FOSSIL REMAINS Nearly complete skulls and minority of skeleton.
ANATOMICAL CHARACTERISTICS Head very broad.
AGE Late Cretaceous, Turonian or Campanian.
DISTRIBUTION AND FORMATION/S Eastern China; Xiaguan.

Mosaiceratops azumai

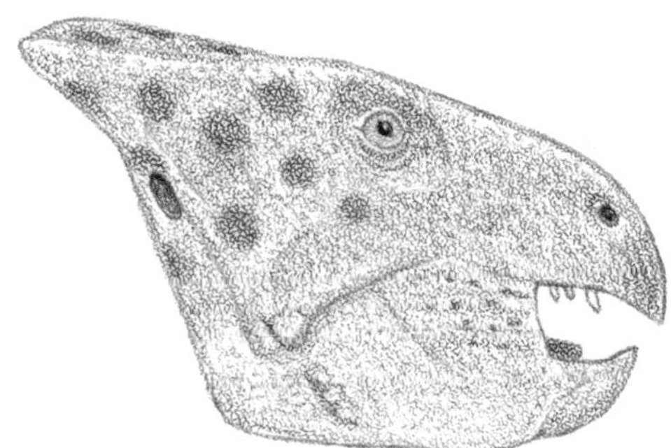

Liaoceratops yanzigouensis

Liaoceratops yanzigouensis

0.5 m (1.7 ft) TL, 2 kg (4 lb)

FOSSIL REMAINS Two skulls, juvenile and adult.
ANATOMICAL CHARACTERISTICS Head deep, frill very short and not broad, teeth near front of upper jaw.
AGE Early Cretaceous, Barremian.
DISTRIBUTION AND FORMATION/S Northeast China; lower Yixian.
HABITAT Well-watered forests and lakes, winters chilly with some snow.

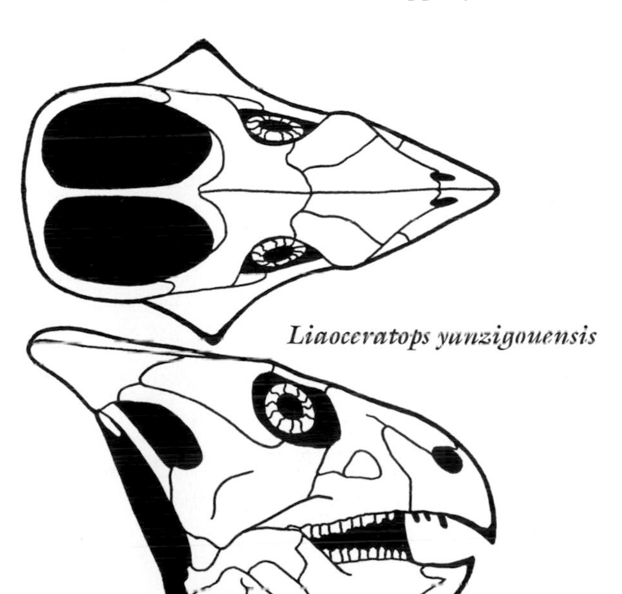

Liaoceratops yanzigouensis

Aquilops americanus

Adult size uncertain

FOSSIL REMAINS Majority of skull, probably juvenile.
ANATOMICAL CHARACTERISTICS Standard for group.
AGE Early Cretaceous, late Aptian to early Albian.
DISTRIBUTION AND FORMATION/S Wyoming; Cloverly.
HABITAT Short wet season, otherwise semiarid with floodplain prairies, open woodlands, and riverine forests.
NOTES Shared its habitat with *Tenontosaurus tilletti*.

Aquilops americanus

Archaeoceratops oshiami
0.9 m (3 ft) TL, 10 kg (20 lb)

FOSSIL REMAINS Skull and partial skeleton.
ANATOMICAL CHARACTERISTICS Head deep, eyes shaded by overhanging rim, frill incipient, teeth near front of upper jaw.
AGE Early Cretaceous, Aptian.
DISTRIBUTION AND FORMATION/S Central China; Xinminbo.
NOTES It is not known whether this is a short-armed biped or longer-armed quadruped.

Cerasinops hodgskissi
2.5 m (8 ft) TL, 175 kg (380 lb)

FOSSIL REMAINS Partial skull and skeleton.
ANATOMICAL CHARACTERISTICS Head deep, frill short, jaw deep. Arm short compared to hind limb, so possibly more bipedal than other protoceratopsids.
AGE Late Cretaceous, Santonian.
DISTRIBUTION AND FORMATION/S Montana; Lower Two Medicine.
HABITAT Seasonally dry upland woodlands.

Helioceratops brachygnathus
1.3 m (4.3 ft) TL, 20 kg (40 lb)

FOSSIL REMAINS Minority of skull.
ANATOMICAL CHARACTERISTICS Lower jaw very deep along entire length.
AGE Late Early Cretaceous or early Late Cretaceous.
DISTRIBUTION AND FORMATION/S Northeast China; Quantou.
HABITAT Well-watered forests and lakes, winters chilly with some snow.
NOTES Shared its habitat with *Changchunsaurus*.

Udanoceratops tschizhovi
4 m (13 ft) TL, 700 kg (1,500 lb)

FOSSIL REMAINS Majority of skull and minority of skeleton.
ANATOMICAL CHARACTERISTICS Head deep, nasal opening large, lower jaw extremely deep and massive. Arm short compared to hind limb, so possibly more bipedal than other protoceratopsids.
AGE Late Cretaceous, Campanian.
DISTRIBUTION AND FORMATION/S Mongolia; Djadokhta.
HABITAT Desert with dunes and oases.
HABITS Healed break in lower jaws implies intense impact combat within species.
NOTES May include *Bainoceratops efremovi*. Prey of *Tsaagan* and *Velociraptor*.

Udanoceratops tschizhovi

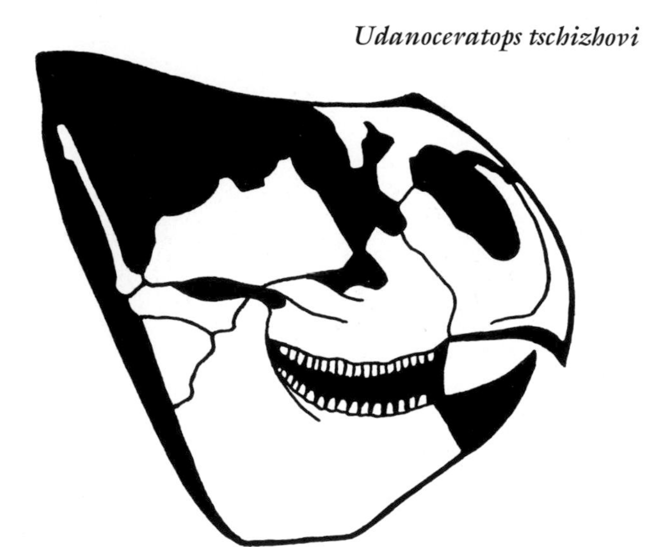

Archaeoceratops oshiami

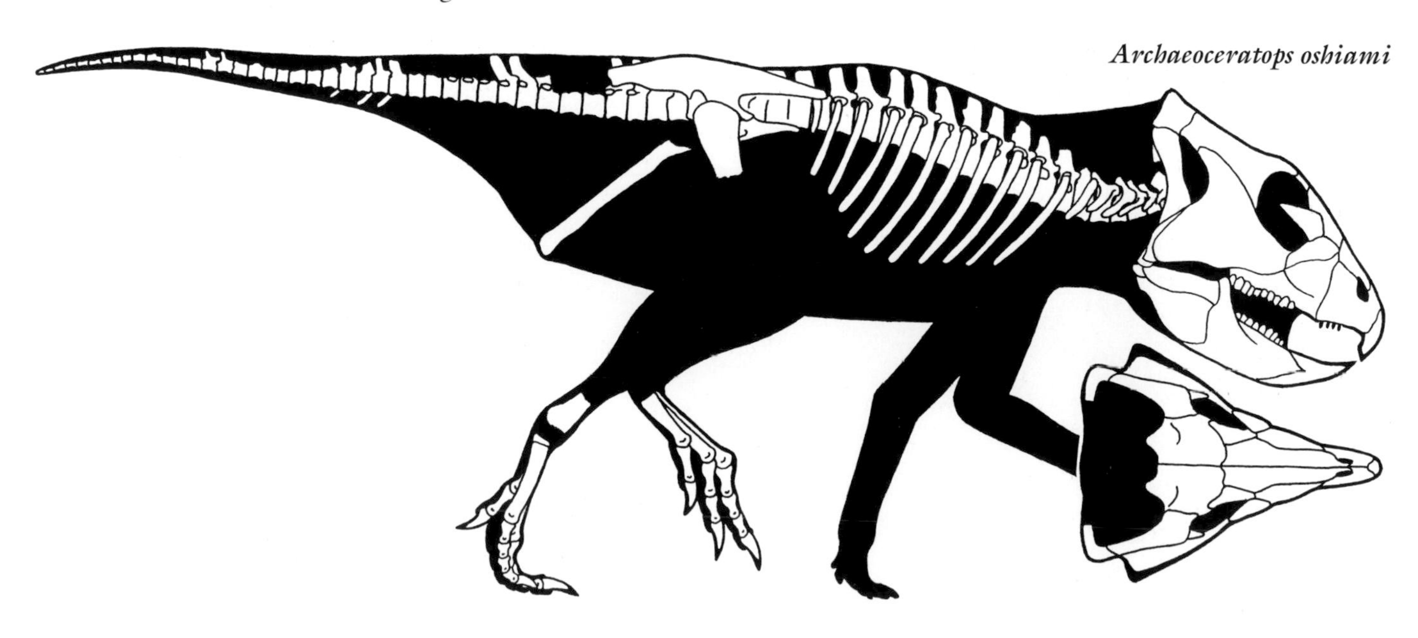

Zhuchengceratops inexpectus
2.5 m (8 ft) TL, 175 kg (380 lb)

FOSSIL REMAINS Minority of skull and skeleton.
ANATOMICAL CHARACTERISTICS Lower jaw extremely deep and massive.
AGE Late Late Cretaceous.
DISTRIBUTION AND FORMATION/S Eastern China; Wangshi Group.
NOTES Shared its habitat with *Sinoceratops*.

Yamaceratops dorngobiensis
0.5 m (1.7 ft) TL, 2 kg (4 lb)

FOSSIL REMAINS Majority of skull and minority of skeletons.
ANATOMICAL CHARACTERISTICS Frill short, fairly broad.
AGE Probably Late Cretaceous, possibly Santonian or Campanian.
DISTRIBUTION AND FORMATION/S Mongolia; Javkhlant.

Auroraceratops rugosus
6 m (20 ft) TL, 1.3 tonnes

FOSSIL REMAINS Nearly complete skull.
ANATOMICAL CHARACTERISTICS Snout short, frill incipient, teeth near front of upper jaw, front of lower jaw shallow, and lower beak pointed.
AGE Early Cretaceous, Aptian.
DISTRIBUTION AND FORMATION/S Northern China; Xinminpu.

Prenoceratops pieganensis (Illustrated overleaf)
1.3 m (4.3 ft) TL, 20 kg (40 lb)

FOSSIL REMAINS Complete skull and skeleton from large number of bones.
ANATOMICAL CHARACTERISTICS Head deep, full extent of frill uncertain, lower jaw very deep.
AGE Late Cretaceous, middle and/or late Campanian.
DISTRIBUTION AND FORMATION/S Montana; Upper Two Medicine.
HABITAT Seasonally dry upland woodlands.

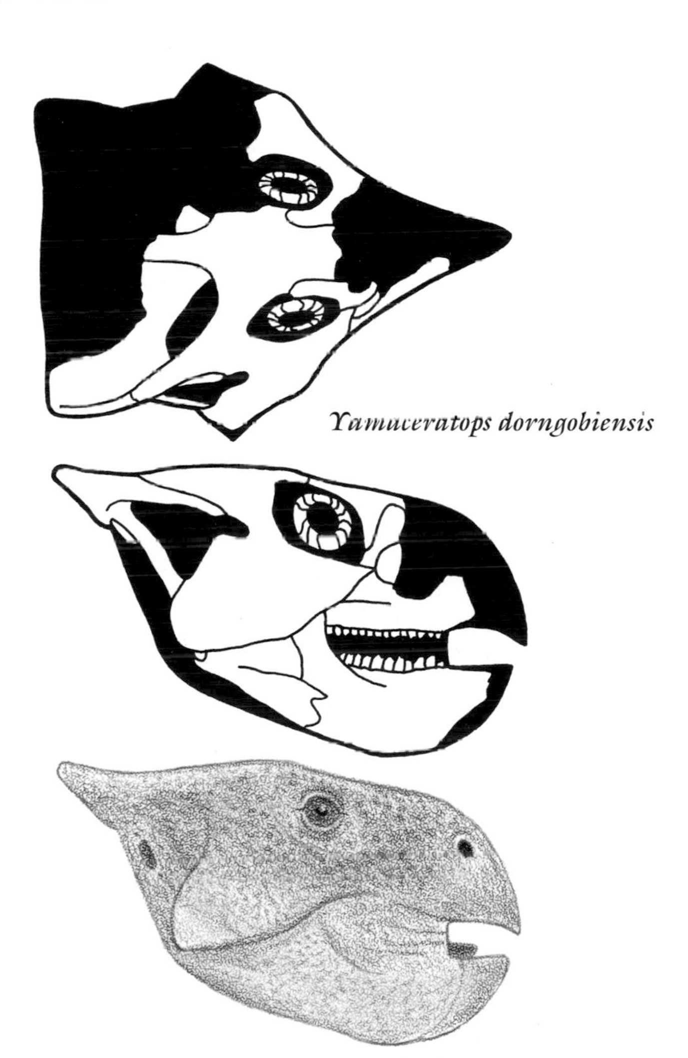
Yamaceratops dorngobiensis

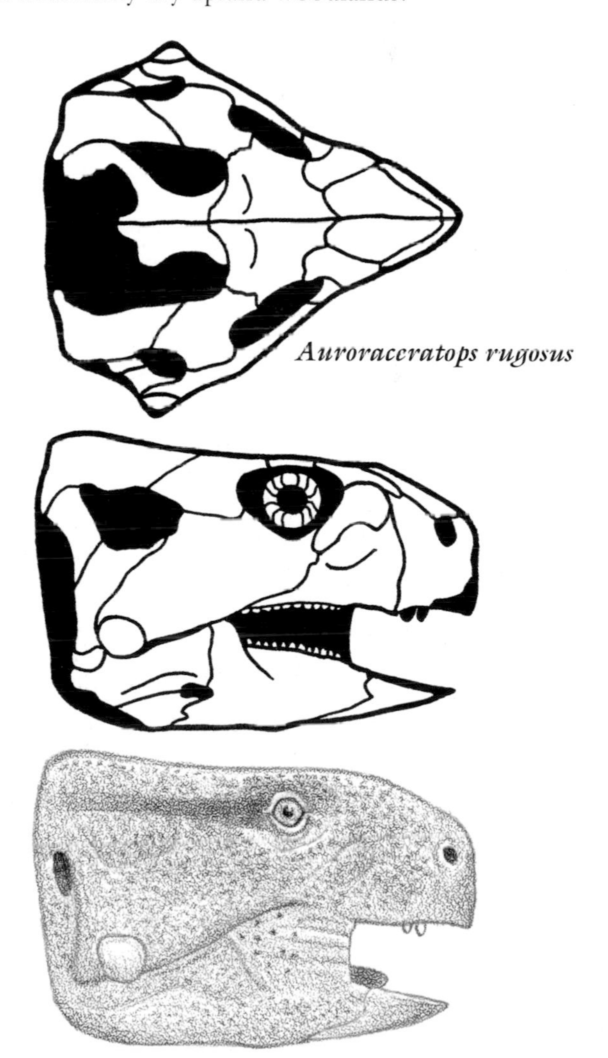
Auroraceratops rugosus

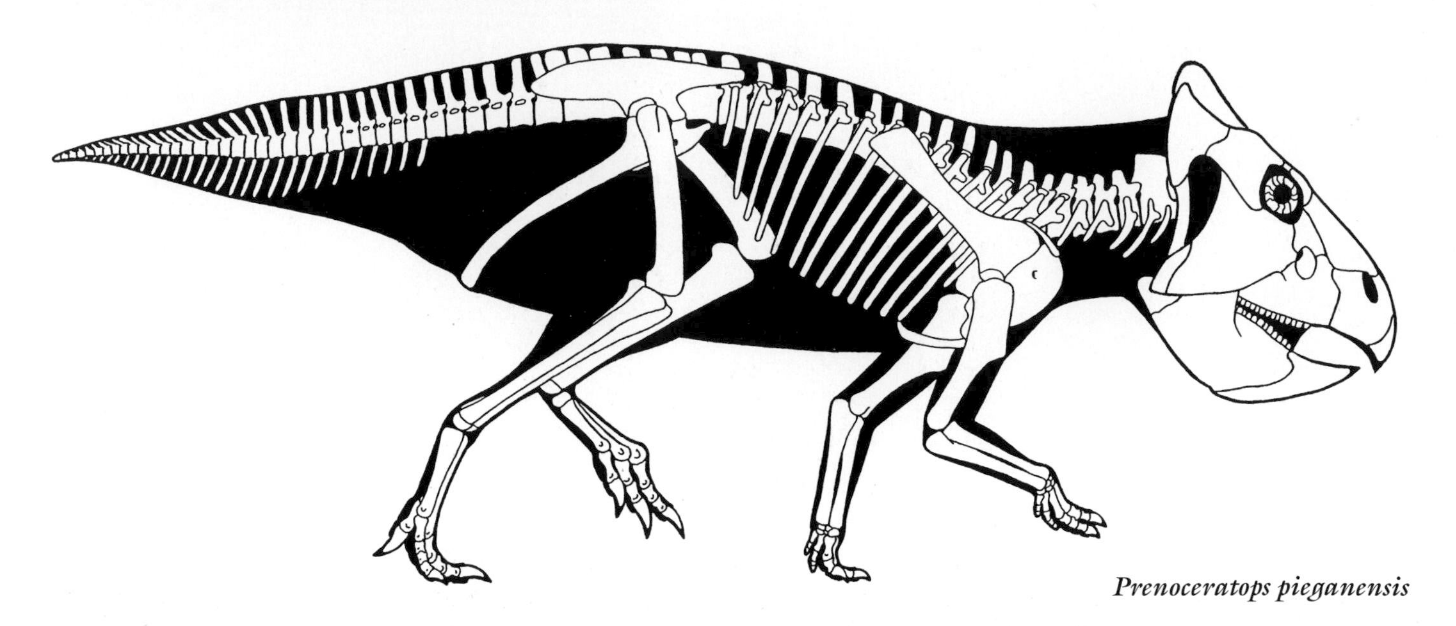

Prenoceratops pieganensis

Leptoceratops gracilis
2 m (6.5 ft) TL, 100 kg (200 lb)

FOSSIL REMAINS Some skulls and skeletons.
ANATOMICAL CHARACTERISTICS Head extremely large, deep, eyes shaded by overhanging rim, frill incipient, lower jaw very deep.
AGE Late Cretaceous, late Maastrichtian.
DISTRIBUTION AND FORMATION/S Montana, Wyoming, Alberta; Hell Creek, Lance, Scollard, levels uncertain.
HABITAT Upland forests.
NOTES May be more than one stratigraphic species.

Montanoceratops cerorhynchus
2.5 m (8 ft) TL, 170 kg (375 lb)

FOSSIL REMAINS Partial skull and skeleton.
ANATOMICAL CHARACTERISTICS Head deep, nasal horn absent, full extent of frill uncertain. Tall vertebral spines over tail form a shallow sail.
AGE Late Cretaceous, latest Campanian and/or early Maastrichtian.
DISTRIBUTION AND FORMATION/S Alberta, Montana; Saint Mary River, lower Horseshoe Canyon.
HABITAT Well-watered, forested floodplain with coastal swamps and marshes.

Leptoceratops gracilis

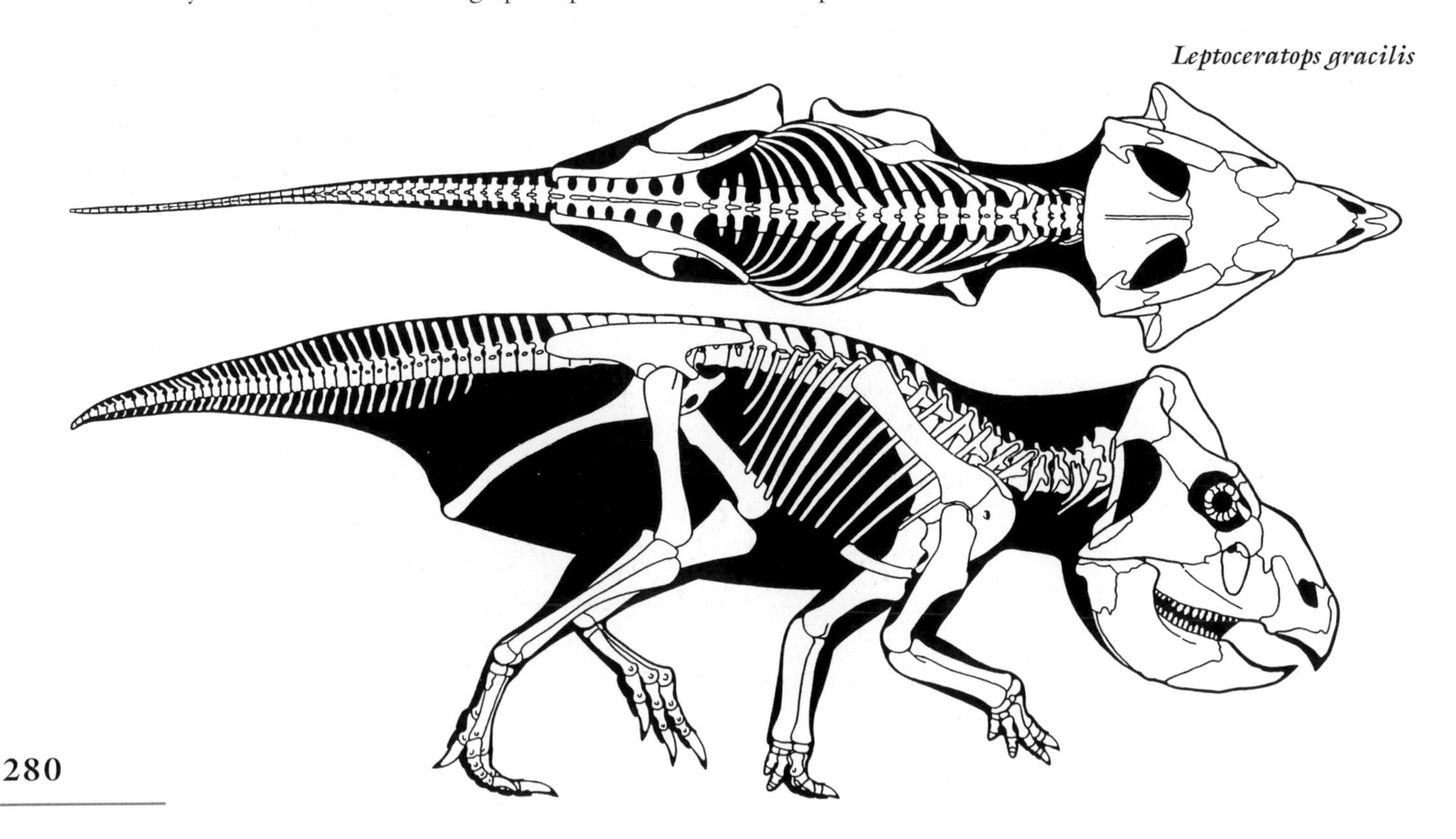

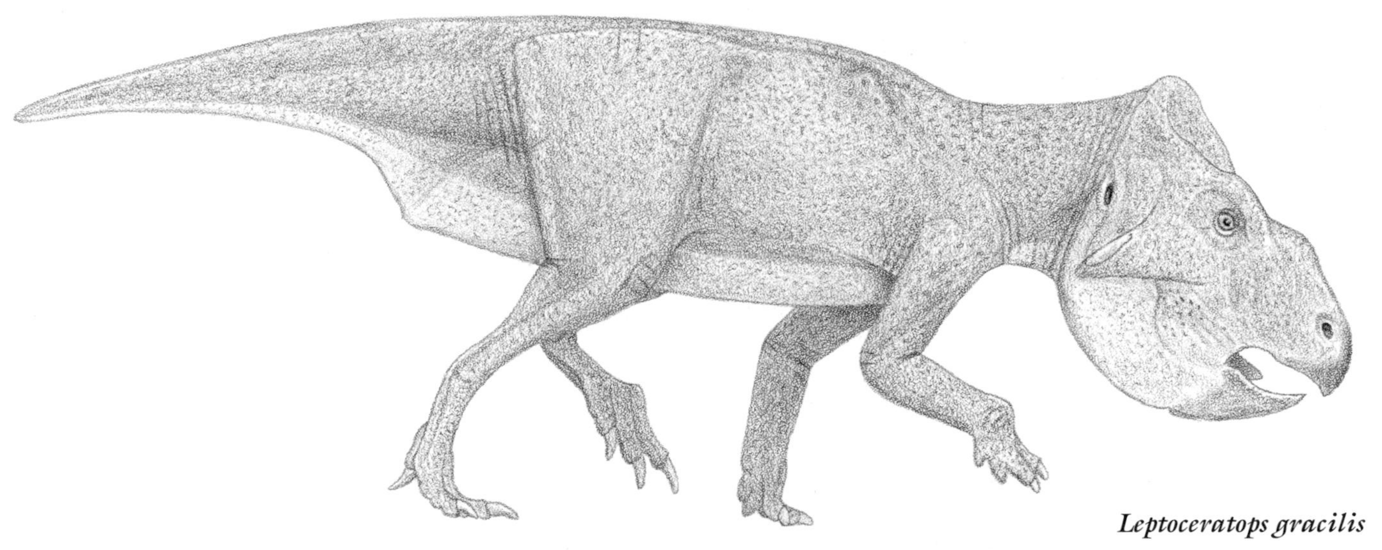

Leptoceratops gracilis

Leptoceratops gracilis

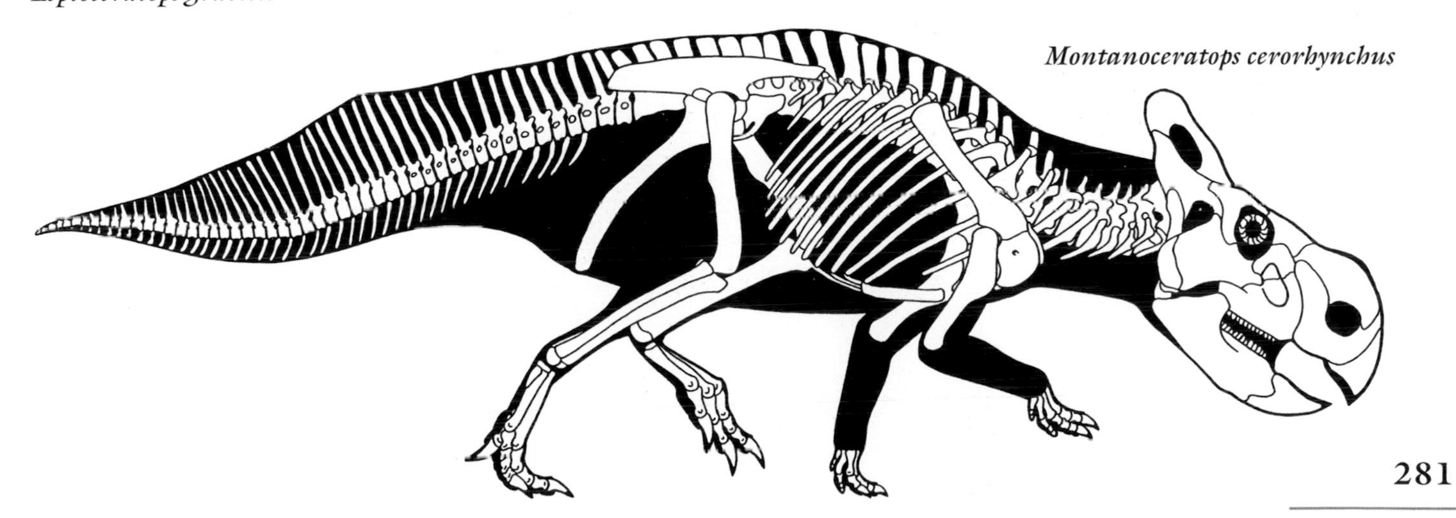

Montanoceratops cerorhynchus

Graciliceratops mongoliensis
Adult size uncertain

FOSSIL REMAINS Partial skull and skeleton.
ANATOMICAL CHARACTERISTICS Head frill short, not broad. Leg long.
AGE Late Cretaceous.
DISTRIBUTION AND FORMATION/S Mongolia; Shireegiin Gashuun.

Protoceratops andrewsi
2.5 m (8 ft) TL, 180 kg (400 lb)

FOSSIL REMAINS Many dozens of skulls and skeletons, many complete, juvenile to adult, completely known.
ANATOMICAL CHARACTERISTICS Head very large and broad, deep, incipient nasal horn, frill large and broad, teeth near front of upper jaw. Tall vertebral spines over tail form a shallow sail.
AGE Late Cretaceous, Campanian.
DISTRIBUTION AND FORMATION/S Mongolia, northern China; Djadokhta, Minhe.
HABITAT Desert with dunes and oases.
HABITS A *Protoceratops* is preserved biting on the arm of a *Velociraptor*, its main enemy.
NOTES The classic protoceratopsid. Shared its habitat with *Udanoceratops*.

Protoceratops hellenikorhinus?
2.5 m (8 ft) TL, 180 kg (400 lb)

FOSSIL REMAINS Several skulls and minority of skeletons.
ANATOMICAL CHARACTERISTICS Head very large and broad, and deep; nasal horn prominent; frill large and very broad.
AGE Late Cretaceous, Campanian.
DISTRIBUTION AND FORMATION/S Mongolia; Bayan Mandahu.
HABITAT Desert with dunes and oases.

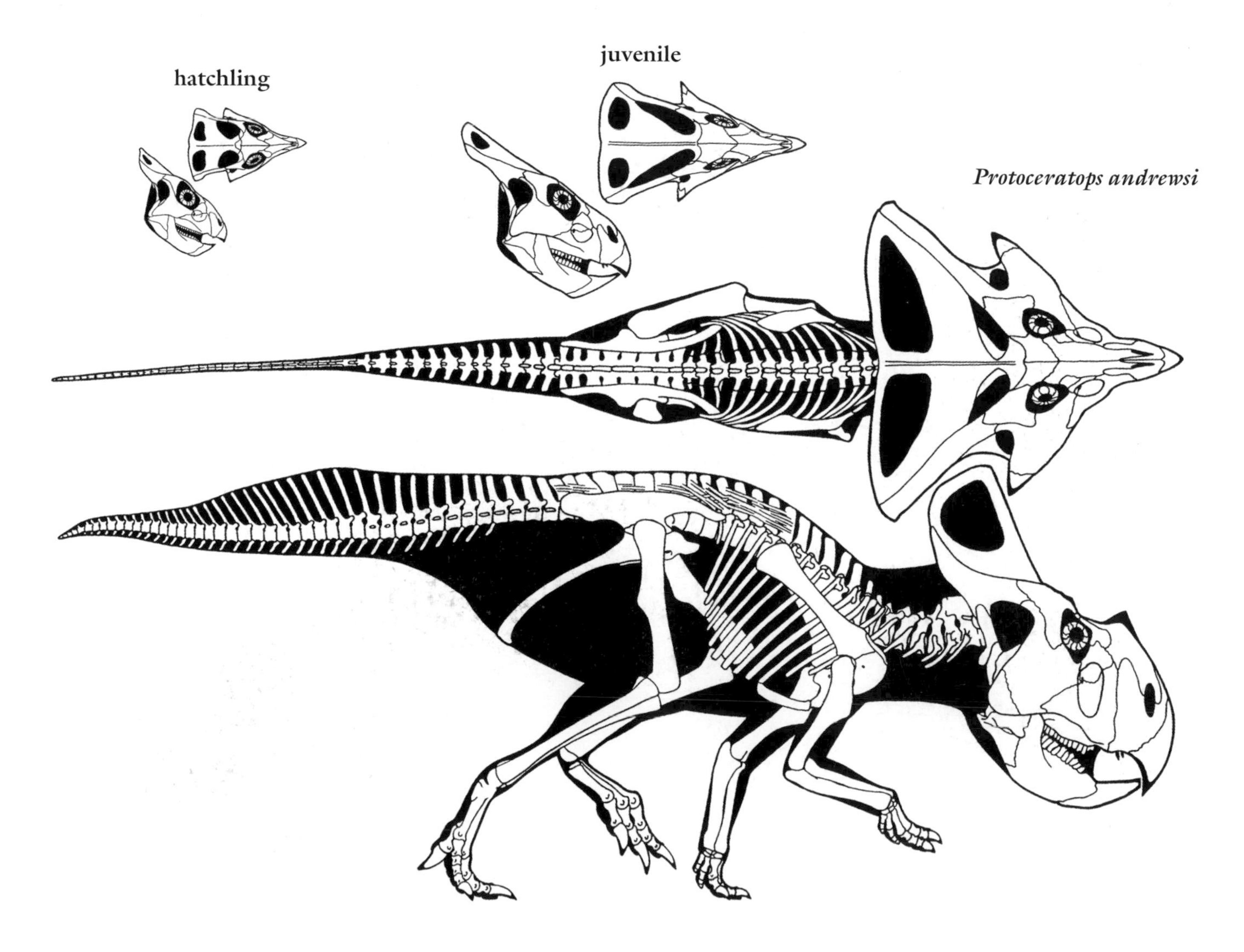

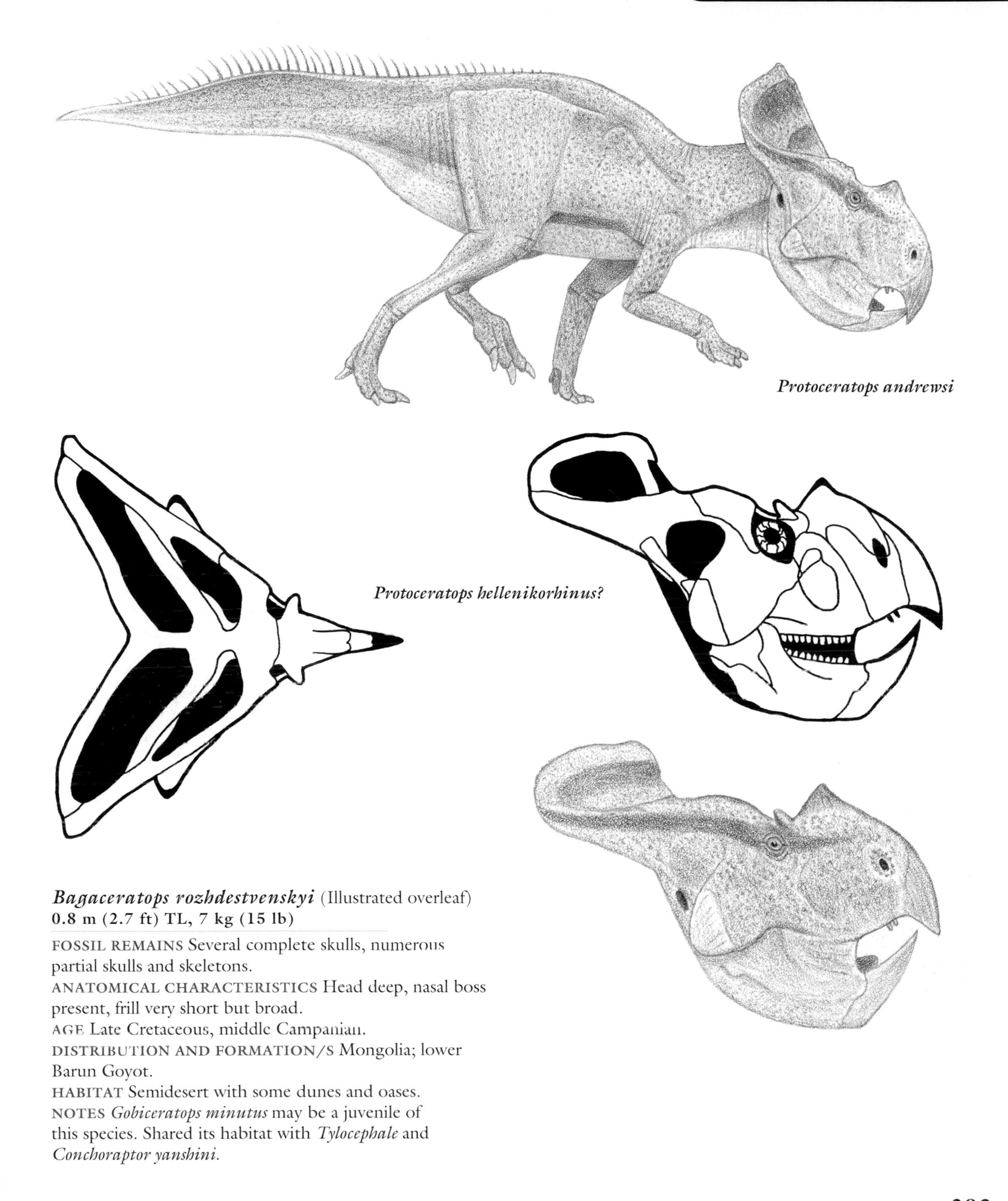
Protoceratops andrewsi

Protoceratops hellenikorhinus?

Bagaceratops rozhdestvenskyi (Illustrated overleaf)
0.8 m (2.7 ft) TL, 7 kg (15 lb)

FOSSIL REMAINS Several complete skulls, numerous partial skulls and skeletons.
ANATOMICAL CHARACTERISTICS Head deep, nasal boss present, frill very short but broad.
AGE Late Cretaceous, middle Campanian.
DISTRIBUTION AND FORMATION/S Mongolia; lower Barun Goyot.
HABITAT Semidesert with some dunes and oases.
NOTES *Gobiceratops minutus* may be a juvenile of this species. Shared its habitat with *Tylocephale* and *Conchoraptor yanshini*.

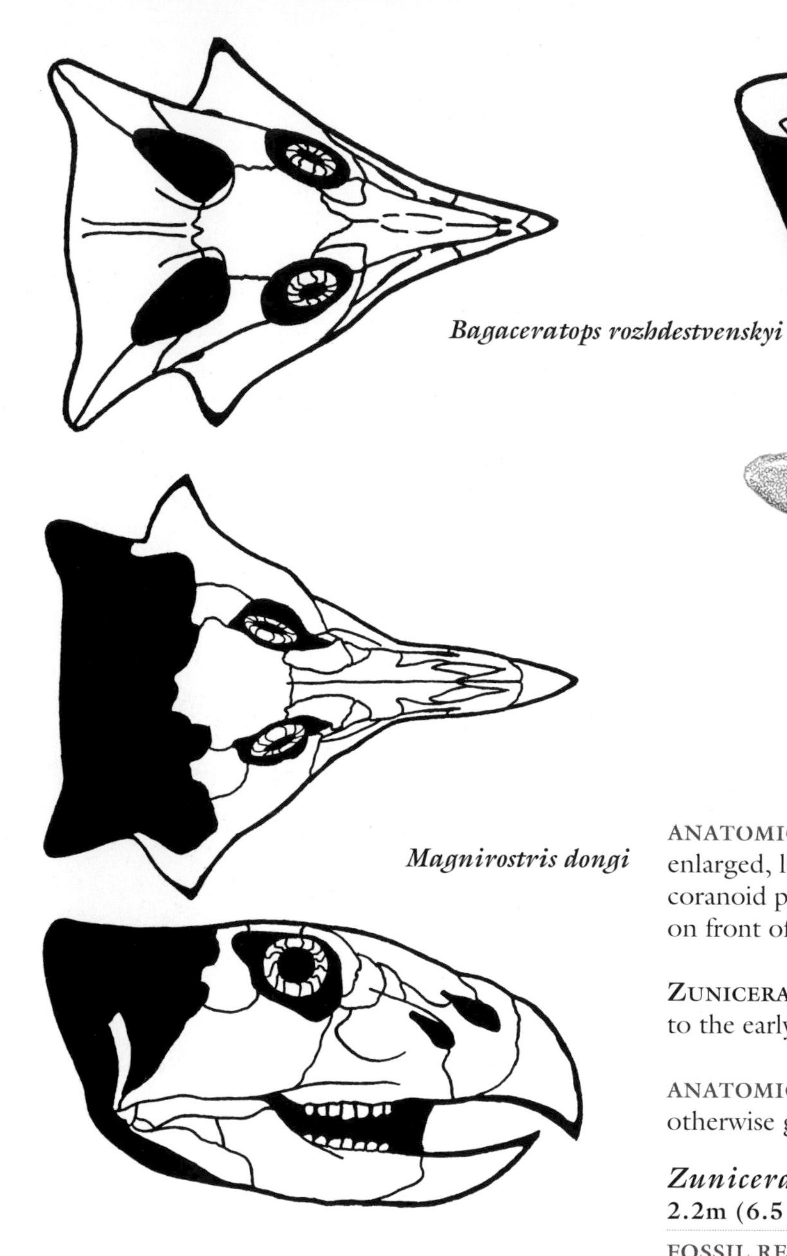

Bagaceratops rozhdestvenskyi

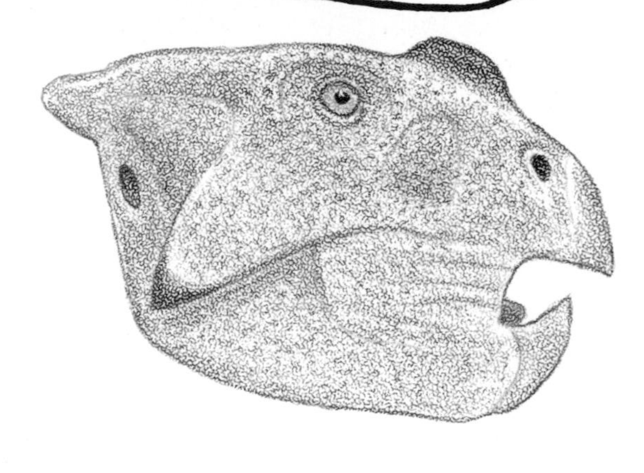

Magnirostris dongi

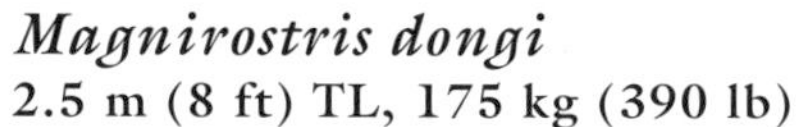

Magnirostris dongi
2.5 m (8 ft) TL, 175 kg (390 lb)

FOSSIL REMAINS Majority of skull.
ANATOMICAL CHARACTERISTICS Head and jaw not as deep as in other protoceratopsids, beak large.
AGE Late Cretaceous, Campanian.
DISTRIBUTION AND FORMATION/S Northern China; Bayan Mandahu.
HABITAT Desert with dunes and oases.

Ceratopsoids Medium-sized to gigantic neoceratopsians limited to the Late Cretaceous of North America and Asia.

ANATOMICAL CHARACTERISTICS Nasal opening enlarged, large nasal and/or brow horns usually present, coranoid process on lower jaw well developed, no teeth on front of upper jaw.

ZUNICERATOPSIDS Medium-sized ceratopsoids limited to the early Late Cretaceous of North America and Asia.

ANATOMICAL CHARACTERISTICS Medium sized, otherwise generally lacking features of ceratopsids.

Zuniceratops christopheri
2.2m (6.5 ft) TL, 175 kg (390 lb)

FOSSIL REMAINS Several partial skulls and skeletons.
ANATOMICAL CHARACTERISTICS Head long, no nasal horn, brow horns large, frill long.

Zuniceratops christopheri

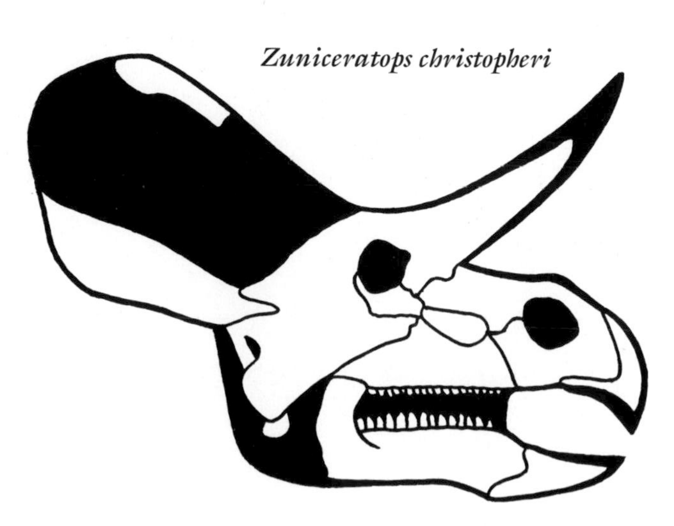

AGE Late Cretaceous, middle Turonian.
DISTRIBUTION AND FORMATION/S New Mexico; Moreno Hill.
HABITAT Coastal swamps and marshes.
HABITS Horns and frills probably used as display organs and weapons during contests within species as well as for defense against theropods.
NOTES Shared its habitat with *Nothronychus mckinleyi.*

Turanoceratops tardabilis
2 m (6.5 ft) TL, 175 kg (400 lb)

FOSSIL REMAINS Minority of a few skulls and skeletons.
ANATOMICAL CHARACTERISTICS Brow horns well developed.
AGE Late Cretaceous, middle or late Turonian.
DISTRIBUTION AND FORMATION/S Uzbekistan; Bissekty.
HABITAT Coastal.
NOTES Shared its habitat with *Levnesovia.*

CERATOPSIDS Large to gigantic ceratopsoids limited to the late Late Cretaceous of North America and Asia.

ANATOMICAL CHARACTERISTICS Fairly uniform except for head adornments. Upper beaks hooked, nasal openings very large, assorted bosses and hornlets on skull, head frills often well developed, main tooth rows long and developed into complex slicing batteries containing hundreds of teeth. Skeletons heavily constructed. First neck vertebrae fused together. Ribs in front of hips contact one another and anterior pubic process of pelvis. Tail sweeps downward, shortened. Trackways show hands farther from midline than hind feet. Fingers shorter. Pelvis very large, indicating exceptional muscle power, toes shorter.

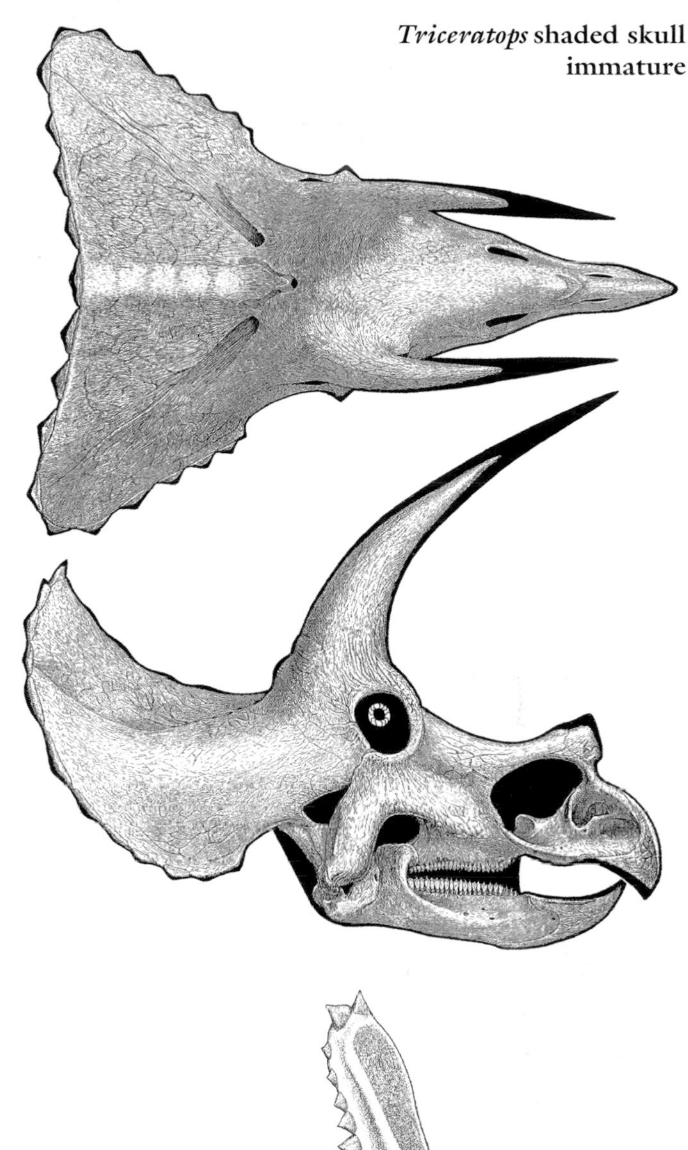
Triceratops shaded skull immature

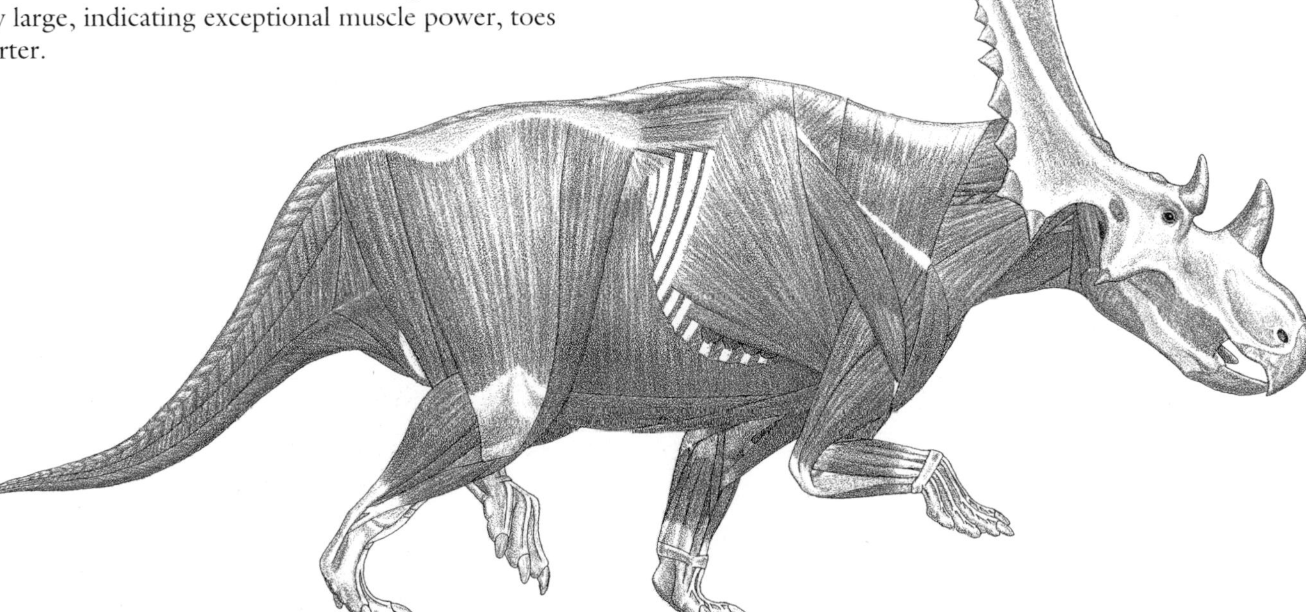
Chasmosaurus muscle study

ONTOGENY Growth rates apparently rapid, probably to reduce exposure to and recover from predation.
HABITS Some species may have fed in shallow waters on occasion; may have scavenged carcasses. Horns and frills used as display organs and weapons during contests within species; ribcage cuirass may have helped protect flanks. Defense may have included rearing like a bear and tilting frill up to intimidate attacker, followed by short fast charge with horns and/or beaks. Frills and in some cases spikes helped protect neck. Single-species bone beds indicate at least some species sometimes congregated in large herds.
NOTES Closest mammal analogs rhinos and giant extinct suids.

Centrosaurines Large ceratopsids limited to the late Late Cretaceous of North America and Asia.

ANATOMICAL CHARACTERISTICS Fairly uniform except for head adornments. Nasal openings subcircular, frill not always strongly elongated, open and subcircular; horns, hornlets, and bosses often exceptionally variable within a species and frequently asymmetrical within an individual. Pelvic ilium slopes down and backward.

Sinoceratops zhuchengensis
5 m (17 ft) TL, 2 tonnes

FOSSIL REMAINS Two partial skulls.
ANATOMICAL CHARACTERISTICS Nasal horn short, brow horns absent, large, upward-curving hornlets on back rim of frill.
AGE Late Late Cretaceous.
DISTRIBUTION AND FORMATION/S Eastern China; Wangshi Group.
NOTES Only known ceratopsid from Asia. Shared its habitat with *Zhuchengceratops*, *Tanius*, and *Tsintaosaurus*.

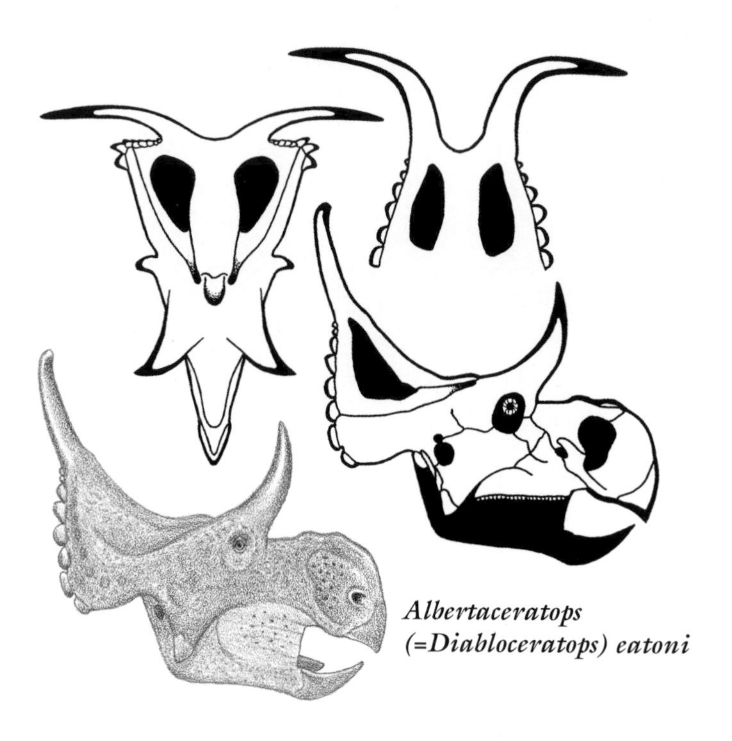
Albertaceratops (=Diabloceratops) eatoni

Albertaceratops (= Diabloceratops) eatoni
4.5 m (14 ft) TL, 1.3 tonnes

FOSSIL REMAINS Majority of skull.
ANATOMICAL CHARACTERISTICS Nasal boss is a low, narrow ridge, brow horns large, frill subvertical, tilted strongly upward, back rim with a pair of long, outward-arcing, slender spikes, small hornlets on rim side.
AGE Late Cretaceous, early Campanian.
DISTRIBUTION AND FORMATION/S Utah; Wahweap.
NOTES Main enemy *Lythronax.*

Albertaceratops nesmoi
5.8 m (19 ft) TL, 3.5 tonnes

FOSSIL REMAINS Majority of a skull, and majority of skeleton from bone beds.

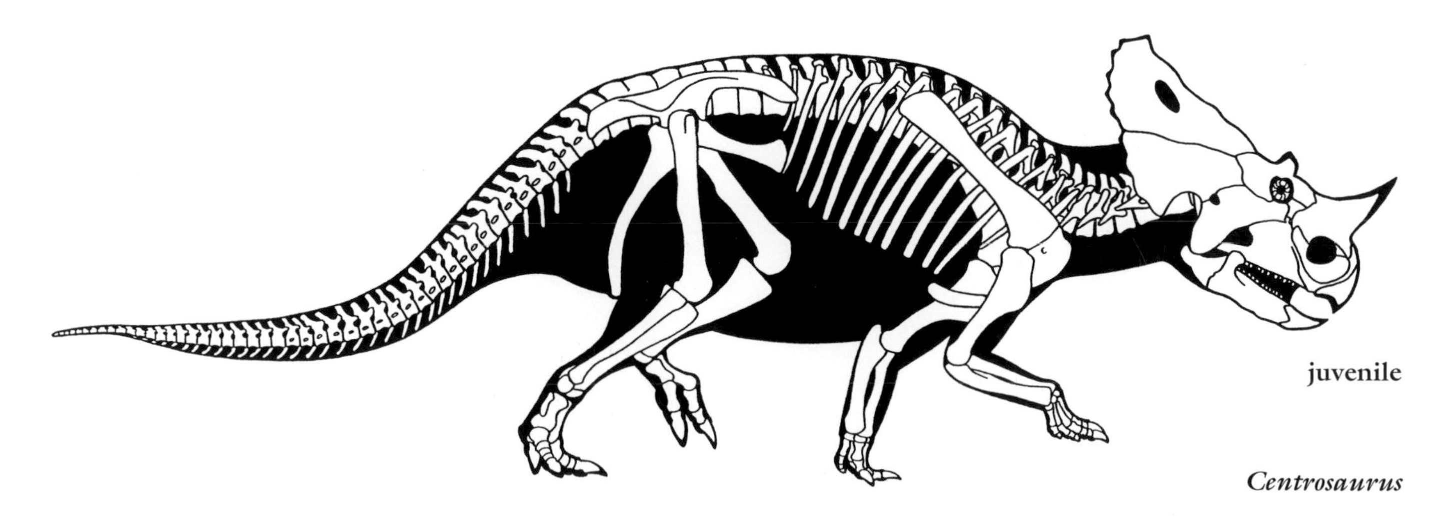
juvenile

Centrosaurus

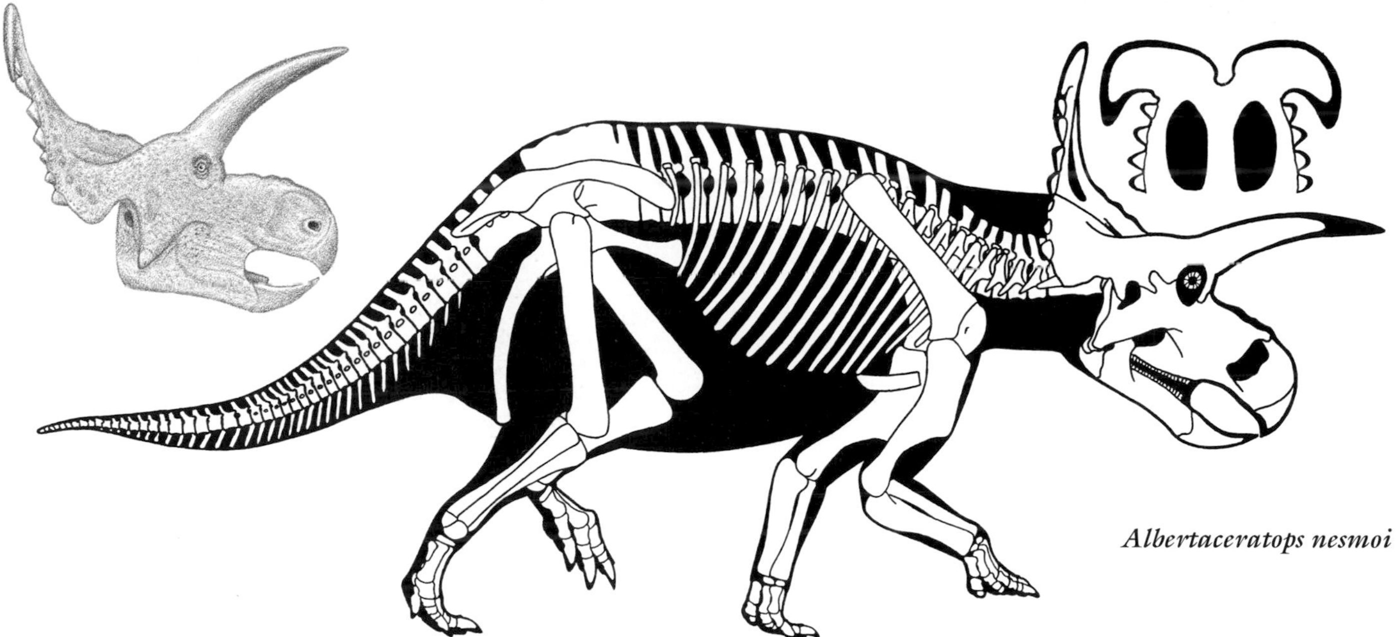

Albertaceratops nesmoi

ANATOMICAL CHARACTERISTICS
Nasal boss is a low, narrow ridge, brow horns large, frill subvertical, back rim with a pair of massive, sideways-arcing spikes, small hornlets on rim side.
AGE Late Cretaceous, middle Campanian.
DISTRIBUTION AND FORMATION/S Alberta; lower Oldman.
HABITAT Well-watered, forested floodplain with coastal swamps and marshes, cool winters.

Avaceratops lammersi
4 m (12 ft) TL, 1 tonne

FOSSIL REMAINS One or two partial skulls and skeleton, immature.
ANATOMICAL CHARACTERISTICS May have well-developed brow horns.
AGE Late Cretaceous, late Campanian.
DISTRIBUTION AND FORMATION/S Montana; Judith River.
HABITAT Well-watered, forested floodplain with coastal swamps and marshes.
NOTES Original skull lacks top, and assignment of second skull with brow horns in this species is uncertain. Relationships of these immature specimens uncertain. Probably the smallest known ceratopsid.

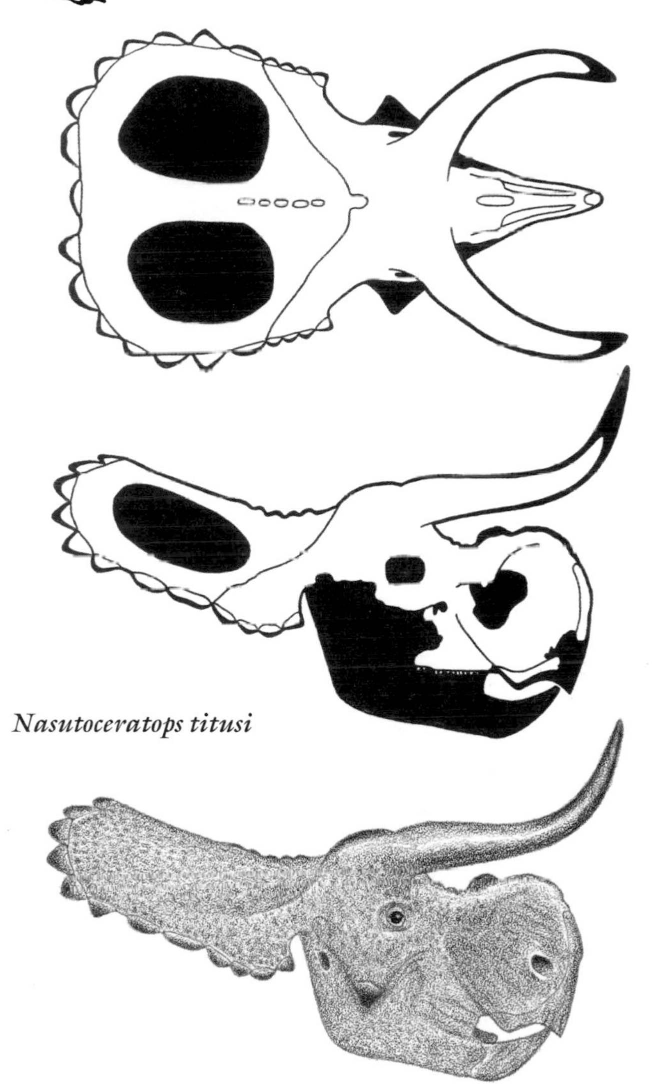

Nasutoceratops titusi

Nasutoceratops titusi
4.5 m (15 ft) TL, 1.5 tonnes

FOSSIL REMAINS Majority of skulls and minority of a skeleton.
ANATOMICAL CHARACTERISTICS Nasal horn/boss barely present, bovid-like brow horns subhorizontal, tips point inward, frill subhorizontal, subcircular, all hornlets small.

AGE Late Cretaceous, late Campanian.
DISTRIBUTION AND FORMATION/S Utah; middle Kaiparowitz.
HABITS Forward-directed brow horns indicate frontal thrusting action.

Wendiceratops pinhornensis

4.5 m (15 ft) TL, 1.5 tonnes

FOSSIL REMAINS Numeorus skull and skeletal elements and majority of skeleton from bone beds.
ANATOMICAL CHARACTERISTICS Nasal horn well developed, brow horns large, subvertical, frill subhorizontal, back rim subcircular, with large, broad, forward-directed hornlets.
AGE Late Cretaceous, middle Campanian.
DISTRIBUTION AND FORMATION/S Alberta; lower Oldman.
HABITAT Well-watered, forested floodplain with coastal swamps and marshes, cool winters.

Centrosaurus (= Coronosaurus) brinkmani

5 m (17 ft) TL, 2 tonnes

FOSSIL REMAINS Bone bed remains.
ANATOMICAL CHARACTERISTICS Nasal horn moderately large and erect, brow horns small and directed sideways, frill subhorizontal, large hornlets project from back rim over openings with bases sprouting minihornlets, innermost rim edge hornlets large and projecting sideways, small hornlets adorn sides of rim.
AGE Late Cretaceous, middle Campanian.
DISTRIBUTION AND FORMATION/S Alberta; lower Dinosaur Park, middle Oldman.
HABITS Vertically directed nasal horn indicates upward thrusting action.
HABITAT Well-watered, forested floodplain with coastal swamps and marshes, cool winters.

Centrosaurus apertus

5.5 m (18 ft) TL, 2.3 tonnes

FOSSIL REMAINS Complete and partial skulls and skeletons, bone beds, completely known.
ANATOMICAL CHARACTERISTICS Large nasal horn either curved forward or vertical, brow horns small, frill subhorizontal, subhorns project from back rim over openings, sometimes strongly asymmetrical in length, innermost rim edge hornlets point toward centerline, small hornlets adorn sides of rim.
AGE Late Cretaceous, late Campanian.
DISTRIBUTION AND FORMATION/S Alberta; uppermost Oldman? lower Dinosaur Park.

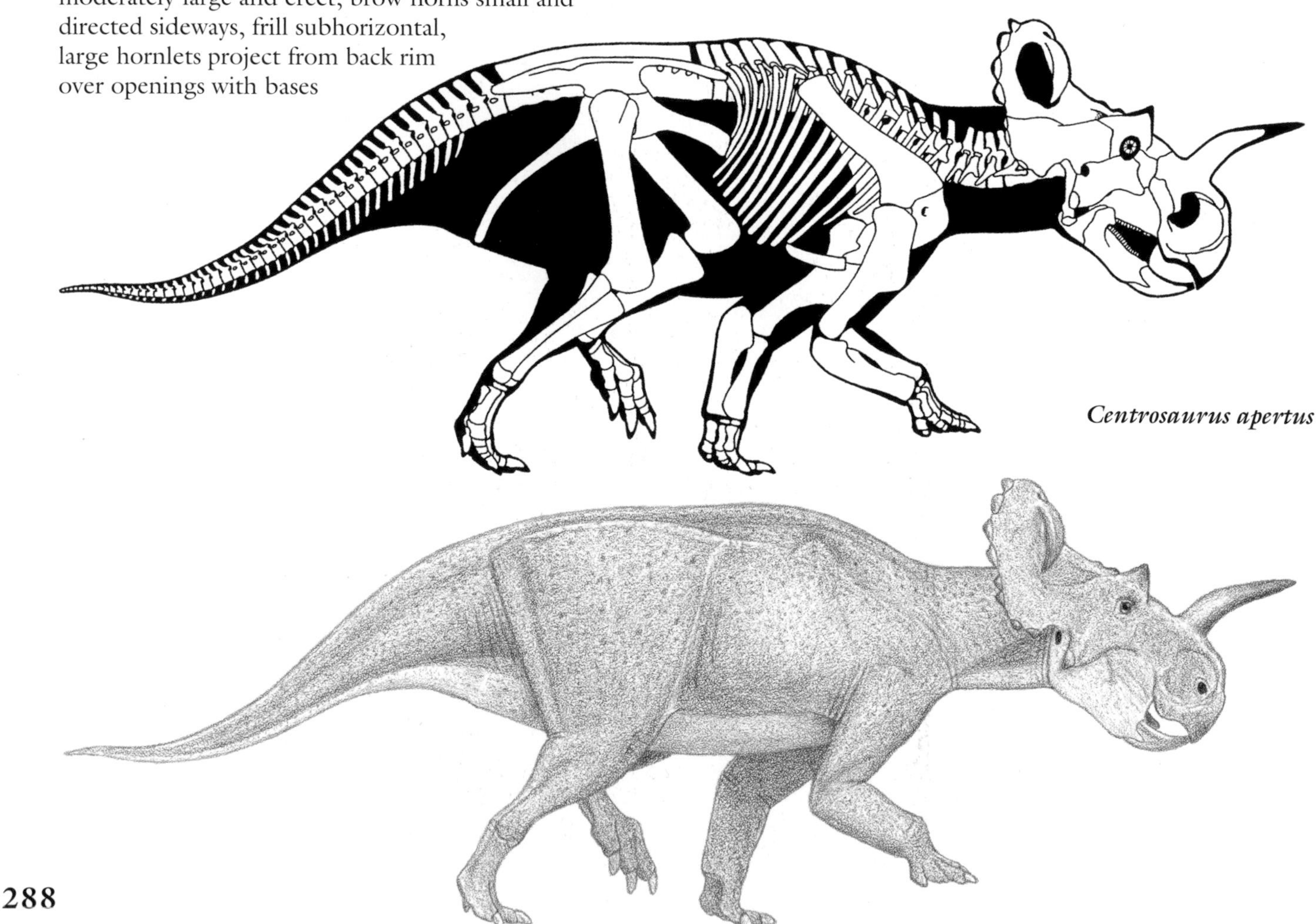

Centrosaurus apertus

HABITAT Well-watered, forested floodplain with coastal swamps and marshes, cool winters.
HABITS Forward-directed nasal horn indicates frontal thrusting action.
NOTES May be the direct descendant of *C. brinkmani.* Remains from the Judith River Formation of Montana named *Monoclonius* are based on inadequate remains and probably belong to *C. apertus* and/or *C. nasicornis.* Shared its habitat with *Chasmosaurus russelli.*

Centrosaurus nasicornis
5 m (17 ft) TL, 2 tonnes

FOSSIL REMAINS Complete skulls and skeletons, bone beds, completely known.
ANATOMICAL CHARACTERISTICS Robustly built. Large nasal horn always vertical, brow horns small, frill subhorizontal, subhorns project from back rim over openings, innermost rim edge hornlets point toward centerline, small hornlets adorn rim.
AGE Late Cretaceous, late Campanian.
DISTRIBUTION AND FORMATION/S Alberta; middle Dinosaur Park.
HABITAT Well-watered, forested floodplain with coastal swamps and marshes, cool winters.
HABITS Vertically directed nasal horn indicates upward thrusting action.
NOTES May be the direct descendant of *C. apertus.* Bone beds indicate sometimes congregated in large herds. Shared its habitat with *Chasmosaurus belli.*

Centrosaurus (= *Spinops) sternbergorum*
4.5 m (14 ft) TL, 1.3 tonnes

FOSSIL REMAINS Partial skulls.
ANATOMICAL CHARACTERISTICS Nasal horn large, brow horns small, subhorns project from back rim over

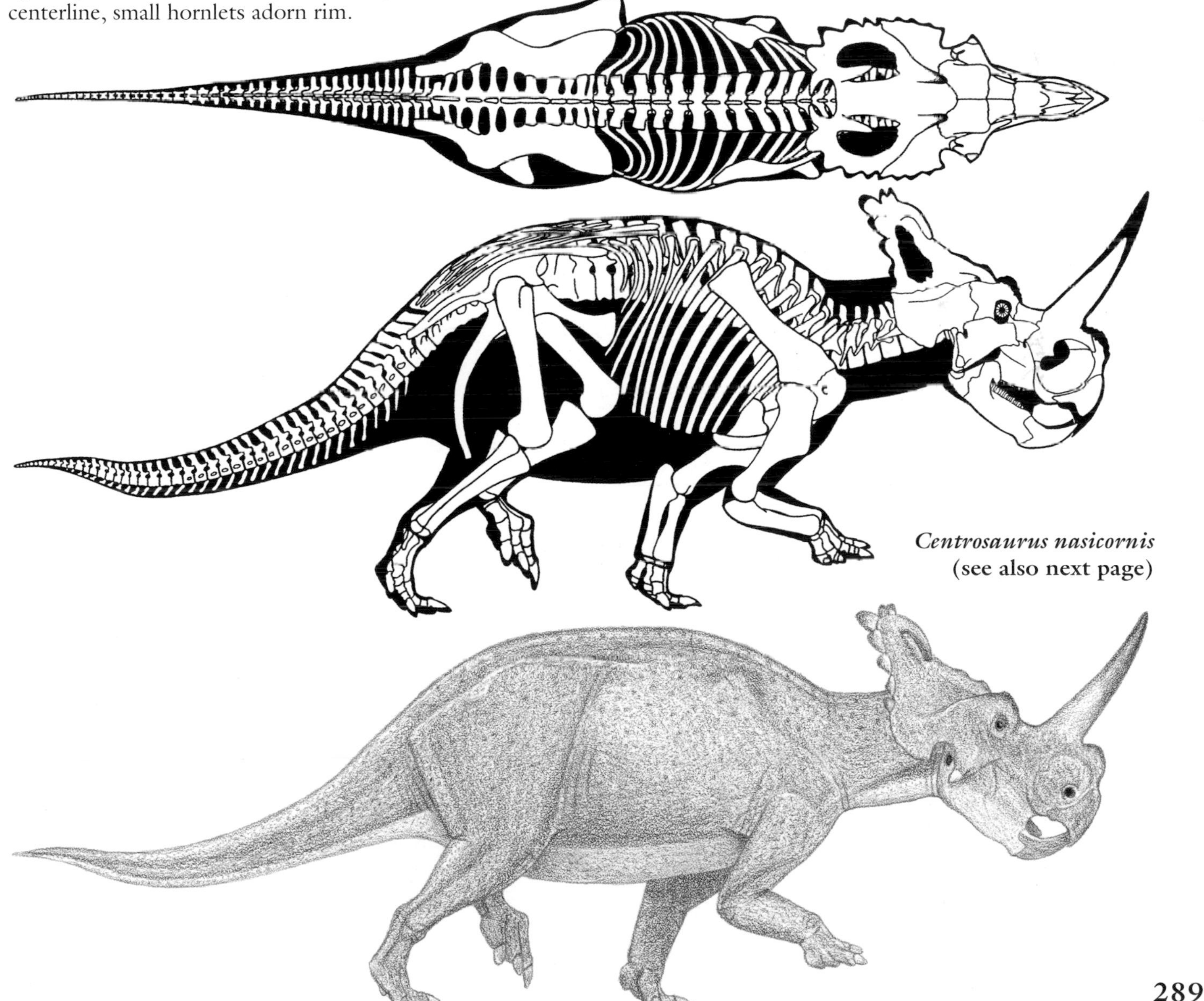

Centrosaurus nasicornis
(see also next page)

Centrosaurus nasicornis

openings, a large spike projects straight back from both sides of rim edge, small hornlets adorn sides of rim.
AGE Late Cretaceous, late Campanian.
DISTRIBUTION AND FORMATION/S Alberta; uppermost Oldman or lower Dinosaur Park.
HABITAT Well-watered, forested floodplain with coastal swamps and marshes, cool winters.
NOTES May be an ancestor of *C. albertensis.*

Centrosaurus (= Styracosaurus) albertensis
5.1 m (17 ft) TL, 1.8 tonnes

FOSSIL REMAINS A few complete and partial skulls and skeletons, bone bed material.
ANATOMICAL CHARACTERISTICS Large nasal horn, brow horns small, frill subhorizontal, rimmed by array of large spikes, innermost projecting sideways, small hornlets on rim side.
AGE Late Cretaceous, late Campanian.
DISTRIBUTION AND FORMATION/S Alberta; upper Dinosaur Park.
HABITAT Well-watered, forested floodplain with coastal swamps and marshes, cool winters.
HABITS Vertically directed nasal horn indicates upward thrusting action.
NOTES May be the direct descendant of *C. nasicornis.* Because the skulls and skeletons of centrosaurines without large brow horns and subhorizontal frills are very similar except for the details of the horns, bosses, and frills, they probably form one genus. Shared its habitat with *Chasmosaurus irvinensis.*

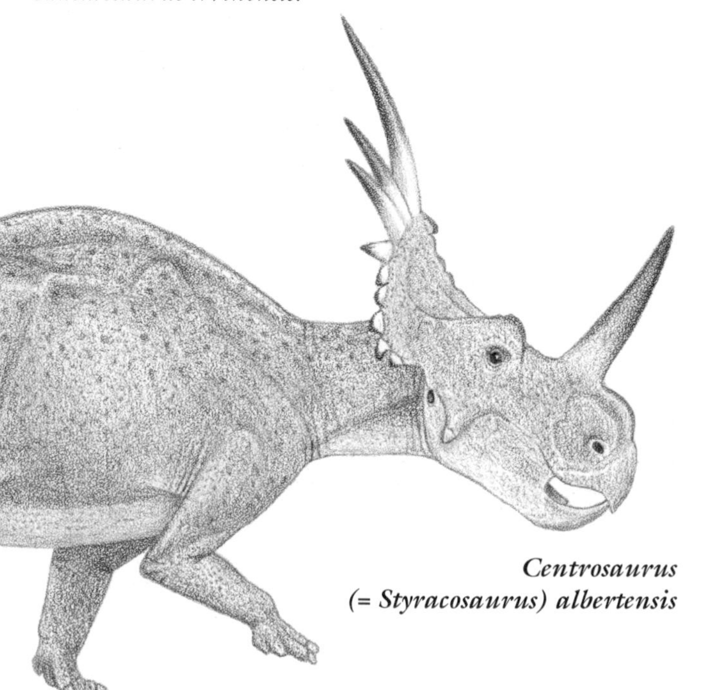

Centrosaurus (= Styracosaurus) albertensis

Centrosaurus (= Rubeosaurus) ovatus
5 m (17 ft) TL, 2 tonnes

FOSSIL REMAINS Numerous skull and skeletal elements.
ANATOMICAL CHARACTERISTICS Large nasal horn, brow horns small, frill subhorizontal, rimmed by array of large spikes, innermost spikes converge inward.
AGE Late Cretaceous, middle and/or late Campanian.
DISTRIBUTION AND FORMATION/S Montana; Upper Two Medicine.
HABITS Vertically directed nasal horn indicates upward thrusting action.
HABITAT Seasonally dry upland woodlands.
NOTES Shared its habitat with *C. procurvicornis.*

Centrosaurus (= Einiosaurus) procurvicornis
4.5 m (14 ft) TL, 1.3 tonnes

FOSSIL REMAINS Several partial skulls and a number of skeletons, juvenile to adult.
ANATOMICAL CHARACTERISTICS Large nasal horn strongly downcurved and deep, brow horns small, frill subhorizontal, two long spikes near middle of back rim, small hornlets on rim side.
AGE Late Cretaceous, middle and/or late Campanian.

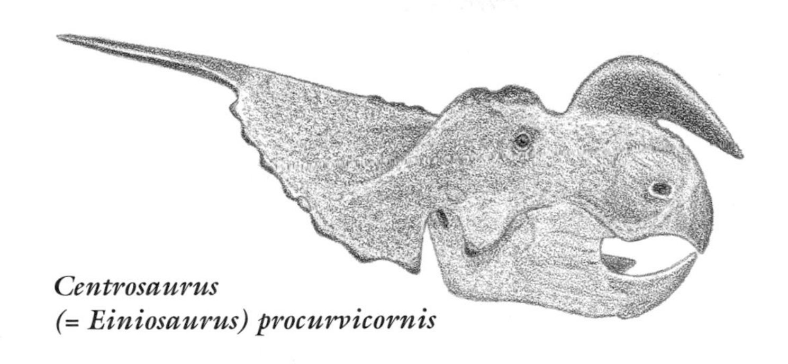

Centrosaurus (= Einiosaurus) procurvicornis

DISTRIBUTION AND FORMATION/S Montana; Upper Two Medicine.
HABITAT Seasonally dry upland woodlands.
HABITS May have rammed other members of species and tyrannosaurids with edge of horn; primary weapon against latter was its beak.
NOTES Taxonomically inadequate juvenile remains named *Brachyceratops montanensis* may belong to this species, alternatively to *C. ovatus.*

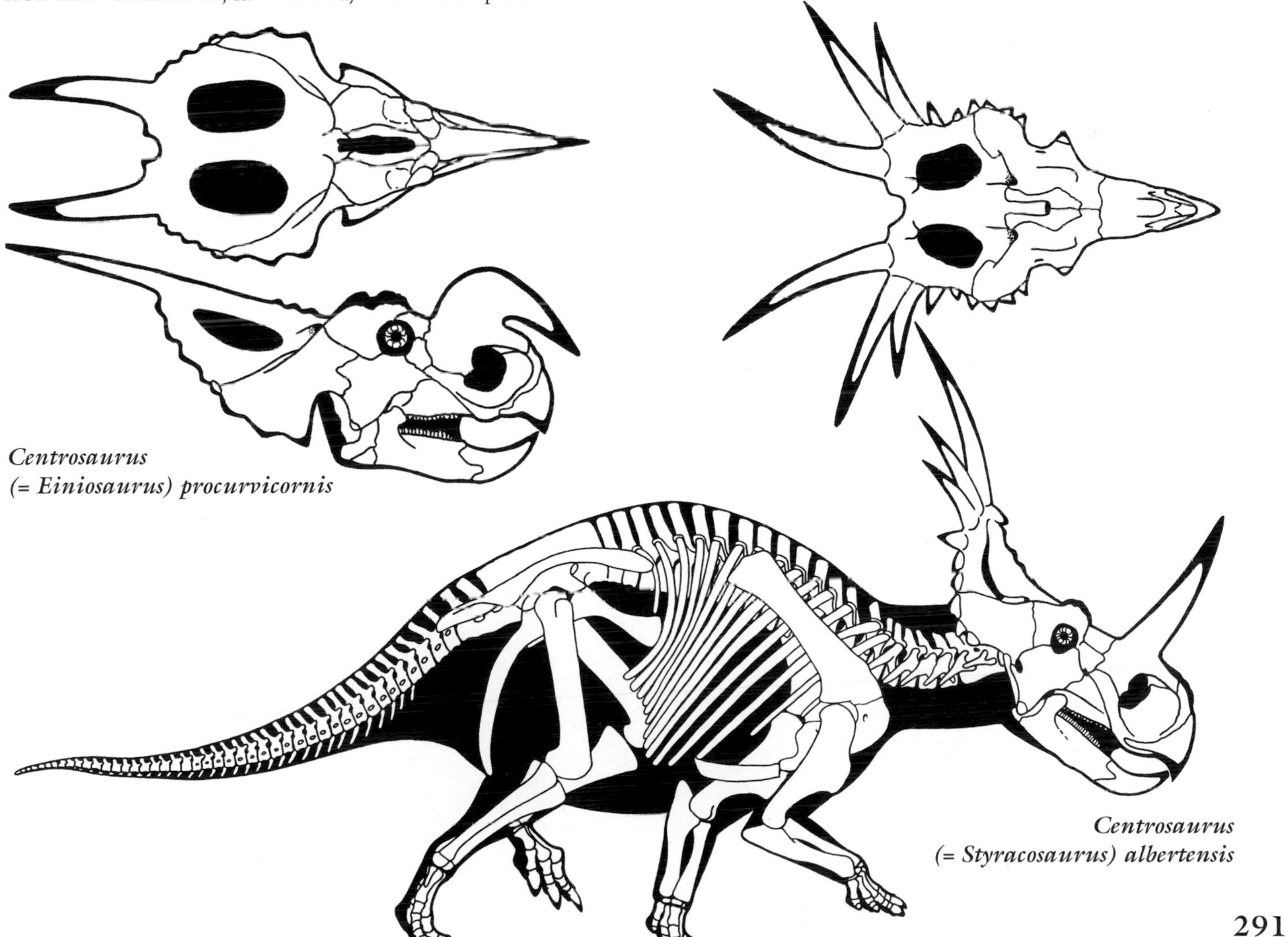

Centrosaurus (= Einiosaurus) procurvicornis

Centrosaurus (= Styracosaurus) albertensis

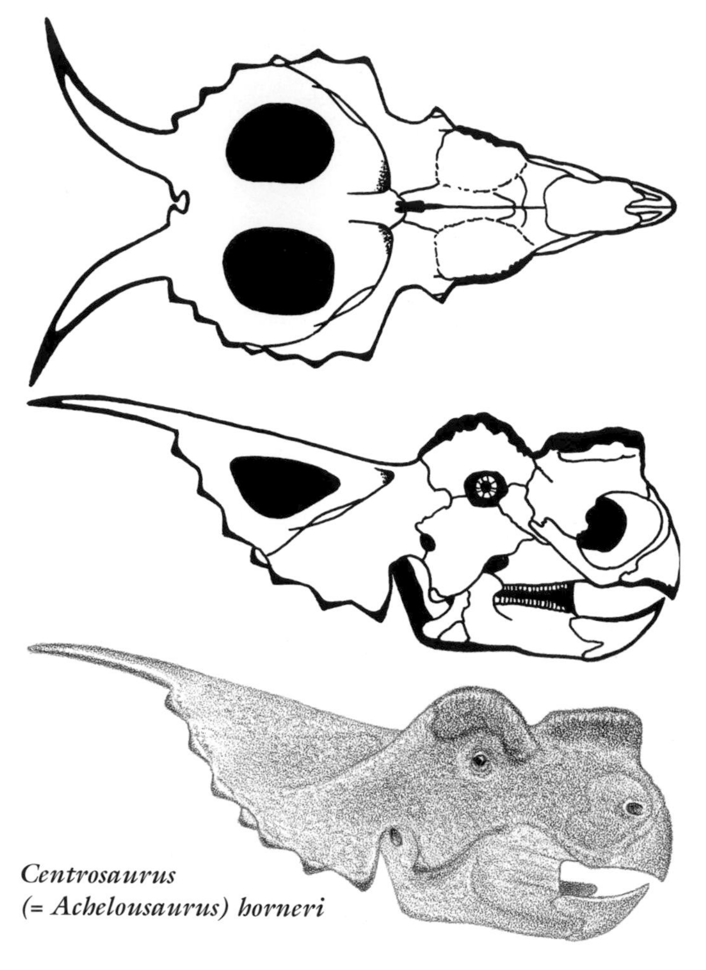

Centrosaurus
(= Achelousaurus) horneri

Centrosaurus (= Achelousaurus) horneri

6 m (20 ft) TL, 3 tonnes

FOSSIL REMAINS Several partial skulls and a partial skeleton.

ANATOMICAL CHARACTERISTICS Nasal and brow bosses present, frill subhorizontal, two long, partly sideways-arcing spikes near middle of back rim, small hornlets on rim side.

AGE Late Cretaceous, middle and/or late Campanian.

DISTRIBUTION AND FORMATION/S Montana; Upper Two Medicine.

Centrosaurus (= Pachyrhinosaurus) lakustai

5 m (17 ft) TL, 2 tonnes

FOSSIL REMAINS A very large number of partial skulls and skeletons.

ANATOMICAL CHARACTERISTICS Nasal and brow bosses replace low juvenile nasal horn and combine into a massive unit in adults, frill subhorizontal, center strut bears irregular short vertical horn/nub/hump(s) in at least one adult morph, two sideways-sweeping spikes on back rim, two small horns along midline point toward each other, small hornlets on rim side.

AGE Late Cretaceous, late Campanian.

DISTRIBUTION AND FORMATION/S Alberta; middle Wapiti.

HABITAT Well-watered, forested floodplain with coastal swamps and marshes, winters cool to cold.

NOTES Presence of keratin horns atop bosses cannot be entirely ruled out in pachyrhinosaurs.

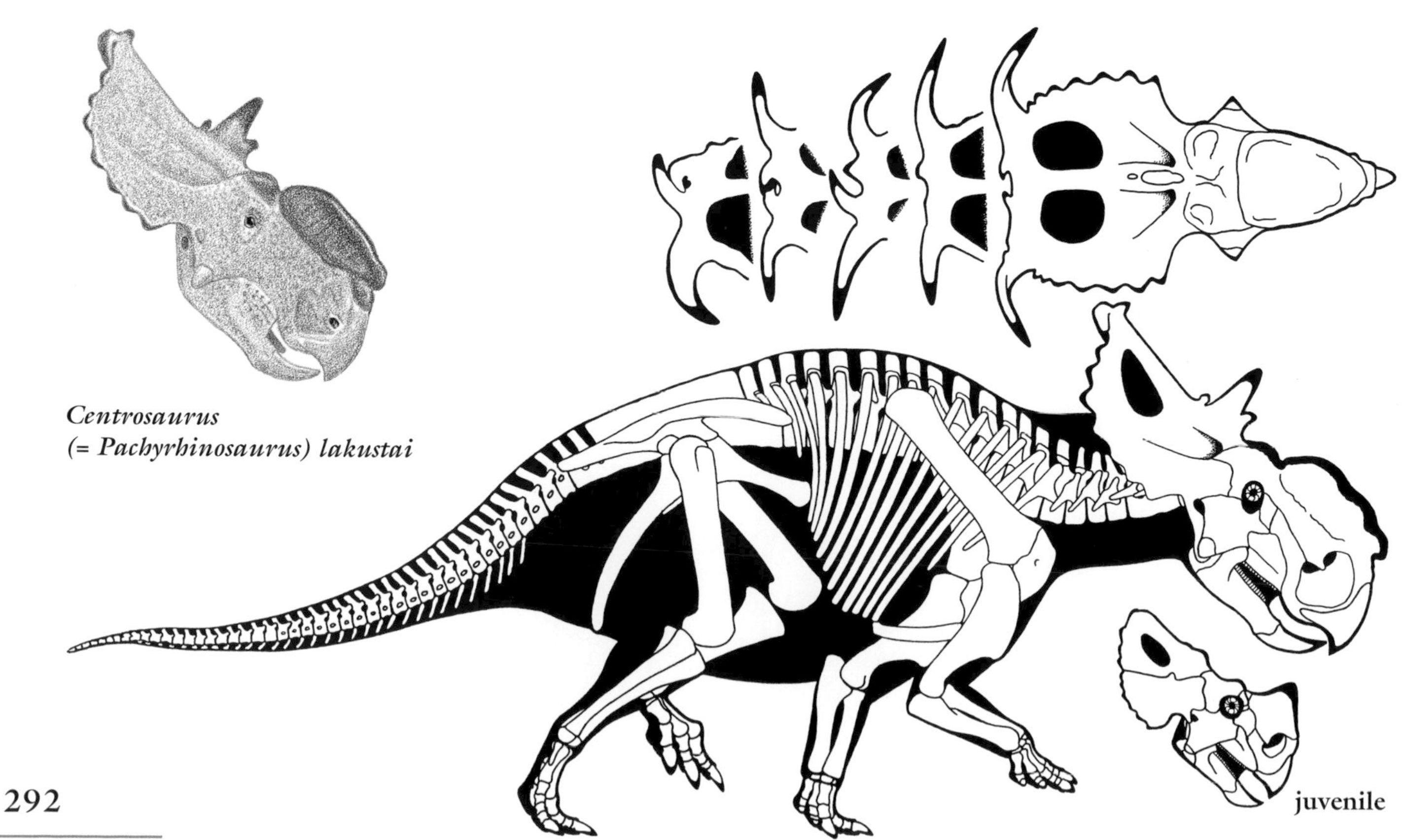

Centrosaurus
(= Pachyrhinosaurus) lakustai

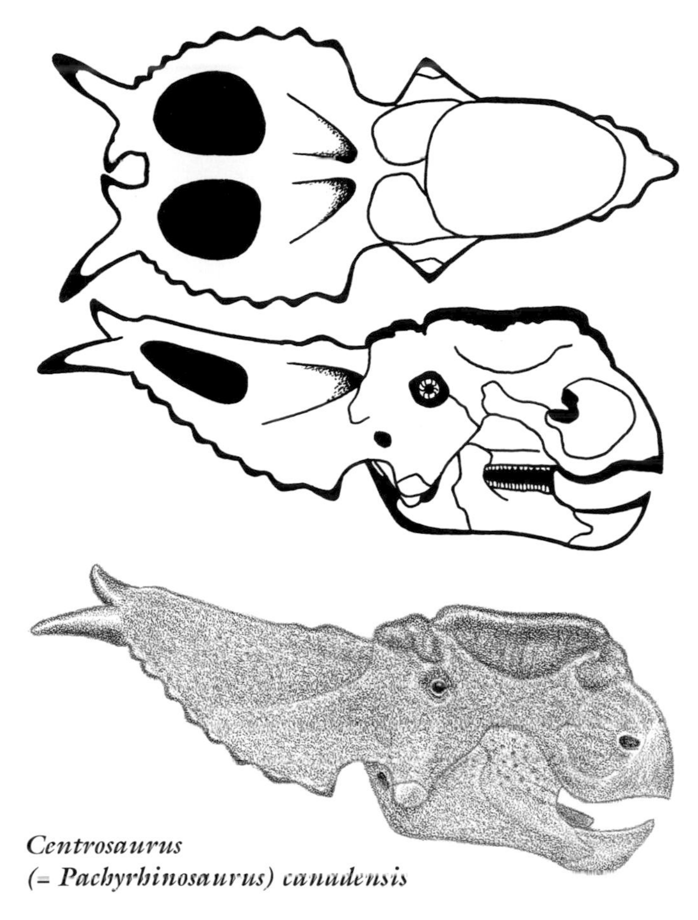

Centrosaurus (= Pachyrhinosaurus) canadensis

Centrosaurus (= Pachyrhinosaurus) canadensis
6 m (20 ft) TL, 3 tonnes

FOSSIL REMAINS A large number of partial skulls and skeletons.
ANATOMICAL CHARACTERISTICS Nasal and brow bosses present in juveniles, combine into a massive unit in adults, frill subhorizontal, two back- and sideways-pointing spikes on back rim, two small horns along midline point toward each other, small hornlets on rim side.
AGE Late Cretaceous, early Maastrichtian.
DISTRIBUTION AND FORMATION/S Alberta; lower Horseshoe Canyon, Saint Mary River.
HABITAT Well-watered, forested floodplain with coastal swamps and marshes, winters cool to cold.
NOTES May be the descendant of *C. lakustai*.

Centrosaurus (= Pachyrhinosaurus) perotorum
5 m (17 ft) TL, 2 tonnes

FOSSIL REMAINS Partial skulls.
ANATOMICAL CHARACTERISTICS Spikes on frill rim robust, small frill hornlets project forward.
AGE Late Cretaceous, middle Maastrichtian.
DISTRIBUTION AND FORMATION/S Northern Alaska; middle Prince Creek.
HABITAT Well-watered coastal woodland, cool summers, severe winters including heavy snows.
NOTES Indicates that polar ceratopsids were not dwarfed compared to those from farther south.

Chasmosaurines Large to gigantic ceratopsids limited to the late Late Cretaceous of North America.

ANATOMICAL CHARACTERISTICS Nasal openings elongated, frill always long and open. Tails generally shorter than those of centrosaurines. Pelvic ilium horizontal.

Mercuriceratops gemini
4 m (13 ft) TL, 1 tonne

FOSSIL REMAINS Small minority of skulls.
ANATOMICAL CHARACTERISTICS Side of frill has a winglet supporting small hornlets.
AGE Late Cretaceous, late Campanian.
DISTRIBUTION AND FORMATION/S Alberta, Montana; lower Dinosaur Park, Upper Two Medicine.

Coahuilaceratops magnacuerna
4 m (13 ft) TL, 1 tonne

FOSSIL REMAINS Partial skull.
ANATOMICAL CHARACTERISTICS Nasal horn low, brow horns very large.
AGE Late Cretaceous, late Campanian.
DISTRIBUTION AND FORMATION/S Northern Mexico; middle Cerro del Pueblo.

Chasmosaurus russelli (Illustrated overleaf)
4.3 m (14 ft) TL, 1.5 tonnes

FOSSIL REMAINS Several complete or partial skulls, partial skeleton.
ANATOMICAL CHARACTERISTICS Nasal horn short, brow horns short at least in adults, frill subhorizontal, back rim very broad and forming a shallow U, corners with large hornlets, small hornlets along side rims. Shoulder withers supported nuchal ligaments to neck and head. Belly broad, hind legs bowed.
AGE Late Cretaceous, late Campanian.
DISTRIBUTION AND FORMATION/S Alberta; at least lower Dinosaur Park.
HABITAT Well-watered, forested floodplain with coastal swamps and marshes, cool winters.
HABITS Short-horned individuals probably relied more on beak than horns for defense.
NOTES May be a sex of *C. belli*.

Chasmosaurus belli (Illustrated on page 295)
4.8 m (16 ft) TL, 2 tonnes

FOSSIL REMAINS A number of skulls and skeletons, completely known.

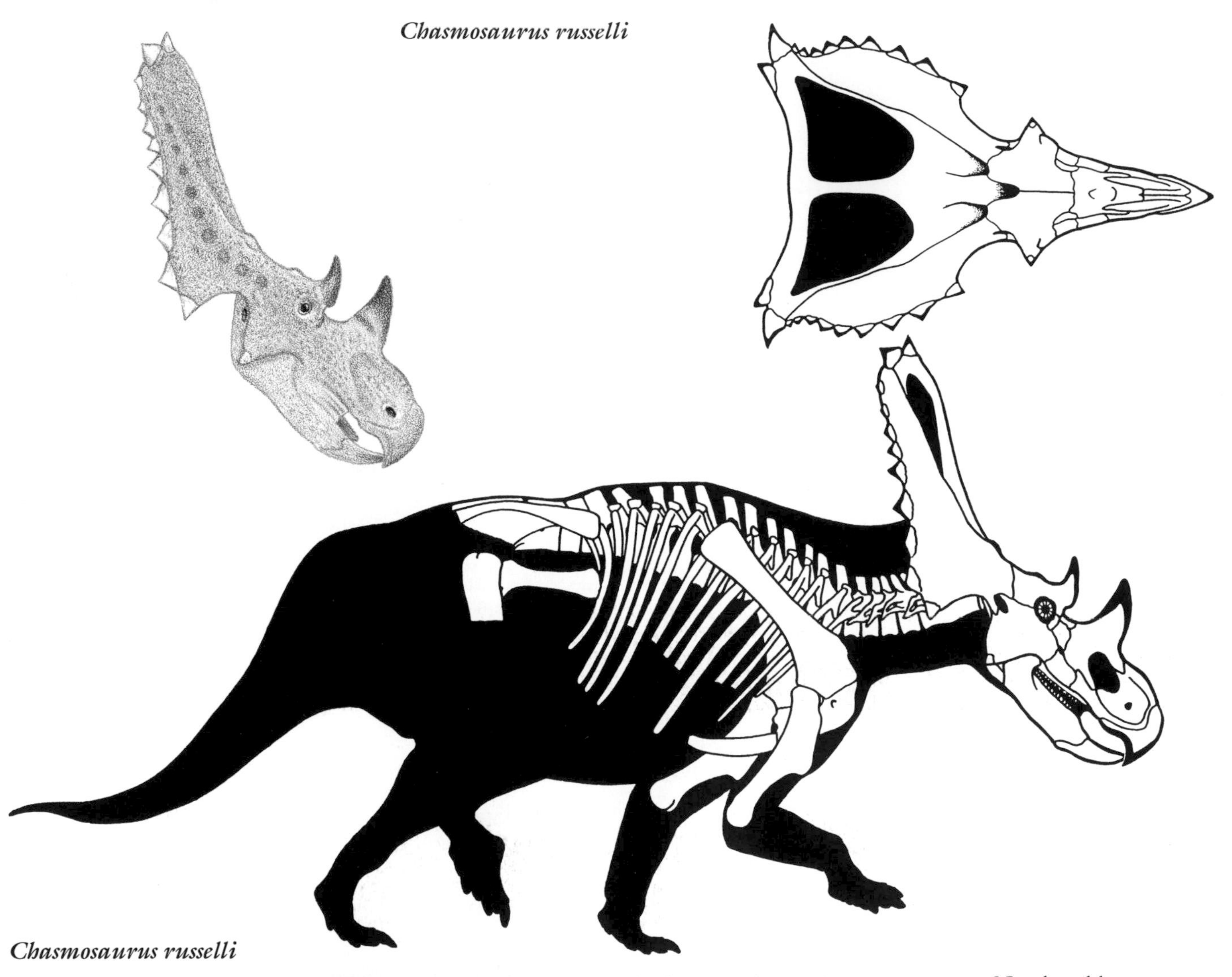

Chasmosaurus russelli

Chasmosaurus russelli

ANATOMICAL CHARACTERISTICS Nasal and brow horns short at least in adults, orientation variable, frill subhorizontal, back rim very broad and forming a shallow V, corners with large hornlets, small hornlets along side rims. Shoulder withers supported nuchal ligaments to neck and head. Belly broad, hind legs bowed.
AGE Late Cretaceous, late Campanian.
DISTRIBUTION AND FORMATION/S Alberta; at least middle Dinosaur Park.
HABITAT Well-watered, forested floodplain with coastal swamps and marshes, cool winters.
HABITS Probably relied more on beak than horns for defense.
NOTES May be the direct descendant of *C. russelli*. Shared its habitat with *Centrosaurus nasicornis*.

Chasmosaurus (= Mojoceratops) perifania

4.5 m (15 ft) TL, 2 tonnes

FOSSIL REMAINS A nearly complete skull and partial skulls.

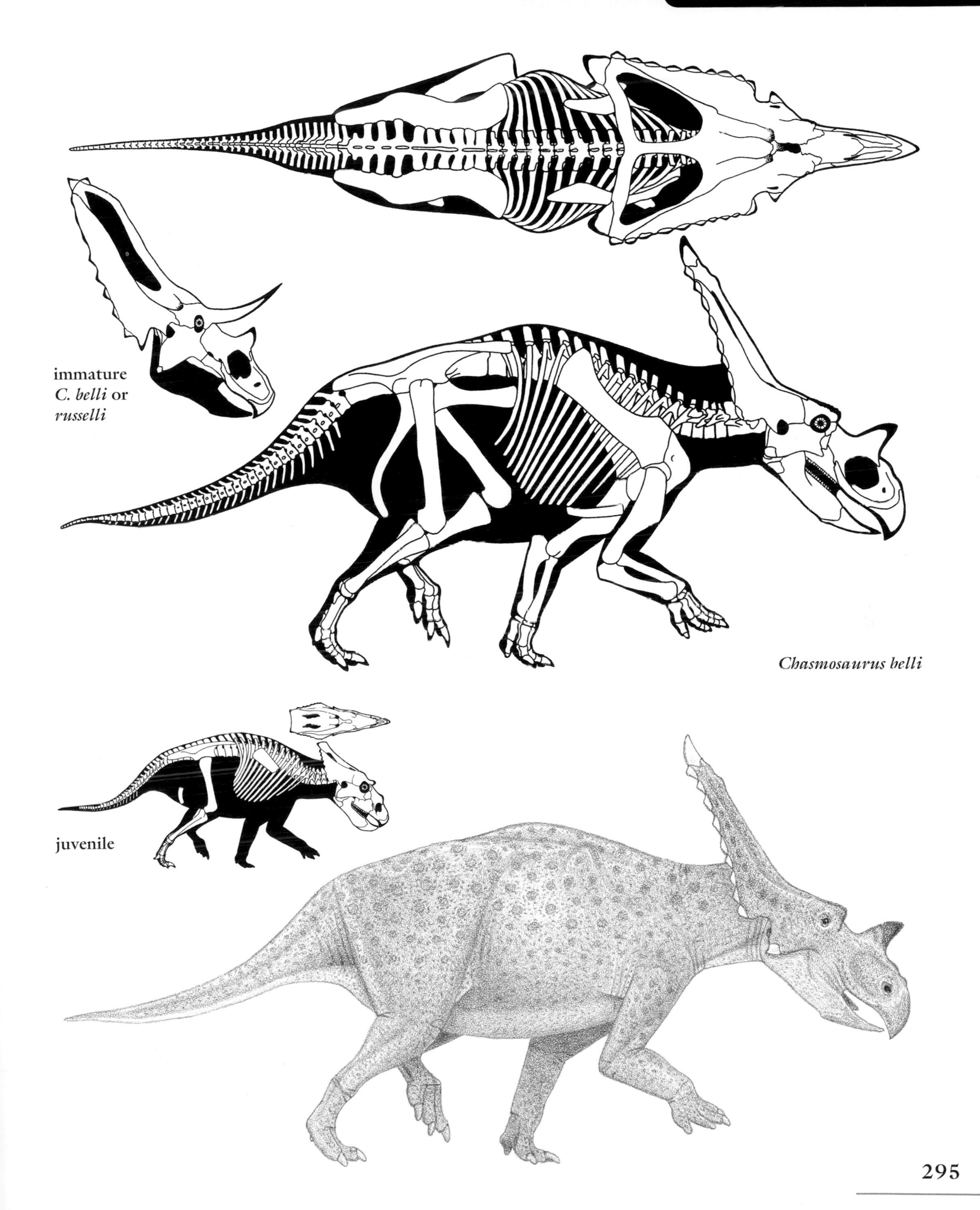
immature
C. belli or
russelli
Chasmosaurus belli
juvenile

ANATOMICAL CHARACTERISTICS Nasal horns short, blunt, brow horns large and directed strongly sideways, frill subhorizontal, back rim very broad and forming a shallow V, corners with large hornlets, small hornlets along side rims.
AGE Late Cretaceous, late Campanian.
DISTRIBUTION AND FORMATION/S Alberta; middle Dinosaur Park.
HABITAT Well-watered, forested floodplain with coastal swamps and marshes, cool winters.
NOTES May be a sexual morph of *C. belli*.

Chasmosaurus (Kosmoceratops) richardsoni
4.5 m (15 ft) TL, 1.2 tonnes

FOSSIL REMAINS A complete skull and partial skulls.
ANATOMICAL CHARACTERISTICS Nasal horns short, long, blunt, brow horns large and projecting sideways, frill not as elongated as in other *Chasmosaurus*, back rim broad, somewhat indented, bearing a row of large, broad, forward-directed hornlets, also forward- and sideways-projecting hornlets at rim corners, small hornlets along side rims.
AGE Late Cretaceous, late Campanian.
DISTRIBUTION AND FORMATION/S Utah; lower and upper Kaiparowits.
NOTES With only frills greatly differing from *Chasmosaurus*, this and *C. irvinensis* probably within former genus; alternatively they form a united genus.

Chasmosaurus (Kosmoceratops = Vagaceratops) irvinensis
4.5 m (15 ft) TL, 1.2 tonnes

FOSSIL REMAINS A few skulls and majority of distorted skeleton.
ANATOMICAL CHARACTERISTICS Nasal horn short, brow horns absent, frill not as elongated as in other *Chasmosaurus*, back rim very broad, straight, bearing a row of large, broad, forward-directed hornlets, small hornlets along side rims. Belly broad and hind legs bowed.
AGE Late Cretaceous, late Campanian.

Chasmosaurus (Kosmoceratops) richardsoni

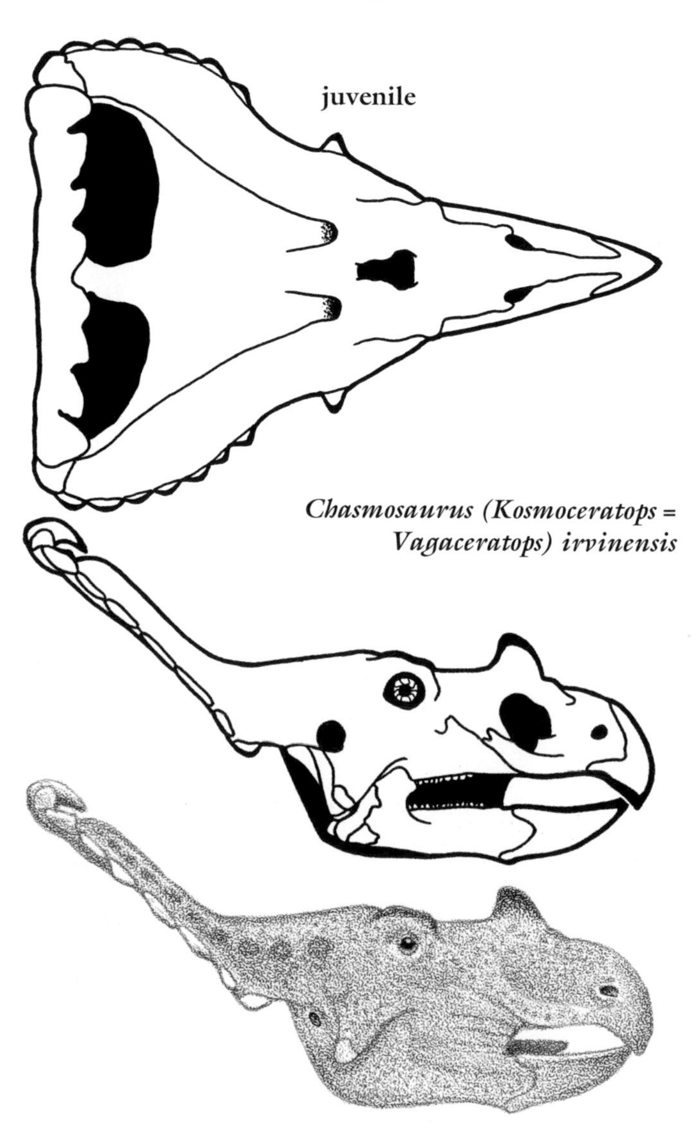

Chasmosaurus (Kosmoceratops = Vagaceratops) irvinensis

DISTRIBUTION AND FORMATION/S Alberta; upper Dinosaur Park.
HABITAT Well-watered, forested floodplain with coastal swamps and marshes, cool winters.
HABITS Probably relied more on beak than horns for defense.

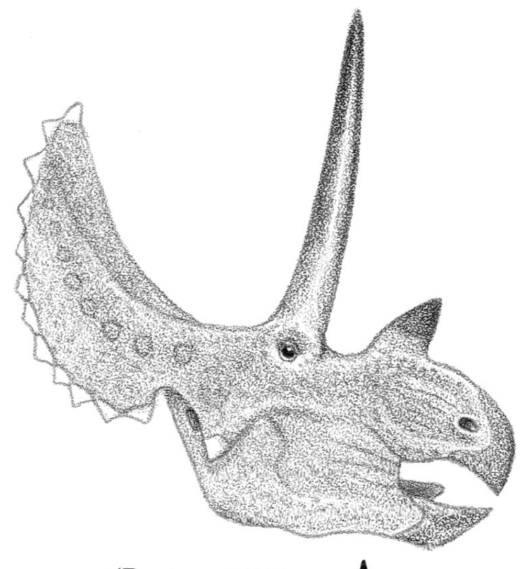

Chasmosaurus (Pentaceratops = Agujaceratops) mariscalensis

Chasmosaurus (Pentaceratops = Agujaceratops) mariscalensis
4.3 m (14 ft) TL, 1.5 tonnes

FOSSIL REMAINS Numerous disarticulated skulls and skeletons, juvenile to adult.
ANATOMICAL CHARACTERISTICS Nasal horn short, brow horns long, frill elongated, back rim not broad, strongly indented, bearing large hornlets, small hornlets along side rims. Belly broad, hind legs bowed.
AGE Late Cretaceous, Campanian.
DISTRIBUTION AND FORMATION/S Texas; Aguja.
NOTES Originally placed in *Chasmosaurus*; the skulls and skeletons of the chasmosaurs are very similar except for the horns and frills, so they probably form one genus. Alternatively, this, *Utahceratops gettyi*, and *Pentaceratops sternbergii* form a united genus.

Chasmosaurus (Pentaceratops = Utahceratops) gettyi
5 m (16 ft) TL, 2 tonnes

FOSSIL REMAINS Partial skulls and skeletons.
ANATOMICAL CHARACTERISTICS Nasal and brow horns short, latter projecting sideways, frill elongated, back

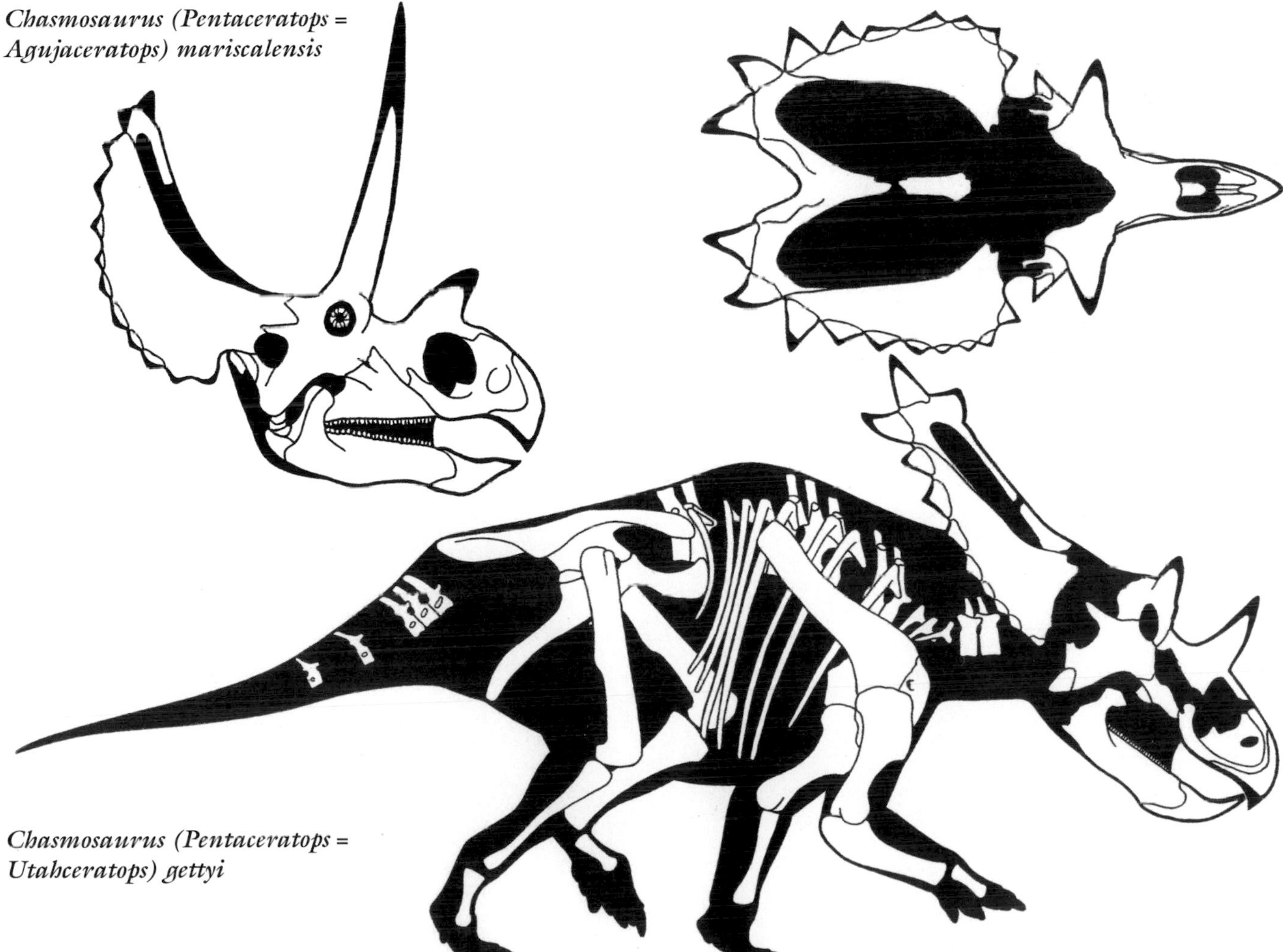

Chasmosaurus (Pentaceratops = Utahceratops) gettyi

rim not broad, strongly indented, bearing large hornlets, innermost projecting vertically, small hornlets along side rims.
AGE Late Cretaceous, late Campanian.
DISTRIBUTION AND FORMATION/S Utah; lower and upper Kaiparowits.

Chasmosaurus (Pentaceratops) sternbergii
5.5 m (18 ft) TL, 2.5 tonnes

FOSSIL REMAINS Several complete or partial skulls, a complete and some partial skeletons, completely known.
ANATOMICAL CHARACTERISTICS Nasal horn long, brow horns long, frill extremely elongated, tilted upward, back rim not broad, strongly indented, bearing large hornlets, innermost projecting vertically, small hornlets along side rims. Shoulder withers supported nuchal ligaments to neck and head. Belly broad, hind legs bowed.
AGE Late Cretaceous, late Campanian and/or early Maastrichtian.
DISTRIBUTION AND FORMATION/S New Mexico; Fruitland, lower Kirtland.
HABITAT Moderately watered floodplain woodlands, coastal swamps and marshes.
NOTES Whether *Pentaceratops* is a distinct genus from *Chasmosaurus* that are very similar aside from horns and frills is problematic. Shared its habitat with *Titanoceratops* and *Nodocephalosaurus.*

Anchiceratops (or *Arrhinoceratops) brachyops*
4.5 m (15 ft) TL, 1.3 tonnes

FOSSIL REMAINS Two nearly complete skulls.
ANATOMICAL CHARACTERISTICS Nasal horn short, projecting forward, brow horns long and directed sideways, frill subhorizontal, all hornlets small.
AGE Late Cretaceous, latest Campanian and earliest Maastrichtian.

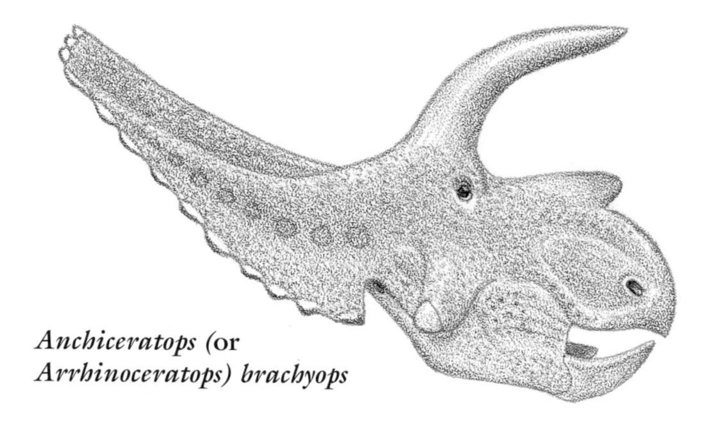

Anchiceratops (or *Arrhinoceratops) brachyops*

DISTRIBUTION AND FORMATION/S Alberta; lower-mid Horseshoe Canyon.
HABITAT Well-watered, forested floodplain with coastal swamps and marshes, cool winters.
NOTES May be a sexual morph of *Anchiceratops*. Prey of *Albertosaurus sarcophagus.*

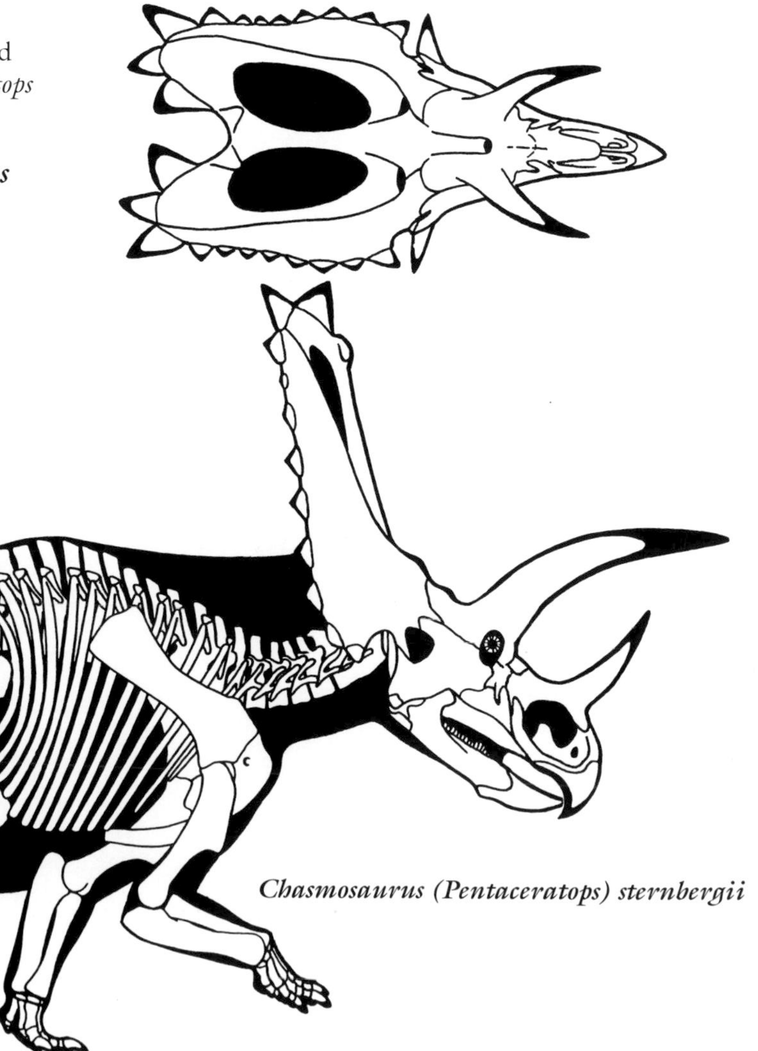

Chasmosaurus (Pentaceratops) sternbergii

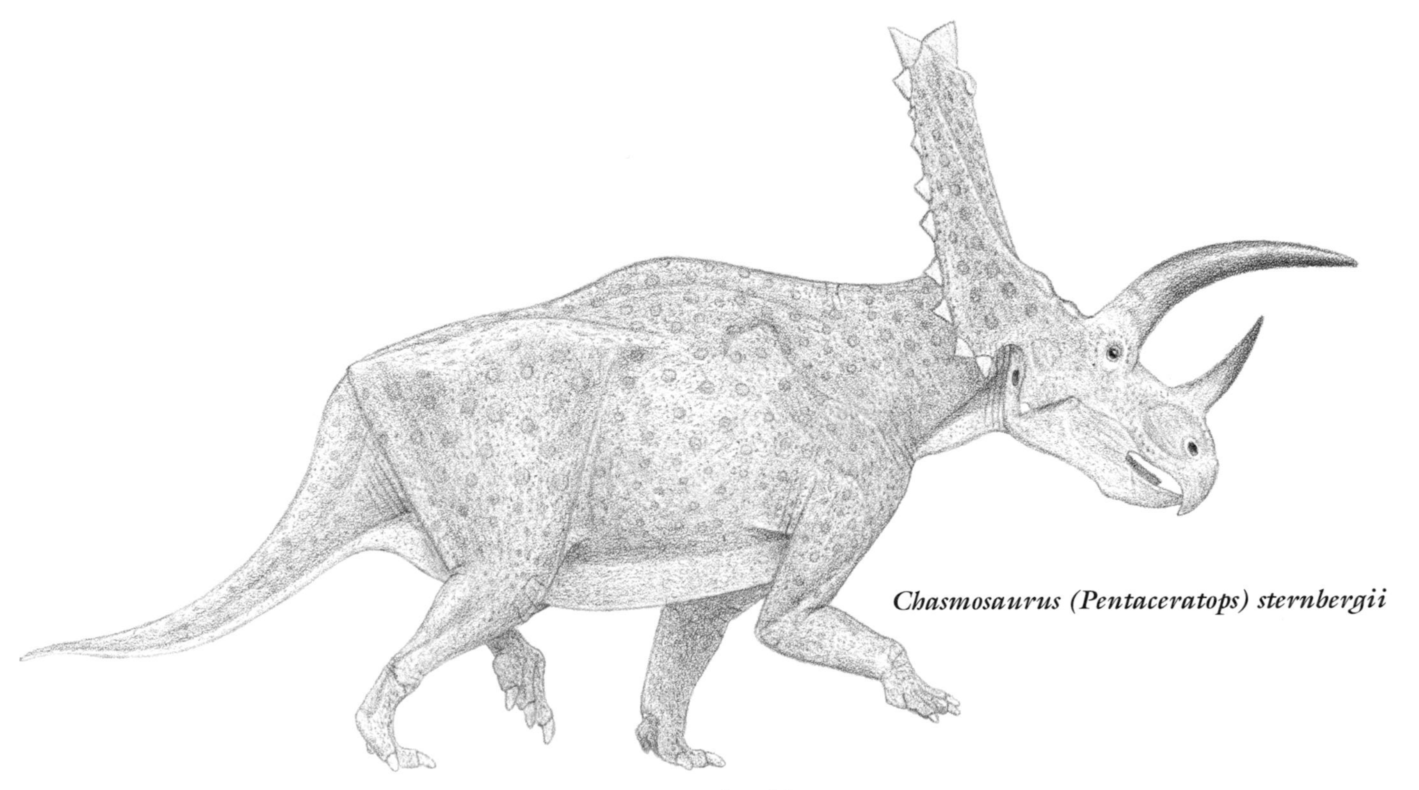
Chasmosaurus (Pentaceratops) sternbergii

Anchiceratops ornatus
4.5 m (15 ft) TL, 1.5 tonnes

FOSSIL REMAINS A few nearly complete skulls and minority of skeleton.
ANATOMICAL CHARACTERISTICS Nasal horn short, projecting forward, brow horns long to very long and directed sideways, frill subhorizontal, back rim not broad, large hornlets astride and atop centerline and along back, small hornlets on rim side.

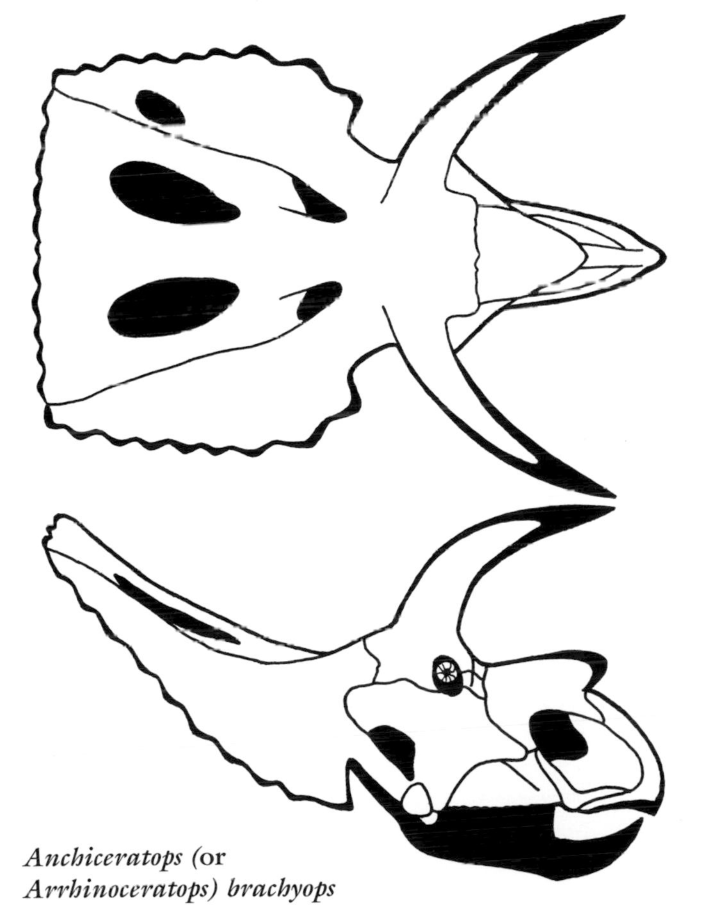
Anchiceratops (or *Arrhinoceratops*) *brachyops*

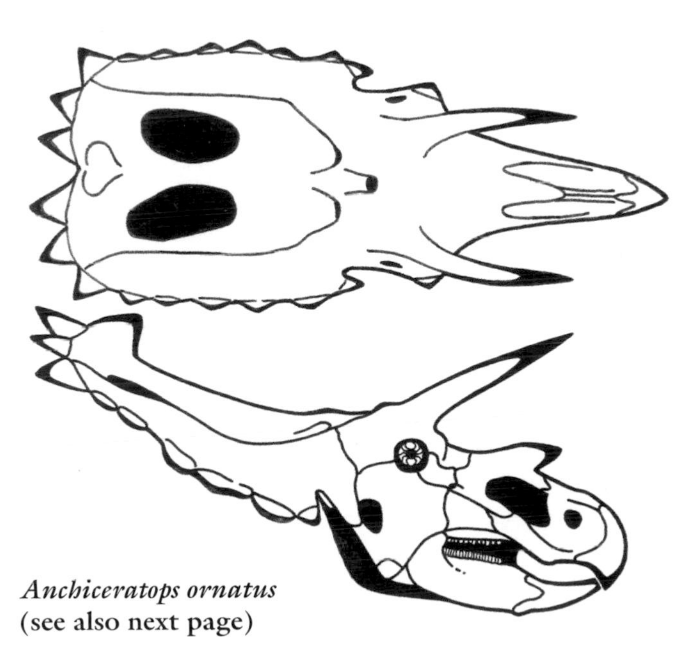
Anchiceratops ornatus
(see also next page)

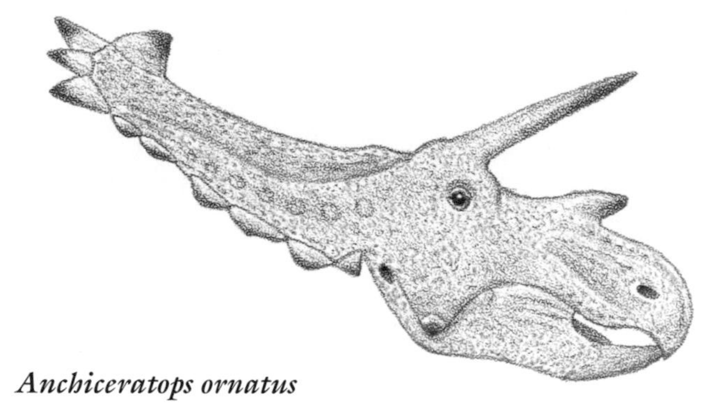

Anchiceratops ornatus

AGE Late Cretaceous, latest Campanian and earliest Maastrichtian.
DISTRIBUTION AND FORMATION/S Alberta; lower-mid Horseshoe Canyon.
HABITAT Well-watered, forested floodplain with coastal swamps and marshes, cool winters.
NOTES The complete skeleton lacking a skull is no longer automatically attributed to this taxon.

Unnamed genus and species?

4.3 m (14 ft) TL, 1.2 tonnes

FOSSIL REMAINS Complete skeleton lacking skull.
ANATOMICAL CHARACTERISTICS Neck elongated. Arm robust.
AGE Late Cretaceous, latest Campanian and/or earliest Maastrichtian.
DISTRIBUTION AND FORMATION/S Alberta; lower-mid Horseshoe Canyon.
HABITAT Well-watered, forested floodplain with coastal swamps and marshes, cool winters.
NOTES Skeleton probably belongs to *Anchiceratops* or less common *Arrhinoceratops*, or may be a distinct taxon.

Triceratopsines Large to gigantic ceratopsids limited to the late Late Cretaceous of North America.

ANATOMICAL CHARACTERISTICS Pelvic ilium slopes down and backward, at least in *Triceratops.*

Regaliceratops peterhewsi

5 m (17 ft) TL, 2 tonnes

FOSSIL REMAINS Majority of skull.
ANATOMICAL CHARACTERISTICS Head and frill short, broad, nasal opening not elongated, nasal and brow horns short, frill openings small, rim adorned by large, stout hornlets, centerline boss on frill.
AGE Late Cretaceous, probably middle Maastrichtian.
DISTRIBUTION AND FORMATION/S Alberta; probably upper Saint Mary River.

Regaliceratops peterhewsi

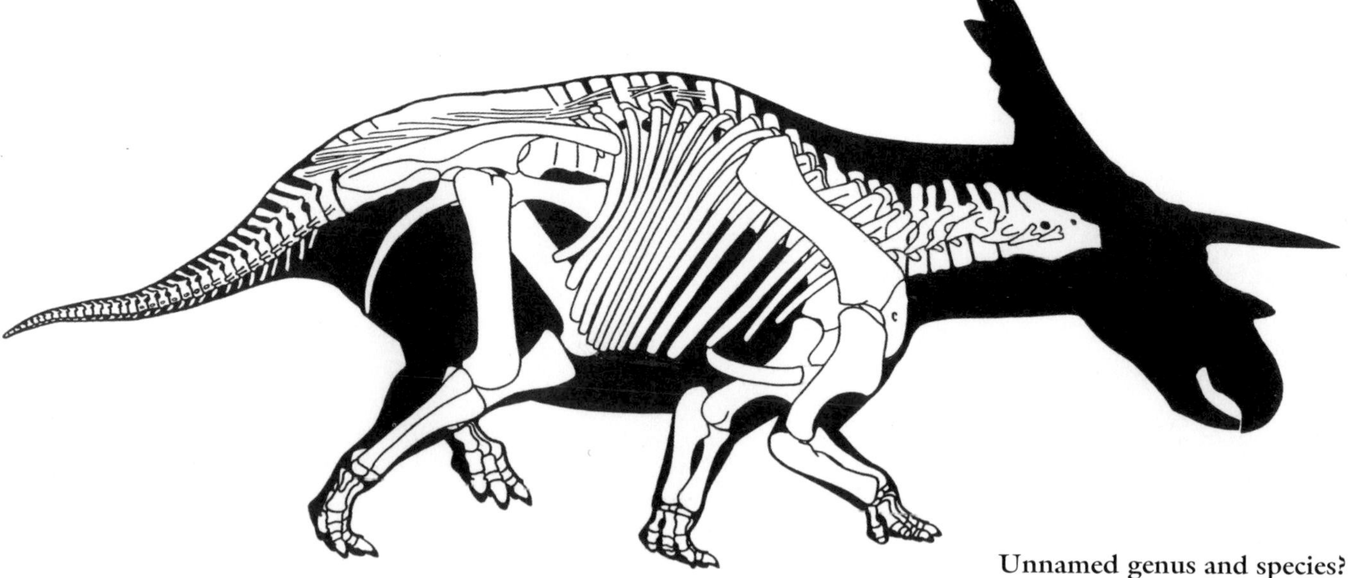

Unnamed genus and species?

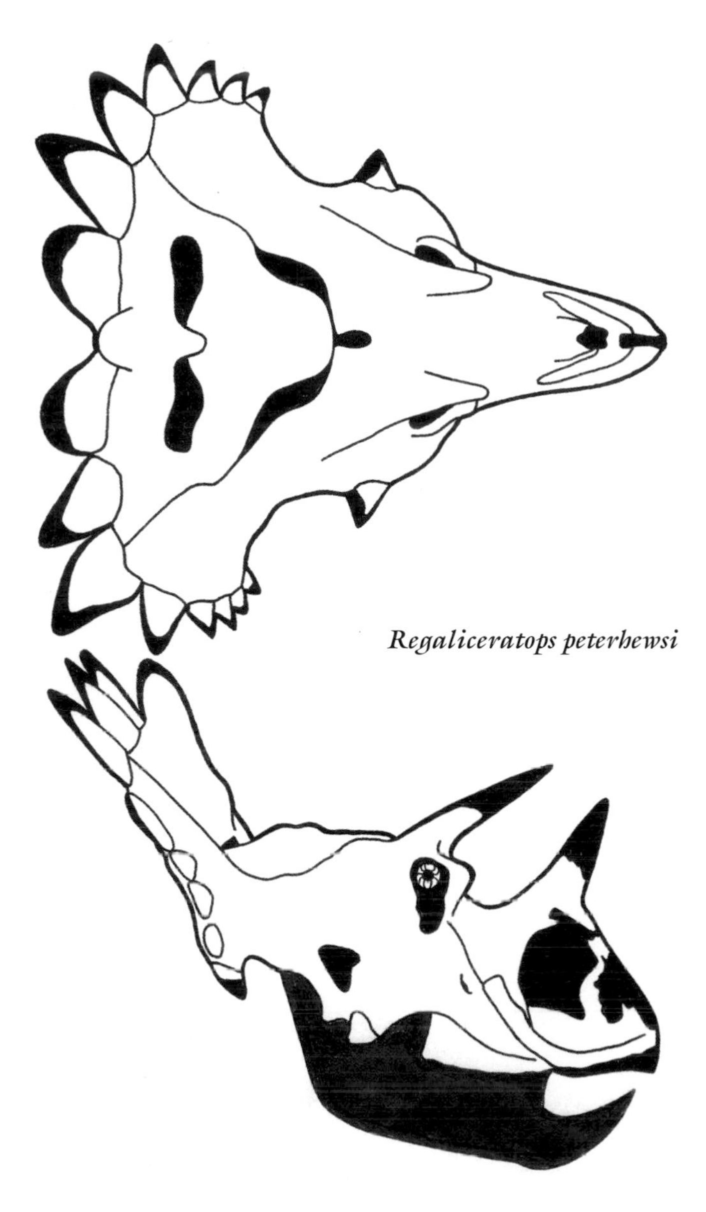
Regaliceratops peterhewsi

Triceratops (= Eotriceratops) xerinsularis
8.5 m (28 ft) TL, 10 tonnes

FOSSIL REMAINS Complete skull and minority of skeleton.
ANATOMICAL CHARACTERISTICS Snout shallow, nasal opening elongated, nasal horn small, brow horns long, frill moderately elongated.
AGE Late Cretaceous, middle Maastrichtian.
DISTRIBUTION AND FORMATION/S Alberta; uppermost Horseshoe Canyon.
HABITAT Well-watered, forested floodplain with coastal swamps and marshes, cool winters.
NOTES Largest known horned dinosaur, largest-headed known land animal. Separation of *Eotriceratops* from very similar *Triceratops/Torosaurus* not warranted; may include *Ojoceratops fowleri* from New Mexico—the latter is *Triceratops*; may be direct ancestor of *Triceratops horridus.*

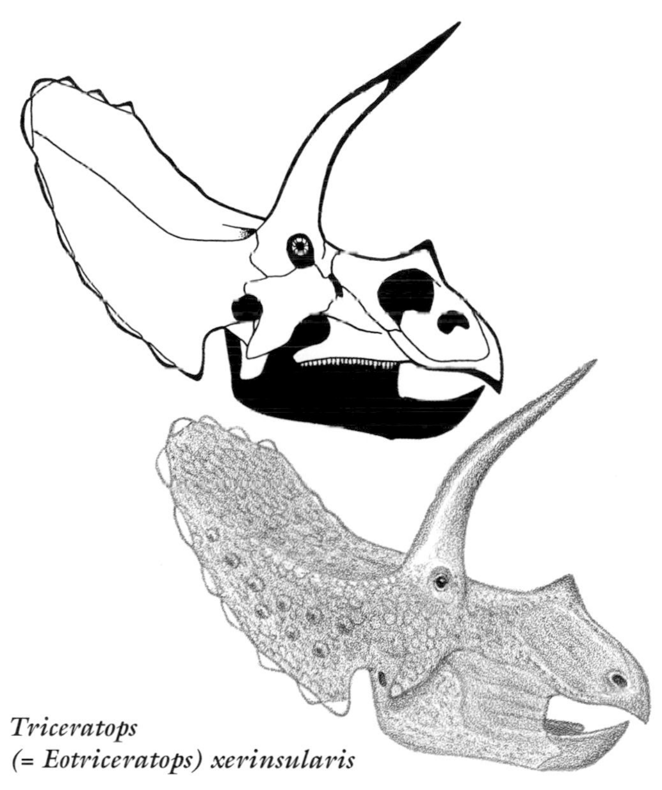
Triceratops (= Eotriceratops) xerinsularis

Titanoceratops ouranos
6.5 m (21 ft) TL, 4.5 tonnes

FOSSIL REMAINS Majority of skull and skeleton.
ANATOMICAL CHARACTERISTICS Nasal opening elongated, nasal horn short, brow horns long.
AGE Late Cretaceous, late Campanian and/or early Maastrichtian.
DISTRIBUTION AND FORMATION/S New Mexico; Fruitland, lower Kirtland.
HABITAT Moderately watered floodplain woodlands, coastal swamps and marshes.
NOTES Originally placed in *Chasmosaurus* (*Pentaceratops*) *sternbergii*; uncertain anatomical features include length of frill, absence or presence of withers; orientation of pelvis precludes restoration.

Triceratops horridus (Illustrated overleaf)
8 m (25 ft) TL, 9 tonnes

FOSSIL REMAINS Numerous skulls and some complete and partial skeletons including juveniles, completely known.
ANATOMICAL CHARACTERISTICS Snout shallow, nasal opening elongated, nasal horn small, brow horns long, shift from arcing backward to forward with maturity; frill probably elongates greatly with maturity as large openings develop in at least one sex, hornlets suppressed and

centerline hornlet lost. Skeleton very robust. Body scales generally larger than on other ceratopsids; largest often have raised cones and are widely spaced in an irregular pattern.
AGE Late Cretaceous, late Maastrichtian.
DISTRIBUTION AND FORMATION/S Dakotas, Wyoming, Montana; lower Lance and Hell Creek.
HABITAT Well-watered coastal woodlands, climate cooler than in latest Maastrichtian, possibly chilly in winter.
NOTES Probably includes *Nedoceratops* (= *Diceratops*) *hatcheri*, at least one adult sex (probably male) is probably *Torosaurus*—assignment of fragmentary remains from more southern regions to latter genus is problematic. Main enemy robust *Tyrannosaurus* species. May be an ancestor of *Triceratops prorsus.*

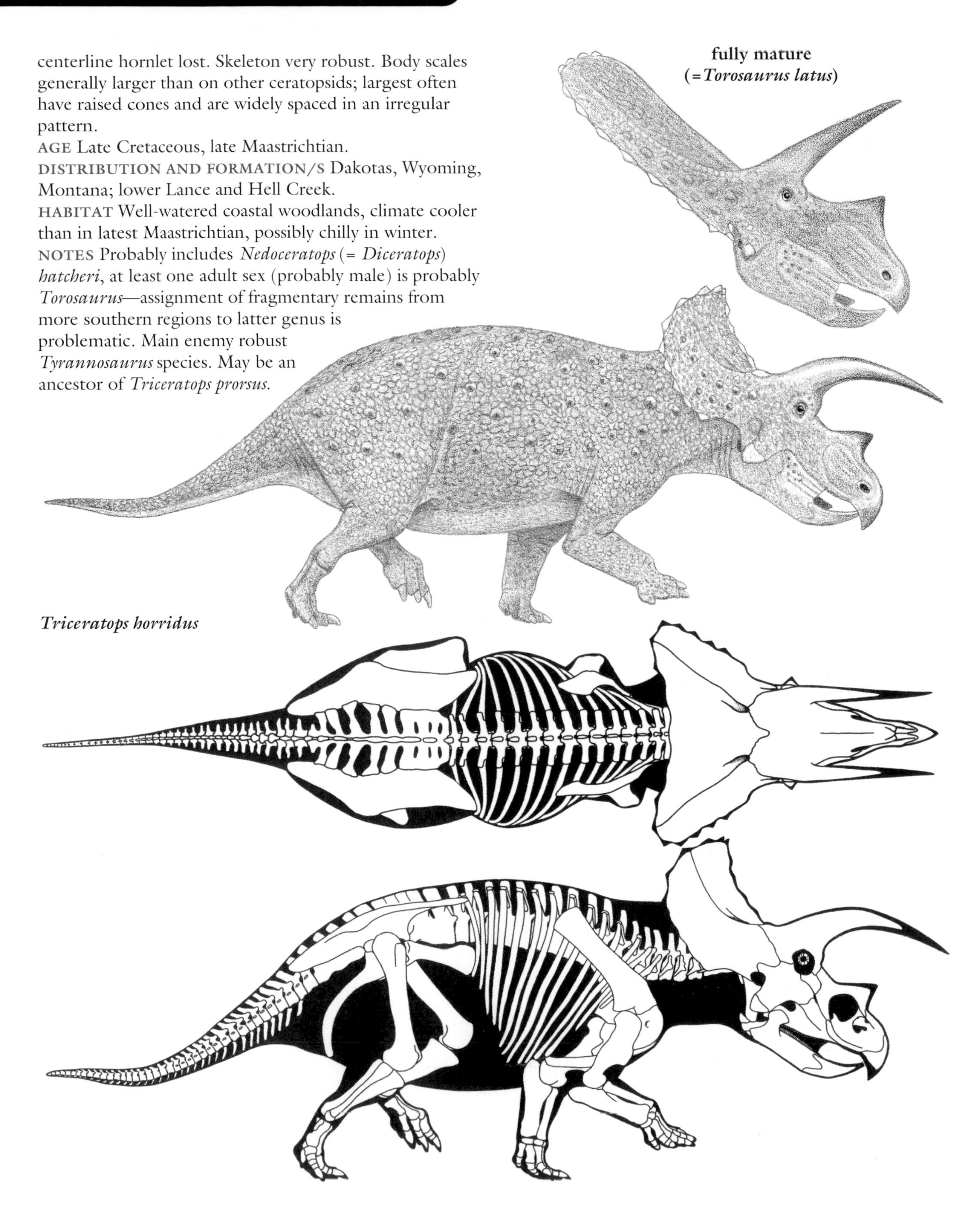

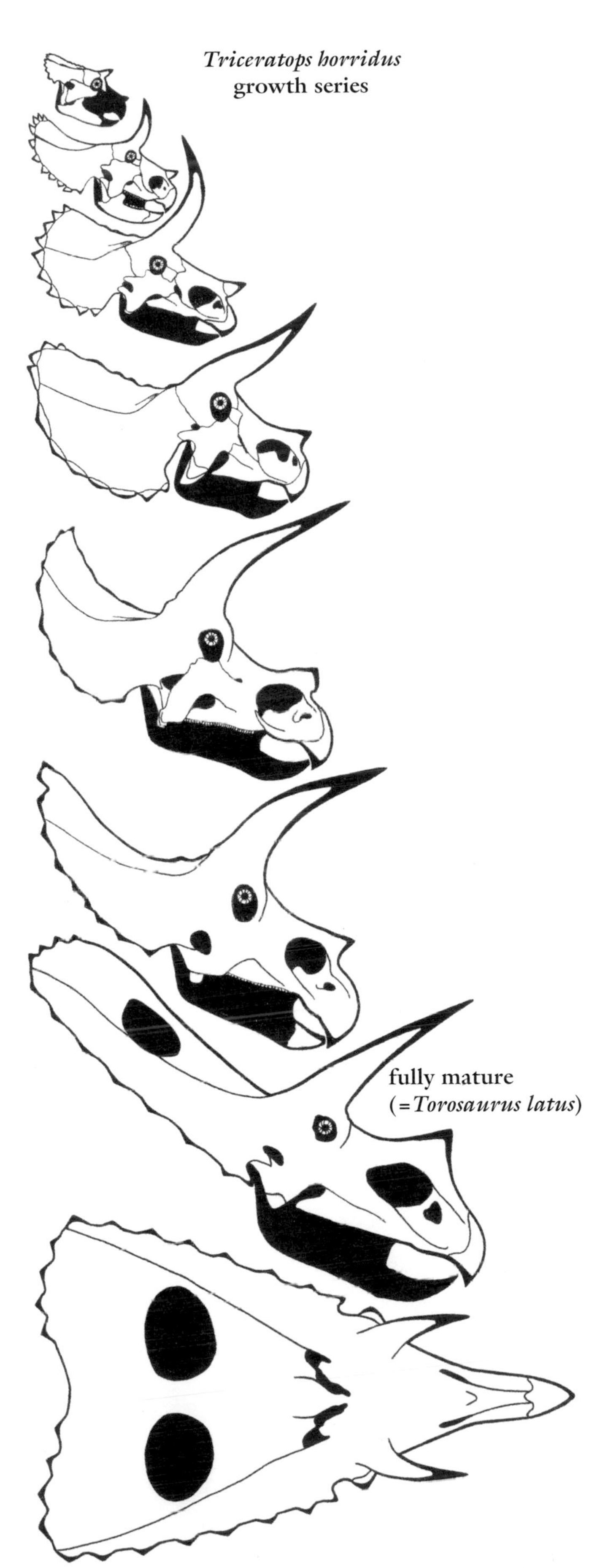

Triceratops horridus
growth series

Triceratops prorsus

8 m (25 ft) TL, 9 tonnes

FOSSIL REMAINS Numerous skulls and skeletal parts.
ANATOMICAL CHARACTERISTICS Snout deep, nasal opening elongated, nasal horn moderately long, brow horns moderately long, frill never elongated; openings develop and hornlets suppressed, centerline hornlet lost when fully mature. Skeleton very robust.
AGE Late Cretaceous, latest Maastrichtian.
DISTRIBUTION AND FORMATION/S Montana, Wyoming, Alberta, Saskatchewan; upper Lance and Hell Creek, Denver, Laramie, Scollard, Frenchman.
HABITAT Well-watered coastal woodlands, climate warmer than earlier in Maastrichtian.
HABITS A bitten-off and healed brow horn confirms aggressive defensive head-to-head combat with its main enemy, *Tyrannosaurus*, either *T. rex* or a more common gracile species.
NOTES *Triceratops* was the most common large herbivore in the Hell Creek, making up two-thirds of the population.

Triceratops prorsus
growth series

ORNITHOPODS

SMALL TO GIGANTIC GENASAUR ORNITHISCHIANS OF THE LATE JURASSIC TO THE END OF THE DINOSAUR ERA, ALL CONTINENTS.

ANATOMICAL CHARACTERISTICS Fairly uniform. Head not greatly enlarged, beak not hooked, eyes large, main tooth rows well developed. Neck S-curved. Trunk and tail stiffened by ossified tendons. Tail moderately long. Strongly bipedal to semiquadrupedal, arm and leg flexed, and latter always long, so good runners. Fingers five to four. Toes four to three. Rib-free lumbar region in front of hips implies a mammal-like diaphragm was present. Large examples scaly; integument of smaller examples uncertain but may have been insulated with fibers.
ONTOGENY Growth rates apparently rapid in at least some examples.
HABITAT Very variable, from tropics to polar winters, from arid to wet.
HABITS Low- to medium-level browsers; some species may have fed in shallow water on occasion. Main defense running, also kicking with feet among medium-sized and larger species.
NOTES The kangaroos, deer, antelope, and cattle of the last half of the Mesozoic, and the most common herbivores of the Cretaceous.

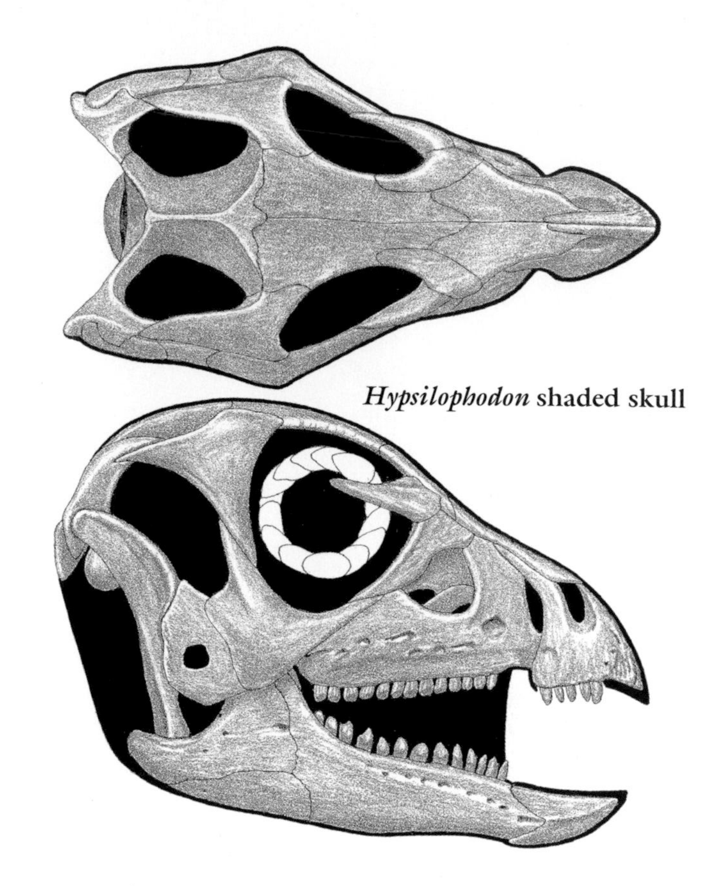

Hypsilophodon **shaded skull**

"HYPSILOPHODONTS"

SMALL TO MEDIUM-SIZED ORNITHOPODS OF THE LATE JURASSIC TO THE END OF THE DINOSAUR ERA, ALL CONTINENTS.

ANATOMICAL CHARACTERISTICS Uniform. Head subtriangular, narrow, beaks narrow, eyes shaded by overhanging rim, teeth at front of upper jaw, main tooth rows well developed. Body and hips fairly narrow, large interrib plates in at least some examples. Tail moderately long. Bipedal except could move quadrupedally at slow speeds. Arm fairly short, so strongly bipedal; hands small, five grasping fingers tipped with small claws. Leg long and usually fairly gracile, so speed potential high, four long toes tipped with blunt claws.
HABITAT Very variable, from tropics to polar winters, from arid to wet.
HABITS Low-level browsers and omnivores, probably picked up insects and small vertebrates. Predominantly terrestrial, probably some climbing ability, well-developed shoulders indicate burrowing ability. Main defense high speed.
NOTES Closest living analogs small kangaroos, deer, and antelope. Burrows attributable to small ornithopods have been found in North America and then-polar Australia. The relationships of these generalized ornithopods are problematic: very possibly not a unified group, and splittable into a number of divisions and subdivisions. Fragmentary Late Cretaceous *Morrosaurus antarcticus* shows presence in Antarctica.

Yandusaurus hongheensis or *Agilisaurus louderbacki*

3.8 m (12 ft) TL, 140 kg (300 lb)

FOSSIL REMAINS Two nearly complete skulls and skeletons and partial remains.
ANATOMICAL CHARACTERISTICS Head small, a few large bladed teeth near front of lower jaws. End of tail fairly deep.
AGE Late Jurassic, Bathonian.
DISTRIBUTION AND FORMATION/S Central China; lower Shaximiao.
HABITAT Heavily forested.
HABITS Bladed teeth facilitated handling small prey items.
NOTES *A. louderbacki and Hexinlusaurus multidens* probably immature examples of this species, but the original specimen of *Y. hongheensis* is very incomplete and much larger than the others, so the first name may be most valid.

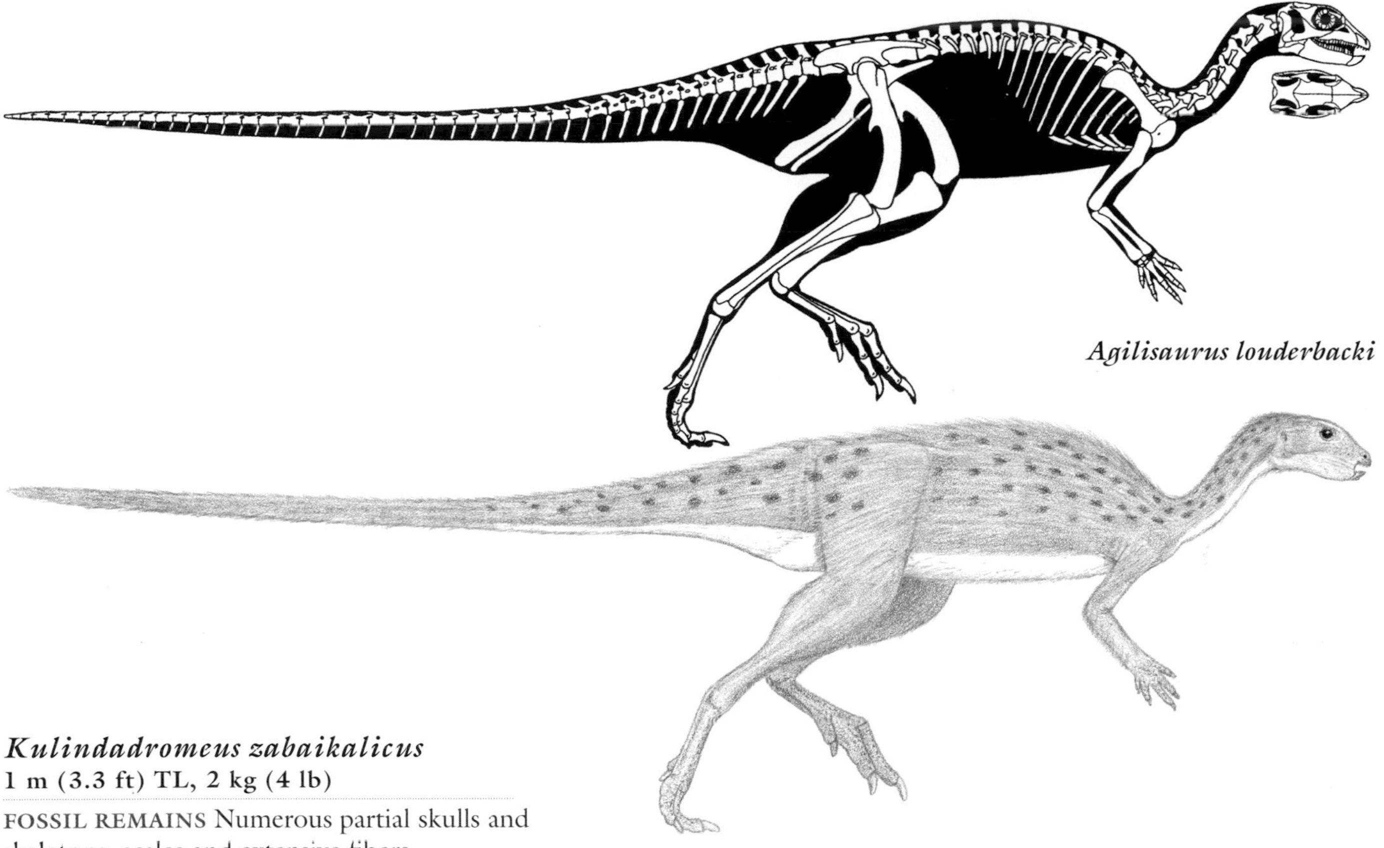

Agilisaurus louderbacki

Kulindadromeus zabaikalicus
1 m (3.3 ft) TL, 2 kg (4 lb)

FOSSIL REMAINS Numerous partial skulls and skeletons, scales and extensive fibers

ANATOMICAL CHARACTERISTICS Much of body covered by short, coarse and fine filaments, latter long on upper shank; small, rounded, nonoverlapping scales on hands and feet, rows of large, overlapping scales on upper half of tail.

AGE Middle or Late Jurassic.

DISTRIBUTION AND FORMATION/S Southern Siberia; Ukureyskaya.

HABITAT Woodlands and lakes.

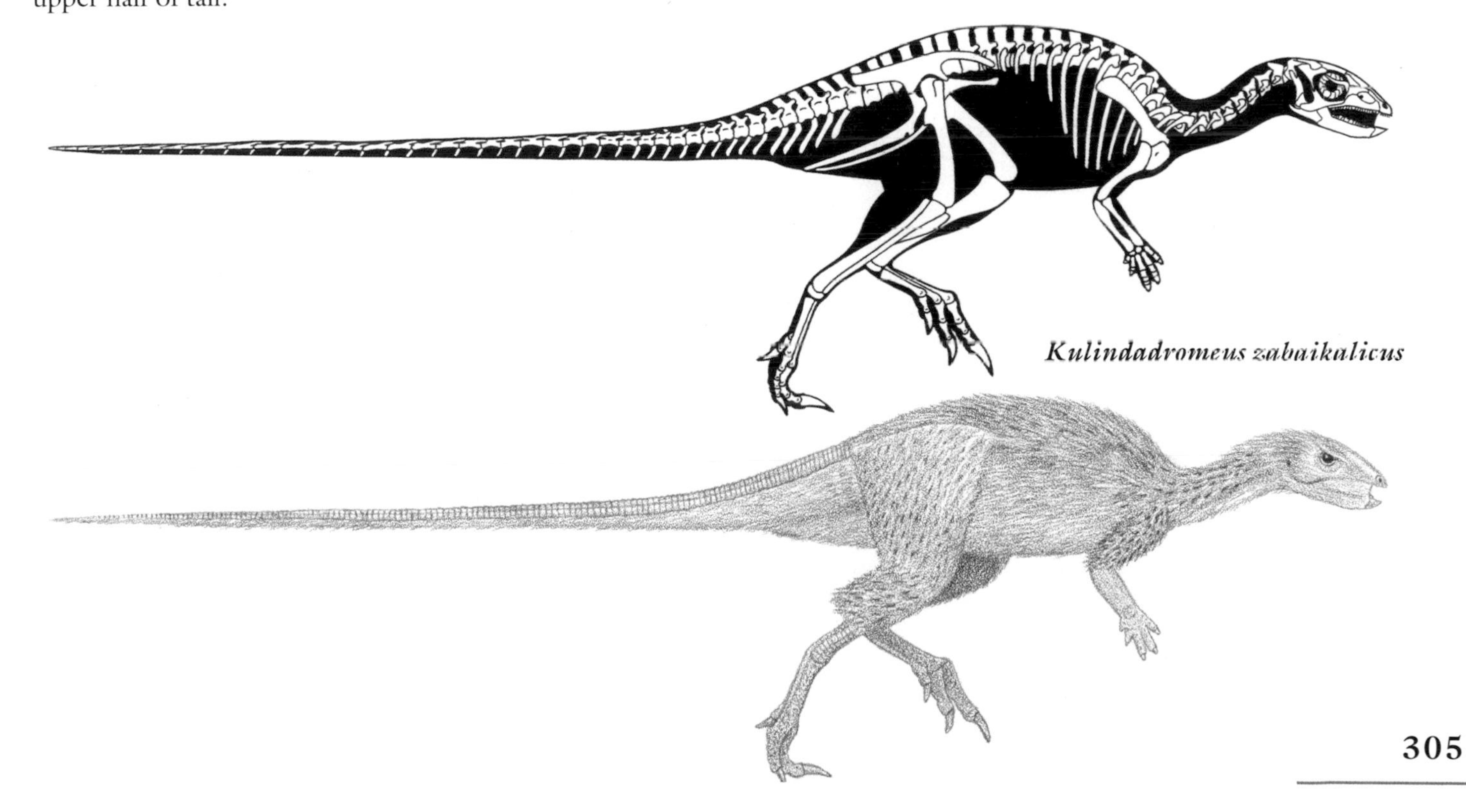

Kulindadromeus zabaikalicus

Drinker nisti
2 m (6.5 ft) TL, 20 kg (45 lb)

FOSSIL REMAINS Partial skeletons.
ANATOMICAL CHARACTERISTICS Insufficient information.
AGE Late Jurassic, middle Tithonian.
DISTRIBUTION AND FORMATION/S Wyoming; upper Morrison.
HABITAT Short wet season, otherwise semiarid with open floodplain prairies and riverine forests.

Othnielosaurus consors
2.2 m (7.5 ft) TL, 30 kg (60 lb)

FOSSIL REMAINS Complete skull and majority of skeletons.
ANATOMICAL CHARACTERISTICS Head small, subrectangular.
AGE Late Jurassic, early Tithonian.
DISTRIBUTION AND FORMATION/S Wyoming, Colorado, Utah; middle Morrison.
HABITAT Wetter than earlier Morrison, otherwise semiarid with open floodplain prairies and riverine forests.
NOTES Once *Othnielia rex*, which is based on inadequate material; it is uncertain whether the skull and the skeleton illustrated here belong to this species or to one or two other taxa.

Jeholosaurus shangyuanensis
Adult size uncertain

FOSSIL REMAINS Two skulls and minority of skeletons, at least one juvenile.
ANATOMICAL CHARACTERISTICS Standard for type.
AGE Early Cretaceous, Barremian.
DISTRIBUTION AND FORMATION/S Northeast China; lower Yixian.
HABITAT Well-watered forests and lakes, winters chilly with some snow.
NOTES Prey of *Sinocalliopteryx* and *Sinornithosaurus zhaoianus*.

Changchunsaurus parvus
1.5 m (5 ft) TL, 10 kg (20 lb)

FOSSIL REMAINS Majority of distorted skull and uncertain part of skeleton.
ANATOMICAL CHARACTERISTICS Skull shallow, upper and lower beaks pointed.
AGE Late Early Cretaceous or early Late Cretaceous.
DISTRIBUTION AND FORMATION/S Northeast China; Quantou.
HABITAT Well-watered forests and lakes, winters chilly with some snow.
NOTES Shared its habitat with *Helioceratops*.

Haya griva
1.6 m (5 ft) TL, 10 kg (20 lb)

FOSSIL REMAINS A few complete to partial skulls and skeletons.
ANATOMICAL CHARACTERISTICS Skull low, front upper teeth large.
AGE Late Cretaceous, probably Santonian.
DISTRIBUTION AND FORMATION/S Mongolia; Javkhlant.

Hypsilophodon foxii
2 m (6.5 ft) TL, 20 kg (45 lb)

FOSSIL REMAINS About a dozen complete and partial skulls and skeletons, juvenile to adult, completely known.
ANATOMICAL CHARACTERISTICS Standard for type.
AGE Early Cretaceous, late Barremian.
DISTRIBUTION AND FORMATION/S Southern England; Wessex.
NOTES The classic hypsilophodont.

Leaellynasaura amicagraphica
3 m (10 ft) TL, 90 kg (200 lb)

FOSSIL REMAINS Minority of skull and skeleton.
ANATOMICAL CHARACTERISTICS Insufficient information.

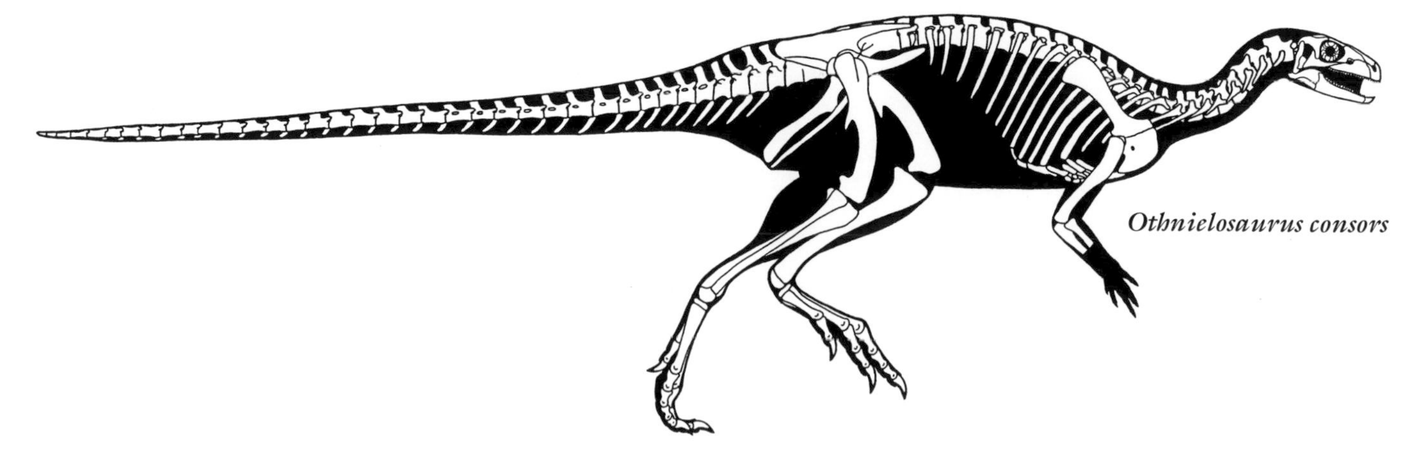
Othnielosaurus consors

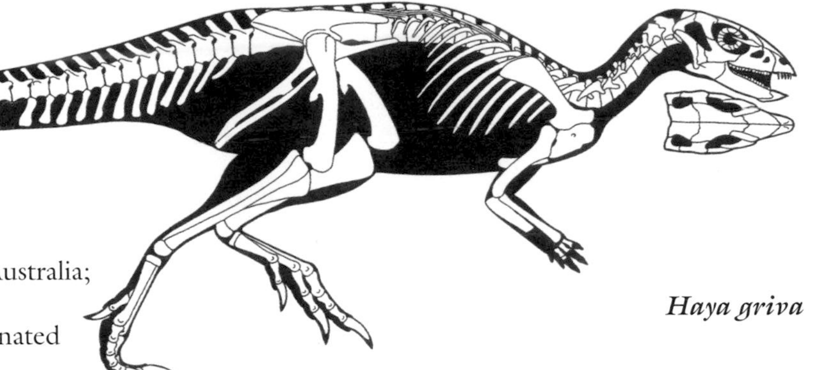

Haya griva

AGE Early Cretaceous, early Albian.
DISTRIBUTION AND FORMATION/S Southern Australia; Eumeralla.
HABITAT Polar forests with warm, daylight-dominated summers and cold, dark winters.

Fulgurotherium australe
1.3 m (4 ft) TL, 6 kg (12 lb)

FOSSIL REMAINS Minority of skeletons.
ANATOMICAL CHARACTERISTICS Insufficient information.
AGE Early Cretaceous, Albian.
DISTRIBUTION AND FORMATION/S Southeast Australia; Griman Creek.
HABITAT Polar forests with warm, daylight-dominated summers and cold, dark winters.

Qantassaurus intrepidus
2 m (6.5 ft) TL, 20 kg (45 lb)

FOSSIL REMAINS Minority of skull.
ANATOMICAL CHARACTERISTICS Insufficient information.
AGE Early Cretaceous.
DISTRIBUTION AND FORMATION/S Southern Australia; Wonthaggi.
HABITAT Polar forests with warm, daylight-dominated summers and cold, dark winters.

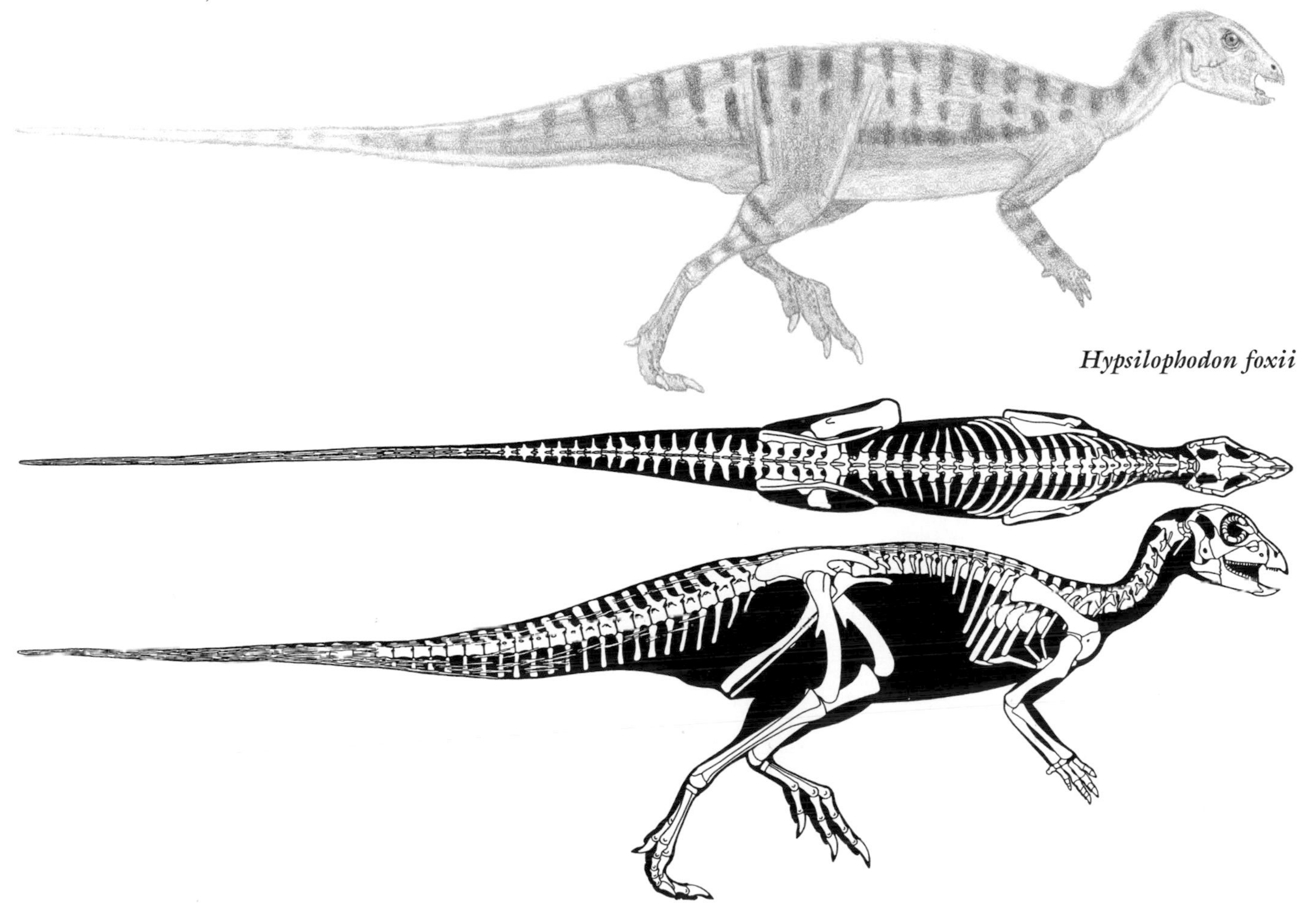

Hypsilophodon foxii

Zephyrosaurus schaffi
2 m (6.5 ft) TL, 20 kg (45 lb)

FOSSIL REMAINS Partial skulls and small portion of two skeletons.
ANATOMICAL CHARACTERISTICS Upper beak strengthened.
AGE Early Cretaceous, middle Albian.
DISTRIBUTION AND FORMATION/S Montana, Wyoming; upper Cloverly.
HABITAT Short wet season, otherwise semiarid with floodplain prairies and open woodlands, and riverine forests.
HABITS Strengthening of snout indicates burrowing.
NOTES Shared its habitat with *Tenontosaurus*. Prey of *Deinonychus*.

Oryctodromeus cubicularis
2 m (7 ft) TL, 20 kg (45 lb)

FOSSIL REMAINS Minority of skulls and skeletons, juveniles and adult, burrows.
ANATOMICAL CHARACTERISTICS Upper beak strengthened. Shoulder blade exceptionally enlarged. Pelvis strengthened.
AGE Late Cretaceous, Cenomanian.
DISTRIBUTION AND FORMATION/S Montana; Blackleaf.
HABITAT Seasonally dry upland woodlands.
HABITS Dug sinuous burrows a few meters long with strengthened beak and arms while bracing with legs.

Orodromeus makelai
Adult size uncertain

FOSSIL REMAINS A number of partial skulls and skeletons.
ANATOMICAL CHARACTERISTICS Upper beak strengthened, large boss on cheek. Shoulder blade exceptionally enlarged.
AGE Late Cretaceous, middle and/or late Campanian.
DISTRIBUTION AND FORMATION/S Montana; Upper Two Medicine, possibly Judith River.
HABITAT Well-watered, forested floodplain with coastal swamps and marshes and drier upland woodlands.
HABITS Strengthening of snout and arms indicates well-developed burrower.

Koreanosaurus boseongensis
Adult size uncertain

FOSSIL REMAINS Minority of skeleton.
ANATOMICAL CHARACTERISTICS Shoulder blade exceptionally enlarged.
AGE Late Cretaceous, Santonian or Campanian.
DISTRIBUTION AND FORMATION/S North Korea; Seonso.
HABITS Strengthening of snout and arms indicates well-developed burrower.

Parksosaurus warreni
2.5 m (8 ft) TL, 45 kg (100 lb)

FOSSIL REMAINS Majority of skull and skeleton.
ANATOMICAL CHARACTERISTICS Shoulder blade exceptionally enlarged, arm robust. Toes unusually long.
AGE Late Cretaceous, middle Maastrichtian.
DISTRIBUTION AND FORMATION/S Alberta; middle Horseshoe Canyon.
HABITAT Well-watered, forested floodplain with coastal swamps and marshes, cool winters.
HABITS Long toes imply *Parksosaurus* was adapted for moving on soft soils near watercourses and in marshlands. Strong shoulders and arms indicate burrowing.

Thescelosaurus neglectus
3 m (10 ft) TL, 90 kg (200 lb)

FOSSIL REMAINS A few complete and partial skulls and skeletons, completely known.
ANATOMICAL CHARACTERISTICS Shoulder blade enlarged, arm robust.
AGE Late Cretaceous, late Maastrichtian.
DISTRIBUTION AND FORMATION/S Colorado, Wyoming, South Dakota; Lance, Hell Creek (levels uncertain), Laramie.
HABITAT Well-watered coastal woodlands.
HABITS Strong shoulders and arms indicate burrowing.
NOTES May include *Thescelosaurus garbanii* and *Bugenasaura infernalis*.

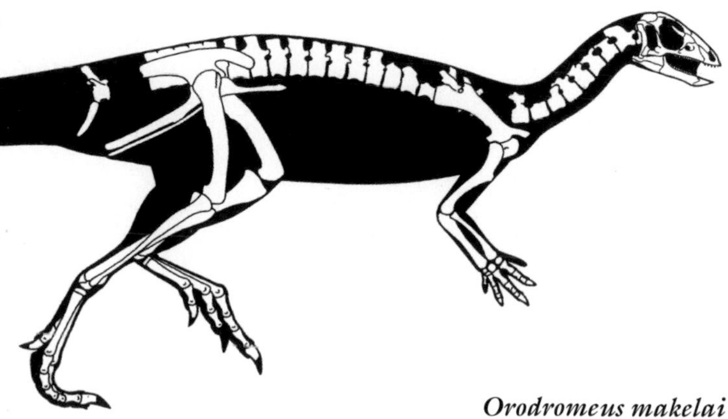

Orodromeus makelai

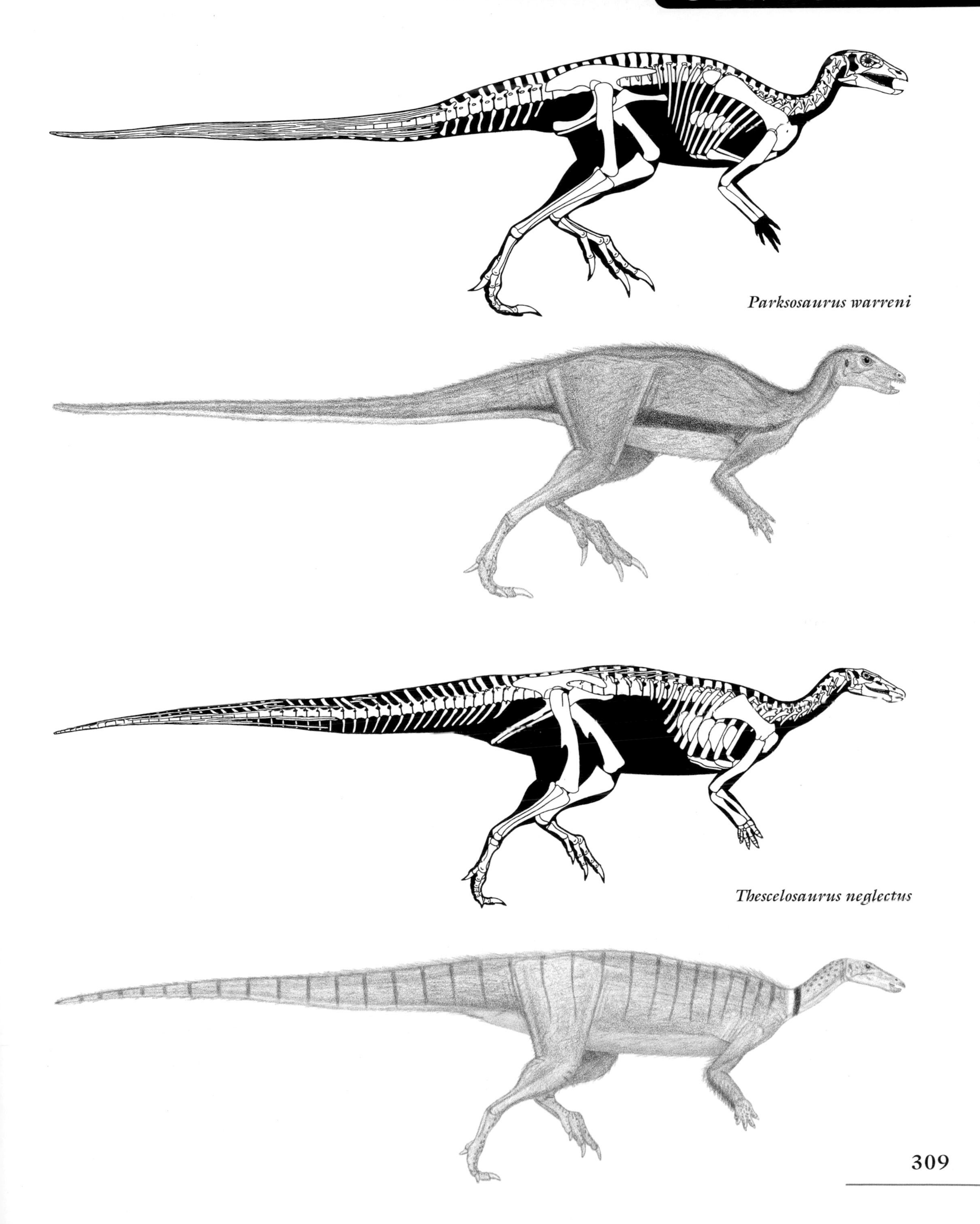

Parksosaurus warreni

Thescelosaurus neglectus

Thescelosaurus assiniboiensis
Adult size uncertain

FOSSIL REMAINS Partial skulls and skeletons.
ANATOMICAL CHARACTERISTICS Shoulder blade enlarged, arm robust.
AGE Late Cretaceous, latest Maastrichtian.
DISTRIBUTION AND FORMATION/S Saskatchewan; Frenchman.
HABITAT Well-watered coastal woodlands.
HABITS Strong shoulders and arms indicate burrowing.
NOTES May be direct descendant of *T. neglectus.*

Gasparinisaura cincosaltensis
1.7 m (5.5 ft) TL, 13 kg (30 lb)

FOSSIL REMAINS Partial skull and skeletons, juvenile to adult.
ANATOMICAL CHARACTERISTICS Standard for type.
AGE Late Cretaceous, late Santonian or early Campanian.
DISTRIBUTION AND FORMATION/S Western Argentina; Anacleto.
NOTES Prey of *Aerosteon* and *Abelisaurus.*

Notohypsilophodon comodorensis
1.3 m (4 ft) TL, 6 kg (12 lb)

FOSSIL REMAINS Minority of skeleton.
ANATOMICAL CHARACTERISTICS Insufficient information.
AGE Late Cretaceous, late Cenomanian or Turonian.
DISTRIBUTION AND FORMATION/S Southern Argentina; lower Bajo Barreal.

Macrogryphosaurus gondwanicus
5 m (16 ft) TL, 300 kg (650 lb)

FOSSIL REMAINS Partial skeleton.
ANATOMICAL CHARACTERISTICS Insufficient information.
AGE Late Cretaceous, late Turonian.
DISTRIBUTION AND FORMATION/S Western Argentina; Portezuelo.
HABITAT Well-watered woodlands with short dry season.
NOTES Prey of *Unenlagia* and *Megaraptor.*

Talenkauen santacrucensis
4.7 m (15 ft) TL, 300 kg (650 lb)

FOSSIL REMAINS Partial skull and majority of skeleton.
ANATOMICAL CHARACTERISTICS Head small.
AGE Late Cretaceous, early Maastrichtian.
DISTRIBUTION AND FORMATION/S Southern Argentina; Pari Aike.
NOTES Prey of *Orkoraptor.*

Anabisetia saldiviai
2 m (6.5 ft) TL, 20 kg (45 lb)

FOSSIL REMAINS Partial skull and skeleton.
ANATOMICAL CHARACTERISTICS Insufficient information.
AGE Late Cretaceous, late Cenomanian or early Turonian.
DISTRIBUTION AND FORMATION/S Western Argentina; Lisandro.
HABITAT Well-watered woodlands with short dry season.

IGUANODONTIANS

SMALL TO GIGANTIC ORNITHOPODS FROM THE LATE JURASSIC TO THE END OF THE DINOSAUR ERA, MOST CONTINENTS.

ANATOMICAL CHARACTERISTICS No teeth on front of upper jaw. Strongly bipedal to semiquadrupedal. Five to four fingers. Four to three toes.
HABITAT Very variable, from sea level to highlands, from tropics to polar winters, from arid to wet.
NOTES Absence from Antarctica probably reflects lack of sufficient sampling.

Gasparinisaura cincosaltensis

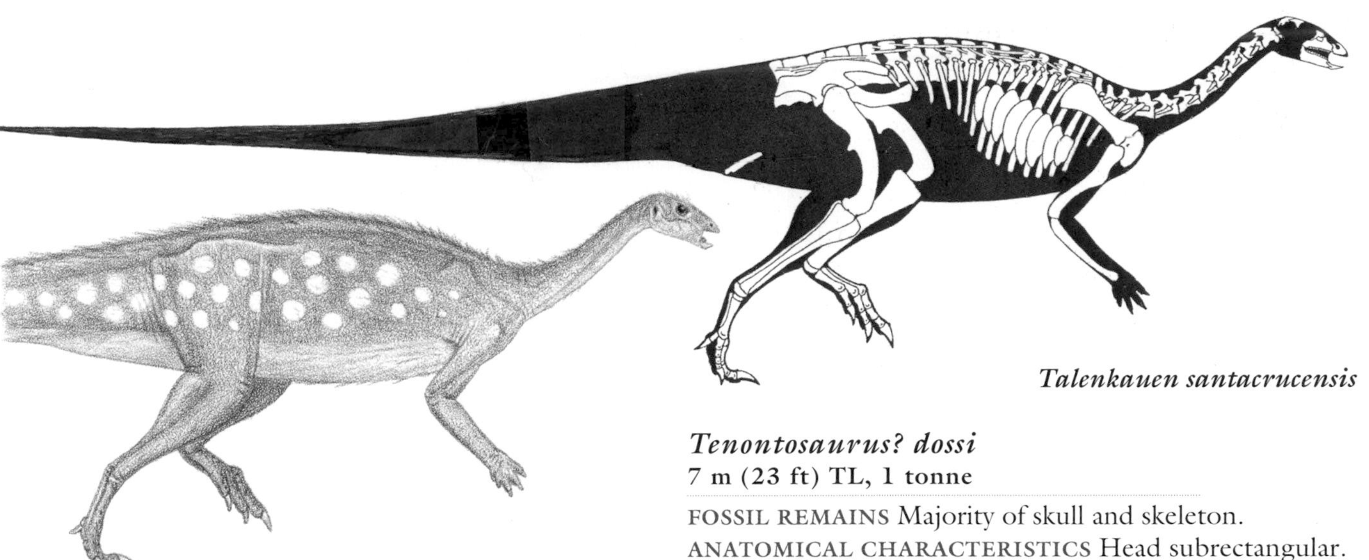

Talenkauen santacrucensis

Tenontosaurs Large iguanodontians limited to the late Early Cretaceous of North America.

ANATOMICAL CHARACTERISTICS Head narrow, beaks narrow, lower beak edge serrated, snout elongated, nasal opening enlarged, eyes shaded by overhanging rim. Body and hips fairly narrow. Tail long, base very deep. Trunk vertebral series downcurved and arm moderately long, so semiquadrupedal; hands short, broad, five grasping fingers tipped with small claws. Four long toes tipped with blunt claws.

HABITS Low- and medium-level browsers.

Tenontosaurus? dossi

7 m (23 ft) TL, 1 tonne

FOSSIL REMAINS Majority of skull and skeleton.

ANATOMICAL CHARACTERISTICS Head subrectangular.

AGE Early Cretaceous, Aptian.

DISTRIBUTION AND FORMATION/S Texas; Twin Mountains.

NOTES Whether this is the same genus as later *T. tilletti* is problematic.

Tenontosaurus tilletti

6 m (20 ft) TL, 600 kg (1,300 lb)

FOSSIL REMAINS Numerous complete and partial skulls and skeletons, completely known.

ANATOMICAL CHARACTERISTICS Head including snout deep, subrectangular.

AGE Early Cretaceous, middle Albian.

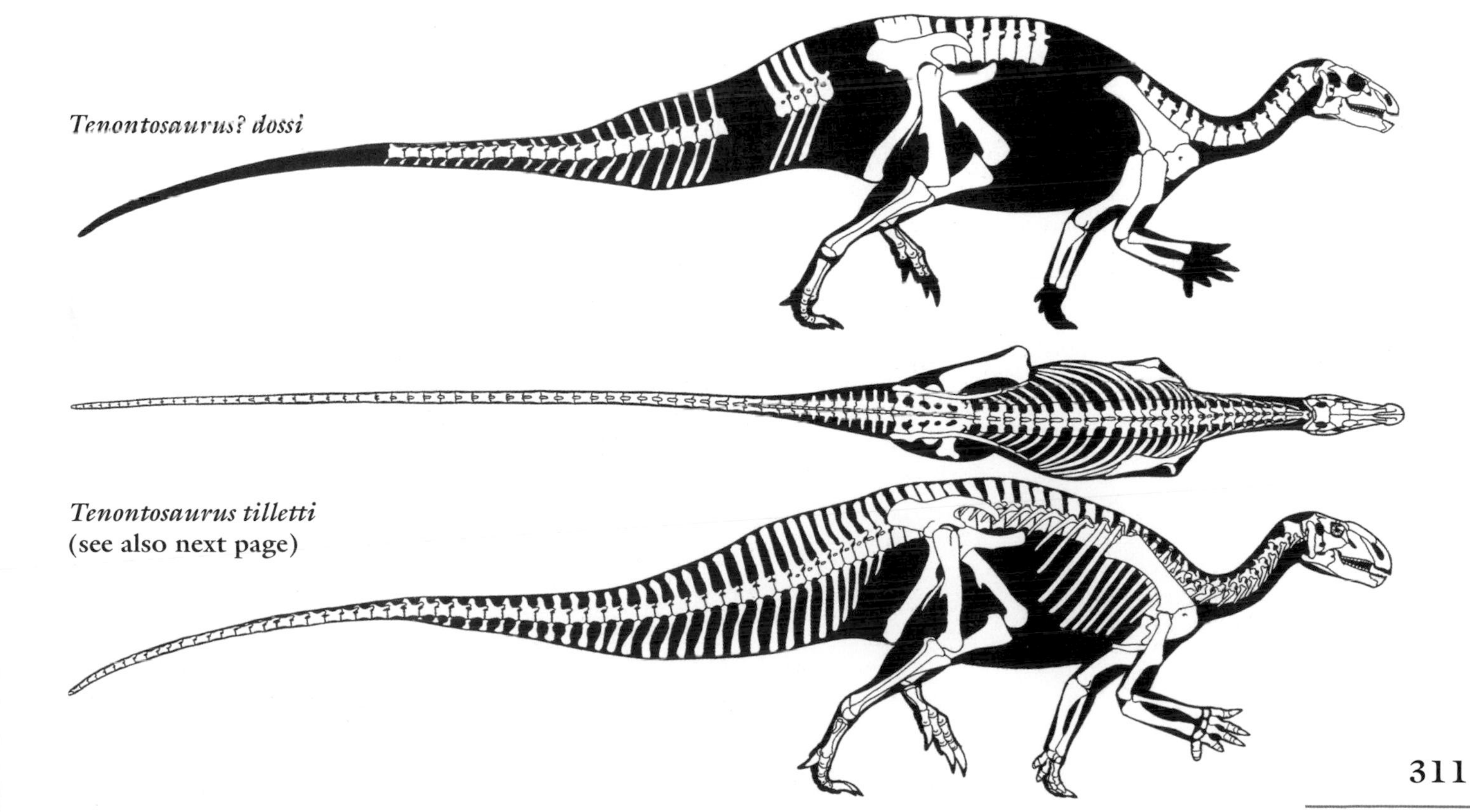

Tenontosaurus? dossi

Tenontosaurus tilletti
(see also next page)

Tenontosaurus tilletti

DISTRIBUTION AND FORMATION/S Montana, Wyoming, Texas; upper Cloverly, Paluxy.
HABITAT Semiarid floodplains to coastal.
NOTES Shared its habitat with *Zephyrosaurus*. Main enemy *Deinonychus*.

Rhabdodonts Medium-sized iguanodontians limited to the late Late Cretaceous of North America.

ANATOMICAL CHARACTERISTICS Head large, broad, subrectangular, heavily constructed, beaks narrow, eyes shaded by overhanging rim, lower jaw deep, teeth large. Skeleton heavily constructed. Body and hips fairly broad. Bipedal except could move quadrupedally at slow speeds. Arm fairly short. Long toes tipped with blunt claws.
HABITAT Forested islands.
HABITS Low- and medium-level browsers, probably able to feed on coarse vegetation.

Rhabdodon priscus
4 m (13 ft) TL, 250 kg (500 lb)

FOSSIL REMAINS Minority of skull and skeletons.
ANATOMICAL CHARACTERISTICS Standard for group.
AGE Late Cretaceous, early Maastrichtian.
DISTRIBUTION AND FORMATION/S France, possibly Spain, Austria, Hungary; Gres de Labarre, Marnes Rouges Inferieures, Gres de Saint-Chinian.
NOTES It is uncertain whether all the remains from the various formations, including those once placed in *Mochlodon suessi*, belong to this species. Some possible *R. priscus* remains are of 6 m individuals.

Rhabdodon (= Zalmoxes) robustus
2.5 m (8 ft) TL, 45 kg (110 lb)

FOSSIL REMAINS Majority of skull and skeletons.
ANATOMICAL CHARACTERISTICS Standard for group.
AGE Late Cretaceous, late Maastrichtian.
DISTRIBUTION AND FORMATION/S Romania; Sanpetru.
NOTES *Z. shqiperorum* probably adult of this species. Shared its habitat with *Telmatosaurus*.

Rhabdodon (= Zalmoxes) robustus

Dryosaurs Small to medium-sized iguanodontians limited to the Late Jurassic and Early Cretaceous of North America, Europe, and Africa.

ANATOMICAL CHARACTERISTICS Highly uniform. Head small, subtriangular, beaks small, partly squared off, eyes shaded by overhanging rim. Body and hips fairly narrow. Arm short so fully bipedal. Hand broad, five short grasping fingers tipped with very small blunt claws. Leg fairly gracile, three long toes tipped with blunt claws.
HABITS Low-level browsers. Main defense running.

Dryosaurus? unnamed species

3 m (10 ft) TL, 100 kg (220 lb)

FOSSIL REMAINS Complete skull and skeleton, other skeletons, nearly completely known.
ANATOMICAL CHARACTERISTICS Standard for group.
AGE Late Jurassic, early Tithonian.
DISTRIBUTION AND FORMATION/S Utah; middle Morrison.
HABITAT Short wet season, otherwise semiarid with open floodplain prairies and riverine forests.
NOTES Usually placed in *D. altus* but probably a different species than the later dryosaur, and differing genera cannot be ruled out. It has been suggested that the largest known specimens are not mature, but absence of larger individuals leaves this possibility unsupported. Shared its habitat with *Camptosaurus aphanoecetes.*

Dryosaurus altus

3 m (10 ft) TL, 100 kg (220 lb)

FOSSIL REMAINS Partial skeletons and skeletal parts.
ANATOMICAL CHARACTERISTICS Standard for group.
AGE Late Jurassic, middle Tithonian.
DISTRIBUTION AND FORMATION/S Wyoming; upper Morrison.
HABITAT Short wet season, otherwise semiarid with open floodplain prairies and riverine forests.

Dryosaurus (= Dysalotosaurus) lettowvorbecki

(Illustrated overleaf)

2.5 m (8 ft) TL, 80 kg (175 lb)

FOSSIL REMAINS A large number of skull and skeletal parts, nearly completely known.
ANATOMICAL CHARACTERISTICS Standard for group.
AGE Late Jurassic, late Kimmeridgian and/or early Tithonian.
DISTRIBUTION AND FORMATION/S Tanzania; middle Tendaguru.
HABITAT Coastal, seasonally dry with heavier vegetation farther inland.

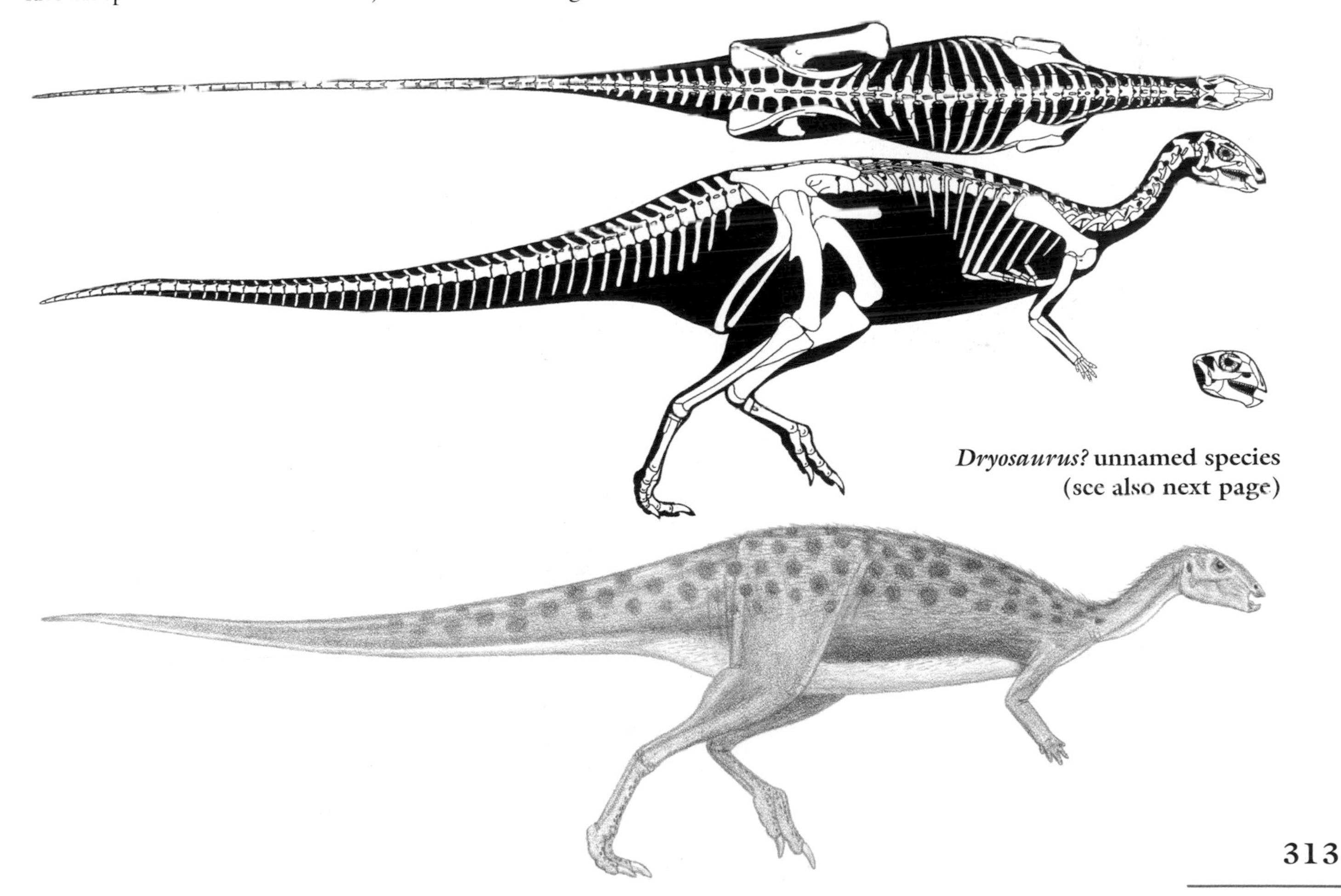

Dryosaurus? unnamed species
(see also next page)

***Dryosaurus?* unnamed species**

NOTES Generic separation from very similar *D. altus* is a classic case of oversplitting. Shared its habitat with *Elaphrosaurus bambergi.*

Valdosaurus canaliculatus

1.2 m (4 ft) TL, 10 kg (20 lb)

FOSSIL REMAINS Small portion of skeletons.
ANATOMICAL CHARACTERISTICS Insufficient information.
AGE Early Cretaceous, Barremian.
DISTRIBUTION AND FORMATION/S Southern England; Wessex.
NOTES The placement of some remains from Romania and Niger in this species or genus is problematic.

Planicoxa venenica

4.5 m (15 ft) TL, 450 kg (1,000 lb)

FOSSIL REMAINS Minority of skeleton.
ANATOMICAL CHARACTERISTICS Insufficient information.
AGE Early Cretaceous, probably Barremian.
DISTRIBUTION AND FORMATION/S Utah; Lower Cedar Mountain.
HABITAT Short wet season, otherwise semiarid with floodplain prairies, open woodlands, and riverine forests.
NOTES Shared its habitat with *Cedrorestes.* Prey of *Utahraptor.*

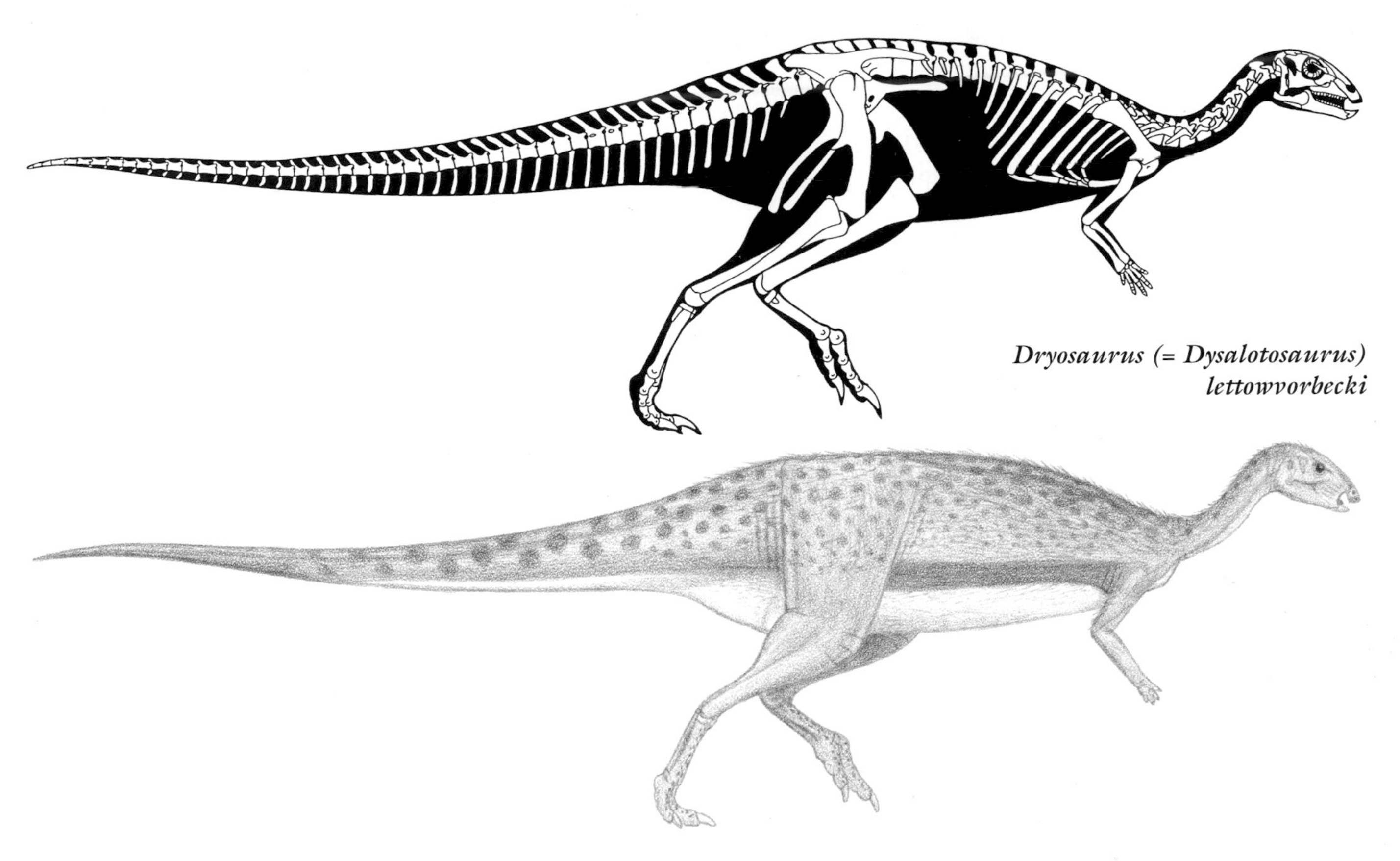

Dryosaurus (= Dysalotosaurus) lettowvorbecki

ANKYLOPOLLEXIA

SMALL TO GIGANTIC IGUANODONTIANS FROM THE LATE JURASSIC TO THE END OF THE DINOSAUR ERA, ALL CONTINENTS.

ANATOMICAL CHARACTERISTICS Head narrow. Strongly bipedal to semiquadrupedal. Thumb claws when present are a spike.
HABITAT Variable, from tropics to polar winters, from semiarid to wet.
HABITS Thumb spikes probably used for competition within the species and for defense against predators.
NOTES At least some ankylopollexians without bony head crests may have had soft-tissue crests.

Camptosaurs

MEDIUM-SIZED TO LARGE IGUANODONTIANS LIMITED TO THE LATE JURASSIC TO EARLY CRETACEOUS OF NORTH AMERICA, EUROPE, AND AUSTRALIA.

ANATOMICAL CHARACTERISTICS Head fairly small and subtriangular, beaks narrow, eyes shaded by overhanging rim. Body and hips fairly broad. Bipedal except could move quadrupedally at slow speeds. Arm fairly short, so strongly bipedal; hands short, broad, five grasping fingers tipped with small claws. Hips deep, four long toes tipped with blunt claws.
HABITS Low- and medium-level browsers.

***Camptosaurus* muscle study**

Camptosaurus dispar
5 m (16 ft) TL, 500 kg (1,000 lb)

FOSSIL REMAINS Majority of skull and skeletons, including juvenile.
ANATOMICAL CHARACTERISTICS Standard for group.
AGE Late Jurassic, late Oxfordian and/or early Kimmeridgian.
DISTRIBUTION AND FORMATION/S Wyoming; lower Morrison.
HABITAT Short wet season, otherwise semiarid with open floodplain prairies and riverine forests.
NOTES Main enemy *Allosaurus fragilis.*

Camptosaurus (Uteodon) aphanoecetes
Adult size uncertain

FOSSIL REMAINS Majority of skeleton.
ANATOMICAL CHARACTERISTICS Standard for group.
AGE Late Jurassic, early Tithonian.
DISTRIBUTION AND FORMATION/S Utah; middle Morrison.
HABITAT Short wet season, otherwise semiarid with open floodplain prairies and riverine forests.
NOTES May be the direct descendant of *C. dispar.* Remains that belong to this or another species indicate individuals approached 2 tonnes.

Cumnoria (or *Camptosaurus*) *prestwichii*
Adult size uncertain

FOSSIL REMAINS Minority of skull and majority of skeleton.
ANATOMICAL CHARACTERISTICS Standard for group.
AGE Late Jurassic, Kimmeridgian.
DISTRIBUTION AND FORMATION/S Eastern England; Kimmeridge Clay.
NOTES Juvenile specimen hinders determination of whether or not this is *Camptosaurus.* Shared its habitat with *Dacentrurus.*

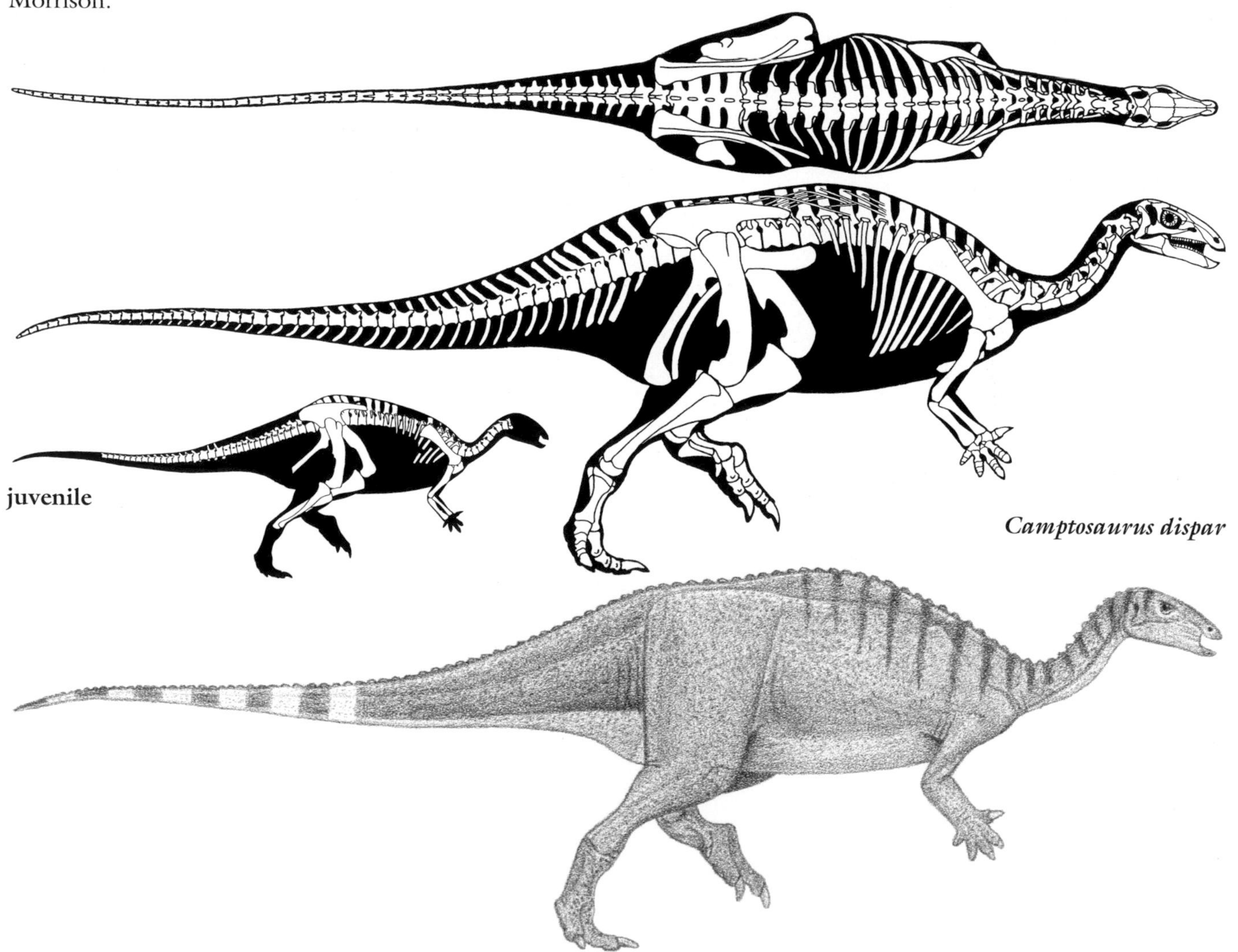

Camptosaurus dispar

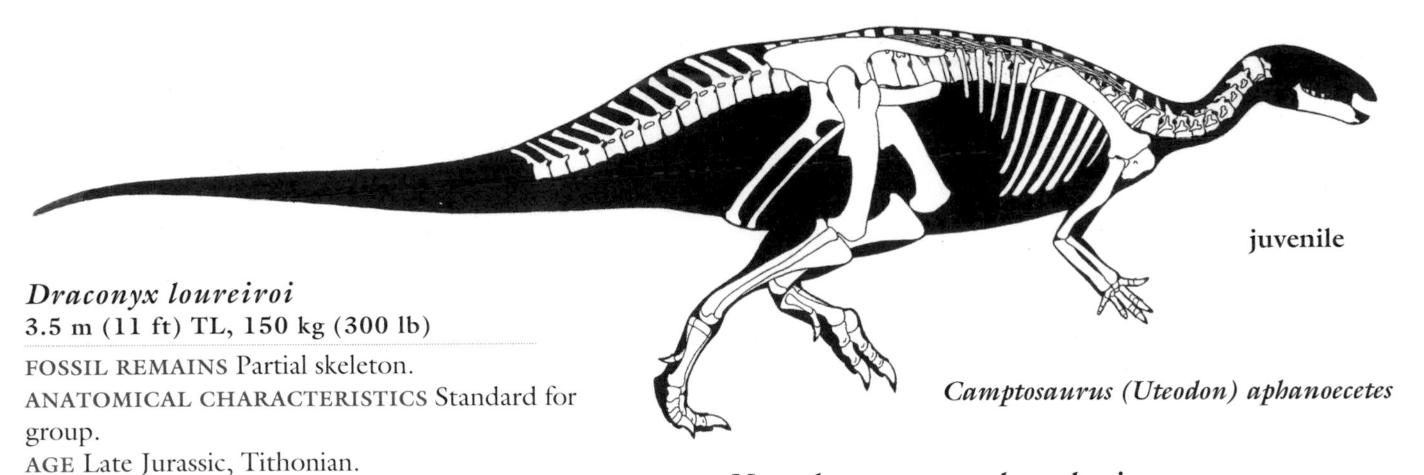

Camptosaurus (Uteodon) aphanoecetes

Draconyx loureiroi
3.5 m (11 ft) TL, 150 kg (300 lb)

FOSSIL REMAINS Partial skeleton.
ANATOMICAL CHARACTERISTICS Standard for group.
AGE Late Jurassic, Tithonian.
DISTRIBUTION AND FORMATION/S Portugal; Lourinha.
HABITAT Large, seasonally dry island with open woodlands.

Ankylopollexia miscellanea

NOTES The relationships of these ankylopollexians are uncertain.

Muttaburrasaurus langdoni
8 m (25 ft) TL, 2.8 tonnes

FOSSIL REMAINS Majority of skulls and partial skeleton.
ANATOMICAL CHARACTERISTICS Head long, fairly shallow and broad; the latter indicates jaw muscles more powerful than in other ankylopollexians, snout elongated, low bulbous crest over snout, nostrils face upward, main teeth rows form slicing batteries. Arm moderately long, so semiquadrupedal.

Muttaburrasaurus langdoni

AGE Early Cretaceous, Albian.
DISTRIBUTION AND FORMATION/S Northeast Australia; Mackunda.
HABITS Possibly omnivores that scavenged carcasses.
NOTES Has been considered a relative of camptosaurs and less plausibly rhabdodonts.

Hippodraco scutodens
4.5 m (15 ft) TL, 400 kg (900 lb)

FOSSIL REMAINS Partial skull and minority of skeleton.
ANATOMICAL CHARACTERISTICS Insufficient information.
AGE Early Cretaceous, probably early Barremian.
DISTRIBUTION AND FORMATION/S Utah; Lower Cedar Mountain.

Theiophytalia kerri
5 m (16 ft) TL, 500 kg (1,000 lb)

FOSSIL REMAINS Majority of skull.
ANATOMICAL CHARACTERISTICS Snout subrectangular.
AGE Early Cretaceous.
DISTRIBUTION AND FORMATION/S Colorado; Purgatoire.
NOTES Until recently this skull was mistakenly thought to be from the much earlier Morrison Formation and was incorrectly used to restore *Camptosaurus* with a deep, subrectangular snout.

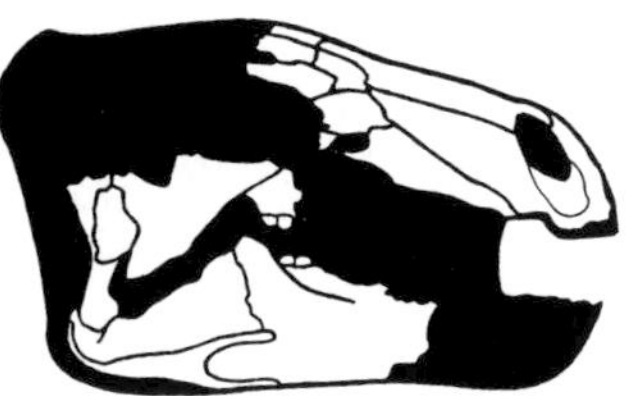

Theiophytalia kerri

Huxleysaurus hollingtoniensis
7 m (23 ft) TL, 2 tonnes

FOSSIL REMAINS Minority of skeleton.
ANATOMICAL CHARACTERISTICS Insufficient information.
AGE Early Cretaceous, early Valanginian.
DISTRIBUTION AND FORMATION/S Southeast England; Wadhurst Clay.
NOTES Originally placed in *Iguanodon*, whose type species has been moved to a much later date; placement in *Hypselospinus fittoni* has not been substantiated.

Iguanodontoids

MEDIUM-SIZED TO GIGANTIC ANKYLOPOLLEXIANS OF THE CRETACEOUS OF THE AMERICAS, EURASIA, AND AFRICA.

ANATOMICAL CHARACTERISTICS Fairly uniform. Head not deep, snout elongated, nasal openings enlarged, upper beaks projecting well below level of upper tooth row, lower beak edges serrated, vertical coronoid projection on lower jaw very well developed, grinding tooth batteries well developed. Head strongly flexed on neck. Neck longer and more slender and flexible than in other ornithopods. Trunk and tail vertebrae stiffened by a dense crisscross lattice of ossified tendons, spines at least fairly tall. Tail deep and flattened from side to side along most of length. Arm length from moderately long to long, so strongly bipedal to semiquadrupedal. Central three fingers short, inflexible, and hooved, outer finger

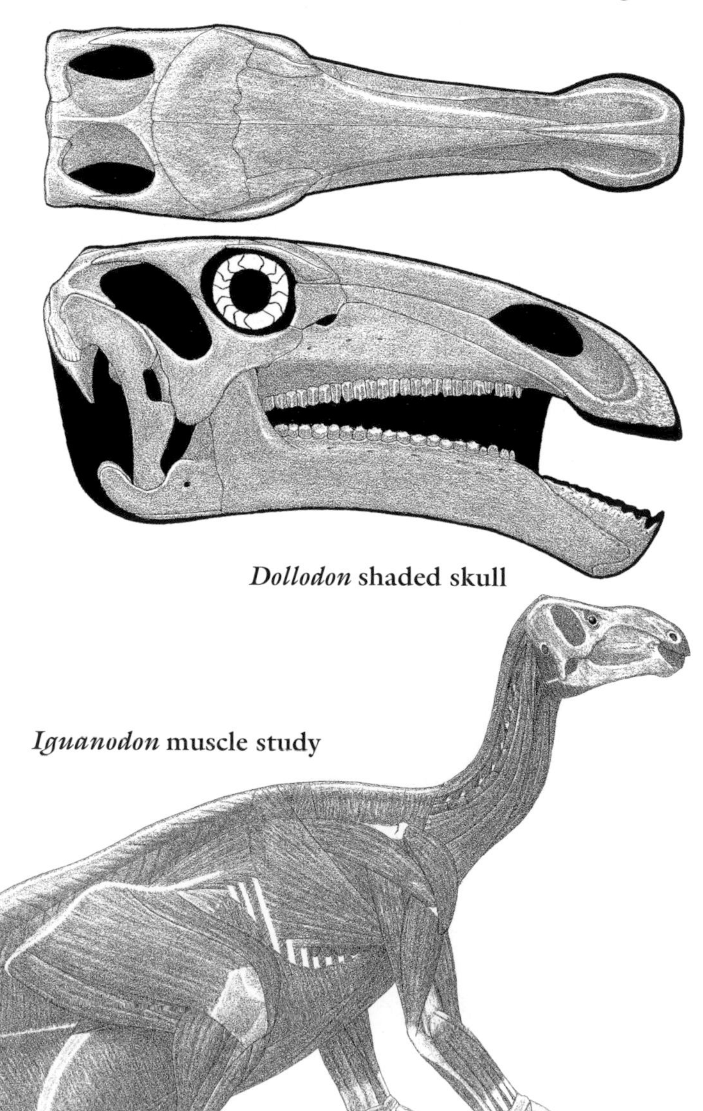

Dollodon shaded skull

Iguanodon muscle study

long, flexible, and divergent, providing a limited grasping ability. Three toes shortened, inflexible, and hooved.
HABITAT Variable, from tropics to polar winters, from seasonally arid to well-watered forests.
HABITS Medium- and low-level browsers and grazers. Tail too stiff to be used as a sculling organ when swimming.
NOTES The oversized cattle of the Cretaceous, and the most common large herbivores of that period.

Iguanodonts Medium-sized to very large iguanodontoids limited to the Early Cretaceous of North America, Eurasia, and Africa.

ANATOMICAL CHARACTERISTICS Fairly uniform. Head fairly shallow, snout long. Eyes usually shaded by overhanging rim. Thumb spikes anchored on heavily built wrist bones.
NOTES Relationships within the group are not well understood; ultimately splittable into a number of subdivisions. Absence from Australia and Antarctica may reflect lack of sufficient sampling.

Barilium dawsoni
8 m (25 ft) TL, 2.5 tonnes

FOSSIL REMAINS Minority of two skeletons.
ANATOMICAL CHARACTERISTICS Insufficient information.
AGE Early Cretaceous, early Valanginian.
DISTRIBUTION AND FORMATION/S Southeast England; Wadhurst Clay.
NOTES Originally placed in *Iguanodon*, whose type species has been moved to a much later date.

Dakotadon lakotaensis
6 m (20 ft) TL, 1 tonne

FOSSIL REMAINS Majority of skull, small portion of skeleton.
ANATOMICAL CHARACTERISTICS Head subrectangular, beak narrow and rounded.
AGE Early Cretaceous, probably Barremian.
DISTRIBUTION AND FORMATION/S South Dakota; Lakota.
HABITS Middle- and low-level browser.
NOTES Incorrectly placed in *Iguanodon*. Shared its habitat with *Hoplitosaurus*.

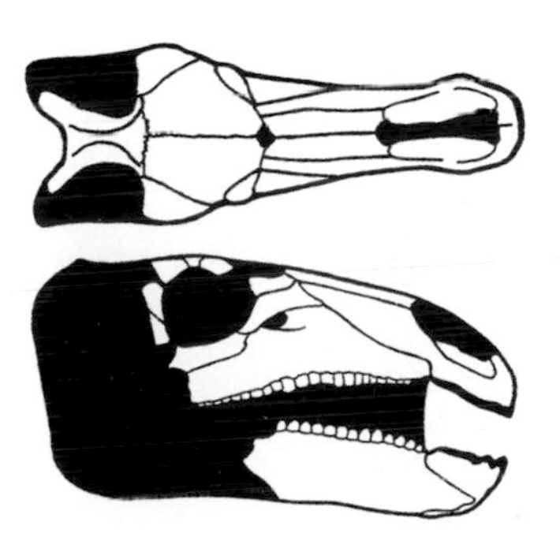

Dakotadon lakotaensis

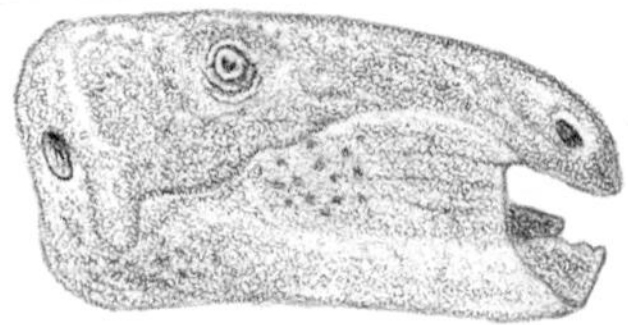

Iguanacolossus fortis
9 m (30 ft) TL, 5 tonnes

FOSSIL REMAINS Minority of skull and skeleton.
ANATOMICAL CHARACTERISTICS Heavily constructed.
AGE Early Cretaceous, probably lower Barremian.
DISTRIBUTION AND FORMATION/S Utah; Lower Cedar Mountain.

Lanzhousaurus magnidens
10 m (35 ft) TL, 6 tonnes

FOSSIL REMAINS Minority of skull and skeleton.
ANATOMICAL CHARACTERISTICS Heavily constructed. Lower jaw fairly deep, teeth extremely large. Modest shoulder withers support nuchal ligaments to neck and head.
AGE Early Cretaceous.
DISTRIBUTION AND FORMATION/S Central China; Hekou Group.
HABITS Middle- and low-level browser, able to consume coarse vegetation.
NOTES Had the largest known teeth of any herbivorous dinosaur.

Lurdusaurus arenatus
7 m (23 ft) TL, 2.5 tonnes

FOSSIL REMAINS Small portion of skull and partial skeleton.
ANATOMICAL CHARACTERISTICS Arm massively constructed, hand short and broad, thumb spike enormous.
AGE Early Cretaceous, late Aptian.
DISTRIBUTION AND FORMATION/S Niger; upper Elrhaz.
HABITAT Coastal river delta.
HABITS Middle- and low-level browser.
NOTES Shared its habitat with *Ouranosaurus* and *Nigersaurus*.

Darwinsaurus evolutionis
Adult size uncertain

FOSSIL REMAINS Minority of skull and skeleton.
ANATOMICAL CHARACTERISTICS Lower jaw shallow, very long gap between beaks and short main tooth rows. Arm massively constructed, upper hand long and narrow, thumb spike enormous.
AGE Early Cretaceous.
DISTRIBUTION AND FORMATION/S Southeast England; uncertain.

NOTES Originally placed in *Iguanodon*, whose type species has been moved to a much later date; placement in *Hypselospinus fittoni* has not been substantiated.

Sellacoxa pauli
8 m (26 ft) TL, 3 tonnes

FOSSIL REMAINS Minority of skeleton.
ANATOMICAL CHARACTERISTICS Insufficient information.
AGE Early Cretaceous, early Valanginian.
DISTRIBUTION AND FORMATION/S Southeast England; lower Wadhurst Clay.
NOTES Placement in *Hypselospinus fittoni* has not been substantiated.

Hypselospinus fittoni
7 m (23 ft) TL, 2.5 tonnes

FOSSIL REMAINS Minority of several skulls.
ANATOMICAL CHARACTERISTICS Insufficient information.
AGE Early Cretaceous, early Valanginian.
DISTRIBUTION AND FORMATION/S Southeast England; Wadhurst Clay.
NOTES Originally placed in *Iguanodon*, whose type species has been moved to a much later date.

Unnamed genus? *galvensis*
Adult size uncertain

FOSSIL REMAINS Numerous partial skulls and skeletons of adults and juveniles.
ANATOMICAL CHARACTERISTICS Insufficient information.
AGE Early Cretaceous, early Barremian.
DISTRIBUTION AND FORMATION/S Spain; Camarillas.
NOTES Placement in *Iguanodon* not yet substantiated.

Iguanodon bernissartensis
8 m (26 ft) TL, 3.2 tonnes

FOSSIL REMAINS Over two dozen complete skulls and skeletons, completely known.
ANATOMICAL CHARACTERISTICS Heavily constructed. Head subrectangular, beak narrow and rounded, lower jaw fairly deep. Arm long, so semiquadrupedal; hand and spike very large, upper hand long and fairly narrow. Foot large.

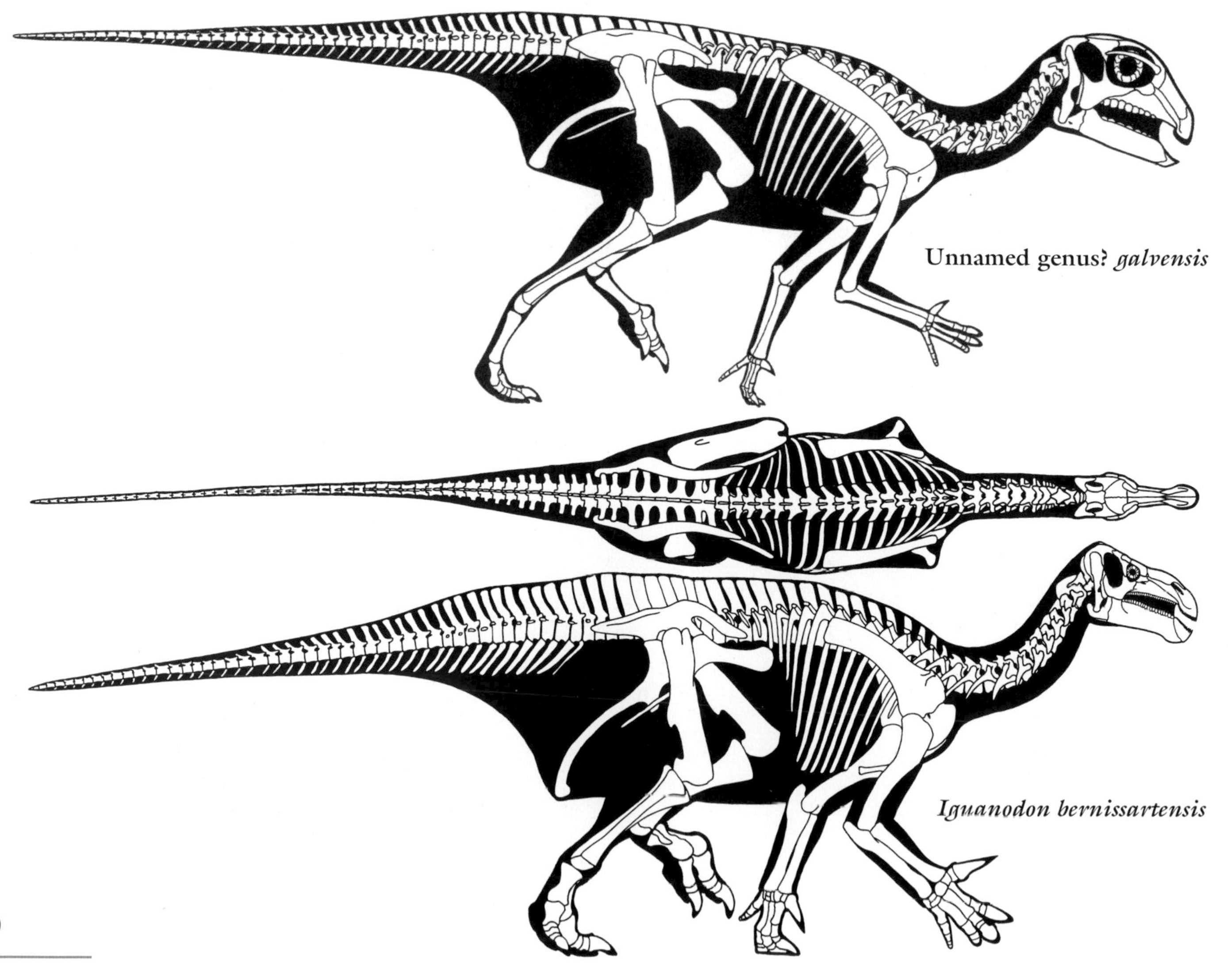
Unnamed genus? *galvensis*

Iguanodon bernissartensis

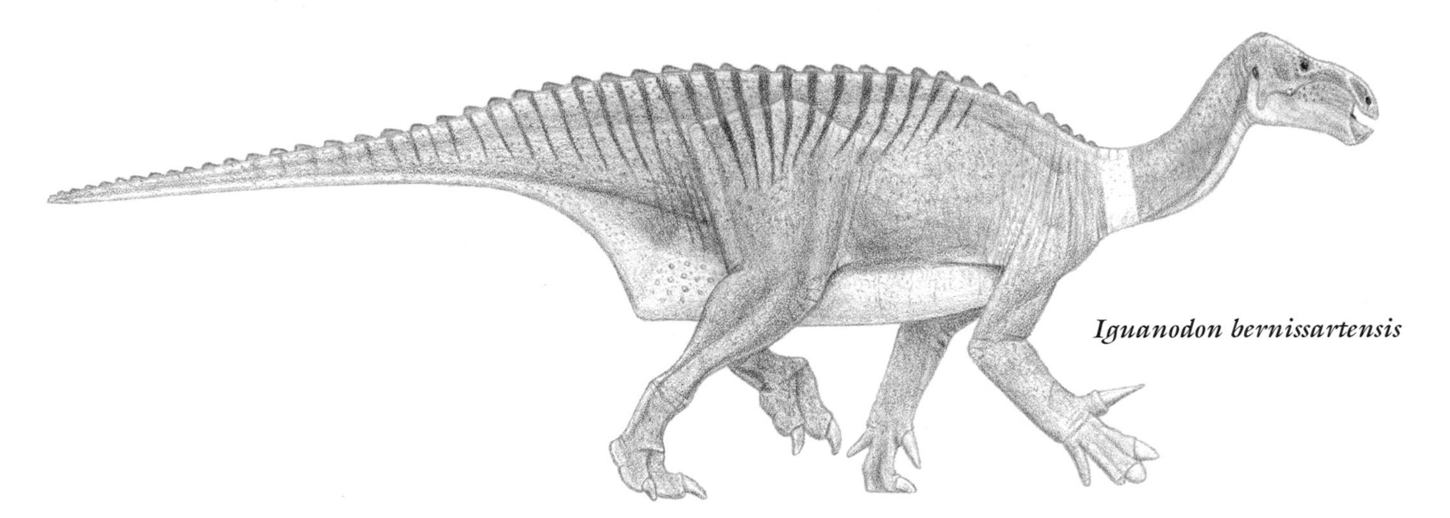
Iguanodon bernissartensis

AGE Early Cretaceous, probably late Barremian, possibly earliest Aptian.
DISTRIBUTION AND FORMATION/S Belgium; Wealden equivalent, level uncertain.
HABITS Middle- and low-level browser.
NOTES The classic iguanodont. Over the years *Iguanodon* became a taxonomic grab bag into which a large number of remains from many places and times were placed. In accord with a decision of the committee that handles such issues, this is now the set of remains that is labeled *Iguanodon*. The original English teeth the genus was based on are much older, from the Valanginian, but are not distinctive. Other remains that probably belong to *Iguanodon* and may or may not belong to this species are known from Germany and England, not common in the latter country. Found in ancient sinkhole fill, shared its habitat with *Dollodon bampingi*.

Dollodon bampingi
6.5 m (21 ft) TL, 1.1 tonne

FOSSIL REMAINS Complete skull and skeleton.
ANATOMICAL CHARACTERISTICS Head shallow,

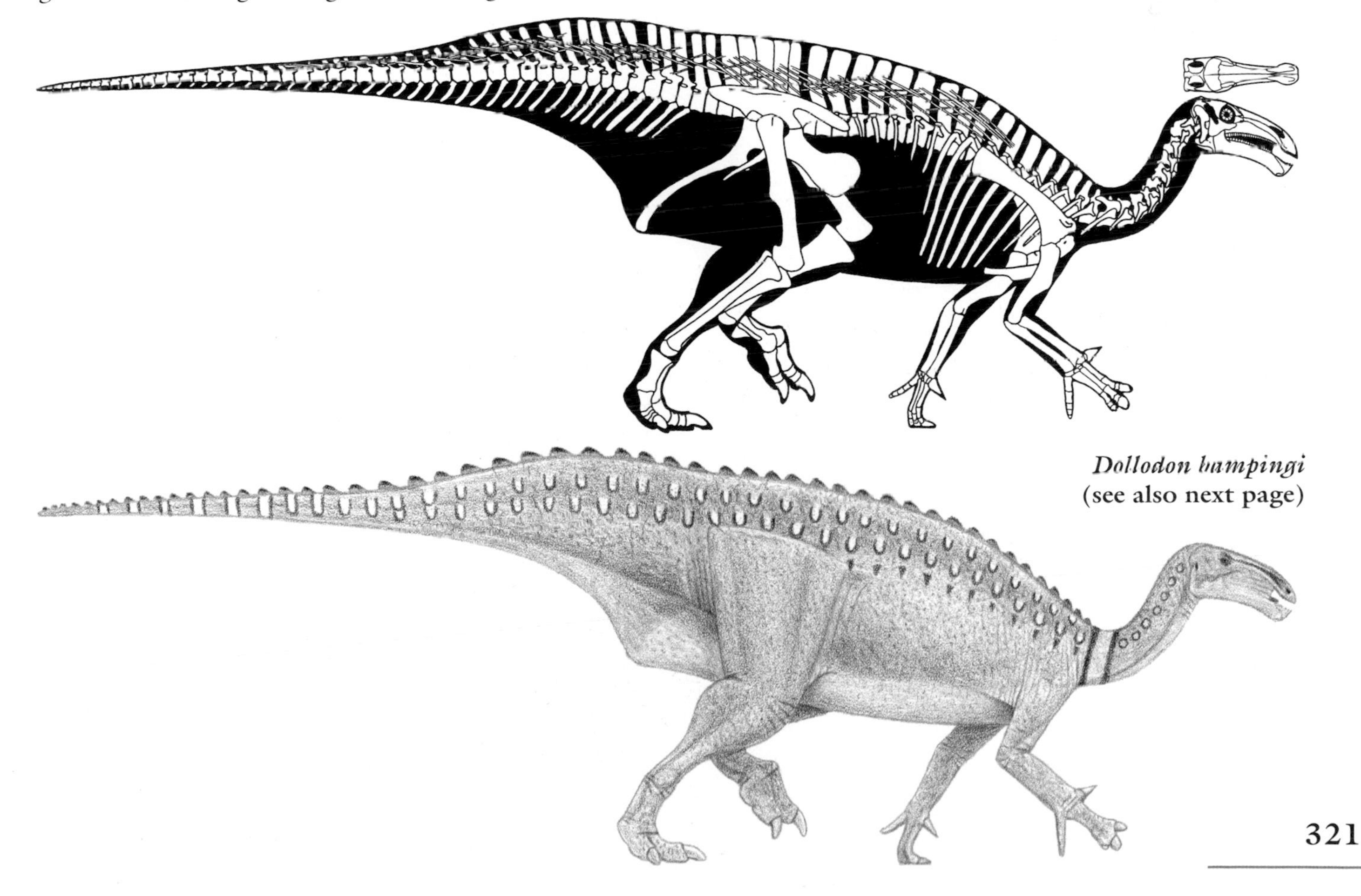
Dollodon bampingi
(see also next page)

Dollodon bampingi

subrectangular, snout very long, beak narrow and rounded, significant gap between beaks and tooth rows. Tall vertebral spines over trunk, hips, and tail form a shallow sail. Arm fairly long, so semiquadrupedal; upper hand long and narrow, thumb claw small.

AGE Early Cretaceous, probably late Barremian, possibly earliest Aptian.

DISTRIBUTION AND FORMATION/S Belgium; Wealden equivalent, level uncertain.

HABITS Middle- and low-level browser.

NOTES Found in ancient sinkhole fill. This iguanodont was confused with the distinct and probably later-appearing *Mantellisaurus*. A few remains imply genus and perhaps species were present in England ("*Iguanodon*" *seelyi* is too fragmentary to be diagnostic).

Mantellisaurus atherfieldensis

Adult size uncertain

FOSSIL REMAINS Nearly complete skull and majority of skeleton, numerous bones.

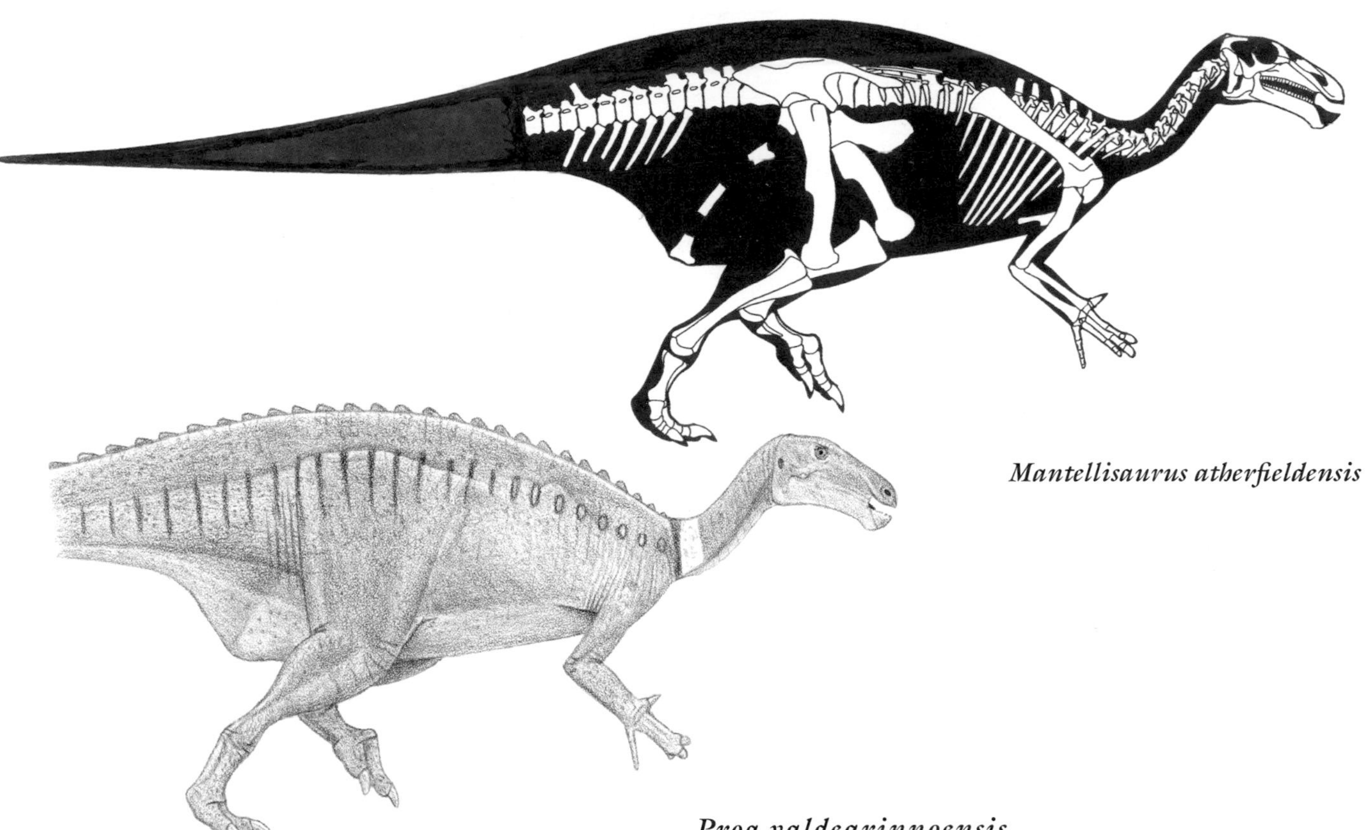

Mantellisaurus atherfieldensis

ANATOMICAL CHARACTERISTICS Snout long, beak narrow and rounded. Bipedal except could move quadrupedally at slow speeds. Arm fairly short, so largely bipedal; upper hand long and narrow, thumb claw small. Toes longer than in other iguanodonts.
AGE Early Cretaceous, earliest Aptian, possibly latest Barremian.
DISTRIBUTION AND FORMATION/S Southeast England; Vectis, probably equivalent level beds in Europe.
HABITS Middle- and low-level browser.
NOTES Incorrectly placed in *Iguanodon*; placement of many incomplete European remains in this taxon is problematic. Shared its habitat with *Polacanthus*.

Mantellodon carpenteri
Adult size uncertain

FOSSIL REMAINS Partial skeleton.
ANATOMICAL CHARACTERISTICS Insufficient information.
AGE Early Cretaceous, early Aptian.
DISTRIBUTION AND FORMATION/S Southeast England; lower Lower Greensand.
NOTES Incorrectly placed in *Iguanodon*; placement in probably earlier *Mantellisaurus atherfieldensis* probably incorrect, although it may have descended from latter.

Proa valdearinnoensis
5.5 m (18 ft) TL, 1 tonne

FOSSIL REMAINS Partial skulls and skeletons.
ANATOMICAL CHARACTERISTICS Robustly built. Beak narrow and rounded.
AGE Early Cretaceous, early Albian.
DISTRIBUTION AND FORMATION/S Spain; Escucha.
HABITS Middle- and low-level browser.

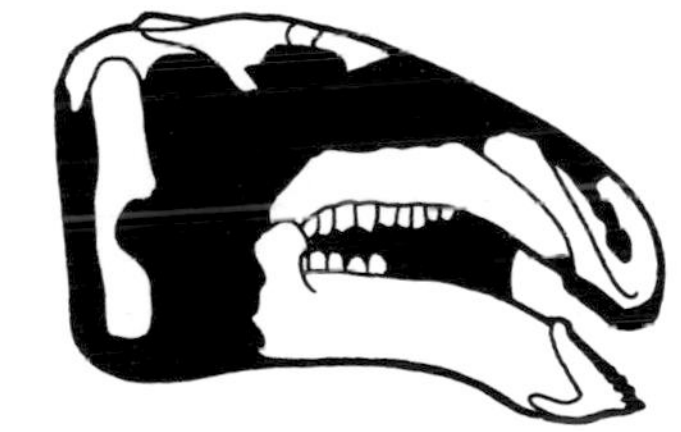

Proa valdearinnoensis

Xuwulong yueluni
5 m (16 ft) TL, 650 kg (1,400 lb)

FOSSIL REMAINS Complete skull and majority of skeleton.
ANATOMICAL CHARACTERISTICS Head subrectangular, beak small. Front half of tail very deep.
AGE Early Cretaceous, Albian.
DISTRIBUTION AND FORMATION/S Central China; Xinminbo Group.
NOTES Absence of limbs precludes skeletal restoration.

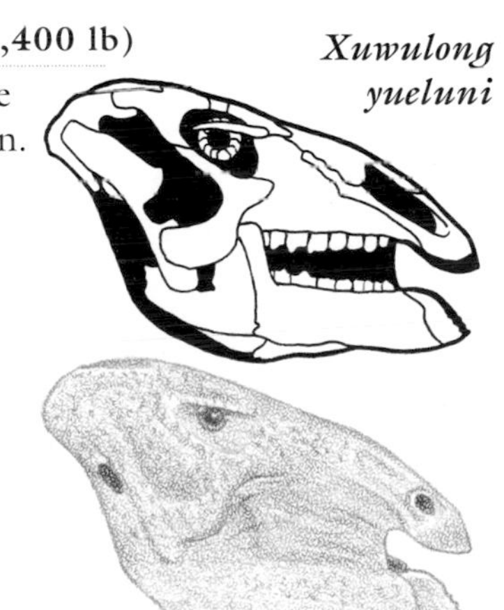

Xuwulong yueluni

Equijubus normani
7 m (23 ft) TL, 2.5 tonnes

FOSSIL REMAINS Complete skull and minority of skeleton.
ANATOMICAL CHARACTERISTICS Head subtriangular, beak narrow and rounded, eyes not shaded by overhanging rim, long gap between beaks and main tooth rows, incipient third tooth in each position.
AGE Early Cretaceous, Albian.
DISTRIBUTION AND FORMATION/S Central China; Xinminbo Group.
HABITS Middle- and low-level browser.

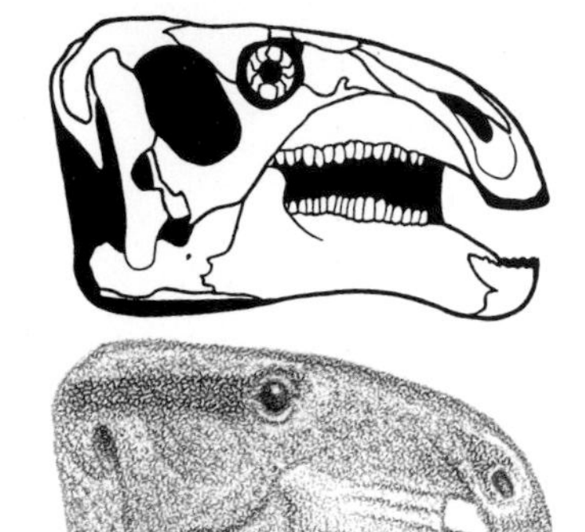

Equijubus normani

Bolong yixianensis
Adult size uncertain

FOSSIL REMAINS Two largely complete juvenile specimens.
ANATOMICAL CHARACTERISTICS Robustly built. Beak rounded. Tail not large. Arm and hand short, so largely bipedal; thumb claw small.
AGE Early Cretaceous, early Aptian.
DISTRIBUTION AND FORMATION/S Northeast China; middle Yixian.
HABITAT Well-watered forests and lakes, winters chilly with some snow.
HABITS Middle- and low-level browser.

Jinzhousaurus yangi
5 m (16 ft) TL, 650 kg (1,400 lb)

FOSSIL REMAINS Complete skull and skeleton.
ANATOMICAL CHARACTERISTICS Robustly built. Beak large, narrow and rounded. Tail not large. Arm and hand short, so largely bipedal; thumb claw small.
AGE Early Cretaceous, early or middle Aptian.
DISTRIBUTION AND FORMATION/S Northeast China; Jiufotang.
HABITAT Well-watered forests and lakes, winters chilly with some snow.
HABITS Middle- and low-level browser.

Fukuisaurus tetoriensis
4.5 m (15 ft) TL, 400 kg (900 lb)

FOSSIL REMAINS Majority of skull.
ANATOMICAL CHARACTERISTICS Skull short and fairly deep.
AGE Early Cretaceous, Aptian or Albian.
DISTRIBUTION AND FORMATION/S Japan; Kitadani.

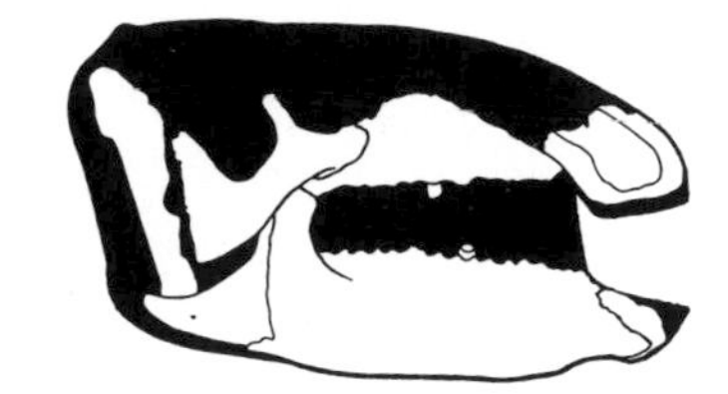

Fukuisaurus tetoriensis

Bolong yixianensis

Jinzhousaurus yangi

Altirhinus kurzanovi
6.5 m (21 ft) TL, 1.1 tonne

FOSSIL REMAINS Nearly complete and partial skulls, minority of skeletons.
ANATOMICAL CHARACTERISTICS Snout deepened into a prominent arched crest, beak narrow and rounded, incipient third tooth in each position. Upper hand long and narrow.
AGE Early Cretaceous, Aptian or Albian.
DISTRIBUTION AND FORMATION/S Mongolia; Huhteeg Svita.
HABITS Middle- and low-level browser.
NOTES Shared its habitat with *Shamosaurus*.

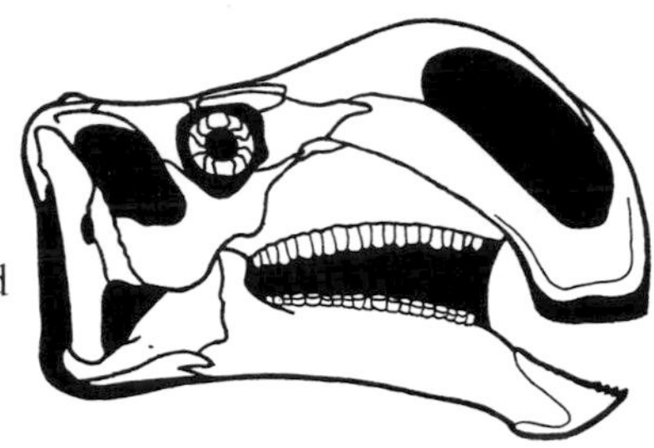

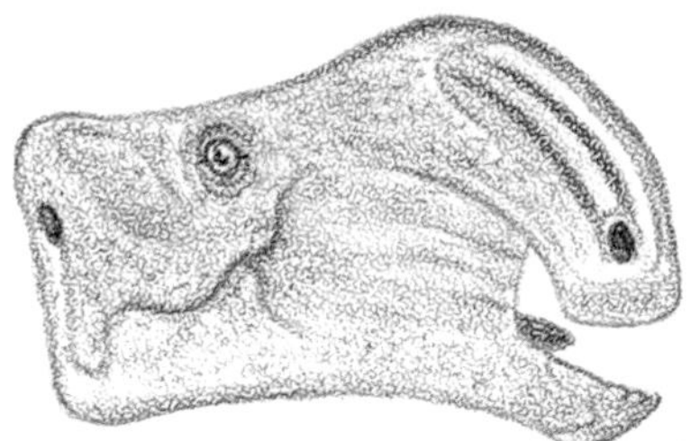

Altirhinus kurzanovi

Ouranosaurus nigerensis
8.3 m (27 ft) TL, 2.2 tonnes

FOSSIL REMAINS Complete skull and majority of two skeletons.
ANATOMICAL CHARACTERISTICS Head shallow, subtriangular, snout very long, beak squared, low midline crest on top of middle of head, beak broad and partly squared off, very long gap between beak and tooth rows. Very tall vertebral spines over trunk, hips, and tail forming very tall finback sail. Arm fairly long, so semiquadrupedal; upper hand short and broad, thumb claw small.
AGE Early Cretaceous, late Aptian.
DISTRIBUTION AND FORMATION/S Niger; upper Elrhaz.
HABITAT Coastal river delta.
HABITS Square muzzle at end of long snout is adaptation for reaching down to and mowing ground cover; also able to browse at low and medium levels.
NOTES Shared its habitat with *Lurdusaurus* and *Nigersaurus*, the latter of which was a competing square-mouthed grazer.

Probactrosaurus gobiensis (Illustrated overleaf)
5.5 m (18 ft) TL, 1 tonne

FOSSIL REMAINS Majority of several skulls and skeletons.
ANATOMICAL CHARACTERISTICS Head subrectangular, beak narrow and rounded, significant gap between beaks and tooth rows, incipient third tooth in each position.

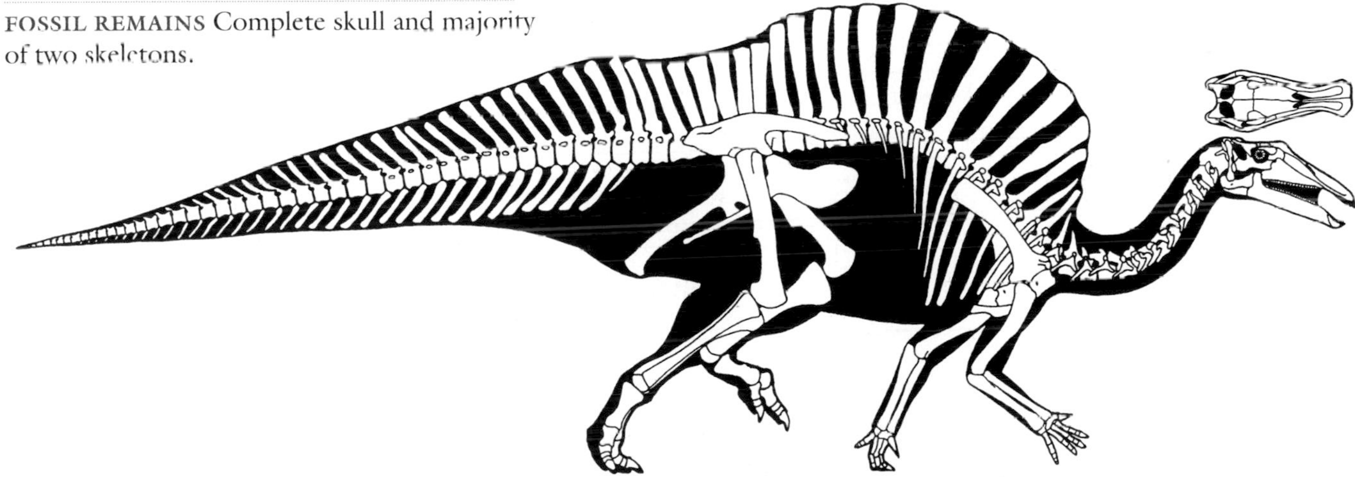

Ouranosaurus nigerensis

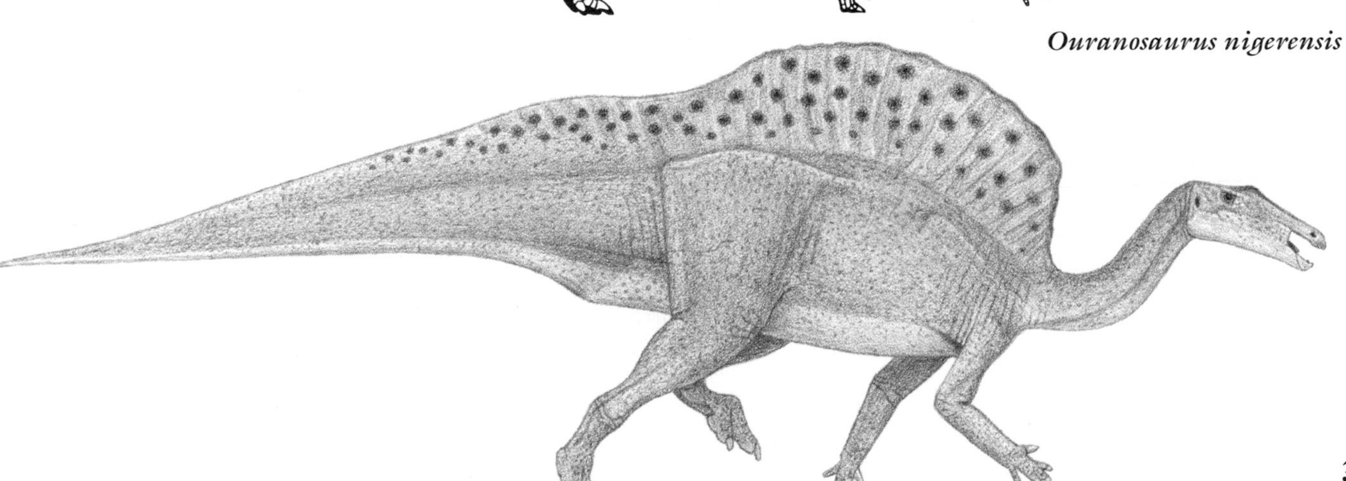

Probactrosaurus gobiensis

Arm long, so semiquadrupedal; upper hand long and narrow, thumb claw small.
AGE Early Cretaceous, Albian.
DISTRIBUTION AND FORMATION/S Northern China; Dashuigou.
HABITS Middle- and low-level browser.
NOTES Probably includes *P. alashanicus*.

Cedrorestes crichtoni
Adult size uncertain

FOSSIL REMAINS Minority of skeletons.
ANATOMICAL CHARACTERISTICS Insufficient information.
AGE Early Cretaceous, probably Barremian.
DISTRIBUTION AND FORMATION/S Utah; Lower Cedar Mountain.
HABITAT Short wet season, otherwise semiarid with floodplain prairies, open woodlands, and riverine forests.
NOTES Relationships problematic; if a hadrosaurid it is the only one known from the Early Cretaceous. Shared its habitat with *Planicoxa*. Prey of *Utahraptor*.

Hadrosaurs Large to gigantic iguanodontoids limited to the Late Cretaceous of the Americas, Eurasia, and Antarctica.

ANATOMICAL CHARACTERISTICS Uniform, especially noncranial skeletons. Eyes usually not shaded by overhanging rim, gap between beaks and dental batteries, at least three teeth in each position forming highly developed grinding pavement including hundreds of teeth. Downcurved front trunk vertebral series supports deep nuchal ligaments to neck and head and lowers shoulders; arm also moderately long, so semiquadrupedal; wrist bones reduced, upper hand elongated and narrow, thumbs lost, leaving at most four fingers. Vertical wrinkles in shoulder region in at least some species.
HABITS Main defense running, using gracile arms to improve speed and turning ability, also kicking with legs. One to two dozen eggs deposited in pit nest, covered by mound of soil.

Kritosaurus muscle study

BASO-HADROSAURS Large hadrosaurs of the Late Cretaceous of the Americas and Eurasia.

ANATOMICAL CHARACTERISTICS Uniform. Head fairly shallow, snout long, beaks narrow to fairly broad, nasal openings large, bony crests absent, gaps between beaks and tooth rows not long.

Protohadros byrdi
Adult size uncertain

FOSSIL REMAINS Majority of skull and minority of skeleton.
ANATOMICAL CHARACTERISTICS Upper beak fairly broad, rounded, projects strongly downward, front of lower jaw deep and curves strongly downward.
AGE Late Cretaceous, middle Cenomanian.
DISTRIBUTION AND FORMATION/S Texas; Woodbine.
HABITAT Coastal river delta.
HABITS May have been more prone to feeding on aquatic plants than other iguanodontoids.

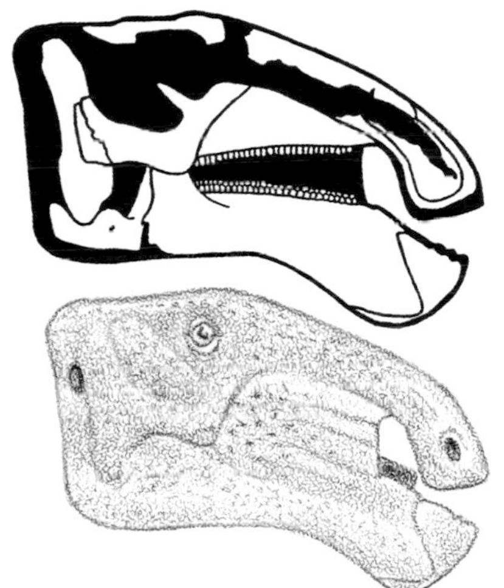

Protohadros byrdi

Tanius sinensis
7 m (23 ft) TL, 2 tonnes

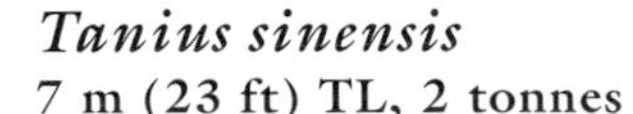

FOSSIL REMAINS Minority of several skulls and skeletons.
ANATOMICAL CHARACTERISTICS Insufficient information.
AGE Late Late Cretaceous.
DISTRIBUTION AND FORMATION/S Eastern China; Wangshi Group.
NOTES Shared its habitat with *Sinoceratops.*

Shuangmiaosaurus gilmorei
7.5 m (25 ft) TL, 2.5 tonnes

FOSSIL REMAINS Minority of skull.
ANATOMICAL CHARACTERISTICS Insufficient information.
AGE Late Cretaceous, Cenomanian or Turonian.
DISTRIBUTION AND FORMATION/S Northeast China; Sunjiawan.

Eolambia caroljonesa
6 m (20 ft) TL, 1 tonne

FOSSIL REMAINS Majority of skull and partial skeletons, juvenile to adult.
ANATOMICAL CHARACTERISTICS Snout elongated, beak fairly broad, partly squared off.
AGE Late Cretaceous, early Cenomanian.
DISTRIBUTION AND FORMATION/S Utah; Upper Cedar Mountain.
HABITAT Short wet season, otherwise semiarid with floodplain prairies, open woodlands, and riverine forests.
HABITS Middle- and low-level browser and grazer.
NOTES Shared its habitat with *Animantarx.*

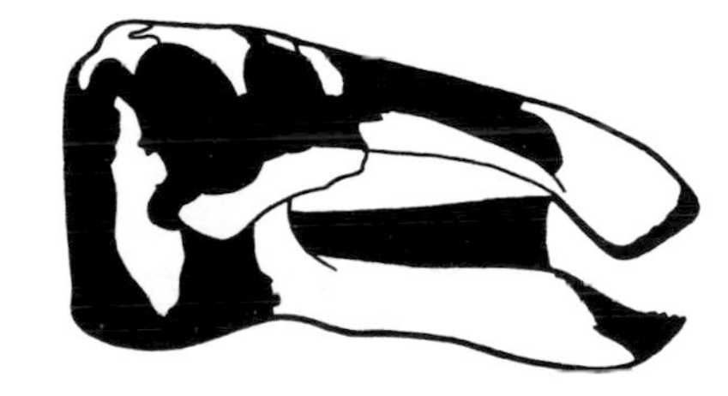

Eolambia caroljonesa

Levnesovia transoxiana
Adult size uncertain

FOSSIL REMAINS Minority of skull.
ANATOMICAL CHARACTERISTICS Insufficient information.
AGE Late Cretaceous, middle or late Turonian.
DISTRIBUTION AND FORMATION/S Uzbekistan; Bissekty.
HABITAT Coastal.
NOTES Shared its habitat with *Turanoceratops.*

Bactrosaurus johnsoni

6.2 m (20 ft) TL, 1.2 tonnes

FOSSIL REMAINS Majority of skulls and skeletons.
ANATOMICAL CHARACTERISTICS Beak narrow, rounded.
AGE Late Cretaceous, probably Campanian.
DISTRIBUTION AND FORMATION/S Northern China; Iren Dabasu.
HABITAT Seasonally wet-dry woodlands.
HABITS Middle- and low-level browser.
NOTES Probably includes *Gilmoreosaurus mongoliensis.* Prey of *Alectrosaurus.*

Telmatosaurus transylvanicus

5 m (16 ft) TL, 600 kg (1,200 lb)

FOSSIL REMAINS A number of partial skulls and skeletons.
ANATOMICAL CHARACTERISTICS Beak narrow, rounded.
AGE Late Cretaceous, late Maastrichtian.
DISTRIBUTION AND FORMATION/S Romania; Sanpetru.
HABITAT Forested island.
HABITS Middle- and low-level browser.
NOTES Small size of most individuals presumed to represent island dwarfism, but some researchers cite larger specimens and higher estimate of size of island as evidence otherwise. Shared its habitat with *Rhabdodon robustus.*

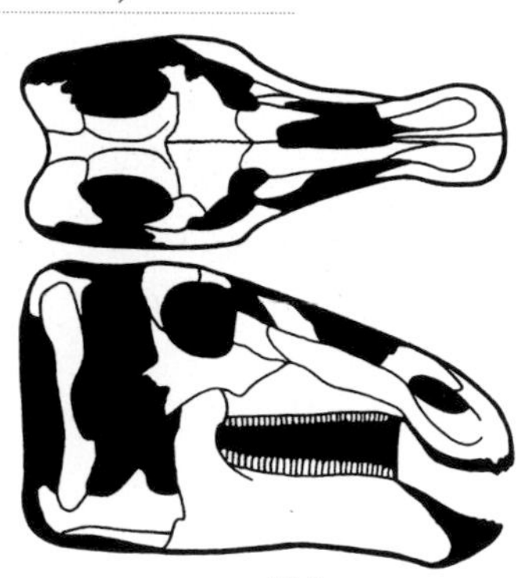

Telmatosaurus transylvanicus

Claosaurus agilis

Adult size uncertain

FOSSIL REMAINS Minority of skull and partial skeleton.
ANATOMICAL CHARACTERISTICS Insufficient information.
AGE Late Late Cretaceous.
DISTRIBUTION AND FORMATION/S Kansas; Niobrara.
NOTES Found as drift in marine deposits.

Huehuecanauhtlus tiquichensis

6 m (20 ft) TL, 1 tonne

FOSSIL REMAINS A number of partial skulls and skeletons.
ANATOMICAL CHARACTERISTICS Beak narrow, rounded.
AGE Late Cretaceous, Santonian.
DISTRIBUTION AND FORMATION/S Central Mexico; unnamed.

Tethyshadros insularis

4 m (13 ft) TL, 300 kg (650 lb)

FOSSIL REMAINS Nearly complete skull and skeleton, some additional material.

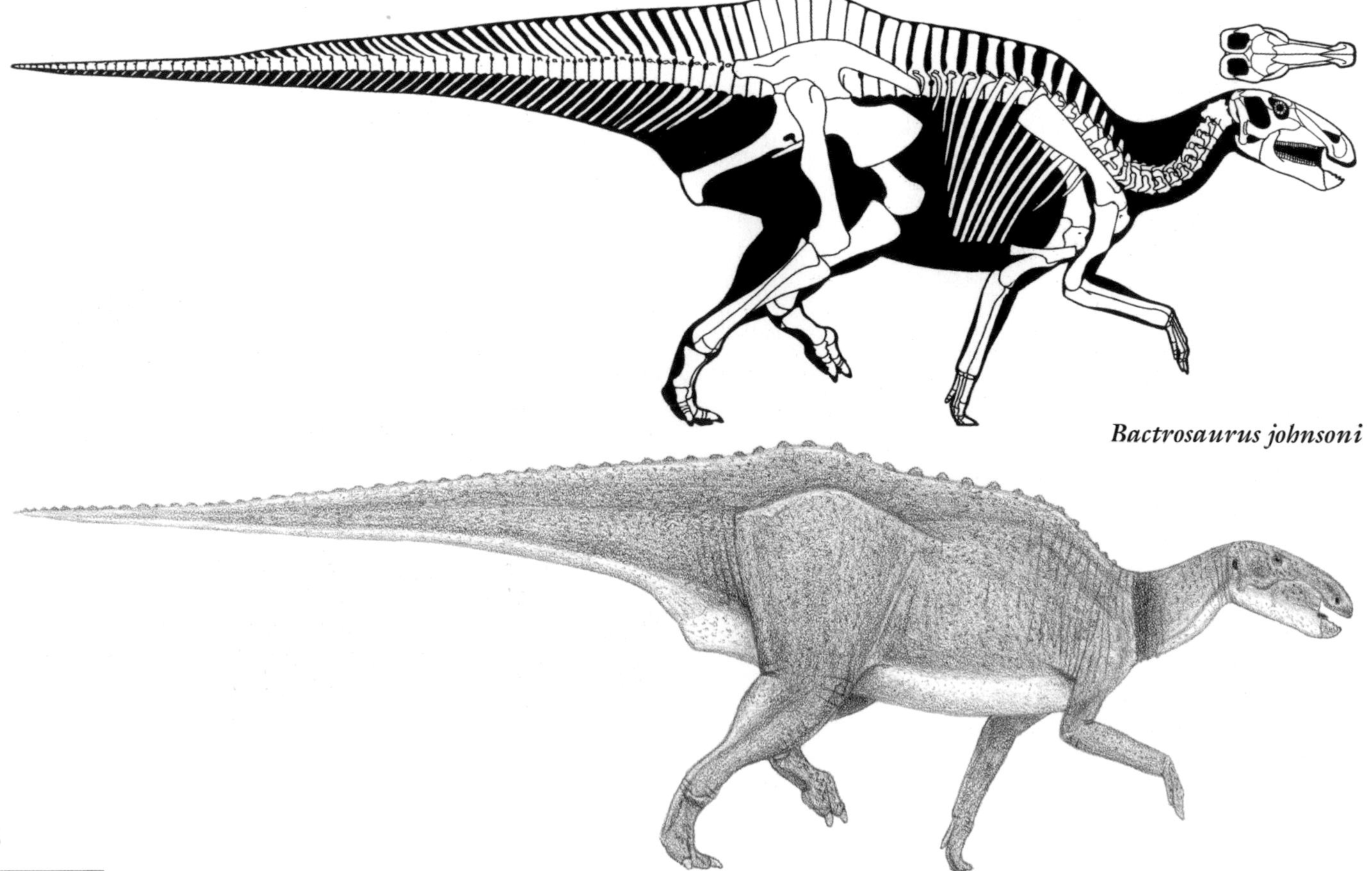

Bactrosaurus johnsoni

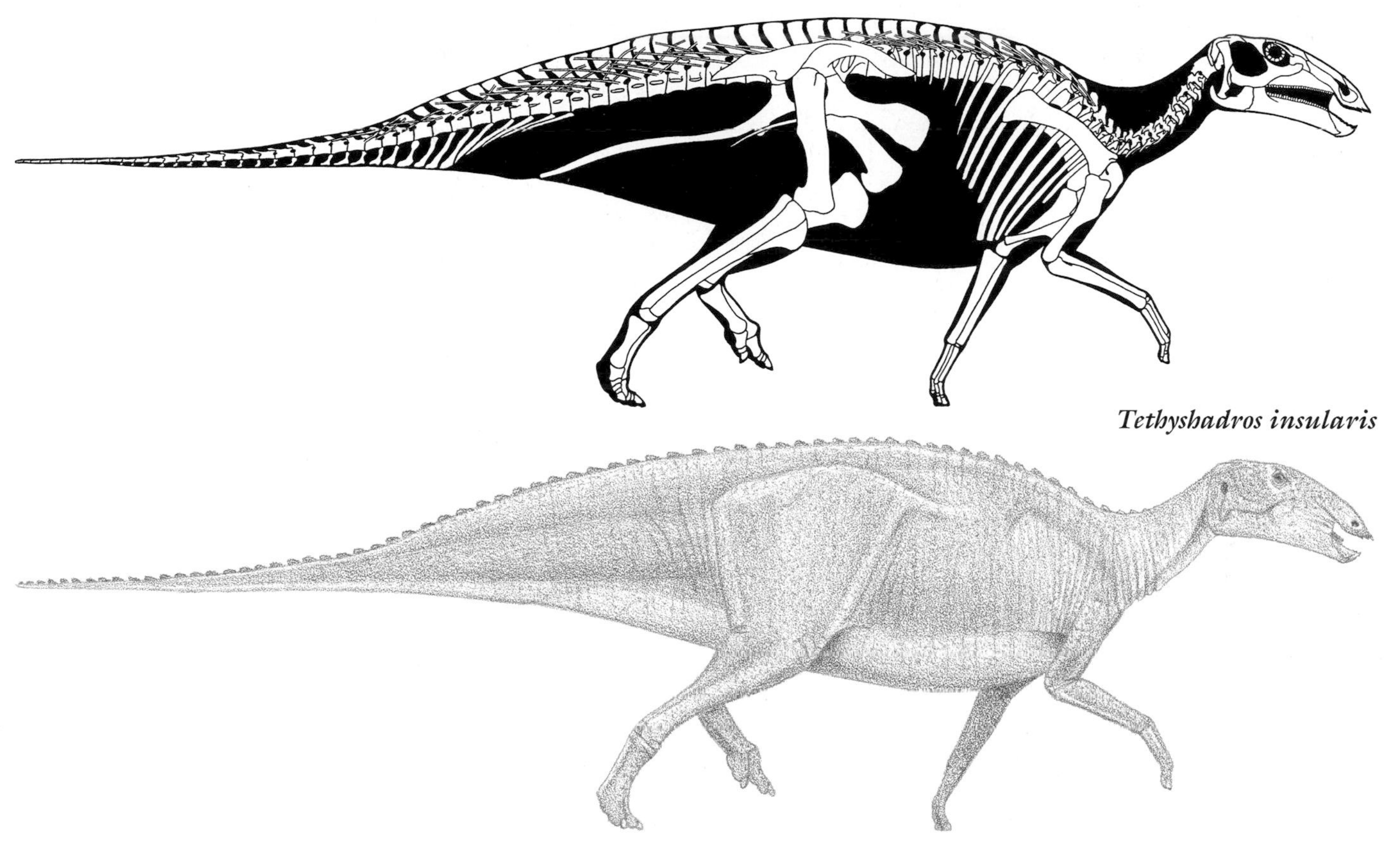

Tethyshadros insularis

ANATOMICAL CHARACTERISTICS Upper beak projecting forward, beak serrations well developed, eyes shaded by overhanging rim. Most of tail slender. Only three fingers. Pelvis elongated backward, shifting large portion of gut behind legs. Limbs short relative to mass, lower leg elongated, toes very short.
AGE Late Cretaceous, late Campanian or early Maastrichtian.
DISTRIBUTION AND FORMATION/S Italy; Liburnian.
HABITAT A large island.
HABITS Although the limbs are the most speed adapted among hadrosaurs yet known, their shortness is contradictory to very high speeds.
NOTES The smallest known hadrosaur, probably an example of island dwarfing. *Tethyshadros* skeleton is the most distinctive among hadrosaurs yet found.

HADROSAURIDS Large to gigantic hadrosaurs limited to the Late Cretaceous of the Northern Hemisphere.

HABITS Head crests when present used for competitive visual and vocal display within species; they did not improve the sense of smell.

Hadrosaurines* or *Saurolophines Large to gigantic hadrosaurids limited to the Late Cretaceous of the Northern Hemisphere.

ANATOMICAL CHARACTERISTICS Very uniform except for head. Head fairly shallow, subrectangular, snout long, nasal openings large.
NOTES Whether this group is hadrosaurines or saurolophines depends on uncertain placement of *Hadrosaurus* relative to other hadrosaurs. This group may be splittable into a number of subdivisions. Among dinosaurs only hadrosaurines matched some sauropods in size.

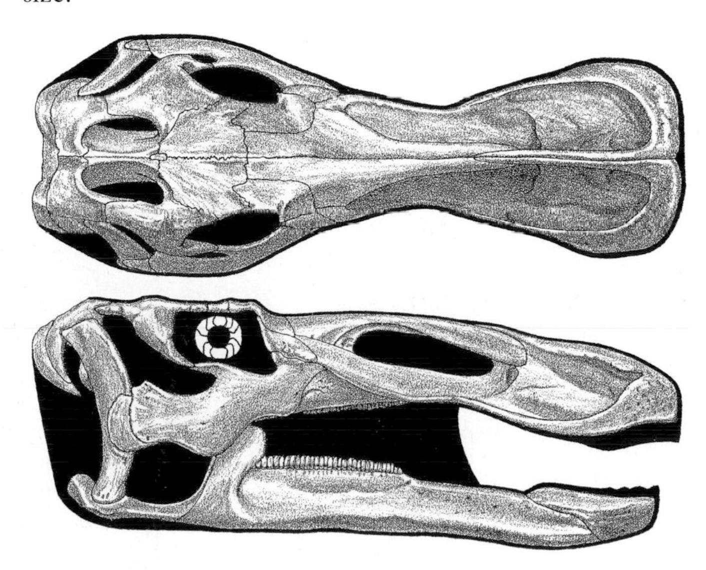

***Edmontosaurus* shaded skull**

Hadrosaurus foulkii
7 m (23 ft) TL, 2 tonnes

FOSSIL REMAINS Minority of skeleton.
ANATOMICAL CHARACTERISTICS Insufficient information.
AGE Late Cretaceous, Campanian.
DISTRIBUTION AND FORMATION/S New Jersey; Merchantville.
NOTES Found as drift in marine sediments.

Wulagasaurus dongi
9 m (30 ft) TL, 3 tonnes

FOSSIL REMAINS Numerous skull and skeletal bones.
ANATOMICAL CHARACTERISTICS Insufficient information.
AGE Late Cretaceous, Maastrichtian.
DISTRIBUTION AND FORMATION/S Northeast China; Yuliangze.

Barsboldia sicinskii
10 m (34 ft) TL, 5 tonnes

FOSSIL REMAINS Minority of skeleton.
ANATOMICAL CHARACTERISTICS Tall vertebral spines over trunk and tail form a low sail.
AGE Late Cretaceous, late Campanian and/or early Maastrichtian.
DISTRIBUTION AND FORMATION/S Mongolia; Nemegt.
HABITAT Well-watered woodland with seasonal rain, winters cold.
NOTES Shared its habitat with *Saurolophus angustirostris*. Main enemy *T. bataar*.

Shantungosaurus giganteus
15 m (50 ft) TL, 13 tonnes

FOSSIL REMAINS Several partial skulls and skeletons.
ANATOMICAL CHARACTERISTICS Snout very long, beak moderately broad, squared off, bony crest absent, lower jaw fairly deep.
AGE Late Late Cretaceous.
DISTRIBUTION AND FORMATION/S Eastern China; lower Xingezhuang.
HABITS Middle- and low-level browser and grazer.
NOTES Includes *Zhuchengosaurus maximus* and *Huaxiaosaurus aigahtens*. Some size estimates are somewhat exaggerated, but this is the largest known ornithischian and rivals some sauropods. Presence of soft-tissue head crest uncertain.

Kundurosaurus nagornyi
7 m (23 ft) TL, 2 tonnes

FOSSIL REMAINS Minority of skull and skeleton.
ANATOMICAL CHARACTERISTICS Insufficient information.

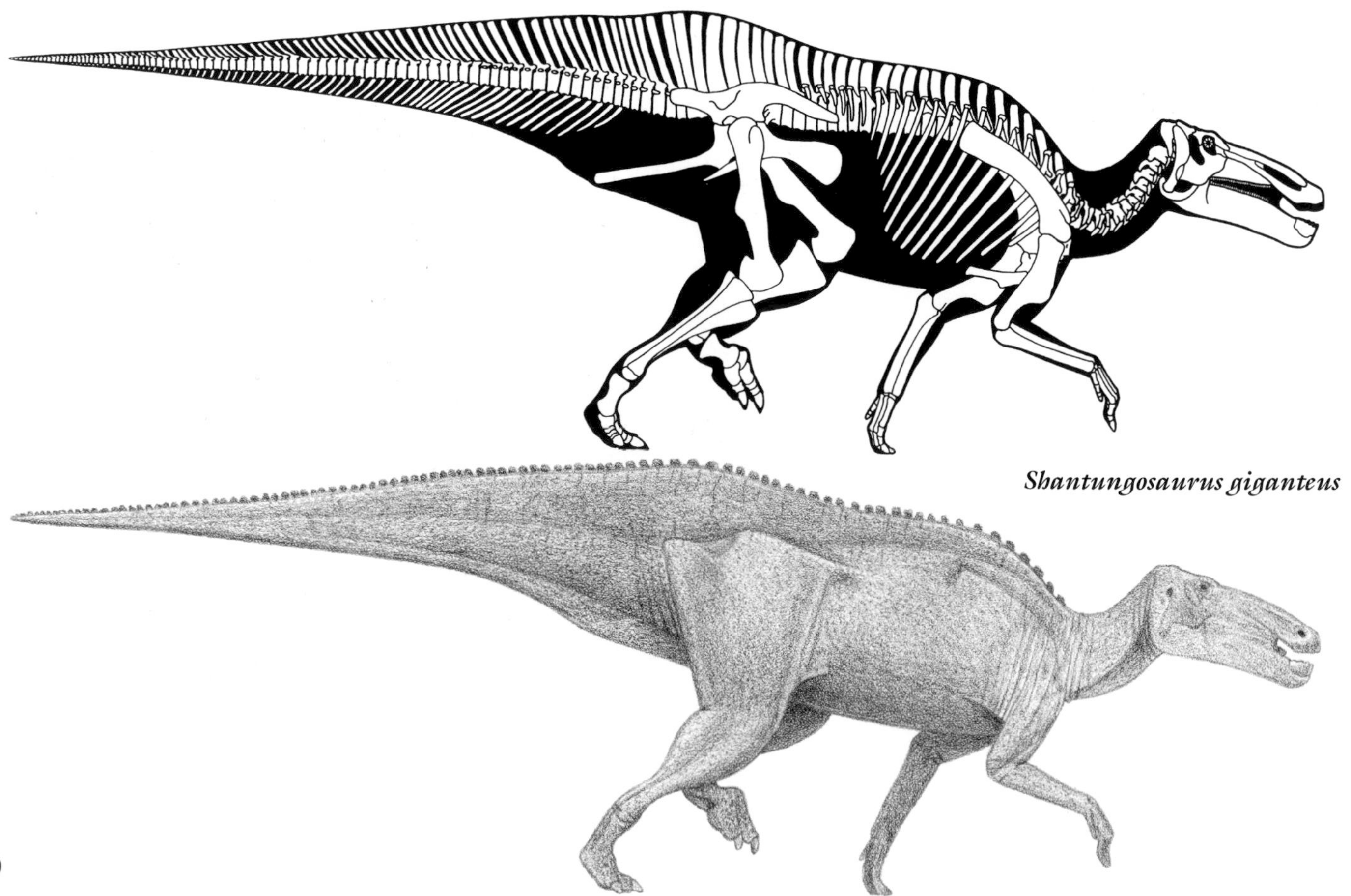

Shantungosaurus giganteus

AGE Late Cretaceous, probably late Maastrichtian.
DISTRIBUTION AND FORMATION/S Eastern Siberia: Udurchukan.

Edmontosaurus (= Ugrunaaluk) kuukpikensis

Adult size uncertain

FOSSIL REMAINS Bone bed, almost all immature.
ANATOMICAL CHARACTERISTICS Insufficient information.
AGE Late Cretaceous, middle Maastrichtian.

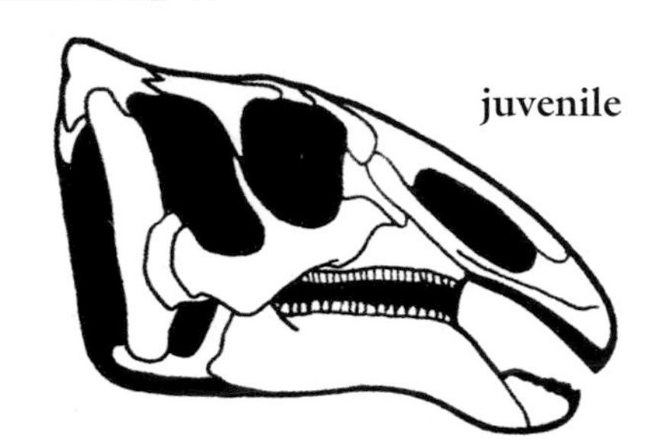

DISTRIBUTION AND FORMATION/S Northern Alaska; middle Prince Creek.
HABITAT Well-watered coastal woodland, cool summers, severe winters including heavy snows.
NOTES Indicates that polar hadrosaurs were not dwarfed compared to those from farther south. Presence of soft-tissue head crest uncertain.

Edmontosaurus regalis

9 m (30 ft) TL, 3.7 tonnes

FOSSIL REMAINS Numerous complete and partial skulls and skeletons.
ANATOMICAL CHARACTERISTICS Head deep in large adults, beak broad, rounded, shovel shaped, bony crest

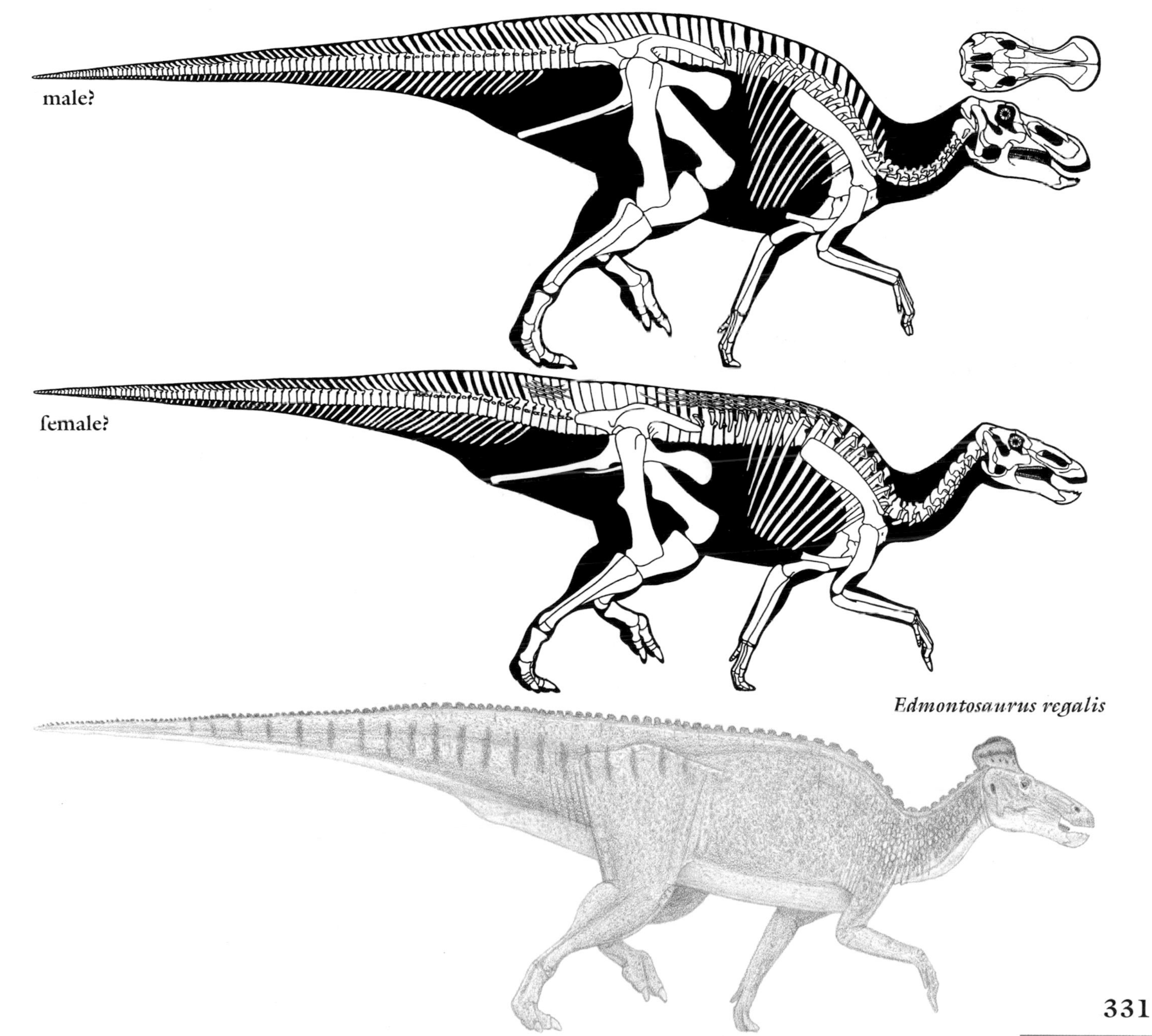

absent. Medium-sized, rounded, subtriangular soft-tissue crest atop back of head, pavement of large, vertically oblong scales adorn base of neck.
AGE Late Cretaceous, latest Campanian and early Maastrichtian.
DISTRIBUTION AND FORMATION/S Alberta; lower Horseshoe Canyon, Saint Mary River.
HABITAT Well-watered, forested floodplain with coastal swamps and marshes, cool winters.
HABITS Middle- and low-level browser and occasional grazer.
NOTES Head crest may have been on one sex (if so, probably male). Main enemy *Albertosaurus sarcophagus*.

Edmontosaurus (= Anatosaurus) annectens
9 m (30 ft) TL, 3.2 tonnes

FOSSIL REMAINS Numerous complete and partial skulls and skeletons including several "mummies," a few juveniles, completely known.
ANATOMICAL CHARACTERISTICS Head low, elongated, snout very long, gap between beaks and tooth rows very long, especially in large adults, beak broad, squared, shovel shaped, bony crest absent. Rectangular serrations along back, each corresponding to a vertebral spine at least along tail, full extent uncertain.
AGE Late Cretaceous, late Maastrichtian.

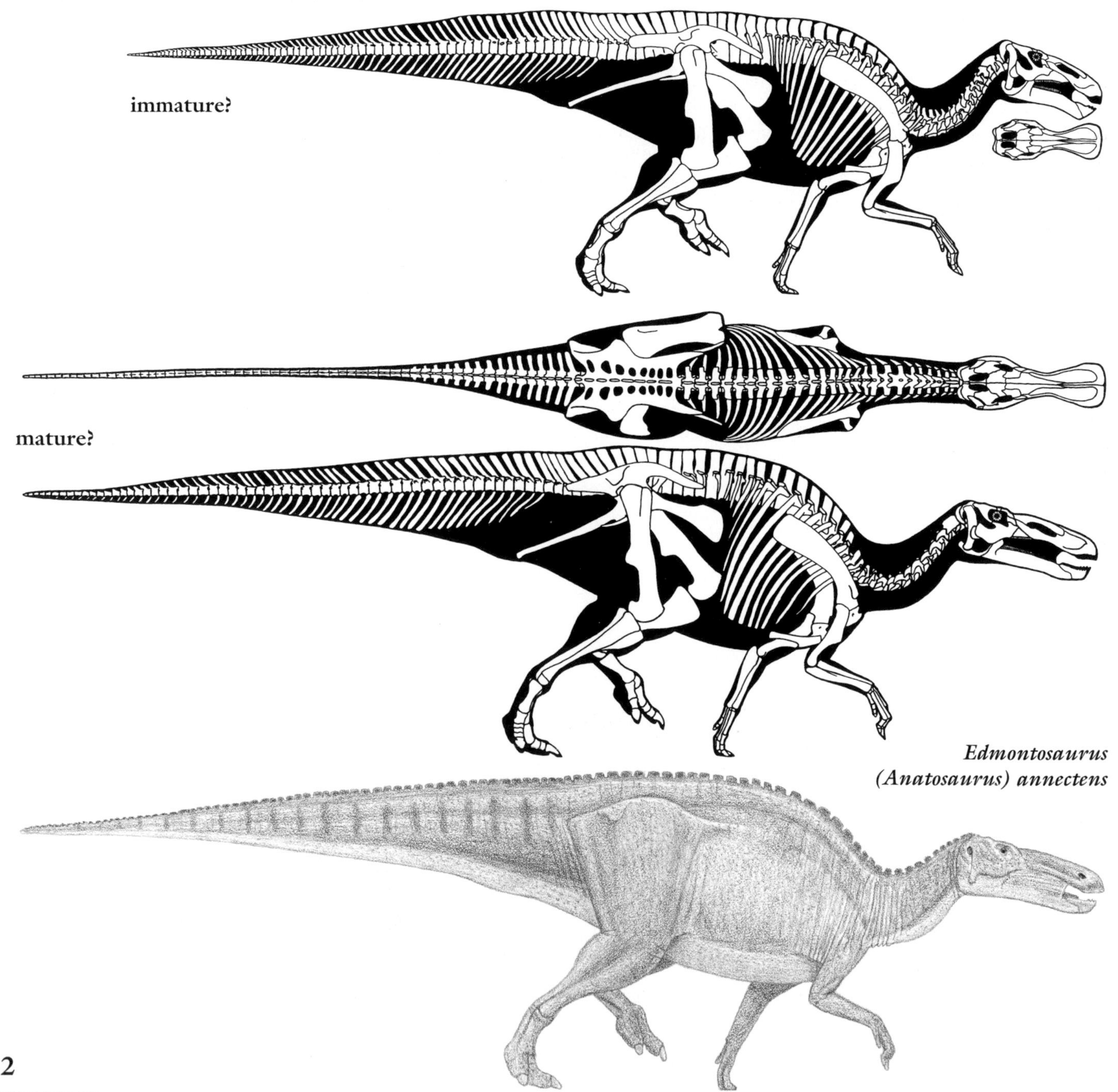

Edmontosaurus (Anatosaurus) annectens

DISTRIBUTION AND FORMATION/S Colorado, Wyoming, South Dakota, Montana, Alberta, Saskatchewan; Lance, Hell Creek (levels uncertain), Laramie, Scollard, Frenchman, etc.
HABITAT Well-watered coastal woodlands.
HABITS Square muzzle at end of increasingly long snout is adaptation for reaching down to and mowing ground cover, reached a maximum in adults; tooth microwear supports consuming grit-covered low plants. Also able to browse at low and medium levels.
NOTES *E. saskatchewanensis* is a juvenile of this taxon. Long-snouted *E.* (*Anatotitan*) *copei* is a distinct species if it is shown to be restricted to the uppermost Lance and Hell Creek. The classic "duck-billed" hadrosaur, this was the known ornithopod most adapted for grazing ground cover, matched only by some square-mouthed sauropods. Presence of soft-tissue head crest uncertain. The common hadrosaur in its habitat; one bone bed may contain tens of thousands of individuals. Healed bite mark of top of tail of one specimen verifies that main enemy was *Tyrannosaurus*; shared its habitat with the even more common *Triceratops*.

Saurolophus (= Lophorhothon) atopus
Adult size uncertain

FOSSIL REMAINS Minority of skull and majority of skeletons.
ANATOMICAL CHARACTERISTICS Shallow transverse crest over orbits.
AGE Late Cretaceous, Campanian.
DISTRIBUTION AND FORMATION/S Alabama, North Carolina; Mooreville Chalk, Black Creek.

Saurolophus (= Prosaurolophus) blackfeetensis
Adult size uncertain

FOSSIL REMAINS Several partial skulls and skeletons, large juveniles.

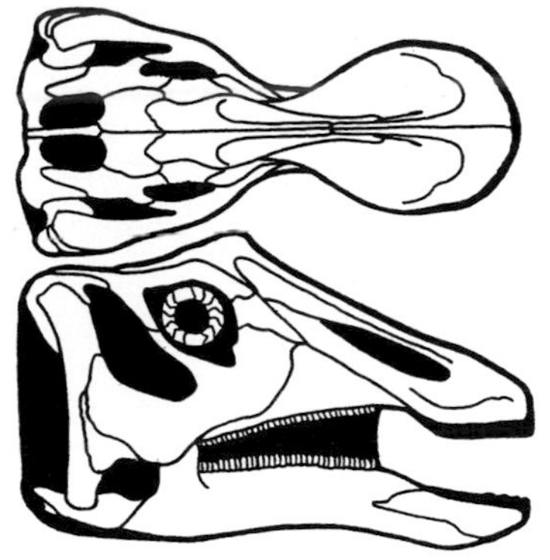

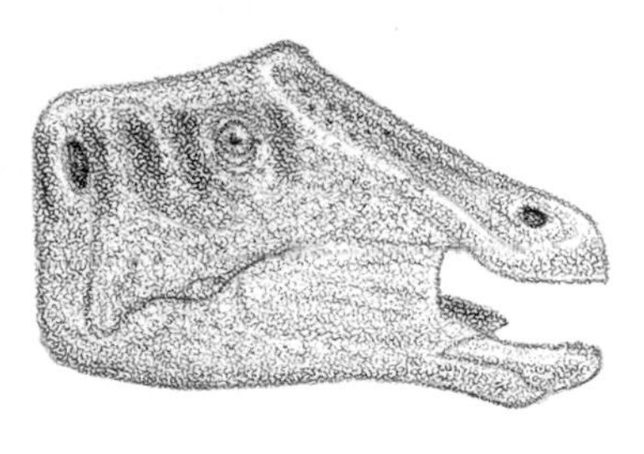

Saurolophus (= Prosaurolophus) blackfeetensis

ANATOMICAL CHARACTERISTICS Beak spoon shaped, shallow transverse crest over orbits.
AGE Late Cretaceous, middle and/or late Campanian.
DISTRIBUTION AND FORMATION/S Montana; Upper Two Medicine.
HABITAT Seasonally dry upland woodlands.

Saurolophus (= Prosaurolophus) maximus
8.5 m (27 ft) TL, 3 tonnes

FOSSIL REMAINS Numerous skulls and skeletons, completely known.
ANATOMICAL CHARACTERISTICS Beak spoon shaped, shallow transverse crest over orbits.
AGE Late Cretaceous, late Campanian.
DISTRIBUTION AND FORMATION/S Alberta; upper Dinosaur Park.
HABITAT Well-watered, forested floodplain with coastal swamps and marshes, cool winters.
NOTES Shared its habitat with *Hypacrosaurus intermedius* and *H. lambei*. Main enemy *Albertosaurus libratus*.

Saurolophus osborni (Illustrated on p. 335)
8.5 m (27 ft) TL, 3 tonnes

FOSSIL REMAINS Two complete skulls and a nearly complete skeleton, almost completely known.

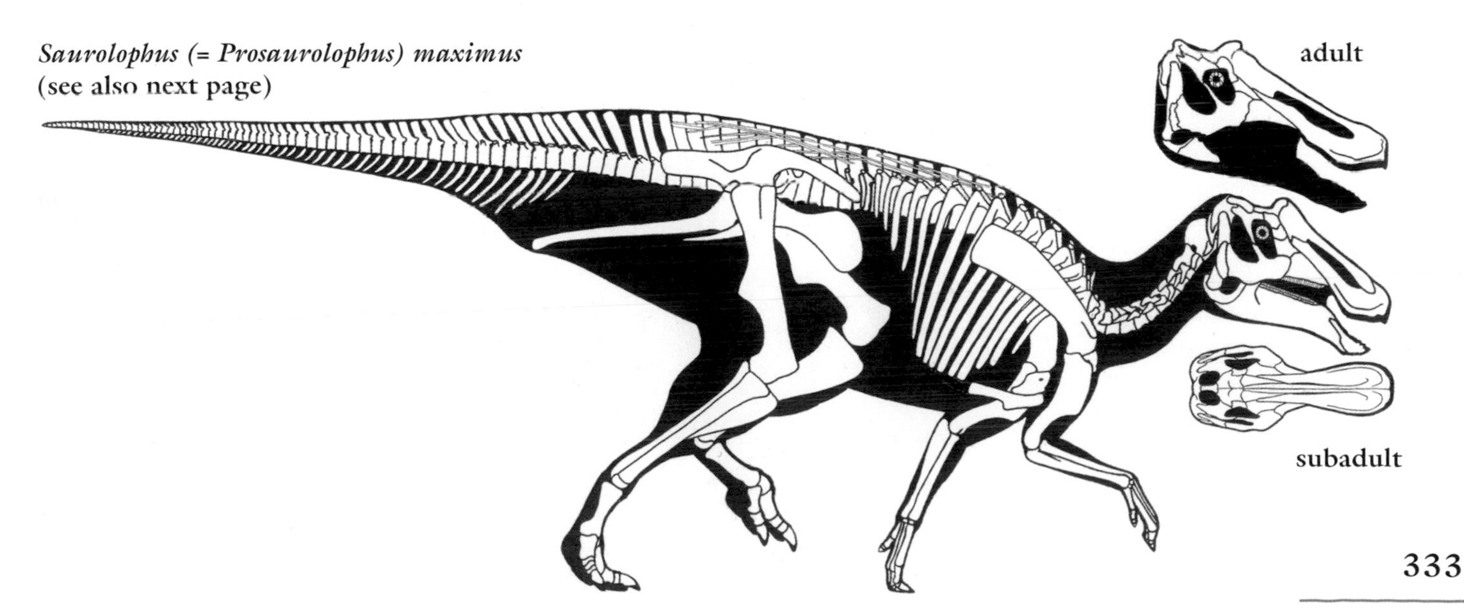

Saurolophus (= Prosaurolophus) maximus
(see also next page)

Saurolophus (= Prosaurolophus) maximus and *Albertosaurus libratus* juveniles (right).

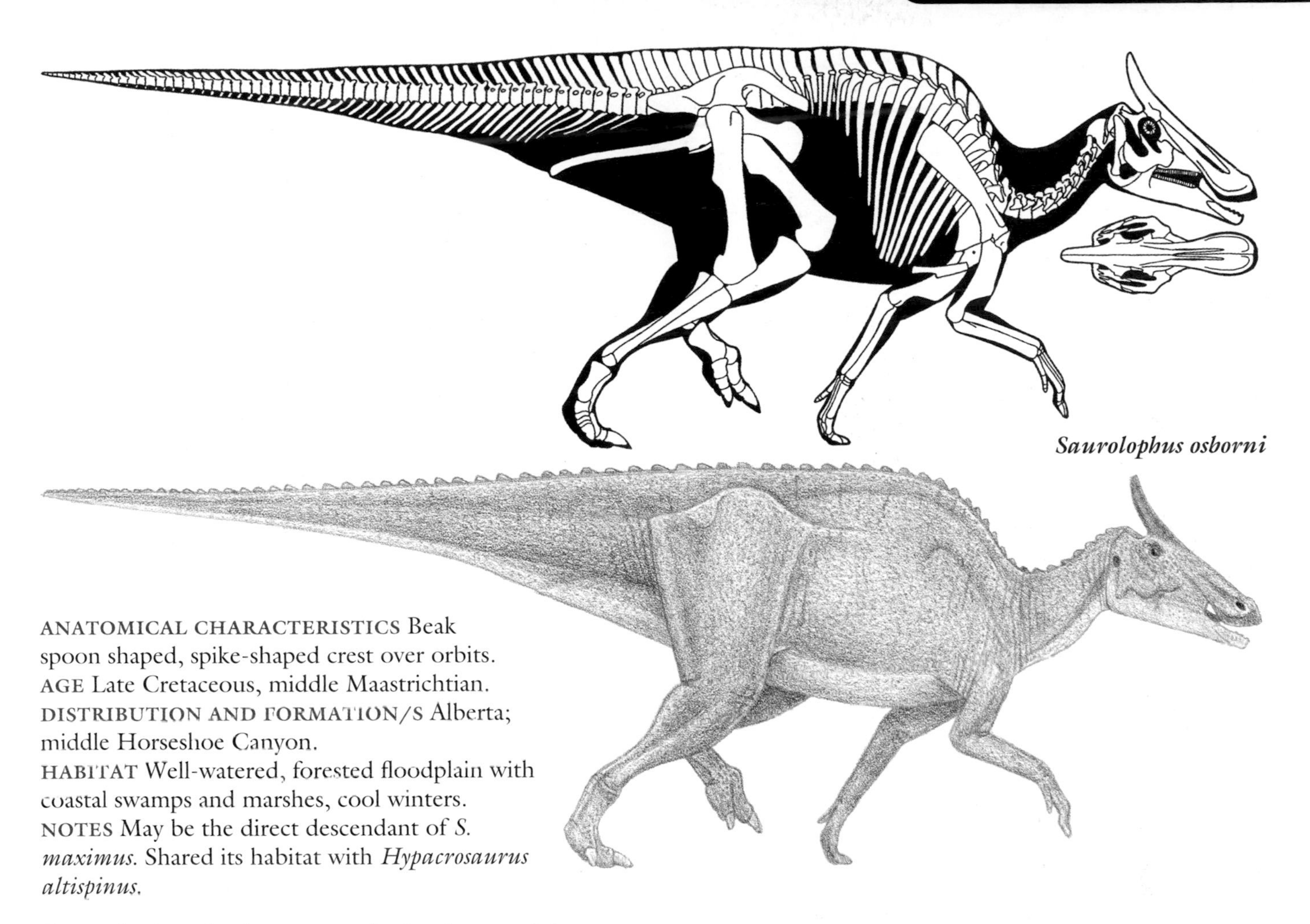

Saurolophus osborni

ANATOMICAL CHARACTERISTICS Beak spoon shaped, spike-shaped crest over orbits.
AGE Late Cretaceous, middle Maastrichtian.
DISTRIBUTION AND FORMATION/S Alberta; middle Horseshoe Canyon.
HABITAT Well-watered, forested floodplain with coastal swamps and marshes, cool winters.
NOTES May be the direct descendant of *S. maximus*. Shared its habitat with *Hypacrosaurus altispinus*.

Saurolophus (= Augustynolophus) morrisi
8 m (26 ft) TL, 3 tonnes

FOSSIL REMAINS Majority of skull and skeleton.
ANATOMICAL CHARACTERISTICS Beak spoon shaped, spike-shaped crest over orbits.
AGE Late Cretaceous, late Maastrichtian.
DISTRIBUTION AND FORMATION/S California; Moreno.

Saurolophus angustirostris
13 m (43 ft) TL, 11 tonnes

FOSSIL REMAINS Complete skull and numerous skeletons, almost completely known.
ANATOMICAL CHARACTERISTICS Beak spoon shaped, spike-shaped crest over orbits.
AGE Late Cretaceous, late Campanian and/or early Maastrichtian.
DISTRIBUTION AND FORMATION/S Mongolia; Nemegt.

growth series

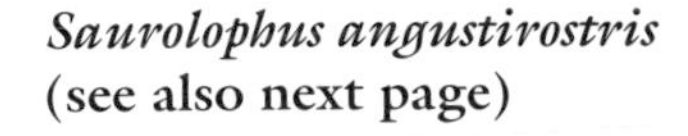

Saurolophus angustirostris
(see also next page)

Saurolophus angustirostris

HABITAT Well-watered woodland with seasonal rain, winters cold.
NOTES Approached *Shantungosaurus* in size. Main enemy *T. bataar*.

Acristavus gagslarsoni
8 m (26 ft) TL, 3 tonnes

FOSSIL REMAINS Majority of skull.
ANATOMICAL CHARACTERISTICS Front edge of upper beak indented, bony crest absent.
AGE Late Cretaceous, early Campanian.
DISTRIBUTION AND FORMATION/S Montana; Lower Two Medicine.

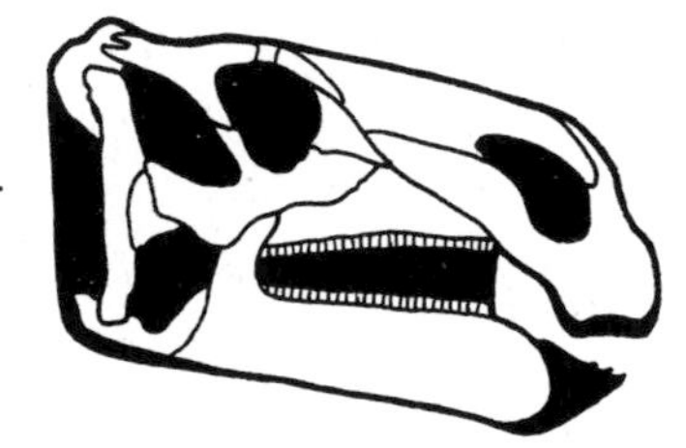
Acristavus gagslarsoni

Maiasaura (or *Brachylophosaurus*) *peeblesorum*
7 m (23 ft) TL, 2.5 tonnes

FOSSIL REMAINS Several complete skulls and majority of skeletons, bone beds, juvenile to adult, numerous nests, completely known.
ANATOMICAL CHARACTERISTICS Beak narrow, squared off, shallow transverse crest over orbits. Spherical eggs 10 cm (4 in) in diameter.
AGE Late Cretaceous, middle and/or late Campanian.
DISTRIBUTION AND FORMATION/S Montana; Upper Two Medicine.
HABITAT Seasonally dry upland woodlands.
HABITS Middle- and low-level browser. Bone beds indicate at least sometimes congregated in large herds. May have nested in colonies, hatchlings apparently remained in nest during first weeks of growth, probably fed by parents. Transverse head crest best suited for frontal displays with head not strongly pitched down.
NOTES May be a subgenus of *Brachylophosaurus*. Shared its habitat with *Saurolophus blackfeetensis*.

Brachylophosaurus (= *Probrachylophosaurus*) *bergei*
10 m (33 ft) TL, 5 tonnes

FOSSIL REMAINS Partial skull and skeleton.
ANATOMICAL CHARACTERISTICS Small horizontal platelike crest over rear of head.
AGE Late Cretaceous, middle Campanian.
DISTRIBUTION AND FORMATION/S Montana; middle Judith River.

Maiasaura (or *Brachylophosaurus*) *peeblesorum* nestling (not to scale)

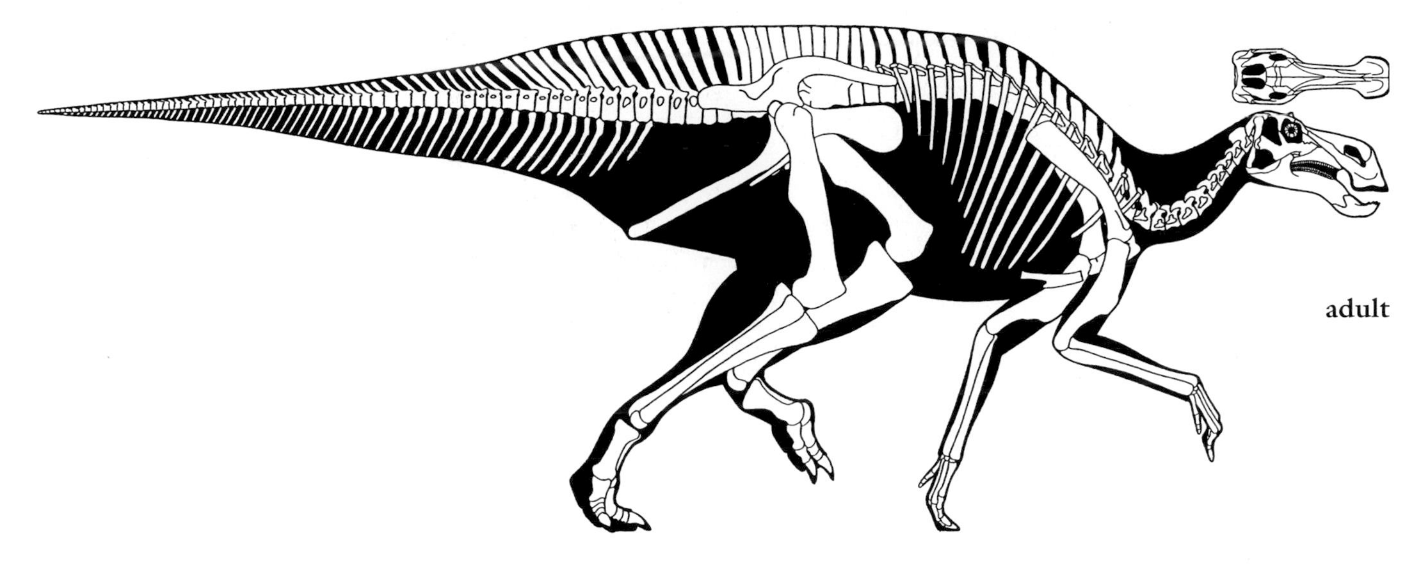

Maiasaura (or *Brachylophosaurus*) *peeblesorum*

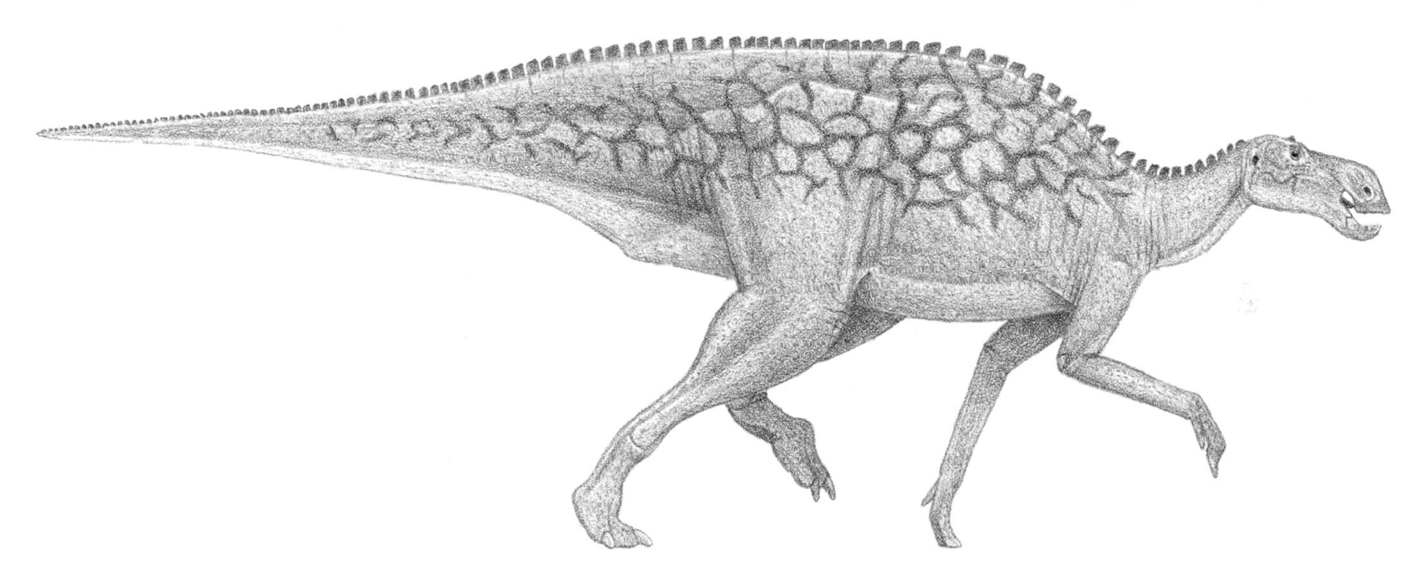

HABITS Transverse head crest best suited for frontal displays with head strongly pitched down.
NOTES The new genus title for this species so similar to *B. canadensis* is a classic example of generic oversplitting. May be direct ancestor of *B. canadensis*.

Brachylophosaurus canadensis (Illustrated overleaf)
11 m (35 ft) TL, 7 tonnes

FOSSIL REMAINS Several complete skulls and skeletons, including mummy, completely known.
ANATOMICAL CHARACTERISTICS Head rather small, beak narrow, squared off, flat, platelike crest over rear of head. Hatchet-shaped serrations along back, each corresponding to a vertebral spine, full extent uncertain.
AGE Late Cretaceous, early Campanian.
DISTRIBUTION AND FORMATION/S Alberta; upper Oldman, lower Dinosaur Park.
HABITAT Well-watered, forested floodplain with coastal swamps and marshes.
HABITS Middle- and low-level browser. Transverse head crest best suited for frontal displays with head strongly pitched down.
NOTES Specimen from the Foremost Formation usually placed in this taxon is probably an earlier species.

Kritosaurus* (or *Gryposaurus*) *latidens
7.5 m (25 ft) TL, 2.5 tonnes

FOSSIL REMAINS Partial skulls and majority of skeleton.
ANATOMICAL CHARACTERISTICS Long, shallow crest ridge over nasal opening moderately developed.

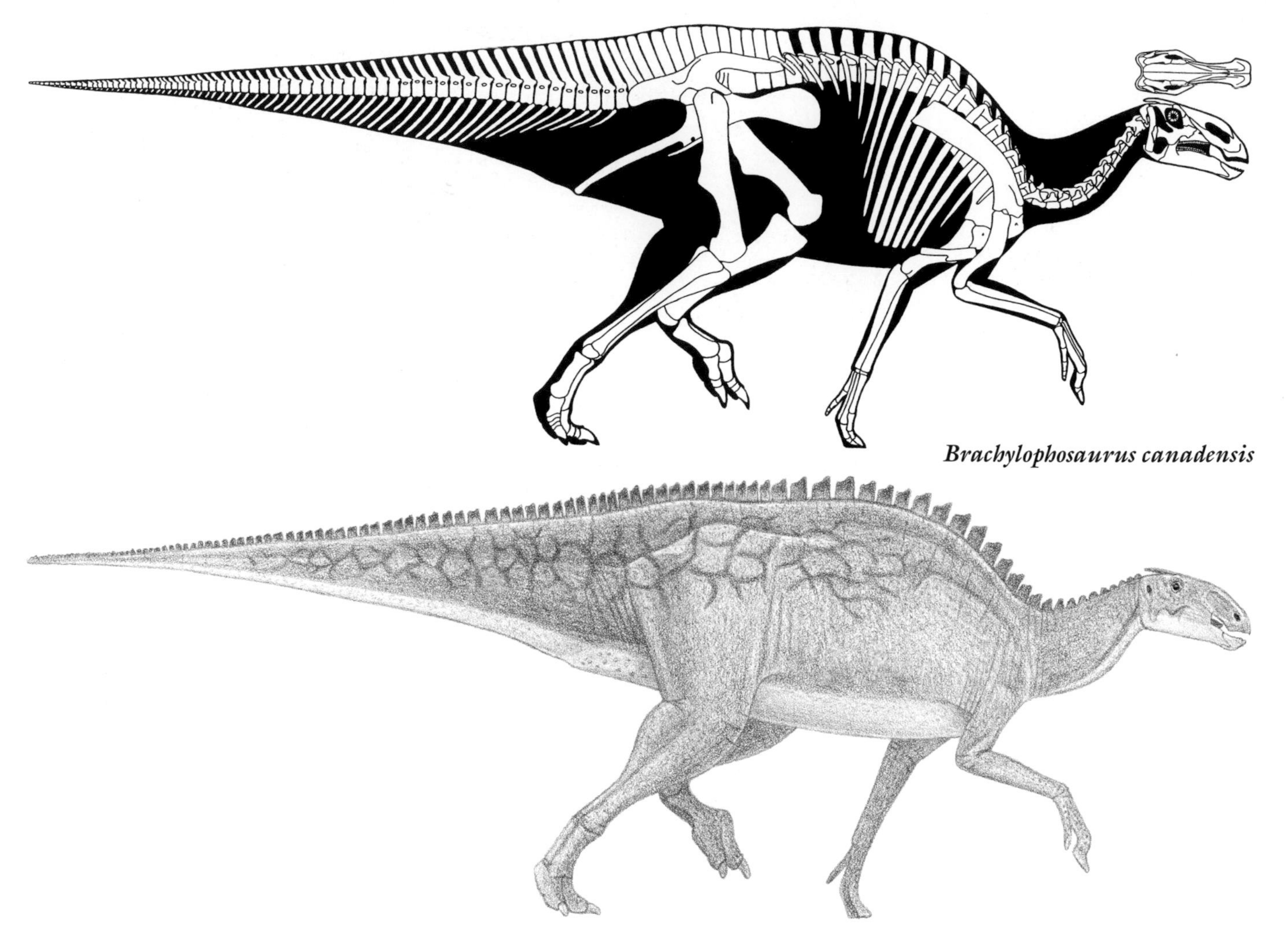

Brachylophosaurus canadensis

AGE Late Cretaceous, early and/or middle Campanian.
DISTRIBUTION AND FORMATION/S Montana; Lower Two Medicine.
HABITAT Seasonally dry upland woodlands.
HABITS Middle- and low-level browser.
NOTES May be ancestral to at least some of the kritosaurs listed below from the same region. Because the skulls and skeletons of kritosaurs are very similar except for the crest, they probably form one genus; *Gryposaurus* may be a subgenus.

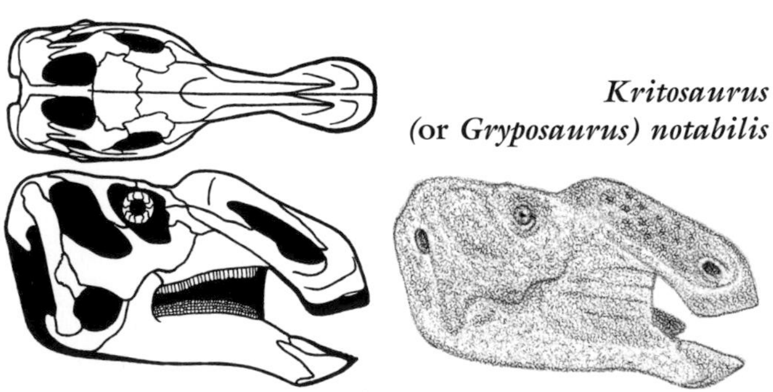

Kritosaurus (or *Gryposaurus*) *notabilis*

Kritosaurus (or *Gryposaurus*) *notabilis*
8 m (26 ft) TL, 3 tonnes

FOSSIL REMAINS A number of skulls, partial skeletons.
ANATOMICAL CHARACTERISTICS Long, shallow crest ridge over nasal opening well developed, beak narrow, rounded. Large conical serrations along back do not correspond to neural spines, full extent uncertain.
AGE Late Cretaceous, late Campanian.
DISTRIBUTION AND FORMATION/S Alberta; lower Dinosaur Park.
HABITAT Well-watered, forested floodplain with coastal swamps and marshes, cool winters.
HABITS Middle- and low-level browser.
NOTES May have been the direct ancestor of *G. incurvimanus.*

Kritosaurus (or *Gryposaurus*) *incurvimanus*
7 m (23 ft) TL, 2.2 tonnes

FOSSIL REMAINS A few skulls and majority of skeleton.
ANATOMICAL CHARACTERISTICS Long, shallow crest

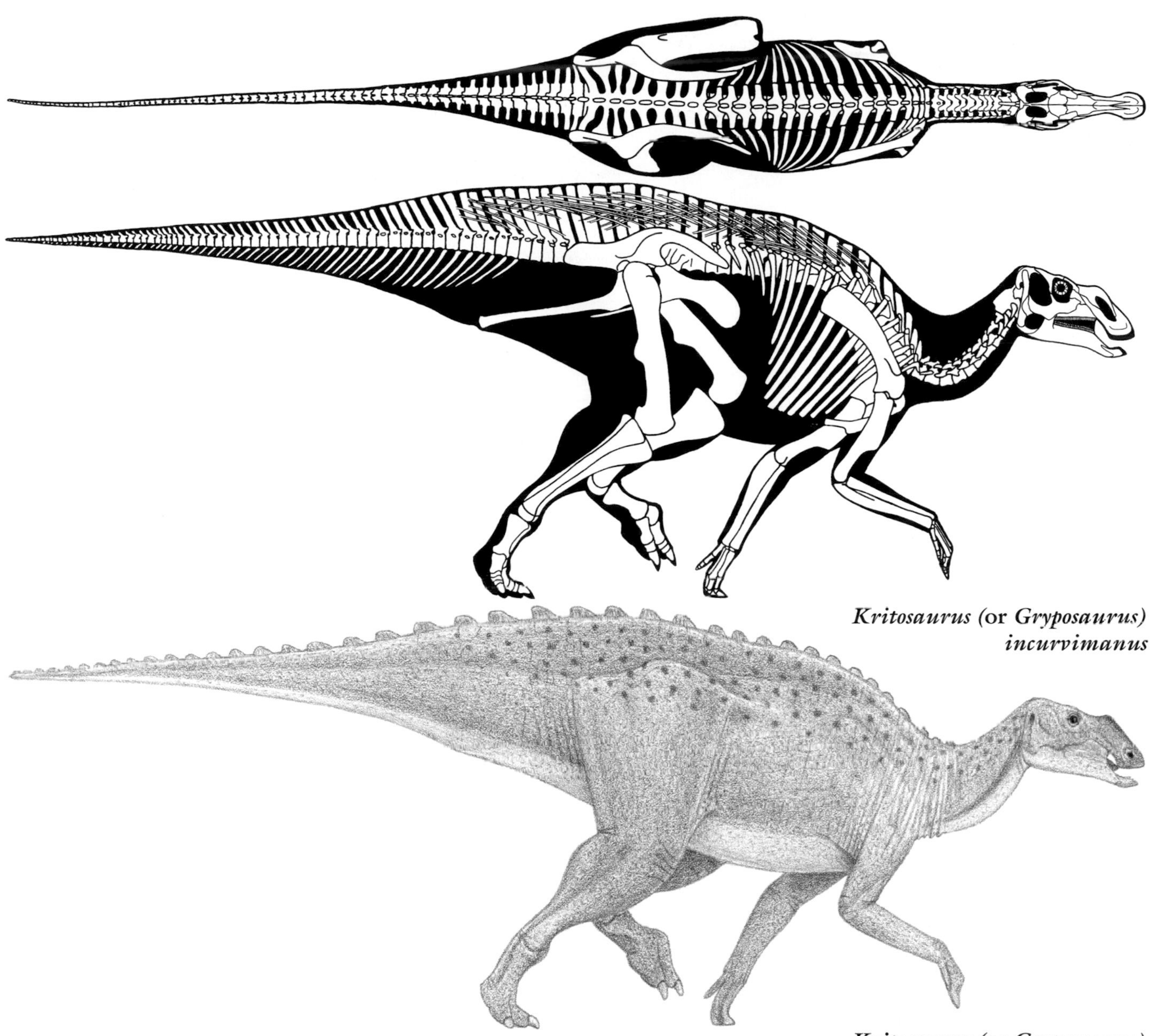

Kritosaurus (or *Gryposaurus*) *incurvimanus*

Kritosaurus (or *Gryposaurus*) *monumentensis*

ridge over nasal opening weakly developed.
AGE Late Cretaceous, late Campanian.
DISTRIBUTION AND FORMATION/S Alberta; middle to upper Dinosaur Park.
HABITAT Well-watered, forested floodplain with coastal swamps and marshes, cool winters.
HABITS Middle- and low-level browser.

Kritosaurus (or *Gryposaurus*) *monumentensis*

8 m (26 ft) TL, 3 tonnes

FOSSIL REMAINS Majority of skull and skeleton.
ANATOMICAL CHARACTERISTICS Head and skeleton heavily built. Head exceptionally deep and vertical; long, shallow crest ridge over nasal opening well developed.
AGE Late Cretaceous, late Campanian.
DISTRIBUTION AND FORMATION/S Utah; Kaiparowits.
HABITS Middle- and low-level browser, depth and strength of head and of jaw muscles indicate ability to process coarse vegetation.
NOTES Shared its habitat with *Nasutoceratops*.

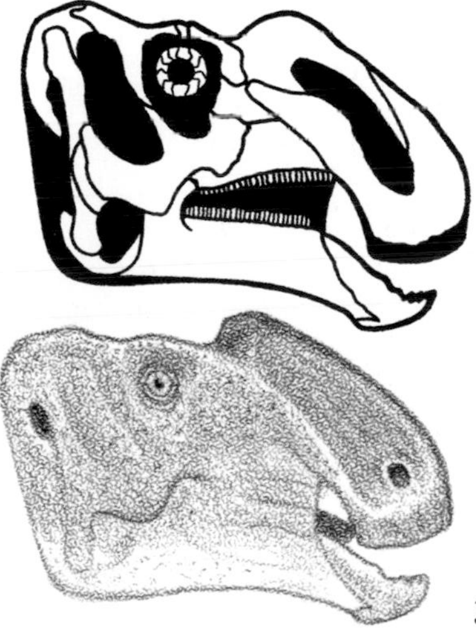

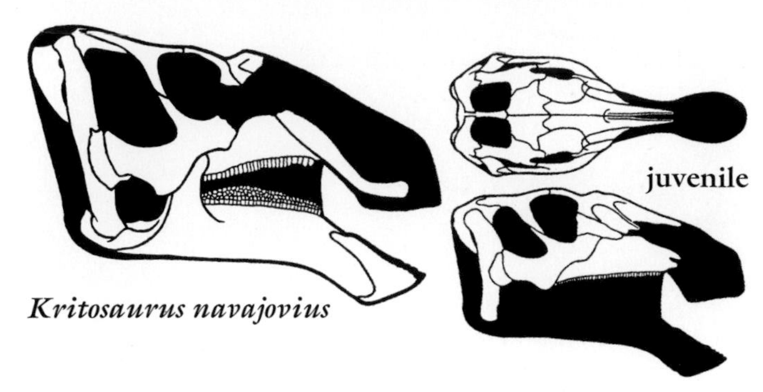

Kritosaurus navajovius

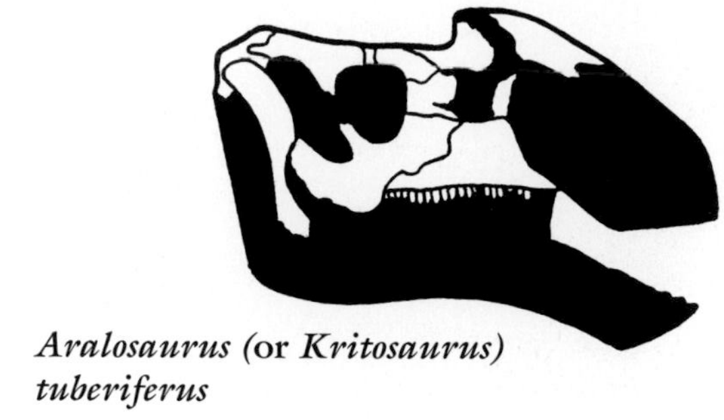

Aralosaurus (or *Kritosaurus*) *tuberiferus*

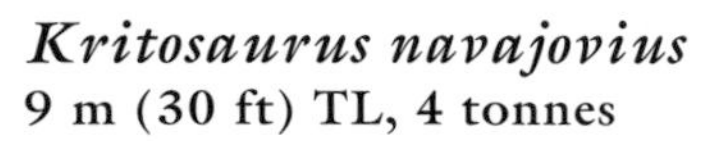

Kritosaurus navajovius
9 m (30 ft) TL, 4 tonnes

FOSSIL REMAINS Partial skull.
ANATOMICAL CHARACTERISTICS Insufficient information.
AGE Late Cretaceous, late Campanian and/or early Maastrichtian.
DISTRIBUTION AND FORMATION/S New Mexico; lower Kirtland.
HABITAT Moderately watered floodplain woodlands, coastal swamps, and marshes.
NOTES *Naashoibitosaurus ostromi* may be an immature member of this species.

Kritosaurus (or *Anasazisaurus*) *horneri*
7.5 m (25 ft) TL, 2.5 tonnes

FOSSIL REMAINS Partial skull.
ANATOMICAL CHARACTERISTICS Long shallow crest ridge over nasal opening with small hook at back end.
AGE Late Cretaceous, late Campanian and/or early Maastrichtian.
DISTRIBUTION AND FORMATION/S New Mexico; lower Kirtland.
HABITAT Moderately watered floodplain woodlands, coastal swamps and marshes.

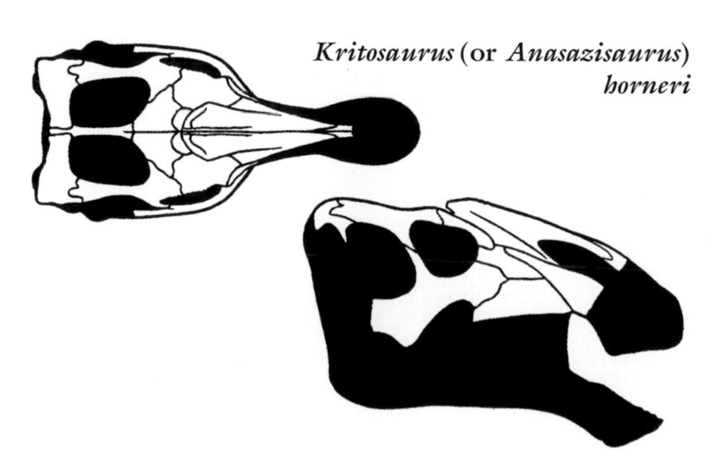

Kritosaurus (or *Anasazisaurus*) *horneri*

NOTES Shared its habitat with *Nodocephalosaurus* and *K. navajovius.*

Aralosaurus (or *Kritosaurus*) *tuberiferus*
Adult size uncertain

FOSSIL REMAINS Partial skull, possibly a large juvenile.
ANATOMICAL CHARACTERISTICS Long, prominent crest ridge over nasal opening with incipient hook at back end.
AGE Late Cretaceous, late Campanian.
DISTRIBUTION AND FORMATION/S Kazakhstan; Beleuta Svita.

Secernosaurus koerneri
Adult size uncertain

FOSSIL REMAINS Minority of skeleton, probably juvenile.
ANATOMICAL CHARACTERISTICS Insufficient information.
AGE Late Cretaceous, late Cenomanian or Turonian.
DISTRIBUTION AND FORMATION/S Southern Argentina; lower Bajo Barreal.
NOTES This and *Willinakaqe* probably form a distinct South American hadrosaur group.

Willinakaqe salitralensis
7 m (23 ft) TL, 2 tonnes

FOSSIL REMAINS Partial skull and skeletons.
ANATOMICAL CHARACTERISTICS Insufficient information.
AGE Late Cretaceous, late Campanian and/or early Maastrichtian.
DISTRIBUTION AND FORMATION/S Central Argentina; lower Allen.

Lambeosaurines Large to gigantic hadrosaurids limited to Late Cretaceous of the Northern Hemisphere.

ANATOMICAL CHARACTERISTICS Very uniform except for head crests, which are always atop back of head and contain looping segments of nasal passages. Head subtriangular, snout slender, beaks narrow and rounded, nasal openings narrow.

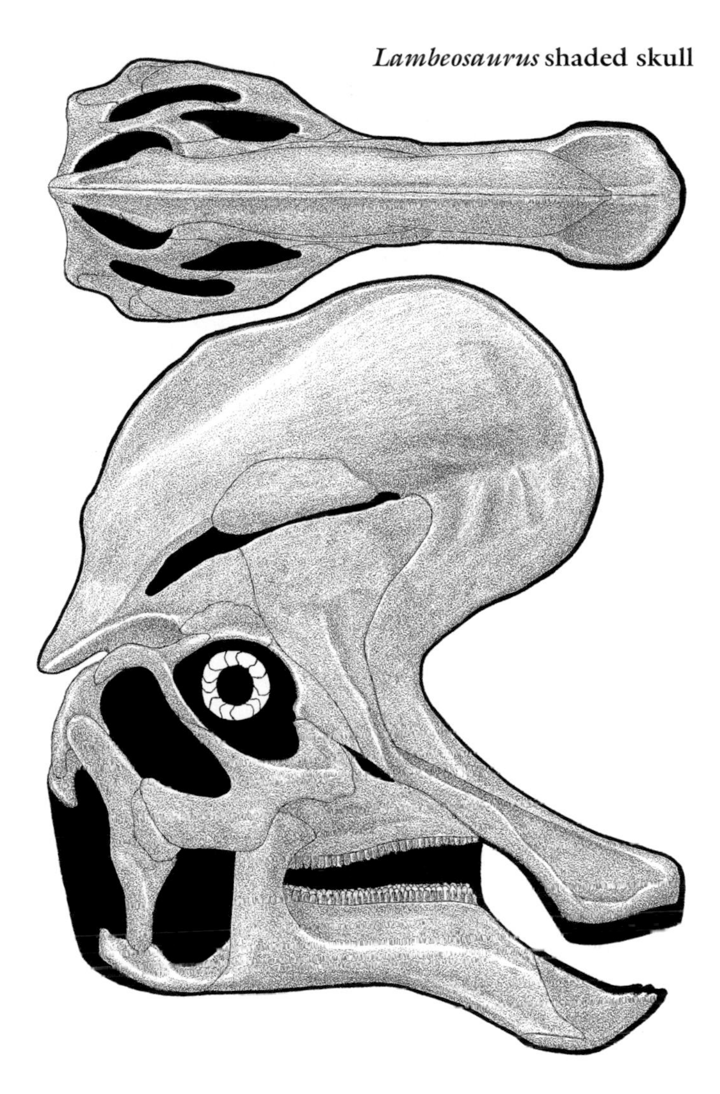
Lambeosaurus shaded skull

HABITS Middle- and low-level browsers. Crests may have been used to help generate vocal sounds for display and communication.

Nanningosaurus dashiensis
7.5 m (25 ft) TL, 2.5 tonnes

FOSSIL REMAINS Minority of skull and skeleton.
ANATOMICAL CHARACTERISTICS Insufficient information.
AGE Late Late Cretaceous.
DISTRIBUTION AND FORMATION/S Southern China; unnamed.

Sahaliyania elunchunorum
7.5 m (25 ft) TL, 2.5 tonnes

FOSSIL REMAINS Numerous skull and skeletal bones.
ANATOMICAL CHARACTERISTICS Insufficient information.
AGE Late Cretaceous, Maastrichtian.
DISTRIBUTION AND FORMATION/S Northeast China; Yuliangze.

Pararhabdodon isonensis
Adult size uncertain

FOSSIL REMAINS Minority of skull and skeletons.
ANATOMICAL CHARACTERISTICS Lower jaw strongly downcurved and unusually broad, indicating head was wider than usual in hadrosaurs.
AGE Late Cretaceous, late Maastrichtian.
DISTRIBUTION AND FORMATION/S Northeast Spain; Tremp.
NOTES Probably includes *Koutalisaurus kohlerorum*. Shared its habitat with *Arenysaurus*.

Arenysaurus ardevoli
6 m (20 ft) TL, 1 tonne

FOSSIL REMAINS Partial skull and skeleton.
ANATOMICAL CHARACTERISTICS Insufficient information.
AGE Late Cretaceous, late Maastrichtian.
DISTRIBUTION AND FORMATION/S Northeast Spain; Tremp.

Amurosaurus riabinini
8 m (26 ft) TL, 3 tonnes

FOSSIL REMAINS Majority of skull and skeleton.
ANATOMICAL CHARACTERISTICS Insufficient information.
AGE Late Cretaceous, late Maastrichtian.
DISTRIBUTION AND FORMATION/S Eastern Siberia; Udurchukan.

Angulomastacator daviesi
Size uncertain

FOSSIL REMAINS Minority of skull.
ANATOMICAL CHARACTERISTICS Snout strongly downturned.
AGE Late Cretaceous, Campanian.
DISTRIBUTION AND FORMATION/S Texas; Aguja.

Parasaurolophus walkeri
7.5 m (25 ft) TL, 2.6 tonnes

FOSSIL REMAINS Several complete or partial skulls, majority of skeletons.
ANATOMICAL CHARACTERISTICS Long, backward-projecting, arced tube crest.
AGE Late Cretaceous, late Campanian.
DISTRIBUTION AND FORMATION/S Alberta, New Mexico?; lower to middle Dinosaur Park, possibly lower Kirtland.

Parasaurolophus walkeri

HABITAT Well-watered forests to north and east, seasonally dry basins to west and south.
NOTES Albertan *P. walkeri* and New Mexican *P. tubicen* are not distinct from one another, and short-crested New Mexican *P. cyrtocristatus* may be a sexual morph or subadult of this species.

Charonosaurus (or *Parasaurolophus*) *jiayinensis*
10 m (23 ft) TL, 5 tonnes

FOSSIL REMAINS Partial skull and majority of skeletons, juveniles to adults.
ANATOMICAL CHARACTERISTICS Insufficient information.
AGE Late Cretaceous, late Maastrichtian.
DISTRIBUTION AND FORMATION/S Northeast China; Yuliangze.
NOTES May have had a tubular crest like that of *Parasaurolophus.*

Tsintaosaurus sphinorhinus
8.3 m (27 ft) TL, 2.5 tonnes

FOSSIL REMAINS Two partial skulls and majority of a few skeletons.
ANATOMICAL CHARACTERISTICS Head crest tall and broad.
AGE Late Cretaceous, probably Campanian.
DISTRIBUTION AND FORMATION/S Eastern China; Wangshi Group.
NOTES Crest used to be thought to be a narrow, vertical tube. Shared its habitat with *Tanius* and *Sinoceratops.*

Tsintaosaurus sphinorhinus

Olorotitan arharensis
8 m (26 ft) TL, 3.1 tonnes

FOSSIL REMAINS Majority of skull and skeleton.
ANATOMICAL CHARACTERISTICS Large, subvertical, fan-shaped crest. Neck longer than in other hadrosaurs. Tail further stiffened.
AGE Late Cretaceous, late Maastrichtian.
DISTRIBUTION AND FORMATION/S Eastern Siberia; Tsagayan.
HABITS Longer neck indicates ability to browse at a relatively high level.
NOTES May be the same genus or species as *Amurosaurus riabinini*.

Olorotitan arharensis

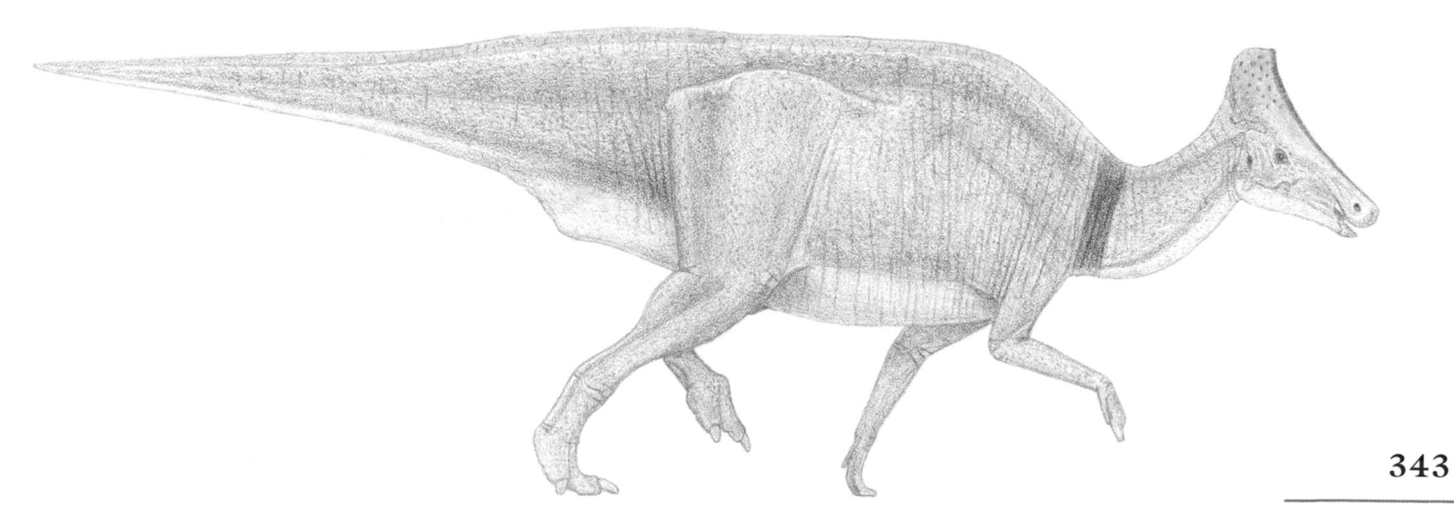

Nipponosaurus (or *Hypacrosaurus*) *sachaliensis*
Adult size uncertain

FOSSIL REMAINS Partial skull and skeleton, juvenile.
ANATOMICAL CHARACTERISTICS Insufficient information.
AGE Late Cretaceous, Santonian or early Campanian.
DISTRIBUTION AND FORMATION/S Sakhalin Island (east of Siberia); Ryugase Group.

Hypacrosaurus (= Velafrons) coahuilensis
Adult size uncertain

FOSSIL REMAINS Nearly complete skull, partial skeleton, juvenile.
ANATOMICAL CHARACTERISTICS Large, semicircular crest atop back of skull.
AGE Late Cretaceous, middle Campanian.
DISTRIBUTION AND FORMATION/S Northeastern Mexico; Cerro del Pueblo.
NOTES Shared its habitat with *Coahuilaceratops.*

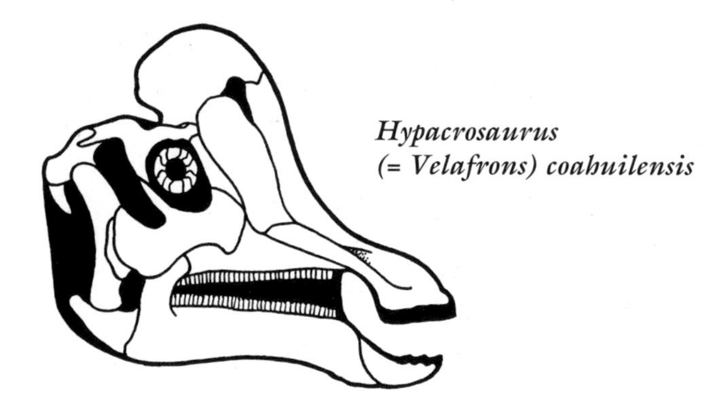
Hypacrosaurus (= Velafrons) coahuilensis

Hypacrosaurus (= Corythosaurus) casuarius
8 m (26 ft) TL, 2.8 tonnes

FOSSIL REMAINS A number of complete skulls and skeletons, including mummies, completely known. Low ribbon frill along back appears to connect to back of head crest.
ANATOMICAL CHARACTERISTICS Large, semicircular crest atop back of skull.
AGE Late Cretaceous, late Campanian.
DISTRIBUTION AND FORMATION/S Alberta; lower Dinosaur Park.
HABITAT Well-watered, forested floodplain with coastal swamps and marshes, cool winters.
NOTES Because the skulls and skeletons of *Hypacrosaurus*, *Corythosaurus*, and *Lambeosaurus* are very similar except for the head crest and height of the neural spines, they probably form one genus. Shared its habitat with *Parasaurolophus walkeri.*

Hypacrosaurus (= Corythosaurus) intermedius
7.7 m (25 ft) TL, 2.5 tonnes

FOSSIL REMAINS Complete and partial skulls and skeletons.
ANATOMICAL CHARACTERISTICS Large, subtriangular crest atop back of skull.
AGE Late Cretaceous, late Campanian.
DISTRIBUTION AND FORMATION/S Alberta; middle and upper Dinosaur Park.
HABITAT Well-watered, forested floodplain with coastal swamps and marshes, cool winters.
NOTES Previously thought to be the female of earlier *H. casuarius*, may be its descendant.

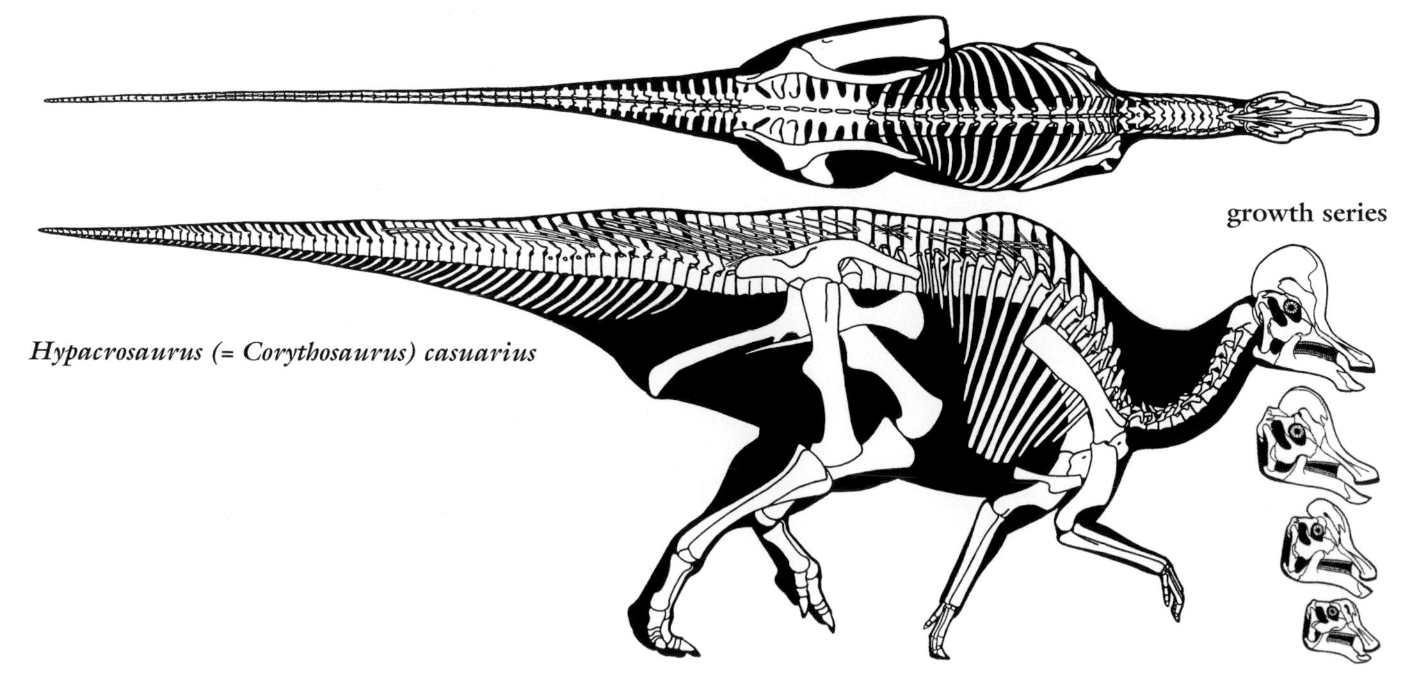

Hypacrosaurus (= Corythosaurus) casuarius

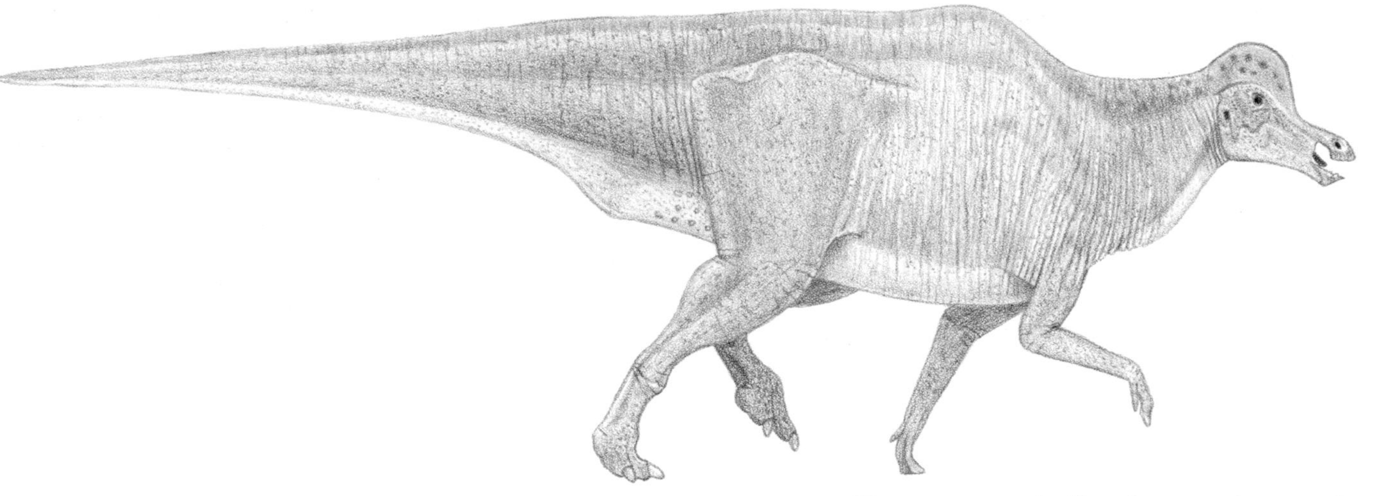

Hypacrosaurus (= Corythosaurus) casuarius

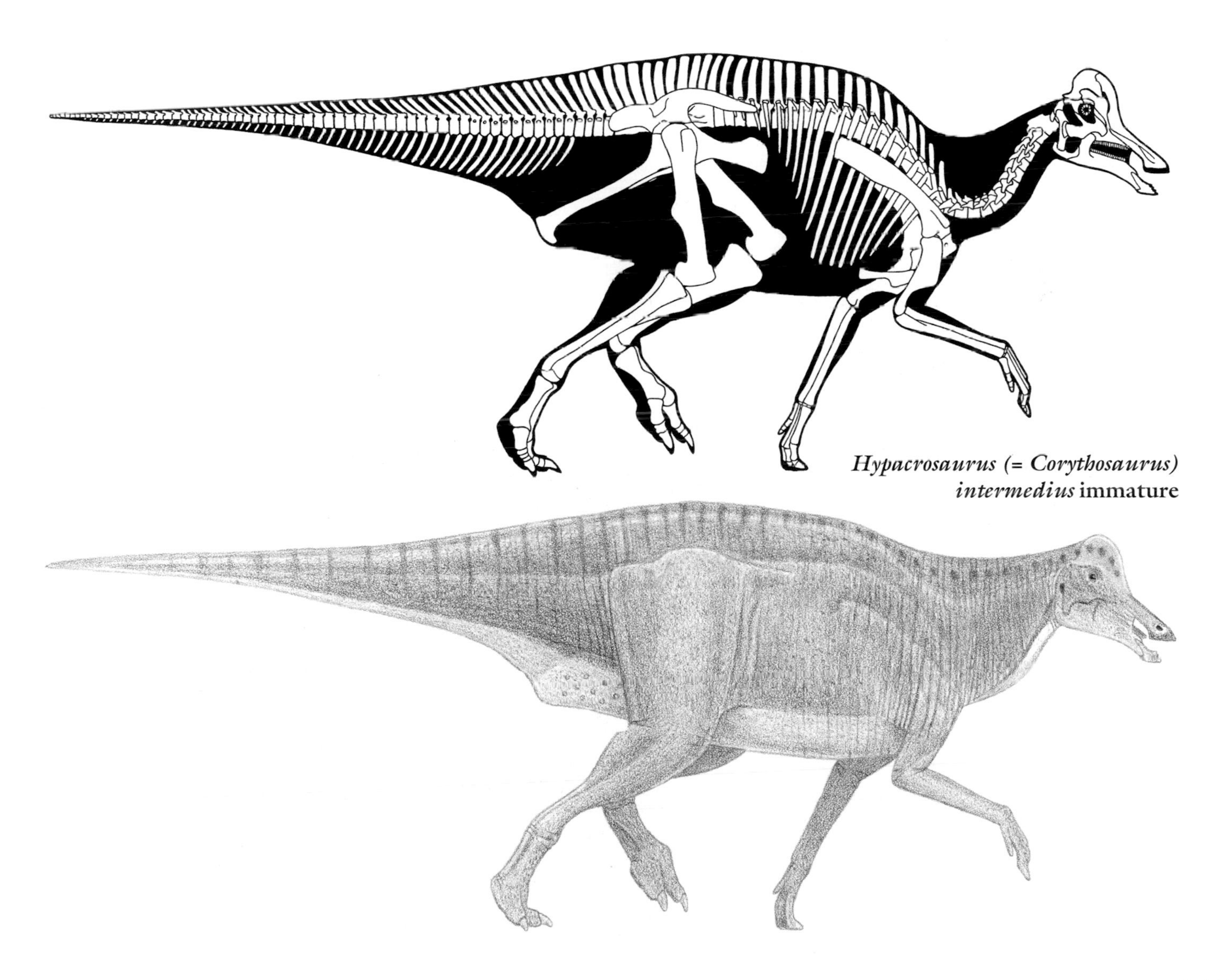

Hypacrosaurus (= Corythosaurus) intermedius immature

Hypacrosaurus stebingeri
8 m (26 ft) TL, 3 tonnes

FOSSIL REMAINS Complete skull and majority of skeletons, juvenile remains including embryos, nests.
ANATOMICAL CHARACTERISTICS Large, semicircular crest atop back of skull. Tall vertebral spines over trunk form a low sail. Eggs spherical, 20 cm (8 in) in diameter.
AGE Late Cretaceous, middle and/or late Campanian.
DISTRIBUTION AND FORMATION/S Montana, Alberta; Upper Two Medicine.
HABITAT Seasonally dry, upland woodlands.

Hypacrosaurus altispinus
8 m (26 ft) TL, 3.4 tonnes

FOSSIL REMAINS Several skulls and partial skeletons.
ANATOMICAL CHARACTERISTICS Large, subtriangular crest atop back of skull. Tall vertebral spines over trunk form a prominent sail.
AGE Late Cretaceous, middle Maastrichtian.
DISTRIBUTION AND FORMATION/S Alberta; middle Horseshoe Canyon.
HABITAT Well-watered, forested floodplain with coastal swamps and marshes, cool winters.
NOTES Shared its habitat with *Saurolophus osborni.* Main enemy *Albertosaurus sarcophagus.*

Hypacrosaurus? laticaudus
9 m (30 ft) TL, 4 tonnes

FOSSIL REMAINS Minority of skull and skeleton.
ANATOMICAL CHARACTERISTICS Tall vertebral spines of trunk and tail form a prominent sail.
AGE Late Cretaceous, Campanian.
DISTRIBUTION AND FORMATION/S Baja California, Mexico; El Gallo.

Hypacrosaurus (Lambeosaurus) clavinitialis
7.7 m (25 ft) TL, 3.3 tonnes

FOSSIL REMAINS A few complete skulls, majority of skeleton.
ANATOMICAL CHARACTERISTICS Large, hatchet-shaped crest atop back of head.

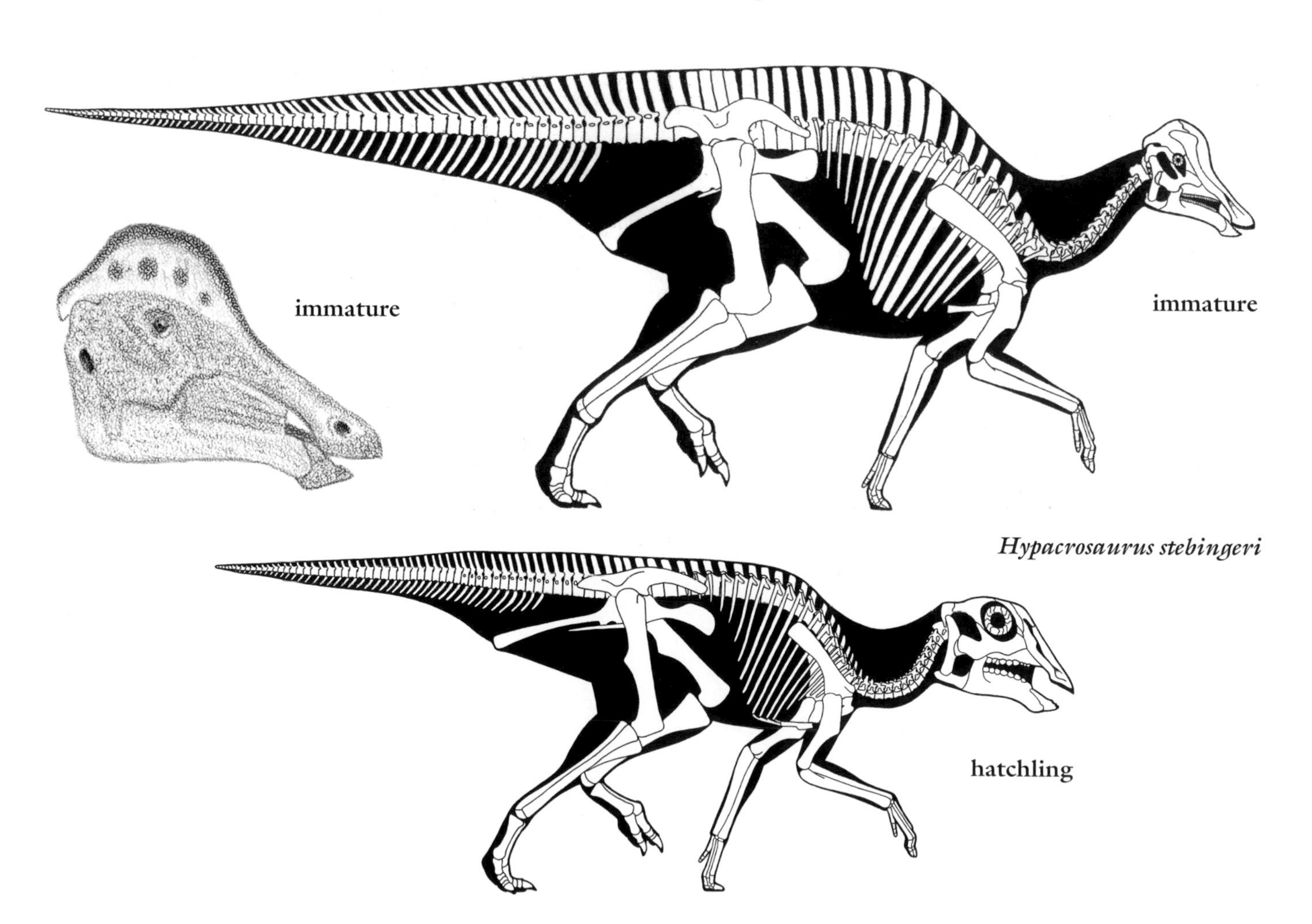

Hypacrosaurus stebingeri

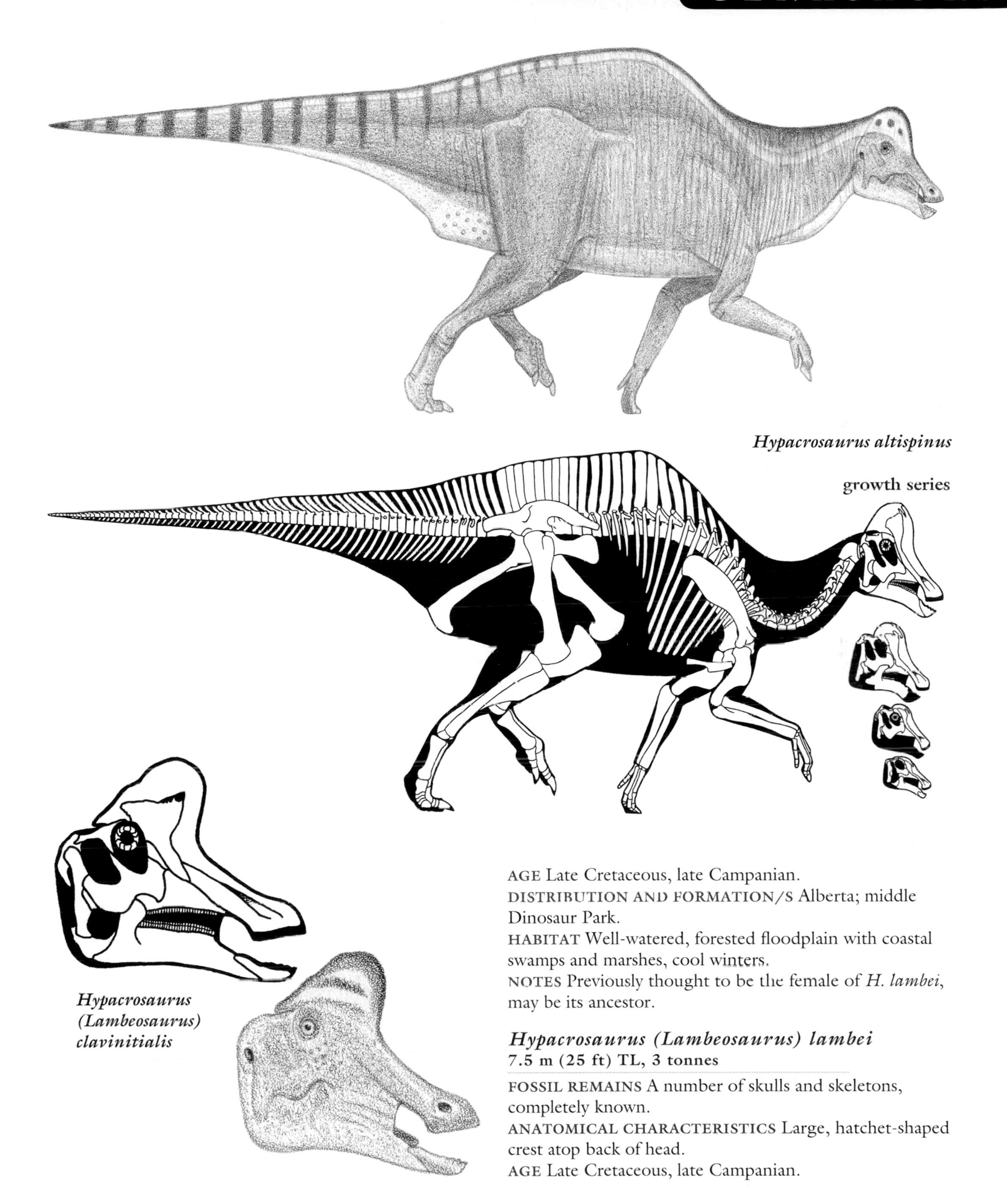

Hypacrosaurus altispinus

growth series

Hypacrosaurus (Lambeosaurus) clavinitialis

AGE Late Cretaceous, late Campanian.
DISTRIBUTION AND FORMATION/S Alberta; middle Dinosaur Park.
HABITAT Well-watered, forested floodplain with coastal swamps and marshes, cool winters.
NOTES Previously thought to be the female of *H. lambei*, may be its ancestor.

Hypacrosaurus (Lambeosaurus) lambei

7.5 m (25 ft) TL, 3 tonnes

FOSSIL REMAINS A number of skulls and skeletons, completely known.
ANATOMICAL CHARACTERISTICS Large, hatchet-shaped crest atop back of head.
AGE Late Cretaceous, late Campanian.

DISTRIBUTION AND FORMATION/S Alberta; middle to upper Dinosaur Park.
HABITAT Well-watered, forested floodplain with coastal swamps and marshes, cool winters.
NOTES Specimens with largest crests may be males.

Hypacrosaurus (Lambeosaurus) magnicristatus
7 m (23 ft) TL, 2.5 tonnes

FOSSIL REMAINS A few skulls, part of skeleton.
ANATOMICAL CHARACTERISTICS Extremely large, oblong crest atop back of head.
AGE Late Cretaceous, late Campanian.
DISTRIBUTION AND FORMATION/S Alberta; uppermost Dinosaur Park.
HABITAT Well-watered, forested floodplain with coastal swamps and marshes, cool winters.
NOTES May be the direct descendant of *H. lambei.*

adult or male

growth series

subadult or female

Hypacrosaurus (Lambeosaurus) lambei

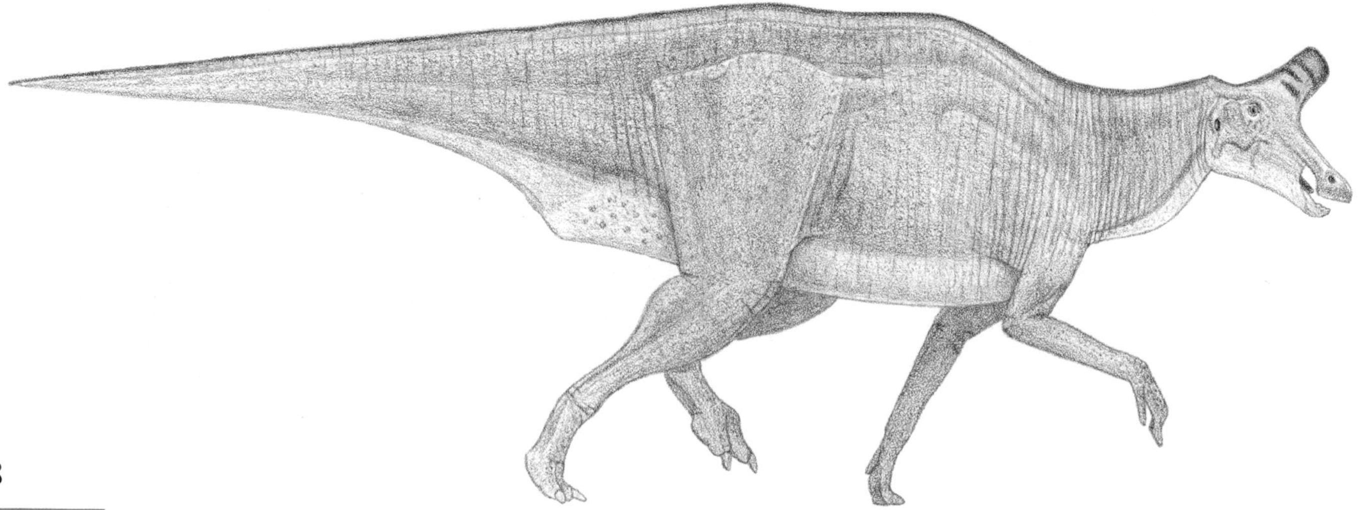

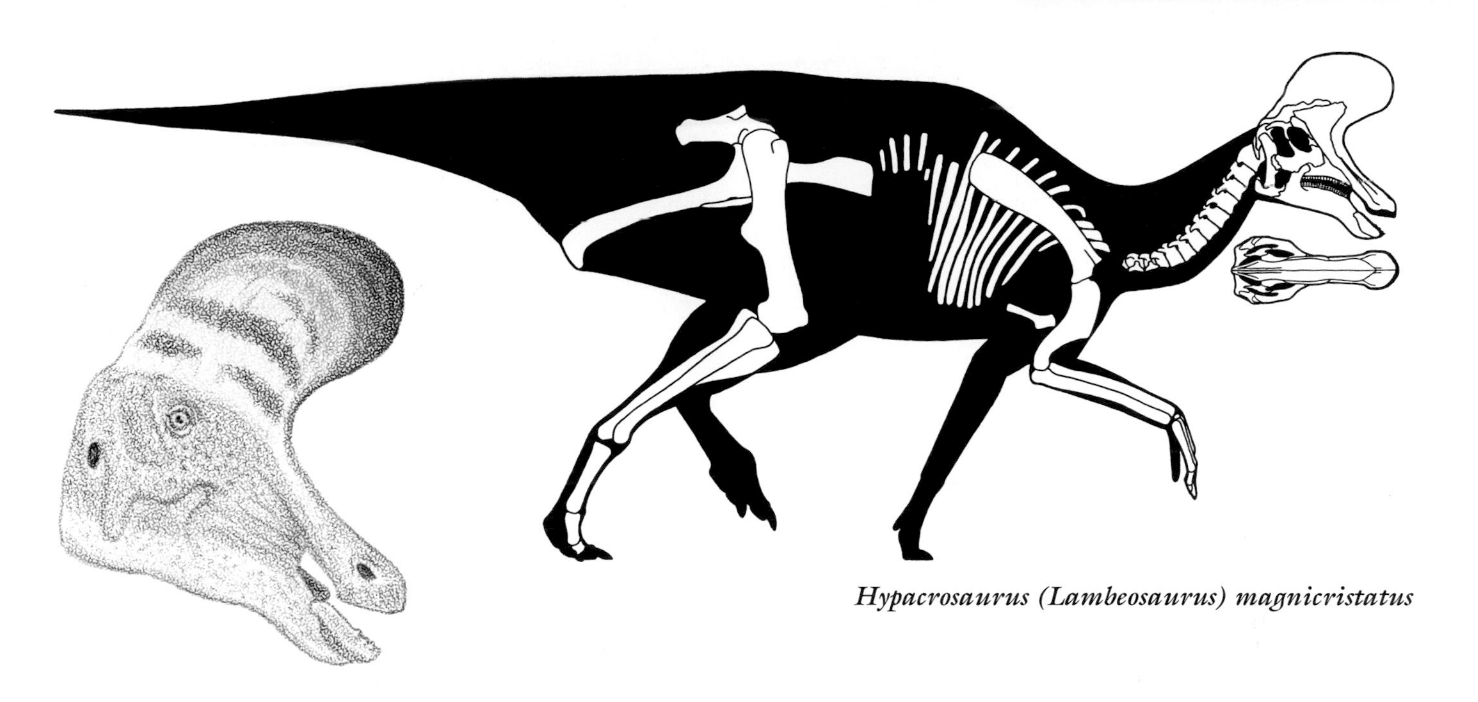

Hypacrosaurus (Lambeosaurus) magnicristatus

Hypacrosaurus (Lambeosaurus) magnicristatus

ADDITIONAL READING

Brett-Surman, M., and J. Farlow. 2011. *The Complete Dinosaur*. 2nd ed. Bloomington: Indiana University Press.

Glut, D. 1997–2012. *Dinosaurs: The Encyclopedia* (including Supplements 1–7). London: McFarland & Company.

Paul, G. S., ed. 2000. *The Scientific American Book of Dinosaurs*. New York: St. Martin's Press.

———. 2002. *Dinosaurs of the Air*. Baltimore: Johns Hopkins University Press.

Weishampel, D., P. Dodson, and H. Osmólska. 2004. *The Dinosauria*. 2nd ed. Berkeley: University of California Press.

INDEX

This index covers dinosaur groups, genera, and species, as well as dinosaur-bearing formations described in the main directory section starting on page 69.

Dinosaur Taxa

Formations

When a formation is cited more than once on a page, the number of times is indicated in parentheses.